Landscape Series

Volume 21

Ramesh Krishnamurthy, Department Landscape Level Planning, Wildlife Institute of India, Chandrabani, Dehradun, India

Quang Bao Le, International Center for Agricultural Research in the Dry Areas, Cairo, Egypt

Yu-Pin Lin, Dept of Bioenvi Systems Engineering, National Taiwan University, Taipei, Taiwan

Benjamin Kofi Nyarko, Dept of Geography & Regional Planning, University of Cape Coast, Cape Coast, Ghana

Henrique Pereira, German Centre for Integrative Biodiversity Research (iDiv), Martin Luther University Halle-Wittenberg, Leipzig, Germany

Alexander Prishchepov, Dept of Geosci & Natural Resource Mgmt, University of Copenhagen, Copenhagen, Denmark

Robert M. Scheller, North Carolina State University, Raleigh, USA

Kalev Sepp, Inst. Agricultural & Environ. Sciences, Estonian University of Life Sciences, Tartu, Estonia

Anton Shkaruba, Inst. of Agricultural & Environ Sciences, Estonian University of Life Sciences, Tartu, Estonia

Ileana Stupariu, Dept. Regional Geography and Environment, University of Bucharest, Bucharest, Romania

Raymond Tutu, College of Humanities, Delaware State University, DOVER, USA

Teiji Watanabe, Fac of Environmental Earth Sci, A-301, Hokkaido University, Sapporo, Japan

Wei-Ning Xiang, University of North Carolina at Charlotte, Charlotte, USA

Qing Zhao, Institute of Applied Ecology, Chinese Academy of Sciences, Shenyang, China

Springer's innovative Landscape Series is committed to publishing high quality manuscripts that approach the concept of landscape and land systems from a broad range of perspectives and disciplines. Encouraging contributions that are scientifically-grounded, solutions-oriented and introduce innovative concepts, the series attracts outstanding research from the natural and social sciences, and from the humanities and the arts. It also provides a leading forum for publications from interdisciplinary and transdisciplinary teams across the globe.

The Landscape Series particularly welcomes contributions around several globally significant areas for landscape research, which are anyhow non-exclusive:

- Climate and global change impacts on landscapes and ecosystems including mitigation and adaptation strategies
- Human Dimensions of Global Change

- Biodiversity and ecosystem processes linked to ecosystems, landscapes and regions
- Biogeography
- Ecosystem and landscape services including mapping, assessment and modelling
- Land System Science
- Regional ecology (including bioregional theory & application)
- Human-Environment Interactions and Social-Ecological Systems & Frameworks (SESF) - including theories, practice and modelling

Volumes in the series can be authored or edited works, cohesively connected around these and other related topics and tied to global or regional initiatives. Ultimately, the Series aims to facilitate the application of landscape research and land system science to practice in a changing world, and to advance the contributions of landscape theory and research and land system science to the broader scholarly community.

Vittorio Ingegnoli

Landscape Bionomics and Planetary Health

A Human Health and Integrated Landscape Ecology Perspective

Springer

Vittorio Ingegnoli
Biomedical, Surgical and Dental Sciences
One-Health Unit
University of Milan
Milan, Italy

ISSN 1572-7742 ISSN 1875-1210 (electronic)
Landscape Series
ISBN 978-3-032-08907-6 ISBN 978-3-032-08908-3 (eBook)
https://doi.org/10.1007/978-3-032-08908-3

This Springer imprint is published by the registered company Springer Nature Switzerland AG
The registered company address is: Gewerbestrasse 11, 6330 Cham, Switzerland

Idemque cum caelum, terras, maria, rerumque omnium naturam perspexerit, eaque unde generata, quo recursura, quando, quo modo obitura, quid in iis mortale et caducum quid divinum aeternumque sit, viderit, Ipsumque ea moderantem et regentem paene prehenderit, seseque non unis circumdadum moenibus, popularem alicuius definiti loci, sed civem totius mundi, quasi unius urbis agnoverit: in hac ille magnificentia rerum atque in hoc conspectu et cognitione Naturae, Dii immortales! quam se ipse noscet! Quod Apollo Praecipit Pythius; quam contemnet, quam despiciet, quam pro-nihilo putabit ea, quae vulgo dicuntur amplissima!"

Marcus Tullius Cicero (De Legibus Chap. XXIII).

Thus, once [man] will have studied heaven, earth, seas and the nature of all the things and will have seen their origin and their possible end, when and how they will perish, what in them is mortal and transient, what divine and eternal, and almost will have touched the Being who governs them, and once [man] will recognize himself as not closed by walls in a limited site, but citizen of the entire world such as a single city: in this magnificence of things, in this spectacle and knowledge of Nature, Immortal Gods, how much he will have known himself such as the precept of Apollo Pytius! How much he will condemn; how much he will despise and consider as nothing what mass reputes as important.

Foreword I

In engineering, as in most technical fields, quick decision-making, guided by a robust protocol, is often required. However, diagnostic and management actions are frequently implemented automatically and based on unquestioned assumptions, thus oversimplifying a reality that is inherently complex. The characterization of the natural environment, however, demands a comprehensive perspective, as it involves numerous abiotic and biotic components, and the relationships among them are highly diverse. When faced with fragmented approaches, constrained by specialized knowledge, there is a clear need for a conceptual and epistemological framework that encourages a holistic view of Nature.

In this context, the concept of Bionomy, introduced by Prof. Ingegnoli, offers a renewed, powerful, and effective perspective on Landscape Ecology. It departs from deterministic approaches, focusing instead on dynamics, transformative processes, and the investigation of the history or evolution of natural systems, which must be studied through the lens of nonequilibrium thermodynamics.

Through the systemic perspective of Landscape Bionomy, the scale factor becomes paramount, highlighting the landscape as a key level of life organization. Landscape possesses unite structural, functional, and dynamic properties, distinguishing it from other units of study, such as ecosystems—redefined by Ingegnoli in his works—, communities, and populations, which become progressively smaller in both spatial and temporal dimensions. In this sense, Prof. Ingegnoli's work has laid a solid foundation for thorough diagnoses of the "health" of landscapes and their components (tesserae, corridors, matrices), like the clinical method used to identify pathologies or alterations from a "normal" state.

In the current era, the Anthropocene, humans, as members of the various hierarchical levels of life organization, have created new systems—ranging from urban to semi-natural and transitional rural environments—and can alter, and indeed do alter, the conservation status of these systems and their surrounding environments to varying degrees.

The human attributes of awareness, rationality, and the desire for knowledge and truth demand that management at the territorial scale focus on improving the state variables of the landscape and its components, once the levels of induced dysfunction have been diagnosed. In short: the assessment of degradation.

The novel contribution of Vittorio Ingegnoli, clearly conveyed in the title of this book, is his demonstration that certain degradations on the landscape scale are at the root of environmental degeneration, with both direct and indirect consequences for human health. These degradations go beyond the more obvious and widely recognized issues, such as air, water, and soil pollution or the extinction of mature vegetation covers. They also involve the loss of crucial ecosystem services like air purification, hydrological regulation, and the direct production of goods, all of which directly impact social and individual health. The aphorism of Spanish philosopher Ortega, from *Meditations on Quixote* (1914), is pertinent here: "I am I and my circumstance; and, if I do not save it, I do not save myself." The landscape is an integral part of that circumstance.

The systemic paradigm of landscape, emphasized by Prof. Ingegnoli in his previous works, continues to shed light on the topic for those familiar with his research. The range of valuable content that readers of this new book will uncover is vast and challenging to summarize in just a few lines. As a forestry engineer primarily focused on teaching Geobotany (the characterization and mapping of vegetation and the study of its dynamics) and Ecosystem and Landscape Ecology, I found Ingegnoli's conceptual and methodological approach through Landscape Bionomy to be an invaluable contribution. It provides structure and clarity to knowledge and intuitions that I aim to pass on to students of natural environment engineering, the future professionals in this field.

The order and systematic approach in the evaluation procedure (LaBiSV) of different vegetation models that make up landscapes are both useful and pedagogical. They enhance understanding of the maturity level assignment of tesserae, a concept developed in Spain by J. Ruiz de la Torre. In fact, this can be linked to the estimation of "bionomic quality" and the calculation of biopotentiality (BTC), which refers to the degree of relative metabolic capacity and the relative ability to maintain antithermic conditions. Prof. Ingegnoli proposes a methodology that evaluates each tessera as part of a broader, hierarchically structured landscape unit, much like how cells, tissues, organs, and apparatuses within the body of an evolved animal form the elements of a higher-level system.

The concepts of human habitat, standard habitat, general landscape metastability, functional networks (or landscape apparatuses), and transformation deficit are key terms defined by Ingegnoli in a diagnostic model that constitutes the core of the first section of this book.

In addition to these revised aspects of Landscape Bionomy, the author delves into the degradation of the environment and planetary health, particularly through an in-depth examination of research linking human health to landscape alterations. He explores the processes involved and proposes specific prevention treatments, closely aligned with this scale. Territorial planning can play a crucial role in maintaining a landscape state that contributes to human health. In the final part of the book, Prof. Ingegnoli dedicates a chapter to specific case studies and concludes with insightful reflections on the importance of the "environmental doctor," the ecoiatrist.

Prof. Ingegnoli exemplifies the wise, multifaceted researcher, whose expertise spans the Natural Sciences, Architecture, and even painting—each grounded in a broad cultural foundation, strong theoretical underpinnings, and a sense of pragmatism. He lives by the principle that *humani nihil alienum*—nothing human is foreign to him—and, equally, nothing in the natural world is beyond his interest. With his modest and humble nature, he seeks not personal gain, but the joy of fully understanding, at least in part, the Clock of the Universe.

César López Leiva

Department of Natural Systems and Resources, ECOGESFOR Research Group, School of Forestry Engineering and Natural Resources, Technical University of Madrid – Universidad Politécnica de Madrid, Madrid, Spain

Foreword II

In the twilight of the Anthropocene, as humanity's footprint on the planet deepens, scientific disciplines capable of recognizing the intrinsic connection between the health of Earth's systems and human health are emerging. Planetary Health represents this epistemological revolution: no longer a medicine isolated from its environmental context, but a holistic vision where human well-being is inextricably linked to the integrity of the ecosystems that host it.

The contemporary landscape presents us with unprecedented challenges. As Prof. Ingegnoli reminds us in this fundamental work, "the Anthropocene, the predominance of man over nature, has produced an environmental crisis" that cannot be understood exclusively in negative terms, since "the destruction of nature has also brought socioeconomic benefits." However, these benefits prove to be "precarious and transitory if obtained by destroying nature." It is precisely here that the urgency for a paradigm shift arises.

Particularly alarming in this context is the emergence of new infectious diseases of zoonotic origin, a phenomenon masterfully described by David Quammen in his work "Spillover." Anthropogenic alterations of ecosystems—deforestation, habitat fragmentation, intensive agricultural expansion, and uncontrolled urbanization—are creating unprecedented ecological interfaces that facilitate the "species jump" of pathogens. As Quammen emphasizes, these are not random events but predictable consequences of our invasion and destructuring of natural balances. Every time we tear through a forest fabric, penetrate deep into pristine ecosystems, or alter the habitat of wildlife, we create opportunities for new viral, bacterial, or parasitic "spillovers," accelerating the emergence of pathogens previously confined to sylvatic cycles.

Landscape Bionomics and Planetary Health emerge as two complementary faces of this paradigmatic transformation. The first rooted in architecture and natural sciences, and the second in the medical-environmental field: both recognize the inadequacy of scientific reductionism and the need to embrace a systemic vision of reality. As Evandro Agazzi acutely points out, "reductionism is today at a standstill because it claims to achieve unity by disregarding variety."

The conceptual innovation proposed in this work transcends traditional disciplinary boundaries. The transition from conventional ecology to landscape bionomics represents a fundamental qualitative leap: no longer isolated ecosystems, but interconnected landscape units; no longer sustainability as mere resource management, but as an ecological "rehabilitation strategy." This approach recognizes the landscape as a "living adaptive complex system," capable of falling ill and healing directly influencing the health of the human communities that inhabit it.

The research presented in this volume demonstrates, with quantitative evidence, how bionomic alterations of the landscape—regardless of pollution—are correlated with the onset of human pathologies. From increased mortality in areas with low bionomic functionality to the incidence of Covid-19 and cancer in relation to environmental degradation: a rigorous scientific framework emerges, establishing causal connections between ecosystem and human health. The recent SARS-CoV-2 pandemic, emblematic of the dynamics described by Quammen, reminds us that ecological balance is not an intellectual luxury but an existential necessity: when we alter the bionomic configuration of the landscape, we not only degrade biodiversity but create ideal conditions for the emergence and spread of new zoonotic pathogens, capable of crossing the species barrier and triggering global health crises.

Particularly illuminating is the exploration of the processes through which the environment influences human health: from ethological alarm signals to chronic stress, from brain-environment neural adaptation to alterations in the gut microbiota. Understanding these mechanisms opens new frontiers for preventive medicine and territorial planning. In this perspective, the prevention of emerging zoonoses cannot be separated from an ecologically aware management of the landscape, which preserves natural balances and minimizes high-risk interfaces between wildlife, domestic animals, and human populations.

As Pope Francis emphasizes in the encyclical Laudato Sì, referenced in the work: "If 'the outer deserts in the world are growing because the inner deserts have become so vast,' the ecological crisis is a call to profound interior conversion." This is the beating heart of Planetary Health: recognizing that the environmental crisis is also an anthropological crisis, requiring not only technical solutions but a profound rethinking of our being in the world.

This volume thus stands as a milestone for a new school of thought that integrates medicine, biology, natural sciences, agronomy, veterinary medicine, territorial planning, and geography in a transdisciplinary approach to planetary health. A fundamental contribution to help humanity respect the environment and, in doing so, preserve its own health. The understanding of zoonotic dynamics described by Quammen integrates perfectly into this vision, highlighting how human health, animal health, and ecosystem integrity are inseparable dimensions of a single systemic reality.

The hope is that, by understanding the complex interrelationships between environmental alterations and human health, we can inaugurate a new era of harmonious coevolution with Earth's living systems, as citizens "not enclosed by walls in a limited place, but citizens of the entire world as of a single city," to quote

Cicero in the passage that eloquently opens this volume. A vision that reminds us that planetary health is not only a scientific necessity but an ethical imperative for our time, in which the prevention of emerging zoonoses becomes emblematic of an integral ecology that simultaneously safeguards biodiversity and human well-being.

Luigi Bonizzi
Alessio Soggiu

Full Prof. of Animal Infectious diseases,
Department of Biomedical, Surgical &
Dental Sciences, One-Health Unit,
University of Milan, Milan, Italy

Preface

About Me and the Environment: Autobiographical Notes

My environmental studies began at the Polytechnic University of Milan in 1963, third year of Architecture, observing the irrepressible urbanization growth of the hinterland of Milan and its senseless destruction of agricultural landscapes, remnant woodland, moors, and historic centers. These studies imposed my first approach to Ecology, not yet taught in our universities; so, I began to study as a self-taught the volume of "Fundamentals of Ecology" by E.P. Odum (1959), and the volume on Vegetation Science "Flora d'Italia" by V. Giacomini and L. Fenaroli (1958).

After my first degree (1967) in Architecture and my military obligations, I won a scholarship for an internship at the Institute of Urban Planning (1970–73, at the Engineer Faculty, Polytechnic) and another scholarship at the Venice University, Cini Foundation, for a high culture course on "Man and the Environment" (1972). In 1980 my first considerations converged in a book "Ecologia e Progettazione" and two years after I opened a course of Principles of Landscape Ecology at the Graduate School of Landscape Architecture at the University of Genoa, as a charged professor. In 1984, I was accepted as a member of the SITE (Italian Scientific Society for Ecology), where I deepened my studies of plant ecology with the help of Sandro Pignatti. Since 1991 I have been a member of IALE (International Association for Landscape Ecology), opening in 1996–97 an Italian Section (SIEP), after the publication of the book "L'ecologia del paesaggio in Italia" (Ingegnoli & Pignatti, 1996).

Perceiving the need to deepen naturalistic knowledge, in 1990 I became a student in Natural Sciences at the State University of Milan, graduated *magna cum laude* in 1995. In 2000–2001, accepting the invitation of R. Levi-Montalcini, I contributed to the book "Frontiers of Life," coordinated by four Nobel Prizes, for the topic "Landscape Ecology."

In 2002 Springer Verlag published my first book on "Landscape Ecology, a Widening Foundation," with the preface by Richard T.T. Forman (Harvard

University) within which, for the first time, the physiology, the growth, and the transformations and the pathologies of different types of landscape were investigated from a biological and quali-quantitative clinical-diagnostic-therapeutic point of view. In the same years, I opened a course of Landscape Ecology, as professor by contract, within the Faculty of Natural Sciences at the State University of Milan, also proponing significant upgrading in vegetation science (Ingegnoli & Pignatti, 2007); in the same times, I participated in the program CON.ECO.FOR, monitoring Italian forests for a European Research Program coordinated by Gottingen University. I have been holding the course of Landscape Ecology until the Academic Year 2015–16.

Between 2011 and 2015, the "widened" Landscape Ecology became "Landscape Bionomics," i.e., the Biological Integrated Landscape Ecology for the training of a physician of ecological systems, firstly in Italian, then in English, always published by Springer Verlag.

Immediately called back to University by Stefano Bocchi, agronomist, since 2016 I have been studying the relationships between the environment and Human Health, regardless of pollution, and I was accepted as a member of SIPNEI (Soc. It. for Psycho-Neuro-Endocrine-Immunology, Rome). My interest has been focusing on the possible influence of the health state of the biologic structure and function of different types of landscape units (LU) on the population active. First results on the relationship between mortality rate (MR) and bionomic functionality (BF) and cancer insurgence in Western Europe seem promising. Recently (2024), the Department of Biomedical, Surgical, and Dental Sciences asked me to further my research on the "One-Health/Planetary Health" sector.

About the Book: Conceptual Notes

It must be underlined that a good part of the crisis in Life Sciences and of the present environmental problems depend on reductionism. As Evandro Agazzi reminds us, reductionism is at a dead end today because it claims to arrive at unity regardless of variety. Many phenomena need to be studied following principles and methods more suitable for the study of complex systems; therefore, to get out of this crisis, being able to combine the historical and cultural heritage of knowledge concerning territories with the employment needs of population, guiding the choices of politicians and administrators toward an advanced idea of protection, conservation, and development, Science needs to change the paradigm, moving toward Holism, at first following the System Theory.

This notable development of systemic disciplines also produces a change concerning interdisciplinarity, which can more easily transform into transdisciplinarity. Recalling what was written by Vitruvius, we can say that the systemic sense (*corpus unum*) has been lost, or greatly weakened, due to reductionism, which at most allowed the convergence of different disciplines toward an objective of study (interdisciplinarity): instead, the disciplines born for a systemic turn following the theoretical bases, and often also the methods, typical of Systems Theory, are spontaneously transdisciplinary.

The need to direct studies toward the Systemic Paradigm has been felt in almost all disciplines, both pertaining to natural and human sciences. This need has pushed toward the development of disciplinary branches of updating: the two most significant disciplines are Landscape Bionomics (LB) and Planetary Health (PH), which were born a few years apart in the architectural/naturalistic and medical/environmental fields, respectively.

As we will see, the structure of this book proves more meaningful than the previous book related to Landscape Bionomics (Ingegnoli, 2015), where the relationship with human health was mainly mentioned. Moreover, this new volume is more essential, contained in a more manageable number of pages (about 330–340 vs. 440 pages).

In addition, many insights and innovations have been occurring in Landscape Bionomics within the last ten years, especially in the fields of human health, sustainability, ecological services, and accounting for certain aspects of natural heritage, and needed to be included. The first of them is to overcome the concept of Sustainability as Sustainable Development—focused only on the management of the resources to postpone their depletion as much as possible and/or on the investigation concerning green energy, production, and connected arguments—, by the principle of Sustainability as Sustainable Restraint; moreover, as "Territorial Regenerative Rehabilitation Strategy," linked to the redefinition of the landscape as a living complex adaptive system (overcoming a sharp distinction between the two concepts of natural and cultural landscape), and to the environmental potentiality of our Planet, even at territorial/regional space/time/information scale. Regenerative Rehabilitation should understand the relationship between the state of health and diseases of a land unit, check its compatibility with the human needs and health, and develop criteria for therapeutic rehabilitation with regenerative function. At present, the definition and monitoring of a Health Protective capacity must lead any political choice and strategy of environmental assessment on territorial planning, development, and management.

I hope that this volume could be used in a new graduated school of Planetary Health, open to medicine, biology, natural sciences, agronomy, veterinary sciences, territorial and urban planning, and geography to help humans to better respect the environment and to preserve its health. Teachers of High School too can find the necessary upgrade for their students. "There is a need for a reform of thought and a reform of education that allows us to face complexities, integrate different knowledge and resist regressions.[…] As long as the sense of the complexity of the world, of life, of human beings and of events remains the prerogative of very few, the world will go from bad to worse," says Morin (2025).

Remember that Pope Francis, in Laudato Sì, underlined: «If *"the external deserts are multiplying in the world, because the internal deserts have become so vast", the ecological crisis is a call to a profound interior conversion» [Benedict XVI, Homily for the Solemn Inauguration of the Petrine Ministry (24 April 2005): AAS 97 (2005), 710].*

N.B.: no part of the text, nor some of the figures, tables, indexes and indicators, flow charts, and models have been created, thought out, or elaborated thanks to any form of artificial intelligence.

Milan, Italy
January 25, 2025

Vittorio Ingegnoli

Acknowledgments The recent evolution of my thinking has been influenced by discussion with—in alphabetic order—Stefano Bocchi (Agronomy, Milano), Luigi Bonizzi (One Health, Milano), Francesco Bottaccioli (SIPNEI, Roma), Richard T.T. Forman (Landscape Ecology, Harvard, Boston), Giuseppe La Torre (Epidemiology, Roma-Sapienza), Francesco Lombardo (Endocrinology, Roma-Sapienza), César Lopez Leiva (Forestry and Ecology, Madrid), Giuseppe Magro (IAIA, Bologna), Domenico Patassini (IUAV, Venezia), Sandro Pignatti (Vegetation Science, Lincei, Roma), Alessio Soggiu (Microbiology, Milano), Marco Sumini (Ingegneria, Bologna), Joan Tort Donada (Geography, Barcelona).

Good discussion also took place with many students at the universities of Milan and Madrid.

Encouragements to persist in my studies, notwithstanding the difficulties to survive as researcher in my Country, have come from many colleagues and friends, especially Domenico Albanese (Industrial Chemistry, Milano), Paolo Scarpazza (Pulmonologist, Vimercate, MB), Andrea Galli (Territorial planner, Ancona).

Many sincere thanks to Antonella Cerri at Springer Milan, Alessandro Ingegnoli, who left us a few years ago, and to all friends reviewing portion of this book. A special appreciation to Elena Giglio Ingegnoli, who reviewed the entire manuscript with an exceptional competence and to Costanza Maria Ingegnoli for her assistance in discussions regarding the topics in Chap. 15.

Main Scientific Recognitions

Below, a summary of the main recognitions received from fellow scientists regarding the studies and discoveries that Ingegnoli was able to carry out to update ecology according to the systemic paradigm. They are listed in alphabetical order, with sincere gratitude for their authoritative scientific discussions and moral encouragement:

Stefano Bocchi [Dr. in Agricultural Science, full Prof. of Agronomy at the State University of Milan], who called Ingegnoli back to the University after the Ecological Landscape Design discipline (degree course Natural Science) was abolished and still collaborates with him in various research fields.

Luigi Bonizzi [Dr. in Veterinary Science, full Prof. of Animal Pathology and One Health at the State University of Milan], who recently invited Ingegnoli as prof.

Fig. 1 Friendly meetings: (Left) Vittorio Ingegnoli and Richard Forman at Harvard University 2017, Boston. (Right) Ingegnoli in a friendly meeting with Sandro Pignatti in Rome, 2019, recently passed away

frequentatore (i.e., visiting professor) at the Department of Biomedical, Surgical & Dental Sciences to study Planetary Health arguments.

Pietro Calissano [Doctor of Medicine, Collaborator of Levi-Montalcini in the USA and Rome, Professor of Neurobiology, past-director of the CNR Neurobiology Center, Rome] and **Rita Levi-Montalcini** (1909–2012) [Doctor of Medicine, Prof. of Biology, Washington University, USA. 1986 Nobel Prize for NGF, past-Director of the Treccani Foundation, Rome], who invited Ingegnoli to write the chapter on *Landscape Ecology* in the volume Frontier of Life (edited by 4 Nobel Prizes), published by Treccani in Rome (1999) and by Academic Press in New York (2001).

Richard T.T. Forman [Dr. in Biology, Professor Emeritus at Harvard University, USA], who met Ingegnoli several times, went on a study trip with him to Tuscany (1998), wrote the presentation of the book *Landscape Ecology, a widening Foundation* (Ingegnoli, 2002), invited him to Harvard in 2017 (Fig. 1).

Hubert Gulink [Dr. in Agrarian Science, Emeritus Prof. of Ecology at KU Leuven, Belgium], who decided to utilize Landscape Bionomics for a comparative research with Ingegnoli, titled: "A Comparison Between Two Rural-Suburban Landscapes from Brussels and Milan" (2015, 2016).

Wolfgang Haber [Dr. in Natural Science, Dr. in Agrarian Science, Prof. Emeritus of Ecology, Past President of the INTECOL. TUM School in Life Science, Technische Universitat Munchen], who invited Ingegnoli to Freising to talk about his studies and recently wrote an email of approval and encouragement.

Marco LaVeglia [Ingegnere idraulico/ambientale, Technical manager AIPO, DTI Lombardia Occidentale, Coordinatore Area Lombardia]. The Interregional Agency for the Po River Basin (AIPO), after having adopted the Landscape Bionomics approach for some consultancies on the ecological compensations related to diversion canals and river expansion basins, has given to Ingegnoli the task of holding a refresher course in landscape bionomics for the technicians of the AIPO headquarters in Parma (2023).

Andrea Lenzi [Dr. in Medicine, President of CNR, President of the National Committee for Biosafety Biotechnology and Life Sciences of the Presidency of the Council of Ministers (CNBBSV), President of National Board of Professors in Life Sciences (Intercollegio), Dep. of Experimental Medicine Chair of Endocrinology, Prof. Emeritus of Endocrinology Sapienza University of Rome] and **Francesco Lombardo** [Dr. in Medicine, full Prof. in Endocrinology, Sapienza Univ., Rome], who firstly asked Ingegnoli to hold a Seminar on the "First Evidence in the Milan Metropolitan Area" of "how Bio-Ecological Alterations of the Environment Damage Human Health, Regardless of Pollution." Co-editor with Ingegnoli and La Torre, of the volume (2022) *Environmental Alteration Leads to Human Disease*, Springer Nature.

Cèsar Lopez Leiva [Dr. in Forestry Engineering, Prof of Ecology, Universidad Politecnica de Madrid], who asked Ingegnoli to supervise a degree thesis on the Botanical Garden of the Polytechnic and then invited him to Madrid to hold two seminars on Landscape Bionomy (2023–24). The collaboration is still in progress.

Zev Naveh (1919–2011) [Dr. in Agronomy, PHD in Ecology, Prof. Emeritus of Landscape Ecology, Technion, Israel Institute of Technology, Haifa], who met Ingegnoli several times, exhorted him to publish (2002) his first volume with Springer and invited Ingegnoli to publish in the *Israel Journal of Plant Sciences* (2005).

Luigi Padovese (1947–2010) [prof. of Patristics at the Pontificia Università Antonianum, Rome, and bishop Apostolic Vicar at Alexandretta, Türkiye], who invited Ingegnoli to refer on "The minority of our mother Earth. Ethical, ecological and economic implications" within the International Seminar on "Minores et subditi omnibus" (2003). Pontifical Athenaeum Antonianum Laurentianum, Rome.

Sandro Pignatti (1930–2025) [Dr. in Biology, former Prof. of Botany at the University of Trieste, Prof. Emeritus at Sapienza, Rome; Academician of the Lincei]. Pignatti has been meeting and exhorting Ingegnoli many times, starting from 1984; he published with him the book *Ecologia del Paesaggio in Italia* (Ingegnoli & Pignatti, 1996), then the article "The impact of the widened Landscape Ecology on Vegetation Science: towards the new paradigm," research for the Accademia dei Lincei (2007), and finally "Cento anni di Fitosociologia" on Braun-Blanquetia (Pignatti & Ingegnoli, 2010).

Mario Raviglione [Dr. in Medicine, Prof. of Global Health at the State University of Milan, formerly Professeur Titulaire of the Global Studies Institute (GSI)—Université de Genève]. Raviglione had been asking Ingegnoli to hold 4 hours of lessons at the II level Master in Global Health, since its inception (2020). Topic: The Intrinsic Relation Between Agricultural Landscape Health and Global Health.

Joan Tort Donada [Dr. of Geography, Prof of Regional Geographic Analysis, Head of Department of Geography (University of Barcelona)], who invited Ingegnoli to hold a Seminar for PHDs in Geography in Barcelona (2023).

Lucia Urbani Ulivi [Dr. in Philosophy, Prof. of Theoretical Philosophy, Catholic University of Sacred Hearth in Milan, and Università della Svizzera Italiana, Lugano], who called Ingegnoli to the research team on Systemic Paradigm, where they collaborated with **Evandro Agazzi**, the well-known past president of the International Society of Epistemology, research that gave rise to the volume *The Systemic Turn in Human and Natural Sciences: A Rock in the Pond* (Springer-Nature, 2019).

Competing Interests The author has no competing interests to declare that are relevant to the content of this manuscript.

Contents

Part III Applications and Conclusion

Part I
Landscape Bionomics: Premise and Contents

Landscape Ecology, A Widening Foundation
(the *foundation of* Landscape Bionomics) *is well named.*
All students and problem-solvers concerned with changing land
should carefully absorb its pages
Richard T. T. Forman
Harvard University,
November 2001

Chapter 1
Environmental Studies and the Systemic Paradigm

Contents

Abstract The Anthropocene, man's domination over nature, has produced an environmental crisis, but this is not understood in an entirely negative sense because the destruction of nature has brought some socioeconomic benefits too. However, we must understand that survival requires a decisive change of direction. The benefits mentioned are precarious and transitory if obtained by destroying nature!

The emergence of the concept of the landscape lead to the discipline of Landscape Ecology, and this science tried to demonstrate the grave perils hidden in nature. Many limitations appeared, however, in Biology (from Genomics to Ecology), from the dominance of the so-called dogma of microbiology to the ambiguous concept of ecosystem. Even geography and landscape ecology, together with general ecology, are of crucial importance in studying the environment but they present many limits. For instance, the definition proposed by the European Landscape Commission (ELC) does not mention the complex adaptive system concept. This failure represents a severe limitation, because it leads to errors in the criteria of application for environmental assessment and territorial design (planning) too.

It must be noted that a good part of the crisis in life sciences depends on reductionism. As Evandro Agazzi reminds us, reductionism is at a dead end today because it claims to arrive at unity regardless of variety. Many phenomena follow principles and methods proper of complex systems and must be studied following

V. Ingegnoli, *Landscape Bionomics and Planetary Health*, Landscape Series 21,
https://doi.org/10.1007/978-3-032-08908-3_1

this approach; therefore, to get out of this crisis, science needs to change the paradigm, moving towards Holism, at first following the System Theory.

As underlined in the Preface, the need to direct studies toward the systemic paradigm have been felt in almost all disciplines, both natural and human sciences. This need has pushed toward the development of disciplinary branches of updating. The two most significant disciplines are landscape bionomics (LB) and planetary health (PH), born a few years apart in the architectural/naturalistic and medical/environmental fields.

Keywords Environmental crisis · Ecological limitations · Landscape · Complex system · Reductionism · Upgrading disciplines · Paradigm shift

1.1 Environmental Crisis

1.1.1 Crisis Salient Features

Many scholars and politicians argue that we are now in the era of the Anthropocene, the predominance of man over nature. This domination has produced an environmental crisis, but this is not understood in an entirely negative sense because unbridled consumerism and the destruction of nature have also brought some socioeconomic benefits: in the last 50 years, on our Planet, poverty has fallen from 90 to 50%, medium life expectancy has increased from 45 to 70 years, cereal production (tons/ha) has more than doubled (Lancet Commission, 2015). It is believed that if man will dominate nature, his increasingly powerful technology will be able to face and overcome the crisis. That is why it is so hard to change direction, yet we must understand that survival requires a decisive radical shift. The aforementioned benefits prove precarious and transitory if obtained by destroying nature!

Let us remember that the extent of the environmental alterations is shocking. The majority of people now recognize the crisis; the mass media debate it and even those who have not studied can understand that:

(a) forests have decreased in the world by 25% in about 130 years,
(b) pollution is widespread,
(c) the countryside no longer has trees (Fig. 1.1) and it is a succession of monocultures,
(d) insects have significantly reduced, especially pollinators,
(e) drinking water is scarce in many areas,
(f) cities have expanded out of all proportion (the conurbations of Milan, Rome, Naples, and Turin contain ¼ of the Italian population),
(g) the historical centers with their cultural heritage are depressed (Fig. 1.1),

Fig. 1.1 Alteration of agrarian and urban landscapes: industrialized rice monoculture near Vercelli (Italy) and the historic downtown of Boston encapsulated in looming towers. (Fig. by Ingegnoli)

(h) landslides and floods have also increased (almost 400% in the last 60 years in Italy),
(i) many serious diseases are on the rise (diabetes-2 and cancer exceeded 165% in the last 30 years), toghether with the rise of the antibiotic resistance,
(j) climate change is now evident. Even if man did not contribute more than 60–70% to the heat increase, it should be noted that the lack of rebalancing capacity of the biosphere due to the decline in ecological efficiency of forests (in addition to their cover decrease) is, in any case, a man's fault.

For some people, the problem of energy should also be added, but here it must be remembered that this problem is too often poorly posed because, cocerning the complex system of Earth, it is not enough to strive for so-called clean energy; it is not an eminently technological or economic problem! It is a reckless increase of entropy in biological systems due to abuse of energy which, even if it were to become truly clean and low-cost, could become too much to be managed by the Planet, thus the greatest danger to Life on Earth. Remember Gandhi's warnings, also taken up by Pope Francis: "We must learn to take a step back and self-limit energy use" (Pope Francis Ist, 2015).

1.2 Importance and Limits of Ecology and Geography

1.2.1 Fundamentals and Limits of Traditional Ecology

After the Second World War, in the 1950s, environmental studies were dominated by the rise of the most important school of Ecology in the world due to Eugene P. Odum (University of Georgia), who published the volume Fundamentals of Ecology (Odum, 1953), reprinted then in 1971 (Odum, 1971). He took up Tansley's concept of *ecosystem (Tansley,* 1935*)*, better-specifying structure and functions and distinguishing the ecology of populations from that of communities. He added the Habitat Approach for the study of aquatic and terrestrial ecological systems and the first mention of the leading applications. The study of the developments of ecosystems is very significant, and an extreme summary proposed by Ingegnoli (Ingegnoli, 1980) is reported in Fig. 1.2.

Many limitations appear, however, in Biology (from Genomics to Ecology), such as the dominance of the so-called dogma of microbiology (Crick, 1970), where the phenotype is still understood for the most part to be directly expressed by the genotype without thinking about discoveries coming from epigenetics and the transmission of information from the environment (Fig. 1.3, left). Furthermore, the concept of *resilience* was favored, forgetting that the most advanced natural systems optimized the opposite concept, that of *resistance* to disturbances and their incorporation (disturbance incorporation). Therefore, they show a *decidedly lower specific biodiversity*: in practice, the more ordered the system, the more its components have efficient connections and are able to oppose or 'incorporate' disturbances; only *resilience requires redundancy*, and therefore a higher biodiversity, because it compensates for the lack of structuring, i.e., organization, with a hyper-abundance of species, or it is configured as a survival strategy when the limits of incorporation of disturbances are exceeded, for example, after a disturbance that destroys a stretch of forest: in the right part of the figure, we can see how the highest biodiversity (edible sp. and total sp.) is found in the most degraded systems, not in the more organized ones (Fig. 1.3. right). It should be noted that even the high biodiversity of tropical forests may be the Earth's response to the disturbance due to the strong heat of the present interglacial, as stated by James Lovelock. In add, one must consider not only the diversity of species but also the heterogeneity/diversity of ecotopes in a landscape and of landscapes within an ecoregion. So, promotion of biodiversity must be carefully related to the health situation of the considered LU.

Moreover, the true concept of ecosystem is ambiguous (Fig. 1.4), as demonstrated by O'Neil et al. (1986), recognizing the impossibility of integrating biotic and functional perspectives. Furthermore, the concept of ecosystem implies a definition based on structures and functions that are not always able to vary together with the scale considered, e.g., from a small ecotope to a large landscape (system of ecosystems); the invariance concerning scale is in contrast with the principle of emergent properties (Ingegnoli, 2002, 2015). It must always be remembered that the scientific concept of the ecosystem is not a mere synonym or a contraction of the term "ecological system".

The conservation of nature excludes man because man's actions always cause an impact on nature, an impact which, however, is often a positive factor, as can be seen by evaluating the so-called Ecosystem Services, provided that this concept can work as well as has been proposed: we will see that other limitations will emerge. The example of Pieve-Emanuele, Milan Metropolitan Area (Table 1.1), a municipality of high urbanization, is given. Note that 60% of its territory is still comprised of vegetation components, therefore photosynthesizing, but only the 15–16% of them are natural, and so counted: the remaining 45% is made up of anthropic vegetation, thus not considered or evaluated, which is scientifically unacceptable!

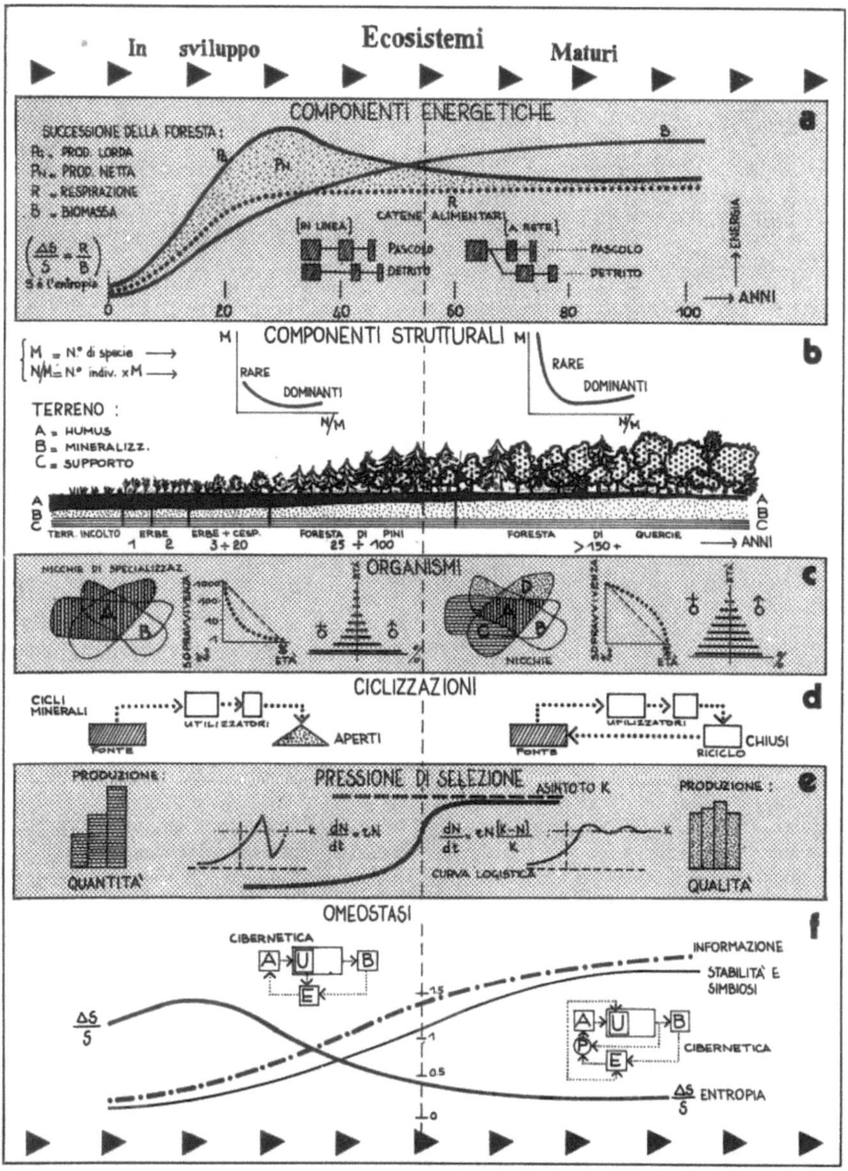

Fig. 1.2 Dynamics of ecosystems and their components according to the School of P.E. Odum (1971) summarized by Ingegnoli in his volume "Ecologia e Progettazione" (1980). Fig. by Ingegnoli, 1980

Note that the principle of emergent properties, although mentioned on the first page of Odum's book is rarely applied in studies. Furthermore, the concept of Landscape and Live Information (sensu Faggin (2022): both symbolic and semantic)

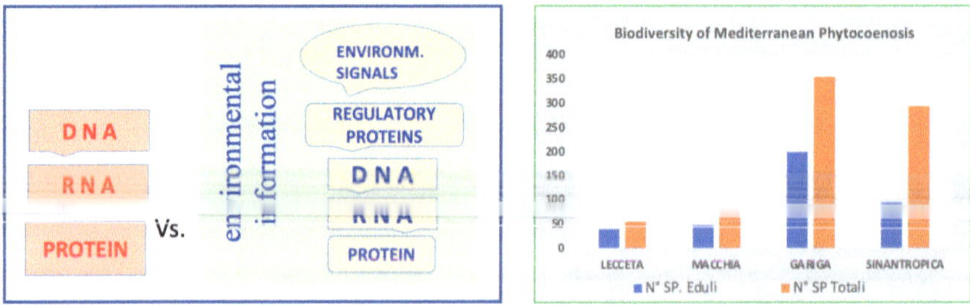

Fig. 1.3 (Left) The dogma of microbiology is not the one supported by F. Crick (1970): the epigenetic control of gene expression, due to DNA methylation, demonstrates that the phenotype is not expressed directly by the genotype. (Right) The most evolved natural systems follow a strategy of resistance stability, not resilience; only resilience needs redundant biodiversity: from Lecceta (sclerophyllous oak forest) to Gariga (Chaparral scrub), we go from 40–50 to 250–350 species; but the pinnacle of the organization is Lecceta. Fig. by Ingegnoli

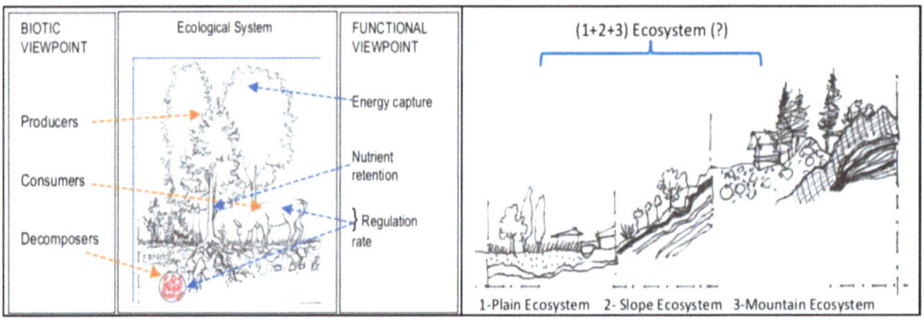

Fig. 1.4 The concept of ecosystem is understood by traditional ecology as independent of space-time dimensions (contrary to reality); furthermore, the ecosystem is not able to integrate biotic and functional criteria… Finally, it is a term abused by journalists and environmentalists/ecologists!. (Fig. by Ingegnoli)

does not appear in this scheme (Fig. 1.5). It is important to underline that any transformation between energy and matter needs a sort of catalysis through an information system, to increase the neg-entropy and to proceed toward ordered forms. Let us remember that when energy is transformed in matter it assumes three spatial dimensions (x, y, z) plus one temporal dimension (t); while if energy is carrying information, it assumes two spatial dimensions (e.g. plane wave) and two temporal dimensions (t_1, t_2) (Manzelli, 1999). Reality is created continuously by the interactions that take place between its "parts" which are not separable, e.g., entanglement and quantum fields Live Information (Faggin, 2022).

Table 1.1 Municipality of Pieve Emanuele

Elements of the landscape	Land cover %	Photosynthetic land cover	Land cover accepted by phytosociology
Dense urbanized	2.50	0.25	0
Cool urbanized	13.28	2.40	0
Industries	14.00	0.14	0
Roads and railways	1.80	0.09	0
Vegetable gardens and gardens	0.75	0.60	0.10
Golf	10,95	8.50	0.60
Public park	1.50	1.35	0.90
Paddies	0.80	0.70	0.10
Arable land	35.00	30.00	1.50
Mowable lawns	5.40	5.00	4.20
Tree-lined and uncultivated rows	6.35	6.00	4.00
Riparian forests	4.50	4.30	4.00
Reforestation	0.50	0.40	0.40
Lambro River	2.40	0.24	0.10
Canals and canals	0.27	0.03	0.10
Total Ha 1310.00	100	60%	16%

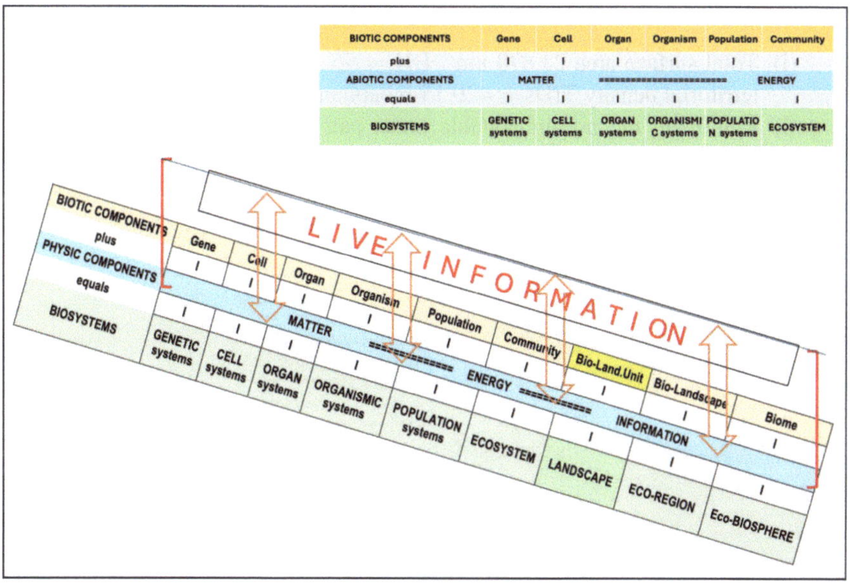

Fig. 1.5 (above) The "Biological Spectrum" sensu E. P. Odum (1953, 1971) ends with the concept of ecosystem. (below) Upgraded Bio-Spectrum, by Ingegnoli. Note the presence of the crucial concepts of Landscape and Live Information (information both symbolic and semantic). Note that any transformation between energy and matter needs a sort of catalysis through an information system

1.2.2 Fundamentals and Limits of Geography

In addition to what was already said in Sect. 1.2, we should note that the ability to address problems on a global scale, the attempts to deal with complex systems and the interest in the relationship between man and nature still form the basis of geographical studies, even if today they are often dominated by statistical economic aspects. The implications of the relationship between urbanization and the rurality of populations, which has been disrupted in recent decades, is an important part of geographical studies, as are many aspects of urban ecology (Fig. 1.6).

As happens with ecology, similar limitations can also be found in geographic studies, often due to the same problems: abuse of the concept of ecosystem at all scales, too rigid concept of sustainability, often improper use of the population density parameter. We will simply remind you that:

(a) Regarding sustainability, it is not a question of managing resources in such a way as to postpone their exhaustion as much as possible, as sustainable development would like, but rather of maintaining the same *environmental potential* for future generations and, therefore, understanding the relationship between pathologies of the area in question and the *state of human health*, verifying their compatibility.

(b) Regarding population density, a distinction needs to be made between that which refers to the entire territory and that which refers to the ecological one, i.e., which refers only to the HH (human habitat), at least excluding physically uninhabitable areas. See an example for the Lombardy region (2015):

   (i) Total surface area 23,870 km^2, HH = 56.20%, inhabitants 10,05 million, territorial density TrDs = 420,15 inhabitants/km^2.

   (ii) Removing the non-inhabitable areas pair to 8.20% (water and glaciers 4.48% + rocks and screes 3.72%), HH = 60.55%, thus the ecological density ED = 458.83 inhabitants/km^2.

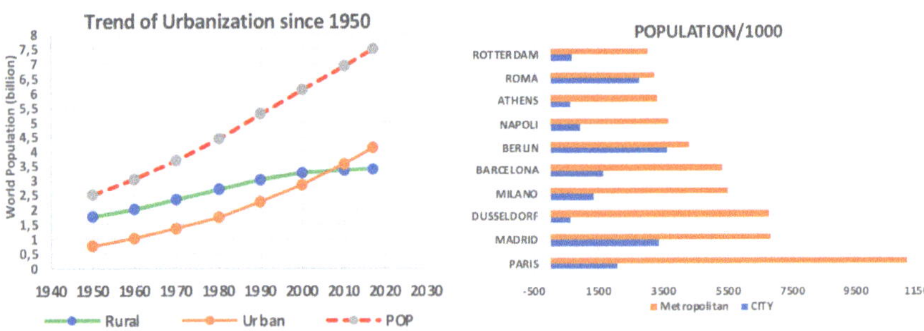

Fig. 1.6 (left) Overtaking of the urbanized population compared to the rural one in 2008. (right) Comparison between some of the major cities in the European Union (2024). Note that the 10 biggest conurbations (53 million inhabitants) measure about 12% of the entire population of European Union

(iii) Removing 8.2% plus the natural habitat areas, pair to 21.87% (natural forests, shrublands, herbaceous formations, swamps), HH = 79.70%, the bionomic density BD = 641.91 inhabitants/km^2, equal to 153% of Territorial Density!

1.2.3 Fundamental and Limits of Landscape Ecology (LE)

1.2.3.1 Man Organizes His Territory: From the First Maps to the Concept of Landscape

Since the Neolithic, man has tried to study his territory while organizing it in a compatible manner with his needs. At the beginning of the Iron Age, graffiti on rocks can be seen representing cultivated and inhabited territories, such as the map of Capo di Ponte in Val Camonica (Brescia, Italy), which still corresponds, after three millennia, to many features of the current landscape (Fig. 1.7); situation similar to the recently discovered carved motifs in the sandstone of the shelter floor within Ségognole 3 shelter (Paris basin, France), that can be compared with major geomorphological features in the present surrounding landscape, potentially becoming "a world-first 3D-model of a Paleolithic territory" (Thiry & Milnes, 2025). Many parts of today's European and circum-Mediterranean territory still show the structure due to Roman centuriation, after two millennia.

In the Greek-Roman period, we began to pass from empiricism to science. In the first century, the disciplines most interested in studying the environment were (in order) Geography, Agronomy, Architecture, and partly Medicine. However, even more interesting is the awareness of scholars to consider these matters in a trans-disciplinary way, as M. Vitruvius Pollio (first century AD) underlines: "*Encyclios enim disciplina uti corpus unum ex his membris est composita…*" ("Interdisciplinarity, as a unique body, is composed by its members") (Vitruvius). Always within the Roman world, the landscape was called *"regio-regionis"* by the

Fig. 1.7 Rural landscape in Val Camonica at Capo di Ponte as it is visible by the place of the graffiti map demonstrating the permanence of the analogous type of rural organization from the beginning of the Iron Age (around 1000 BC) (from Ingegnoli, 2015)

agronomist Columella (first century AD), underlining its geographical aspect, differentiating it from the "prospectus" of the landscape painter, called *"pictor topiarius qui regiones formas pingit";* therefore, landscape and panorama were very distinct.

The confusion between these two concepts had been starting with the landscape painters of the Renaissance when the prevalence of reductionism began to be irresistible, bringing undoubted immediate advantages but also distortions in knowledge. It was necessary to wait for a bio-geographic scholar of the caliber of Alexander Von Humboldt (1840) to distinguish the importance of the concept of Landschaft as *"Der Totalcharakter einer Erdgegend"* once again, that is, the total character of a given territory.

In the next generation, another German biogeographer, Ernst Haeckel (1866), proposed the term Ecology for the field of study of species-environment relationships, and this was undoubtedly a helpful step. However, he placed the landscape outside the field of the natural sciences. The importance of the landscape was only re-proposed in 1922 by Sigfried Passarge (1922), again a German geographer and opponent of the American morphological school of W. M. Davis, who tried to give a new orientation to geography, specifying the scientific concept of Landschaft as something having its proper *physiology*. In the same years, the landscape concept was underlined by the birth of Josias Braun-Blanquet's Phytosociology (Braun-Blanquet, 1926), who founded the well-known Montpellier school.

1.2.3.2 From the Concept of Landscape Trough Landscape Ecology to Landscape Bionomics

A certain number of geographers and territorial planning architects, parallel to Odum's publications, realizing that the concept of landscape had not been considered in ecology, proposed to remedy this, remembering that, otherwise, the spatial distribution in the territory of populations and communities (including man) could not have been analyzed properly.

Carl Troll wrote a significant article: *"Die geographische Landshaft und ihre Erforshung"* (Troll, 1950); Vittorio Ingegnoli published an article "Ecologia Territoriale e Progettazione" (1971)[1] and later the volume "Ecologia e Progettazione" (1980)[2] (Ingegnoli, 1971, 1980), which attempted to complete some aspects of E.P. Odum ecology with landscape analysis; R.H. Mac Arthur proposed a *Geographical Ecology* (Mac Arthur, 1972); H. Leser (Germany) wrote a volume with the explicit title of *Landschaftsokologie* (Leser, 1976). All pioneering works are not to be overlooked, but the true founders of Landscape Ecology were Naveh & Lieberman, Haifa (Naveh & Lieberman, 1984) and Forman & Godron (Harvard and Montpellier, 1981, 1986) followed again by Forman (Forman & Godron, 1981, 1986; Forman, 1995). Zev Naveh's approach is transdisciplinary

---

[1]Territorial Ecology and Planning.

[2]Ecology and Planning.

theoretical and systemic-functional and capable of dealing, at least in part, with complexity; that of Richard T.T. Forman is more pragmatic-applicative and morpho-functional, well linked to the structures of the territory and useful for territorial planning.

As stated by Risser (Risser, 1983), Landscape Ecology is the "synthetic intersection of many related disciplines" with a well-established cultural-scientific structure that focuses on multiscale feedback between spatial patterns and ecological processes (Forman, 1995; Naveh & Lieberman, 1984; Turner et al., 2001). Spatial patterns are quantified by a metrics toolkit, such as patch size, connectivity, and shape, allowing inferences about the dominant processes and scales operating over time. Landscape Ecology, differently from other ecologies, "focuses on the interrelations in -and across- a landscape mosaic, consisting of multiple ecosystems" (Wu, 2012); to better identify the scientific core of landscape ecology, Wu indicated ten key research topics as follows: (i) pattern-process-scale of the landscape; (ii) connectivity and fragmentation; (iii) scale and scaling; (iv) spatial analysis and modelling; (v) land use and land use change; (vi) landscape history and legacy effect; (vii) landscape and climate change interactions; (viii) ecosystem services; (ix) landscape sustainability; (x) accuracy assessment and uncertainty analysis.

During the last decades, many landscape ecologists contributed to developing the science of landscape ecology in terms of key concepts, methods of theory formulation used, pathways of development in the formulation of theories, range of applicability in time and space, and tools for spatial analysis. In their overview of theoretical contributions, Van Eetvelde and Christensen (2023) found 32 theories and four pathways of theory development.

The new discipline called Landscape Bionomics was developed starting from theories belonging to Landscape Ecology: the scientific approach, principles, methodologies, and goals of Landscape Bionomics have been thoroughly described mainly in the three volumes published by Springer in 2002, 2011 and 2015 (Ingegnoli, 2002; 2011, 2015). Forwarding the first volume (Ingegnoli, 2002), Forman writes: "The diverse European and North American approaches and perspectives have finally coalesced. A vibrant palette of methods, both for analytic understanding and to evaluate landscapes for appropriate land uses, is lucidly revealed. An impressive range of theories and simple, useful equations for landscape ecology are laid out. A promising approach to understanding landscape "pathology," or degradation, is introduced. New case-study applications in Italy, Africa, and Europe are presented. The "widening foundation", whereby landscape-ecology principles are increasingly used in fields related to land use, is illustrated from urban planning and conservation biology to agricultural land and economic/ecological sustainability." The third book (Ingegnoli, 2015), still published by Springer, describes and updates the theoretical foundations, with some paragraphs on human health, discipline-specific terms, and applications, and provides new case studies.

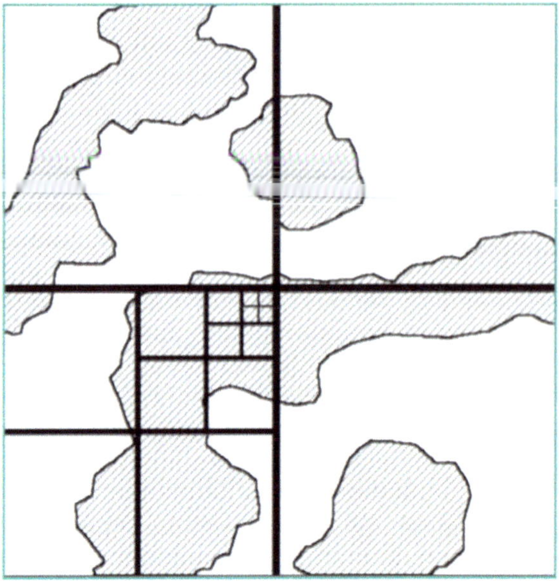

Fig. 1.8 Example of distribution of tree-lined spots framed in a series of grids obtained by dividing the side of the square into two parts each time. The significance value of the attribute per cell was 30% (from: Ingegnoli, 2002).

1.2.3.3 Limits of the Discipline of Landscape Ecology

Landscape Ecology (LE) has added interesting innovations to General Ecology: for example, the use of fractal geometry, i.e. the geometry that appears to be closest to that underlying natural systems, given that Euclidean geometry only considers dimensions integers (0, 1, 2, 3, 4… respectively for the concepts of point, line, surface, volume, space-time…), while the shapes of the natural elements are irregular, with "fractional" dimensions (Mandelbrot, 1975).

The study of the landscape has many applications, including evaluating the distribution methods of the attributes under examination in a territory (Fig. 1.8).

Furthermore, the LE pursued to progress toward a systemic study of complexity. The systemic study is important (Antrop & Van Eetvelde, 2017) because even an in-depth study of the individual ecological communities present in a landscape unit does not allow one to understand the behavior of the "community system" formed by them. Only by starting from the unity of the system toward its parts credible research can be carried out.

Unfortunately, the definition of the landscape made its own by the European Landscape Commission's (ELC) does not mention the concept of complex adaptive systems at all. This silence is a severe limitation, leading to errors in the environmental assessment and territorial design (planning) application criteria. In Fig. 1.9, the definitions of the landscape are reported to observe that the one of ELC does not consider the Principle of Emergent Properties of von Bertalanffy (Von Bertalanffy, 1968) and is, therefore, scientifically incorrect.

Several limitations are also found in the LE. Some limits cited for General Ecology have yet to be exceeded (many researchers still talk about ecosystems or

traditional	*advanced*
The landscape means an area, as perceived by people, whose character is the result of the action and interaction of natural and/or human factors	**The landscape means a complex system, existing as a biological entity, whose character and behaviour are more than the result of the action and interaction of natural and human components**
following the European Landscape Convention (2004)	*following Ingegnoli V. in: Baltimore D, Dulbecco R, Jacob F, Levi-Montalcini R (eds.), Frontiers of Life (2001)*

Fig. 1.9 Contrasting definitions of landscape in Europe

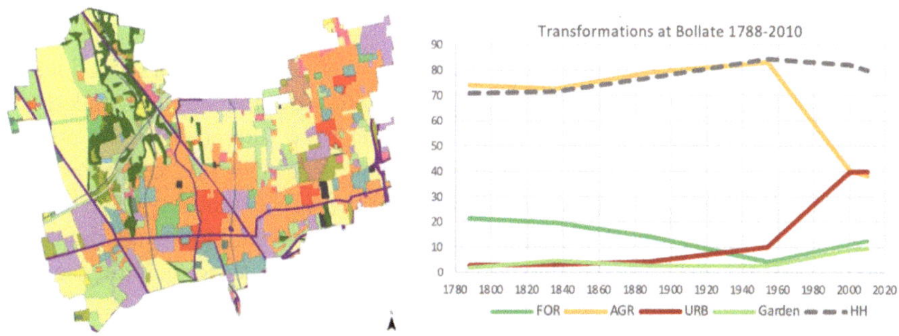

Fig. 1.10 Municipality of Bollate (Milan): transformations from the end of the eighteenth century. As of 2010. Even by applying the principles of the landscape ecology, it is not possible to evaluate the ecological state of its territory: was it better or worse in the past than today? (Fig. by Ingegnoli)

biodiversity incorrectly). The structural references of the landscape have gone beyond the ecosystemic-reductionist one (one ecosystem at a time) but have stopped at the concept of eco-mosaic. Above all, it must be noted that the indices and measures introduced by landscape ecology help evaluate the distribution of components of the territory. However, they do not allow us to understand the structure of a landscape as a complex system and, therefore, how to evaluate any alterations of a landscape unit under consideration. The LE cannot answer some basic questions, such as: (i) How can the territorial transformations -that have occurred to date- be interpreted and evaluated? (ii) Is it possible to evaluate the ecological significance and "direction" of the future transformations?

Looking at Fig. 1.10, we witness in the municipality of Bollate (MI) an intense urban change after 1950, but also a regrowth of woods and gardens. Some questions are inevitable:

1. Can the Landscape Unit (LU) support vast urban growth?
2. In the case of ecological alteration, how to measure ecological dysfunction?
3. Can the increase in grasslands counterbalance the decrease in cultivated fields?
4. Is the decreased forest cover also influenced by decreased ecological efficiency?
5. How do we compare the ecological state of this LU with the surrounding LUs?
6. How can the crucial relationship between the biological state of the environment and human health be studied?

If we want to move forward to study the complexity of environmental systems, we must respect the important scientific discoveries brought by general ecology, geography, and landscape ecology, but also overcome their limits. It is essential to review the development of epistemology.

1.3 The Importance of Paradigms in Science

1.3.1 From Reductionism to Complex Systems

It must be noted that a good part of the crisis in science depends on reductionism. The reductionist method reduces every phenomenon to the simplest underlying structure, believing that the simplest is the physical one; this method was applied to all fields of science. As the eminent epistemology Evandro Agazzi reminds us (Agazzi, 2014, 2019), reductionism is dead today because it claims to arrive at unity regardless of variety. Many phenomena need to be investigated following principles and methods that are more suitable for studying complex systems; therefore, science needs to change paradigms to get out of this crisis.

To avoid reductionism, each object of study can no longer advance through separate sectors but must consider the context, meta-disciplinary schemes, and emergencies. These aims mean studying for "complex systems"and not misleading us, as Galileo did, that scientific knowledge could completely neglect ontology and metaphysics. An oversight like this is dangerous, as it creates a model of knowledge that renounces the essence of reality and truth as absolute. Let me confirm this thought throw the words of Faggin (2022): "The new paradigm, instead, arises from the need to overcome the dichotomy between nature and culture, organism and consciousness, limiting reductionism and studying intertwined fields".

In his book "The Structure of Scientific Revolutions", Thomas Khun (Khun, 1962) was the first scientist to underline what would soon become an axiom for the whole world: the concept of "paradigm". Even in the Sciences, the paradigm shift is linked to the historical and cultural evolution of man, and in our Western World, we can highlight the main changes that have occurred from protohistory up to the present day, as very briefly summarized in Table 1.2.

It is observed that (a) the duration over time of the main paradigms is progressively decreasing, especially in the last two millennia, (b) the transition from the Middle Ages to the Modern Age shows less drastic changes than in the recent transition from reductionism to systemic studies, (c) conventional historical periods are modified.

In summary, the old scientific paradigm was predominantly reductionist, anchored to reversible processes, to the Darwinian "battle for existence", leaving little room for transdisciplinarity. The new paradigm, however, derives from the need to overcome the dichotomy between nature and culture, limiting reductionism and studying complex systems. The new paradigm is characterized by holism, symbiosis and cooperation, irreversible processes, and transdisciplinarity.

Table 1.2 Sequence of the main scientific paradigms in the history of the West (51 centuries)

N	Times	Centuries	Main charactyers	Main paradigms
I	Protohistoric	XXX–XV a.C	First civilizations. Writing (Sumerian). Metallurgy (bronze) and hydraulics. First studies in geometry and medicine.	Civilization allows us to give organized answers to the questions posed by both technology and patient care
II	Ancient	XV–IV a.C.	Alphabet. Persian mathematicians and astronomers, Egyptian doctors. Early Greek philosophers. Thales and Pythagoras.	The concept of 'intelligibility' appears, the theoretical form from which the order of nature can be deduced philosophically and mathematically
III	Classic	IV a.C.–VII	Founding principles of the major scientific branches. Aristotelian logic. School of Alexandria. Euclid and Archimedes. Justinian's codes.	The principle of experimentation appears but governed more by reason than by empiricism. Hypotheses are emerging that will be taken up much later.
IV	Medieval	VII–XV	The advance of science with scholasticism, Thomas Aquinas and Bacon. First universities and hospitals.	The controlled scientific experiment (Bacon); importance of empiricism and birth of science in the modern sense
V	Modern	XV–XX	Galileo and the first fundamental laws. Newton and gravitation. Reductionism and technology, exceptional development of all sciences.	A theory must always be verified with experimental tests and the tool for analyzing the results is mathematics
VI	Current	XX–...	Einstein, quantum physics and general relativity. Systems theory, holism, computer science, epigenetics...	Need to overcome the dichotomy between nature and culture, limiting reductionism and studying complex systems

1.3.2 The Need of Upgrading Disciplines

The scientific world, including the medical one, has proposed new areas and research associations capable of considering the "Systemic Turn" (sensu Urbani-Ulivi) (Urbani-Ulivi, 2019). First, I refer to the meetings promoted in 2015 by the "Rockefeller Foundation-Lancet Commission on Planetary Health" on Lake Como in Bellagio (Lancet Commission, 2015). But it must be underlined that 2015 was an exceptional year for environmental protection because the *UN Agenda 2030 for Sustainable Development commitment to sustainability*, signed at the UN by 193 countries, also began at that time. The volume "Landscape Bionomics, Biological-Integrated Landscape Ecology", published by Springer by Ingegnoli, was released in 2015 too.

In the same year, in parallel to the scientific world, the religious spheres expressed in the Encyclical "Laudato Sì" (Pope Francis Ist, 2015) that we, humans, are intrinsically part of nature:

> When we talk about "environment" we also refer to a particular relationship: that between nature and the society that inhabits it. This prevents us from considering nature as something separate from us or as a mere framework of our lives. We are included in it; we are part of it and we are penetrated by it. (…) It is essential to seek integral solutions, which consider the interactions of natural systems with each other and with social systems. (Laudato Si', 139)

It was then made official by the following Catholic-Orthodox meeting. The need to direct studies toward the systemic paradigm has been felt in almost all disciplines, both of natural and human sciences. This upgrading has pushed towards the development of disciplinary branches of updating. As regards the environment and life sciences, we can mention:

- Epigenetics (Bottaccioli & Bottaccioli, 2020; Carey, 2012) to overcome the limits of neo-Darwinism and link micro and macro-scales.
- Agroecology (Barberi & Bocchi, 2018; Gliessman, 2015) to combat the damage caused by agricultural industrialization.
- Bionomics and Landscape Bionomics, LB (Ingegnoli, 2002; Ingegnoli, 2011; Ingegnoli, 2015), capable of significantly increasing the systemic development of ecology and allowing a better study of the environment/health relationship.
- Planetary Health (Ingegnoli et al., 2022; Myers & Frumkin, 2021) capable of highlighting the relationships between the environment, pollution, and health.
- Information Medicine (Laszlo & Biava, 2019), the revolutionary cell-reprogramming discovery that reverses cancer and degenerative diseases.
- Urban Ecology (Forman, 2014, 2019), capable of studying the environmental state of cities and its consequences on the population.

The two most significant disciplines are Landscape Bionomics (LB) and Planetary Health (PH), which were born a few years apart within the architectural/naturalistic and within medical/environmental fields, respectively.

This notable development of systemic disciplines also produces a change concerning interdisciplinarity (Council of Europe, 2013), which can more easily transform into transdisciplinarity. Recalling what was written by Vitruvius (see Sect. 1.1), we can say that the systemic sense (*corpus unum*) has been lost or greatly weakened due to reductionism, which at most allowed the convergence of different disciplines towards an objective of study (interdisciplinarity). Instead, the disciplines born for a systemic turn following theoretical bases and, often, also methods typical of Systems Theory are spontaneously transdisciplinary.

1.4 Landscape Bionomics and Planetary Health: A First Framework

As a summary of this introductive chapter, allow me to introduce a logical diagram from which we can see that the man-environment relationships brought to a wide ecological crisis, which can be studied and controlled following both Traditional Ecology and System Theory: however, their integration is strongly limited by reductionist residues in Ecology, while it is possible through the new concepts of life and environment formulated as the scientific paradigm shift desired by the two mentioned disciplines of Landscape Bionomics (LB) and Planetary/One Health (PH) with concrete positive results (Fig. 1.11).

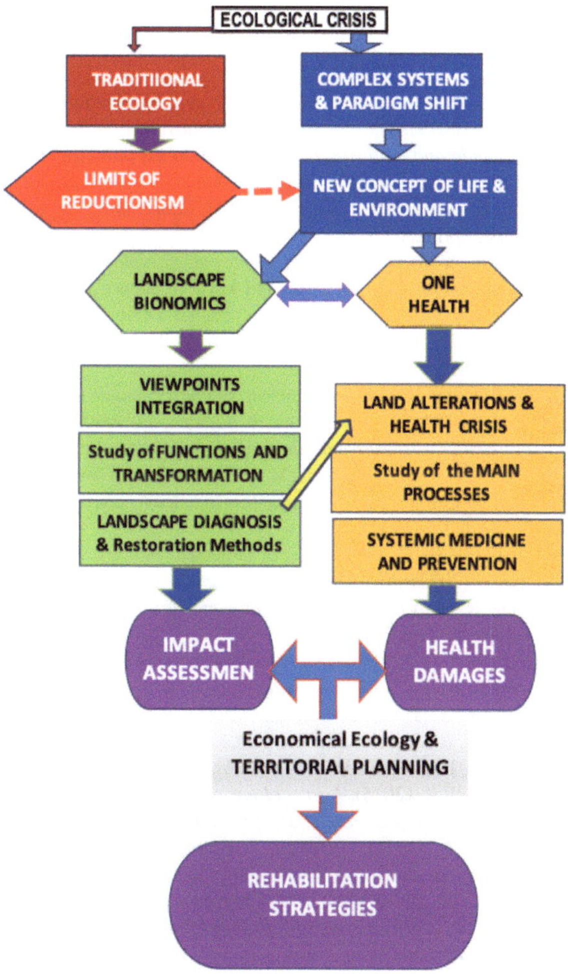

Fig. 1.11 The logical diagram showing the need to develop the two interacting environmental disciplines of Landscape Bionomics and Planetary Health

The contents of LB start from integrating the separated fields of study of traditional ecology, – developing the structure, functions, transformations, and evaluations of landscape units (LU) and their natural and anthropic components -, to arrive at the elaboration of a diagnosys of the health state of the investigated LU, being able to find and quantifying their alterations. With this path, we enter the discipline of PH, which, having ascertained the state of health of the LUs under examination, highlights the processes linking the environment and human health and proposes the first demonstrative studies of this correlation. Only through this path can we develop new ecological territorial planning methods capable of considering human health. The relationships between impact assessment and health damages, through Territorial Planning and economical ecology permit to project the Rehabilitation Strategies.

References

Agazzi, E. (2014). *Science, metaphysics, religion*. Franco Angeli.

Agazzi, E. (2019). *Systemic thinking: An introduction*. In L. Urbani-Ulivi (Ed.), *The systemic turn in human and natural sciences. A rock in the pond* (pp. IX–XVII). Springer.

Anati, E. (1985). *I Camuni*. Jaca Book.

Antrop, M., & Van Eetvelde, V. (2017). *Landscape perspectives, The Holistic Nature of Landscape*. Springer.

Barberi, P., & Bocchi, S. (2018). *Agroecology and organic agriculture: Opportunities for innovative agronomic research* (invited keynote paper). In Conference: 47th congress of the Italian Society of Agronomy (SIA), Marsala, 12–14 September 2018.

Bottaccioli, F., & Bottaccioli, A. G. (2020). *Psycho-neuro-endocrine-immunology and science of integrated care*. Edra.

Braun-Blanquet, J. (1926). *Pflanzensoziologie: Grundzüge der Vegetationskunde*. Springer.

Carey, N. (2012). *The epigenetics revolution: how modern biology is rewriting our understanding of genetics, disease, and inheritance*. Iron Books Publ.

Columella, L. J. M. (Ist. cent.) *De re Rustica*. Loeb Classical Library W Heinemann LTD, .

Council of Europe. (2013). *European landscape convention, 7th council of Europe conference*. Palais de l'Europe.

Crick, F. (1970). Central dogma of molecular biology. *Nature, 227*, 561–563.

Faggin, F. (2022). *Irriducibile. La coscienza, la vita, i computer e la nostra natura*. Mondadori.

Forman, R. T. T. (1995). *Land mosaics: the ecology of landscapes and regions*. Cambridge University Press.

Forman, R. T. T. (2014). *Urban ecology, science of cities*. Cambridge University Press.

Forman, R. T. T. (2019). *Towns, ecology, and the land*. Cambridge University Press.

Forman, R. T. T., & Godron, M. (1981). Patches and structural components for a landscape ecology. *BioScience, 31*, 733–740.

Forman, R. T. T., & Godron, M. (1986). *Landscape ecology*. John Wiley & Sons.

Gliessman, S. R. (2015). *Agroecology, the ecology of sustainable food systems*. CRC Press.

Haeckel, E. (1866). *Generelle Morphologie der organismen* (Vol. 2). Georg Reimer.

Ingegnoli, V. (1971). Ecologia territoriale e progettazione: significati e metodologia. In: L'ingegnere di fronte ai problemi della sopravvivenza umana. In *European congress of FEANI* (pp. 398–400). Collegio Ingegneri.

Ingegnoli, V. (1980). *Ecologia e Progettazione* (pp. 1–311). Cusl.

Ingegnoli, V. (2002). *Landscape ecology: A widening foundation* (p. XXIII+357). Springer.

Ingegnoli, V. (2011). *Bionomia del paesaggio. L'ecologia del paesaggio biologico-integrata per la formazione di un "medico" dei sistemi ecologici* (p. XX+340). Springer.

Ingegnoli, V. (2015). *Landscape bionomics. Biological-integrated landscape ecology* (p. XXIV + 431). Springer.

Ingegnoli, V., Lombardo, F., & LaTorre, G. (2022). *Environmental alteration leads to human disease: A planetary health approach* (Sustainable development goals series. Foreword by Andrea Lenzi) (p. XIV+337). Springer-Nature.

Khun, T. (1962). *The structure of scientific revolutions* (p. 318). The University of Chicago Press.

Lancet Commission. (2015). Safeguarding human health in the Anthropocene epoch: Report of the Rockefeller Foundation-lancet commission on planetary health. *The Lancet, 386*, 1973–2028.

Laszlo, & Biava. (2019). *Information medicine: The revolutionary cell-reprogramming discovery that reverses cancer and degenerative diseases*. Healing Arts Press.

Leser, H. (1976). *Landschaftsokologie*. Ulmer.

Mac Arthur, R. H. (1972). *Geographical ecology*. Princeton University Press.

Mandelbrot, B. (1975). *Les objets fractals: forme, hazard et dimension*. Champs, Science.

Manzelli, P. (1999). Il cervello, la percezione, il colore. Didattica Delle Scienze, n. 202, Firenze.

Myers, S., & Frumkin, H. (2021). *Planetary health: Protecting nature to protect ourselves*. Island Press.

Naveh, Z., & Lieberman, A. (1984/1994) *Landscape ecology: theory and application*. Springer XXVII+360.

O'Neill, R. V., De Angelis, D. L., Waide, J. B., & Allen, T. F. H. (1986). *A hierarchical concept of ecosystems*. Princeton University Press.

Odum, E. P. (1953). *Fundamentals of ecology*. WB Saunders.

Odum, E. P. (1971). *Fundamentals of ecology*. Saunders.

Passarge, S. (1922). *Landschaft und Kulturentwicklung in unseren Klimabreiten*. Friederichsen.

Pope Francis Ist. (2015). *Laudato Sì*. Encyclical.

Risser, P. G. (1983). *Landscape ecology, directions and approaches*. Illinois University.

Tansley, A. G. (1935). The use and abuse of vegetational concepts and terms. *The Ecologist, 16*, 284–307.

Thiry, M., & Milnes, A. (2025). Palacolithic map engraved for staging water flows in a Paris basin shelter. *Oxford Journal of Archeology, 44*(1), 2–26.

Troll, C. (1950). Die geographisched Landschaft und ihre Erforschung. *Studium Generale, 3*, 163–181.

Turner, M. G., Gardner, R. H., & O'Neill, R. V. (2001). *Landscape ecology in theory and practice. Pattern and process*. Springer.

Urbani-Ulivi, L. (2019). *The systemic turn in human and natural sciences. A rock in the pond*. Springer.

Van Eetvelde, & Christensen. (2023). Theories in landscape ecology: An overview of theoretical contributions merging spatial, ecological and social logics in the study of cultural landscapes. *Landscape Ecology, 38*(12), 4033–4064.

Vitruvius P. M. (I sec). De Architectura Lib. I-X.

Von Bertalanffy, L. (1968). *General system theory, foundations, development, and applications*. George Braziller.

Von Humboldt A (1840) Kosmos.

Wu, J. (2012). Key concepts and research topics in landscape ecology revisited: 30 years after the Allerton Park workshop. *Landscape Ecology., 28*, 1–11.

Chapter 2
Complex Systems and the Birth of Bionomics

Contents

Abstract A Summary Notes on the concept of system and on the System Theory (by Alessandro Ingegnoli) is presented.

Then a synthesis on Complex Systems will be exposed, defining hierarchical, dynamic, and dissipative systems, arriving at the general behavior of non-equilibrium systems, metastability, and disturbances, with important upgrading in the interpretation of landscape transformation and the role and importance of disturbances within it. Complexity can be understood as an attribute of a system containing information that is difficult to understand. In the case of complex systems made up of complex subsystems, we can also talk about levels of complexity and hyper-complexity.

After reviewing the main concepts related to the System Theory, we can pass to an upgrading of traditional ecology through the new ecological discipline of Landscape Bionomics. The levels of biological organization cannot be limited to cells,

V. Ingegnoli, *Landscape Bionomics and Planetary Health*, Landscape Series 21,
https://doi.org/10.1007/978-3-032-08908-3_2

organisms, populations, and communities. They must also include the more strictly speaking ecological systems: landscape, ecoregion, and the Earth as a living system (Gaia).

Note that traditional general ecology does not integrate the four criteria of study (space configuration, biotic, functional, and cultural-economic). In contrast, it is their integration that allows us to recognize them as complex adaptive systems as they are. A comparison between traditional ecology (TE) and landscape bionomics (LB) can be seen in Table 2.2. The differences are numerous (only the first 30 have been listed here) and often substantial: for example, the concepts of community and ecosystem, which occupy the same vast range of scales, while they contrasts with the principle of emergent properties.

The landscape is a complex system of ecocoenotopes. This observation is less easy to understand than one might think because, in traditional ecology, the reference unit is the ecosystem, which, as we have underlined in the previous chapter, is an ambiguous concept.

Keywords System theory · Complex system · Advanced ecology · Landscape bionomics · Biological spectrum · EIA (Environmental Impact Assessment) · SEA (Strategic Environmental Assessment)

2.1 Summary Notes on Complex Systems

by Alessandro Ingegnoli[1] (2011a)

2.1.1 Introduction

The notion of a system has remote origins. The term derives from the Greek "susthma" (to be together, meeting) through the Latin *systema* and is used to indicate both a complex of philosophical doctrines and physical theories (for example, the Ptolemaic system); both physiological entities (for example, circulatory system) and a set of physical elements (for example a gear system) and, recently, both physical and algorithmic elements (for example a computer system).

In its correct meaning, it indicates a set that has common characteristics. These characteristics can be the similarity of the components (as in an insular system) or their functional integration (as in the circulatory system). This second aspect is considered in the scientific field.

---

[1] The here reported introduction on the System Theory (Sects. 2.1.1, 2.1.2, 2.1.3, 2.1.4, and 2.1.5) was written in 2011 for my book "Ingegnoli V *Bionomia del Paesaggio*. Springer, Berlin, Milano", by my brother Alessandro (1945–2020), Electronic Engineer, victim of Covid-19 sanitary restrictions.

In the literature, in systems theory, there are multiple definitions of system:

1. Any aggregate of abstract physical elements in which cause-effect relationships can be identified (Aicardi, 2004).
2. Any identity that can be analyzed and broken down.
3. A mathematical model of a physical or artificial phenomenon includes its causes and effects and the mathematical relationship that binds them (Turchiano, 2000).
4. A pair $\Sigma := \{V, R\}$ where V represents the set of variables and R represents the set of relations between the variables (Monaco, 2011).
5. An entity (physical, generally) that responds to an action, exercised through an input, with a reaction called output.

Definitions, sometimes contradictory, tend to highlight the epistemological (1, 2), modelling (3), mathematical (4) or more technical (5) aspect. But, considering what has been said, we can accept as a general definition of the system, in the scientific sense, "*any object of study which, made up of different elements interconnected and interacting with each other or with the external environment, reacts or evolves with own laws in a unitary and organic way*".

2.1.2 The System Theory

If we want to understand the great importance that Systems Theory has assumed, we must start from the potentially almost infinite number of its theoretical and practical applications. This is thanks to the extreme generality of the theoretical approach which allows us to treat as a "system" a very large class of "objects" whether about *physics* (mechanical, fluid dynamic, thermodynamic, optical, circuital, electronic), or about *biology* (tissues, organs, organisms, ... landscape units, landscape districts) but also economic sciences and social sciences.

The main purpose of systems theory (Rinaldi, 1973) is to investigate the methodologies for studying oriented dynamic systems and the class of linear and stationary systems, both continuous time and discrete time. Since it is important to associate the phenomena with a quantitative description, the system to be studied is associated with a model that facilitates its analysis. This model aims to include, in an organic way, cause-effect relationships and to provide the tools for mathematical analysis. The model, however, is always an approximation of the phenomenon under examination and this approximation can be created in a sophisticated way depending on the specific needs and on the difficulties of mathematical treatment. Before the spread of electronic calculators, there were also analog circuit models (analog calculators, created using so-called operational amplifiers) which allowed important progress in the study of systems that were difficult to study through the normal means of mathematical analysis.

However, it was the advent of computers that revolutionized the field of studying system models because special programs, even very complex ones, were developed for simulation and calculation. Through programming techniques, results of great importance have been obtained, both theoretical and experimental, in the study of the properties of systems in both the time and frequency domains.

The mathematical method, whose origin can be attributed to the studies of Galileo, Newton, Leibnitz, Euler, Lagrange, Laplace and so on, of the greatest mathematical physicists, remains the main method for the study of systems. A mathematical model is an application, created using the tools of mathematics, which aims to represent, in the most faithful way possible, a physical phenomenon or a real process from a quantitative and qualitative point of view. It is usually created to provide predictions on the behavior of the system, judge its characteristics and describe the probable evolution of a physical or artificial phenomenon based on initial data (input) and returning final data (output). Its effectiveness can be measured by comparing the data from the calculations with the actual results observed. Mathematical models, complex, are now used in all fields: from meteorology to economics, from physiology to sociology, from traffic to ecology.

An important class of mathematical models is constituted by equations or systems of differential equations, ordinary or partial differential, obtainable starting from the so-called balance equations, for physical systems (mechanical, thermodynamic, electrical, etc.), but also used in economics and other fields. A set of differential equations can describe, for example, the structure of a building and the forces exerted on it, on this basis, the designer can predict the efforts or stresses to which it will be subjected. Another example is the differential equations used to study electronic circuits.

The solution of the differential equations that create the model is not always easy, indeed, in most cases it is necessary to resort to numerical resolution methods because it is impossible to find an analytical method, for which specific systems theory methods have been developed. It is interesting to note that different phenomena can be represented by the same mathematical model, therefore even the same abstract system can be associated with different phenomena. For example, the model that provides the frequency, for small oscillations, of the simple pendulum is that for an oscillating circuit composed of capacitance C and inductance L.

2.1.3 *The Physical Systems*

The physical systems, whose applications are the most studied because they are the ones that best lend themselves to theoretical investigation, can be divided into some important classes.

(i) The first distinction is between linear systems and non-linear systems. A system is linear when it depends linearly on the state variables and the input variables, non-linear if even just one of these dependencies is not linear. Furthermore,

(ii) systems can be stationary and non-stationary; stationary means that the system parameters are constant as time varies; non-stationary means that the system parameters are variable over time. Then,

(iii) there are static systems, i.e. those systems for which the response depends only on the instantaneous values of the input (for example, combinational circuits) and dynamic systems where the adjective "dynamic" indicates that the system evolves over time (for example, sequential circuits) which are the most studied. The main purpose of systems theory, in fact, is precisely to predict the characteristics of evolution over time in a complete and exhaustive way.

(iv) Another important division is that between continuous-time systems and discrete-time systems. For a continuous time-system, i.e., with time considered as a variable that changes continuously (like the infinite points of a straight line), the mathematical model is represented by a system of differential equations. For a discrete time one, i.e., where time is considered to vary in a quantized way by assuming multiple values of a given unit of measurement (for example, the second or the hour etc.) according to the natural numbers (i.e. $t = 0, 1, 2$), the mathematical model is made up of difference equations.

It is also necessary to mention the division between feedback systems (also called "closed loop") and non-feedback systems (called "open loop").

(v) Feedback systems constitute a class of great importance not only for their widespread technical use, but also for their fundamental theoretical implications. In a feedback system, part of the output signal is brought back to the input via an appropriate connection, thus constituting a closed loop (hence the name). Feedback allows oriented dynamic systems to consider what comes out of the system to modify its characteristics; the value of the variable exiting the system is brought back so that it can act, going back to the input of the system. There are two types of feedback, the so-called positive feedback and negative feedback. In positive feedback, the output signal is brought back in such a way as to "reinforce" the input signal. In this way we tend to increase the value of the output but, usually, the system becomes unstable. In negative feedback, in which the output signal is brought back in such a way as to subtract from the input signal, the value of the output tends to decrease, but on the other hand several extremely interesting (and very useful in the technical) properties appear. The stability of the system tends to increase, furthermore there is a significant improvement in the quality of the output signal (i.e., reduction in distortion).

(vi) In open loop systems, however, the value of the manipulable variable is determined within the system using mathematical algorithms (generally table reading). These systems are called predictive because no verification is carried out on the value.

It is important to add for completeness that a generic system can be *deterministic* or *stochastic*. It is deterministic when it is possible to determine the sequence of outputs, knowing the sequence of inputs and the starting state of the system itself. It is stochastic when this prediction is impossible either due to the presence of disturbances or due to the excessive complexity of the system itself (for example meteorological forecasts).

2.1.4 A New Theoretical Approach

There are also particular systems based on concepts that have recently emerged in the field of new theoretical approaches. We mention, among the most interesting:

- *fuzzy systems:* which allow the modeling of complex processes that cannot be well represented through conventional mathematical models because they are based on data without precise contours such as social, economic, managerial, medical diagnostic or perception-based systems human. The fuzzy model is based on groups of rules that make up an algorithm that determines the outputs of a process, through a finite number of steps based on the history of inputs and outputs.
- *neural systems:* formed by networks of elements (artificial neurons) which, unlike usual computer processes, based on a series of subsequent deductions based on binary logic, imitates the functioning of the most complex system of living beings, the nervous system. Neuronal networks can be made *capable of learning*, that is, they can modify their behavior so that their response to certain input stimuli gradually approaches a set of desired responses. They find numerous applications in robotics and image recognition.
- *Linearization.* The systems that occur are, almost all, essentially non-linear. However, for applicative purposes, the trick of hypothesizing them to be linear at least within a certain interval is often resorted to. That is, the effects of nonlinearity are considered negligible in a first approximation and approximate mathematical models are prepared that describe the system as if it were linear (this is called linearization). A linear system provides responses directly proportional to the stresses received and can generally be broken down into the sum of independent parts. However, if we find ourselves in the presence of *non-linearity*, the various components of a system *interact* with each other to make it *impossible to their separation*. Through a linear mathematical model, the system under examination can be represented as a polynomial function, whose coefficients are independent or with negligible mutual interactions, generally "easy" to calculate.

2.1.5 A Generic Dynamic System

A generic dynamic system can be represented as a "black box" with the inputs defined by the $a(t)$ signals and the outputs by the $y(t)$ signals. The state of the system is described by a set of variables, called "state", indicated with the letter $x(t)$, which define the situation the system is in at a certain moment in time.

The inputs act on the state of the system by modifying, over time, the values of the state variables. The values of the system outputs depend, in a complex way, on the system state variables and on the values of the inputs.

Three different equivalent and interchangeable modeling of an oriented dynamic system are theoretically possible:

- input-state-output model which highlights the internal state of the system, the disturbing or forcing causes that act on it, i.e., the inputs, and finally the output.
- input-output model (auto-regressive moving average model) which directly links the inputs and their derivatives with the outputs and their derivatives, "hiding" the state variables.
- model (valid for linear time invariant systems) based on the transfer function in the domain of the Laplace transform, the Fourier transform or the Z transform. Important properties of a directed linear dynamical system are represented by observability and reachability. A system is said to be observable if, for any possible sequence of states and inputs, the current state can be determined in a finite time using the system's outputs and inputs. The controllability of the system depends on observability. A system is defined as reachable if, by manipulating external causes, it can be made to reach the desired operating conditions. Stabilization depends on reachability.

A dynamic system is a system that evolves, both the input and the output change over time. In general, the output at time *t* does not depend only on the system input at the same time, but on a character, called *state*, which represents the history of the system inputs. To describe the variations of the system it is necessary to know at any moment a set of variables, called state variables, as well as the stresses that the system receives from the outside, or inputs to the system.

To create a mathematical model of a given continuous-time system, it is typically necessary to set up a system of differential equations that link the inputs, the state variables, and their derivatives in time. The resolution of this system of equations allows us to understand the functioning of the system and to find the values of the outputs:

$$dx(t)/dt = f[x(t), x_0, a(t), t] \; y(t) = h(x(t), x_0, u(t), y(t))$$

Where: $dx(t)/dt$ is the derivative in t of $x(t)$ and $x(t)$, x_0, $a(t)$, $y(t)$ are column vectors respectively composed of:

- state variables, functions of time t (in general, cannot be fixed or observed);
- state variables at the initial time t_0.
- input variables, functions of time t (which are acted upon, depending on the controllability characteristics of the system to modify the trajectory of the state). Usually there are also input variables, called disturbances, which represent spurious random elements acting on the system.
- outputs, i.e., the variables, functions of time t, from which the estimate of the state can be deduced, depending on the observability characteristics of the system.

The functions f and h depend on t. In this case the set of times coincides with the set R of real numbers and the dynamic system is called continuous.

If, however, the dynamic system has discrete time, the set of times coincides with the set Z of relative numbers so that the differential equations are replaced by difference equations.

2.2 Most Utilized Complex Systems

2.2.1 Main Types of Complexity

Complexity can be understood as an attribute of a system containing information that is difficult to understand (Ruelle, 1991). In the case of complex systems made up of complex subsystems, one also talks about levels of complexity and hyper-complexity. The highest level is called the "proper level of reality." A level of complexity is lower the greater the number of similar parts and the smaller the number of different properties that must be attributed to those parts.

By defining a system as a set of elements that exchange interactions with each other, we will notice that the main types of systems can present the following types of complexity:

1. Organized simplicity: few components, well-defined interactions.
2. Disorganized complexity: many components, random interactions (or similar).
3. Organized complexity: medium components, organic interactions.
4. Irreducible complexity: interactions so linked to form the unity that the removal of even just one part eliminates its functionality (Behe, 2012).

In cybernetics, systems are also distinguished as "adaptive": given a system S, we say that an event E is a stimulus for S if, and only if, the probability $P(S \rightarrow S'/E)$ that a system S shows a change (in elements or processes) upon the occurrence of an event E, is greater than the previous probability that S changes independently of E, therefore: $P(S \rightarrow S'/E) > P(S \rightarrow S')$.

It should be noted that, from a mathematical point of view, organized and irreducible complexities (points 3 and 4) are the most difficult systems (or not at all) to formalize, so it will be necessary to use models to study their behavior, as will be seen.

2.2.2 Hierarchical Systems

The hierarchy of a system depends on the irreversibility of time, which leads to a sequence of scalar processes because the flow of time is not homogeneous. Prigogine (Prigogine et al. 1972; Prigogine 1996) noted that entropic time (which indicates the evolution of systems) is very differentiated in our universe. A landscape, for example, comprises elements that follow different space-time scales (e.g., prairie and forest). The salient characteristics of a hierarchical system can be reported as below:

- *Links between entities:* The distance between levels decreases the bond forces between the components. Note that integration between the parts of a system does not require very close links between them but the presence of some connection (e.g., ecological network).
- *Relative frequency*: At high hierarchical levels, the processes occur with lower frequency (i.e., with a greater return time), which gives nature regularity.
- *Context:* This low frequency allows the high levels to be a context and constraint for the lower levels.
- *Containment*: In a non-nest hierarchy (i.e., dissimilar to Russian dolls) the high levels are distinct entities that do not contain the low levels. Therefore, if a system changes the hierarchical criterion, its structure must change (for example, food chains).
- *Constraints*: Higher levels have fewer constraints. These constraints are to be understood in a passive sense. They are important because they allow predictions.

Let me give an example of hierarchical systems in nature. Figure 2.1 (Ingegnoli, 2002) exposes the hierarchy of levels of biological organization highlighted in the sequence: organ, organism, community/ecosystem, ecotope/landscape unit. This series is represented respectively by a leaf, a plant, a wooded patch, and a vegetated ecotope. Each representation is linked to a graph with the frequency of photosynthetic production, a phenomenon that is increasingly slower with the increase in the hierarchical scale.

We will also observe that the lower levels explain the origin of the level of interest (i.e., they allow an internal description). In comparison, the higher levels explain the meaning of the level of interest (i.e., they allow the understanding of the characters derived from the translated conditions). Therefore, the concept of constraint is better

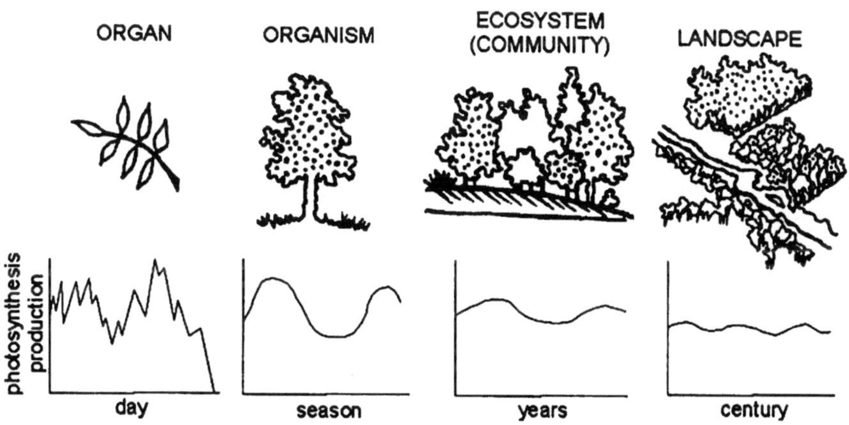

Fig. 2.1 Hierarchical levels of ecological systems. Photosynthesis in different scale systems. Note that a lower process speed corresponds to the behavior of a high-level system and vice versa. (From: Ingegnoli (2002))

than that of limiting factors in ecology. Note that there is a link between constraint and information: in a landscape, we can find sets of conditions (constraint fields) that give basic information on the limits of the transformation dynamics. So, the rupture of these fields alters the information content of a landscape unit.

2.2.3 Dynamic Systems

As previously noted, by knowing the values $x(t_1)$ and $a(t_1, t_2)$, i.e. the state of the system, therefore the response, at instant t_2, can be calculated. A dynamic system can be defined by six variables related to two functions:

1. *Input*: parametric input variables $a(t) \in A$ and input functions $a(.) \in \Omega$
2. *Output*: parametric output variables $y(t) \in Y$ and output functions $y(.) \in \Gamma$
3. *State variables*: $x(t) \in X$
4. *Time variables*: $t \in T$
5. *Function of the state transformation*: $\varphi\ [t, t_0, x(t_0), a(.)]$ from which $x(t) = \varphi\ [t, x(t_0), a(.)]$
6. *Function of output transformation*: $\eta\ [t, x(t)]$, from which $y(t) = \eta\ [t, x(t)]$

The couple state-time (x, t) is important because the set X, T is that of events, i.e., the history of the system. Given an instant t, an initial state $x(t_0)$, an input variable $a(.)$, the transition function $X(t) = \varphi\ [t, t_0, a(.)]$ is univocally determined, and called Movement of the system (Fig. 2.2) (Ingegnoli, 2002). Movement and trajectory of a dynamic system: X represents the field of states (state of the phases).

For example, Ax and Bx (Fig. 2.3) (Ingegnoli, 2002) are, respectively, a set of structural and a set of functional parameters: the *trajectory* is the projection of the movement in each *state field* X and can be used for checks on the bionomics transformation of the system itself, for example, following a project. An example of the application of dynamic systems theory in the study of the ecological state of a landscape is now presented.

In Fig. 2.4 (Ingegnoli, 2015), the model of control of the movement of the eco-landscape system of the Venice lagoon is shown. The structural ratio of sand-banks/tidal area (B/T) is measured on the abscissa, which is very significant for the lagoon; in ordinate, the average territorial bio-potentiality of vegetation (BTC) is estimated in Mcal/m^2/year. Note that going from the year 1900 (B/T = 28.5) to the

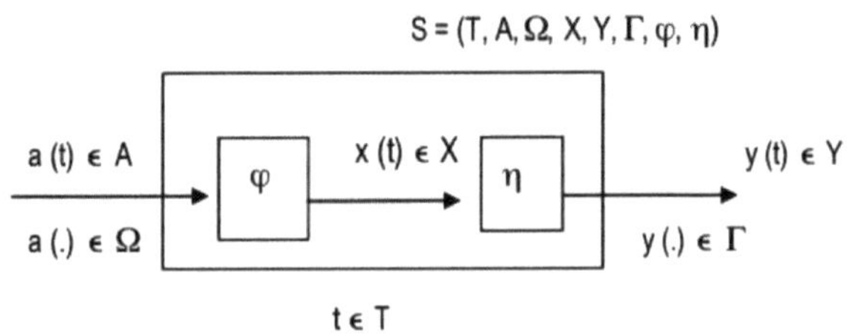

Fig. 2.2 Theoretical representation of a dynamic system, defined by six sets of variables. (From Ingegnoli (2002))

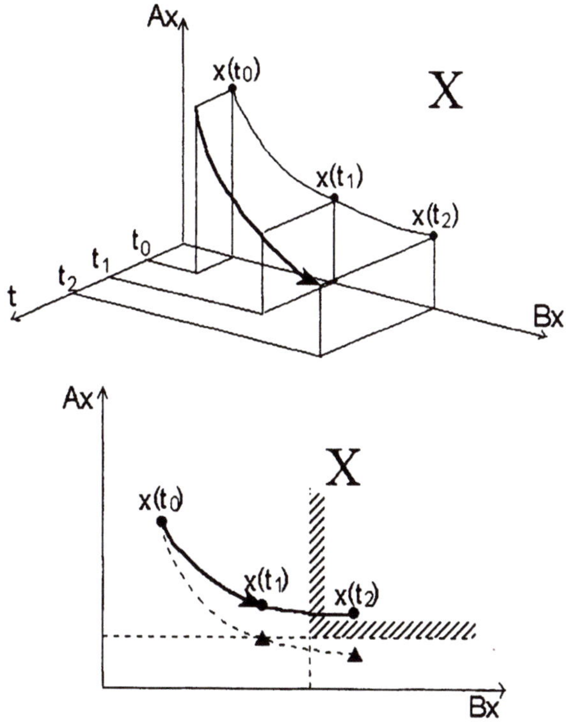

Fig. 2.3 Movement and trajectory of a dynamic system. X represents the field of the states of the system (state of the phases). The Ax and Bx axes are respectively (for example) a set of structural and a set of functional variables. The trajectory is the projection of the movement onto the state field. Note that the area delimited by hatching identifies limits that serve to control the dynamics. (From: Ingegnoli (2002))

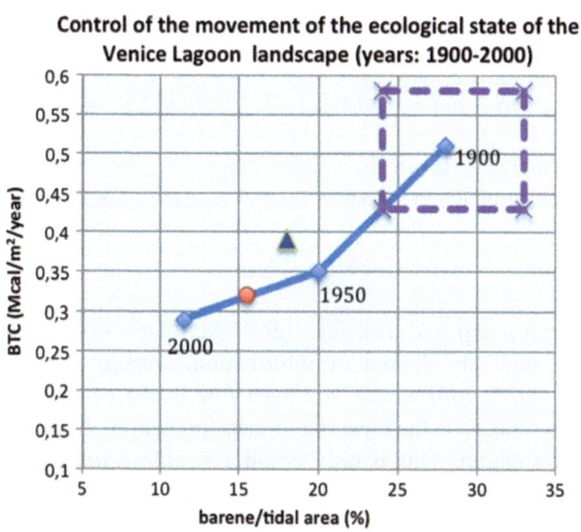

Fig. 2.4 Example of control of the movement of the landscape of Venice. The field of the states of the system (state of the phases) is done by a structural variable (barene/tidal area) and a functional one (biological territorial capacity of vegetation, BTC). The violet square represents the field of biological normality of the system. Two recovery plans (the year 2015) are compared: the blue scenery (through the blue triangle) is the best even if it does not follow back the pattern of transformation, as the red circle does (Ingegnoli, 2015)

year 2000 (B/T = 11), the system left the biological normality field (top right rectangle). In addition, to give a precise idea of the quality of the transformation of the general ecological state of the landscape under examination, this graph also allows us to verify the projection of the movement in the two intervention hypotheses (a = red circle) and (b = blue triangle), showing different distances from the normal range (dotted square).

2.2.4 Dissipative Systems

Why do photosynthesis processes have the greatest responsibility for energy transfer in biological systems? Living systems must be open, to allow the use of free energy *F*, being:

$$F = E–TS\ (where: E = total\ energy, S = entropy,\ \ T = temperature)$$

In an open system we must consider two *entropy flows*: d_eS, due to exchanges with the environment, and d_iS due to *irreversible* processes *within the system*. The second term (d_iS) must have a positive sign (due to the second law of thermodynamics), but the first (d_eS) have not a defined sign. So, the Clausius-Carnot inequality (in Fermi) (Fermi, 1972) becomes:

$$dS = d_eS + d_iS\ (since\ d_iS > 0)$$

In a period in which the system is *stationary* ($dS = 0$), we find:

$$d_eS + d_iS = 0\ and\ d_eS < 0\ (being\ d_eS = –d_iS)$$

Let us remember that *negentropy* flow means "*increasing the order of the system*". In an evolutionary process, when a system reaches a state of lower entropy (new order)

$$S(t_1) < S(t_0)$$

it means that it can *pump out disorder*. Remember that a system in homogeneous equilibrium is completely devoid of information; *entropy* therefore becomes a *measure of how much information has been lost* in the *transformation* and vice versa "*Negentropy*" takes on the meaning of measurement of the *information content* within the transformation. This is only possible in *non-equilibrium conditions*, in *dissipative* systems. That is why the amount of entropy "pumped out" is called negentropy.

A dissipation energy, which allows work, must for example be coupled with the transformation of a system from a state A_0 *to* A_1. The process capable of carrying out this transformation is called *operator (Op)* representing an action rule on a *given function*. Given $A_1 = (Op)\,A_0$, the transformation will be

$$A_1 = [(Op)A_0] \cup (ew \rightarrow ed)$$

where: ew = useful energy, ed = dissipated energy.

If the state of a system represents an eigenfunction (auto-function) for a certain operator, the system does not undergo any other changes. This stationary state is called an *attractor*: a stable condition in a dynamic system. An example: a patch of vegetation capable of holding back a conoid becomes an attractor for the establishment of a forest.

2.3 General Behavior of Non-equilibrium Systems

2.3.1 Metastability and Disturbances

The term metastability refers to a system oscillating around a relatively stable attractor but able also to move towards another attractor. The concept derives from organic chemistry but was underlined by Forman and Godron (1986) as indispensable in landscape studies. Metastability changes the traditional concept of ecological equilibrium, given that a dissipative system must stay far from thermodynamic equilibrium. Note that metastability is not a compromise between stability and instability: high or low metastability depends both on the distance from a position of maximum stability and the height of the local stability thresholds of the system (i.e. far from equilibrium) (Fig. 2.5) (Ingegnoli, 2015). A system may reaches a new level of organization after a phase of instability and the transition to a new metastable level.

Ecological systems with low metastability have good resilience but low resistance to disturbances, while those with high metastability have good resistance. For example, a meadow patch has greater resilience than a forest one.

For systemic reasons the role of disturbances in an ecological system changes too. A set of constraints of an ecological system is related in a one-to-one manner to the *strategy of resistance* to a regime of disturbances, of which the majority can be *incorporated*. That is, the structure of the system can tolerate disturbances, and, sometimes, some can be useful for maintaining order and/or to overcome limitations due to the biological potential of the components (at a lower level), processes that should never be understood in a static sense.

The involvement and the role of disturbances in the process of bifurcation will be demonstrated too. Other limits, such as environmental constraints, come from higher levels.

Therefore, we can talk about the incorporation of disturbances when the organization of the system exercises control over some environmental aspect that is impossible to control at a lower level of organization. If the disturbances are not "off the scale", therefore they can be incorporated, they can be counted among the forces structuring an ecological system. The need to review the methods of both

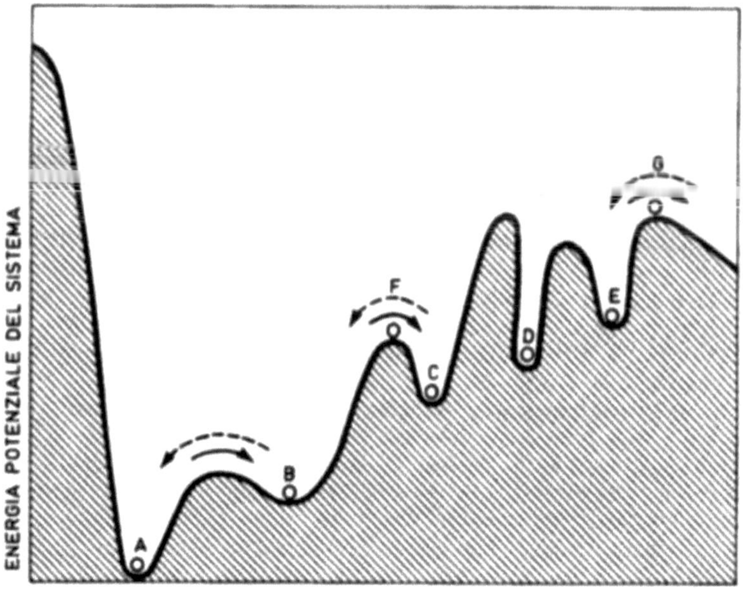

Fig. 2.5 Simple model of metastability. The so-called "roller coaster" model shows how the behavior of a marble ball, comparable to an ecological system, is stable in A, metastable in B, C, D, E, unstable in F, G. Greater metastability does not depend only on the potential energy, but also on the instability or stationarity thresholds (D > E)

environmental impact assessment (EIA) and strategic environmental assessment (SEA) must be underlined here because an elimination of disturbances that do not consider what is mentioned above can become seriously counterproductive.

2.3.2 *Landscape Transformations*

This "systemic approach" allows us to highlight a general model of the following type: (i) thermodynamic conditions can determine an "attractor," which represents a minimum dissipation condition for a system in its field of existence; (ii) possible macro-fluctuations (for example, due to energy accumulation or disturbances) produce an instability that shifts the system towards a new state of order; (iii) this new state, in turn, allows an increase in dissipation and brings the system towards new thresholds to reach a new "attractor". This implies that the so-called "operators" are also important in addition to the attractors, i.e., the transformation processes relevant to the systems under consideration.

This general model, which Ilya Prigogine (Prigogine et al., 1972; Prigogine, 1996) calls "order through fluctuations," manifests itself in all systems characterized by energy flows. These fluctuations are evident above all in living systems, from the

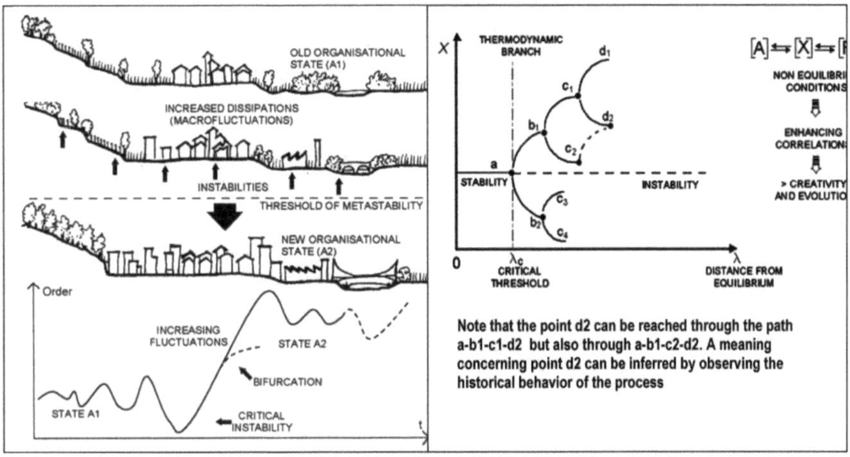

Fig. 2.6 (**a**) An example of transformation of a landscape (from agricultural to urban) following Ilya Prigogine's concept: order through fluctuation. See text for explanation. (From: Ingegnoli (2002)). (**b**) Successive bifurcations in a non-equilibrium system (Ingegnoli, 2015)

physiological ones of the organism (for example, muscle contraction, metabolic rhythms) to the ecological ones of communities or landscapes (for example, ecological succession, the transformation of landscapes) (Fig. 2.6). We must also observe that, once the instability threshold has passed, the system is faced with "bifurcations," that is, the possibility of reordering itself concerning more new attractors, perhaps due to disturbances (Fig. 2.6b) (Ingegnoli, 2015).

Figure 2.6a (Ingegnoli, 2002) schematizes the transformation of an agricultural landscape into a suburban one. From the organization's old state (A1), through an increase in dissipation, the system reaches a critical threshold, which can be represented as a branching point, transforming into the new state of the organization (A2), which means achieving a greater level of order. An increased flow of energy and information (for example, due to the construction of food industries and subsequent influx of inhabitants) produces macro-fluctuations in the local organization; therefore, produces some instabilities (for example, abandonment of crops, loss of riparian vegetation, new buildings, new bridges and roads…). Such instabilities lead to an increase in energy dissipation; the system becomes more challenging to maintain in the old state; once a threshold is reached (for example, the prevalence of urban structures over rural ones), the system changes organization and becomes a city, requiring different management. The system makes a "choice" based on often unpredictable information at each bifurcation. This process exemplifies why we talk about irreversibility too and thus overcomes the reductionist interpretation.

The second law of thermodynamics, noting that energy can only be transmitted with an increase in entropy, indicates the existence of irreversible processes that imply an "arrow of time".

2.4 Bionomic Renewal of Ecological Discipline

2.4.1 Real Systems and the Different Partial Viewpoints

After reviewing the main concepts related to the System Theory, we can pass to an upgrading of traditional ecology through the new ecological discipline of Landscape Bionomic, as underlined in Par. 1.7.

The levels of biological organization cannot be limited to cells, organisms, populations, and communities (*Biological Spectrum* sensu Odum (1971)). However, they must also include the more strictly speaking ecological systems: landscape, ecoregion, and the Earth as a living system (Lovelock & Margulis, 1974), as it is shown in Table 2.1, and Fig. 1.5. Note that the 4 criteria of study (space configurational, biotic, functional, and cultural-economic) are not integrated by traditional general ecology, whereas their integration allows us to recognize and identifies the six levels as complex adaptive systems (last column on the right) as they are.

Life and Environment form a single system, expressed in the space-time-informational dimensions of the biological spectrum. The conditions that allow life are necessary inside and outside a particular entity, distinguishable but homeorhetic, such as the organism or the landscape unit. Bionomics, or the Theory of the Organization of Life on Earth, transforms the fundamental principles of traditional ecology in the awareness that the hierarchical levels of the biological spectrum are types of complex biological systems, each relating to its scale of space-temporal information (Ingegnoli & Giglio, 2005; Ingegnoli, 2002, 2011b, 2015; Pignatti, 2015).

Table 2.1 Hierarchic levels of biological organisation on the Earth: real systems and the different partial viewpoints

Scales	Viewpoints				Real Systems[5]
	Space[1] Configurational	Biotic[2]	Functional[3]	Cultural-Economic[4]	
Global	Geosphere	Biosphere	Ecosphere	Noosphere	*Geo-eco-bio-noosphere*
Regional	Macro-chore	Biome	Biogeographic system	Regional Human systems	*Ecoregion*
Territorial	Chore	Set of communities	Set of Ecosystems	District Human systems	*Landscape*
Local	Micro-chore	Community	Ecosystems[a]	Local Human systems	*Ecocoenotope*
Stationary	Habitat	Population	Population niche	Cultural/Economic....	*Meta-population*
Singular	Living space	Organism	Organism niche	Cultural Economic operator	*Meta-organism*

1 = Not only a topographic criterion, but also a systemic one; 2 = Biological and general-ecological criterion; 3 = Traditional ecological criterion; 4 = Cultural intended as a synthesis of anthropic signs and elements; 5 = Types of living entities really existing on the earth as spatio-temporal-information proper levels

[a]The term ecosystem may be available following the functional viewpoint but, if we refer to the whole system, it must be identified as Ecocoenotope

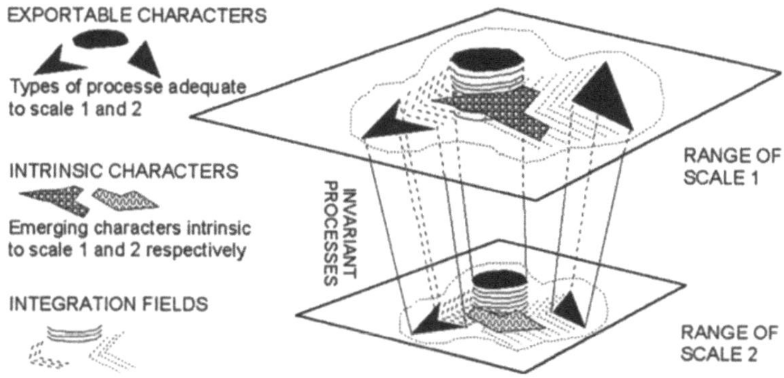

Fig. 2.7 Differences between specific and invariant characteristics of two hypothetical life organization systems. Moving from the first to the second level of organization, the next presents new specific characters, even if the invariant ones remain the same. A different correlation is found between the characters, which can be visualized by hypothetical fields that represent their relationships differently depending on the scale. (From: Ingegnoli (2002))

The "landscape" as the biological organization of a territory, i.e., a "complex adaptive living system, made up of integrated, interacting and co-evolved natural—living and non-living—and human components" is the fulcrum hierarchical level that expresses the transition between medium and large scales.

Remember that the levels in the hierarchy are distinguishable from each other because they operate at distinctly different speeds (Fig. 2.1). There are interactions of various kinds between the various scale levels (Fig. 2.7) (Ingegnoli, 2002): moving from the one to the next level of organization, the next presents new specific characters, even if the "exportable/invariant" ones remain the same. A different correlation is found between the characters, which can be visualized by hypothetical fields that represent their relationships differently depending on the scale.

The concept of constraint shows the behavior of an ecological system as limited by the behaviors of its components and by the environmental constraints imposed by higher levels of the organization. Therefore, there is a link between constraint and information. The alteration (or even loss) of important components of a given level can interrupt the transmission of information between subsequent levels, generating dangerous dysfunctions.

2.4.2 *Comparison Between Traditional Ecology (TE) and Landscape Bionomics (LB)*

As will be seen in Table 2.2, the differences between Traditional Ecology (TE) and Landscape Bionomics (LB) are numerous (even if they are not all listed here) and often substantial.

Table 2.2 Differences between traditional ecology and landscape bionomics

Arguments	Traditional ecology (TE)	Landscape bionomics (LB)
1- Biological Spectrum	(*Sensu Odum* 1971): Biotic elements (gene to community) + Abiotic components (matter & energy) = Biosystems.	(Biotic): *Landscape* and Eco-region included. (Abiotic): even *information* included.
2- Criteria of study	Separated: Spatial, Biotics, Functionals, Anthropics; at most interdisciplinary	Hierarchically integrated as necessary to study real systems and transdisciplinary
3- Ecosistem	Concept misunderstood respect to Tansley definition; misused as independent of space-time scales, often as an ecological system in a generic sense.	Concept not compatible with the Systemic Paradigm, which prefers the concept of ecocoenotope
4- System development	Succession by predefined evening stages and reversibility of the Degradation and Restoration processes to the status *quo ante.*	Development according to the bifurcations of Non-Equilibrium Thermodynamics. Irreversibility of Degradation/Restoration to the status *quo ante. Strategic processes of rehabilitation*
5- Biodiversity	Above all, understood as specific and proportional to the degree of organization	Not proportional to the degree of organization and related to tiles and ecotopes. The most evolved natural systems follow a *resistance* stability strategy, not a *resilience* one; *only resilience needs redundant biodiversity.*
6- Resilience	Prevailing concept over that of resistance	Resistance more important than resilience (appropriate is better than excessive)
7- Landscape	Understood as an aspect of the combination of anthropic and natural factors	Level of organization of life, complex system with its own characteristics
8- Structure of a territory	Ecomosaic, bidimensional structure, made up of tiles of natural and human elements	Ecotissue, multidimensional structure of a complex system, made of interacting and interrelated elements and of their reverberation and redundance
9- Parameters and measures	Sector measures, almost always reductionist; also considered valid for well-structured formations (e.g. forests)	Proper State functions and systemic parameters required for the study of complex formations
10- Ecological space	Space understood as a set of overlapping fields of existence of species	Intended space greater than the set of overlap of the fields of existence of the species, due to the addition of emergent properties

(continued)

Table 2.2 (continued)

Arguments	Traditional ecology (TE)	Landscape bionomics (LB)
11- Vegetation	Usually studied through methodologies still linked to Geobotany and (in Italy) Phytosociology; not including anthropic vegetation	Studied through new proper parameters and analysis criteria for complex systems. The LaBiSV method, the Bio-potentiality of vegetation, various models, etc. including anthropic vegetation
12- Fauna	Very in-depth studies, but above all by species, by habitat, by community.	Addition of the relationships with the bionomic functions of state and the concepts of Standard Habitat and landscape apparatus to the LE methods
13- Ecology and culture	Conventional notes in separate application chapters	Ecological control of culture attributed to cybernetic ethological and bionomic processes
14- Fluvial efficiency	Parameters by sector: Chemical analyses, BOD, Macro-benthos, Alveo data, etc.	RCE/IFF and similar. The surrounding territories are analyzed, as well as the excretory (EXR) and connective (CON) systems
15- Conservation	α, γ, β Indices (specific biodiversity) and similar. Control of endangered species, etc. Opposition between human needs and needs of nature	As in TE, adding the $\tau \times$ BTC analysis. Man is included in nature: attention to anthropized areas.
16- Perimeters and borders	By type of species to be protected, often independent of the geo-morphology of the places	Protection area boundaries: geomorphological and boundary data added. Bionomics 'radii of influence' considered too
17- Agro-ecology	Improvements in the species to be cultivated, in the reduction of pollutants, in soil processing	As for the ET, with more attention to the network of rows etc. In add, respect for every scale interrelationship and bionomic control of agricultural landscape units
18- Territorial planning	Conventional methods for zoning, prevailing socio-economic needs, without obligation of historical processes analysis and with reduction of ecological concepts to protect natural biotopes	Obligation to add analysis of Historical processes and systemic studies of bionomic processes to the usual themes (urbanization, cultivation, industries, natural protection areas, etc.)
19- Strategic environmental assessment	Impact assessment by thematic sector. Mitigation and compensation measures, even distant from the interventions. The socio-economic strategic directions are on par with the ecological ones.	Strategic directions based on health/human health relationship in the landscape unit. Clinical-diagnostic evaluation in a quantitative and bionomic sense.

(continued)

Table 2.2 (continued)

Arguments	Traditional ecology (TE)	Landscape bionomics (LB)
20- Sustainability	Sustainability as Sustainable Progressive Development. Guarantee children the same environment used by their fathers. 17 equivalent SD Goals (UN). Ecologically balanced transformations limited to the subject of intervention.	Sustainability as self-limitation, not of progressive development. Bionomic Carrying Capacity $\sigma = SH/SH^*$ Guarantee children the rehabilitated environmental potentialities used by their fathers. 17 Non-equivalent SDGs: education and ethics paramount. Ecological transformations balanced helping the ecological state of surroundings and of the LU.
21- Ecosistem services	Limited to humans. Supporting, regulating, provisioning, cultural. Modest variations in PES (Payment of Ecosystem Services) compared to the market.	Open to all components of biological systems. Added: Health Preserving and Protective. Related to landscape systems. Notable variations in PES.
22- Genetics	Still mostly linked to neo-Darwinism, albeit correct. Most diseases still depending on genetic disorders. Epigenetics often kept on the sidelines.	Rejection of the dogma of biology (DNA): the phenotype does not depend only on the genotype. Importance of Epigenetics that recognizes the crucial environment-genome links.
23- Resource consumption	National ecological surplus or deficit measured per person (ecological footprint). Production area necessary to regenerate the consumed resources (ha/capita)	More parameters involved: bionomic efficiency and landscape apparatuses, Minimum Standard Habitat SH* both for humans and animals; multidimensionality and multifunctionality of the ecocoenotopes in a territory.
24- Environment vs. human health	Only three categories of environmental alteration considered, with pollution prevailing. Internal processes, specific and general external factors.	Addition of a fourth category: the dysfunctions of landscape units, measured according to bionomics.
25- Space/time dynamics	Rarely considered; usually limited to a few years and to the site under consideration, or only months, in relation to reductionist measures for specific sectors	From regional to local scale with times of at least a few decades and for at least two past states, to detect the methods of change
26- History	Little appreciated, as studies are often limited to classical thermodynamics	Unavoidable, as closely linked to non-equilibrium thermodynamics, so the current state is not entirely understandable without history

(continued)

Table 2.2 (continued)

Arguments	Traditional ecology (TE)	Landscape bionomics (LB)
27- Information content	Limited to the concepts of heterogeneity and Shannon-Weaver dominance	Linked to the concept of landscape metastability and biodiversity of landscape elements, measurable with the g-LM indicator
28- Ecological networks	Created to address fragmentation problems. designed as natural infrastructures	Linked to undisturbed Areas of Strategic Importance, measured as for Landscape Ecology. Importance of the concept of PCA (potential core areas).
29- Transformation deficit	Marginal and underdeveloped concept, given the little importance of the history of ecological systems.	Important quantifiable concept, strategic to direct rehabilitation, to be related to a period of recognizable normality.
30- Cybernetics	Feedback controls in application modeling between social needs and the environment	The capacity for cultural direction due to (unconscious) information processes of bionomic systems also considered.

2.4.3 Premises to the Study of the Structure of the Landscape

It has been noted that the landscape is a system that results from integrating a set of biotic communities (including the human one) and the ecosystems they formed, as well as their spatial configuration at a territorial scale (Cora). The landscape is, therefore, a complex system of ecocenotopes. This observation is less easy to understand than one might think because, in general ecology, the reference unit is the ecosystem, which, as we have underlined, is an ambiguous concept. For many scholars, community, and ecosystem occupy the same vast range of scales, but this contrasts with the principle of emergent properties. Precisely for this reason, the need to consider both the intrinsic and propagable (invariant) characteristics that each level of the organization of life on Earth presents (Fig. 2.7) was underlined.

The structural characteristics of the landscape present in classical landscape ecology are referred to two models (Fig. 2.8) (Ingegnoli, 2002): the so-called fragmentation model, or traditional ecological mosaic, and the variegated model, or ecological mosaic with variable geometry (Samways, 1996).

These models represented two opposing visions: one ecosystemic and one species-specific. However, they don't express the real complexity in studying the landscape as a biological system. Furthermore, it is necessary to keep in mind that the main components of a landscape's structure are not only the natural, semi-natural, and anthropic (agricultural or urbanized) ones; there are also peculiar integrative configurations (Forman & Godron, 1986) such as agricultural fields with scattered farmhouses and residual woodlands. Every biological system has a

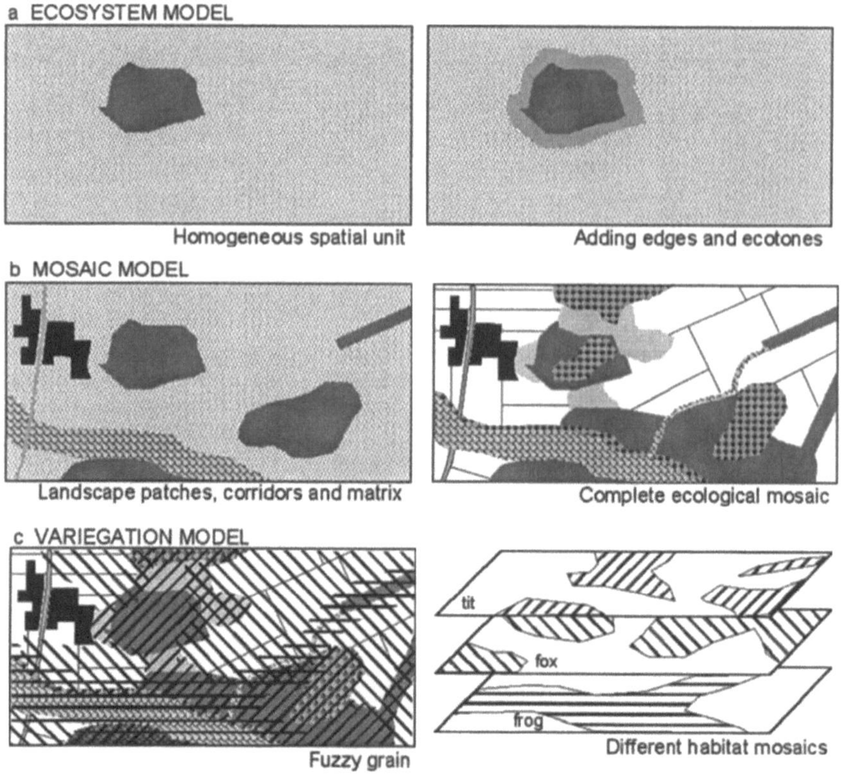

Fig. 2.8 Main structural models of the landscape; (**a**) ecosystem model; (**b**) ecological mosaic (or fragmentation) model; (**c**) variegated pattern. Each model can be represented in two versions, the second of which is more detailed. (From: Ingegnoli (2002))

very complex structure, made up of well-defined functional sets (for example, mitochondria at the cellular level or a wooded patch at the landscape level) in a context of substrates varying in space-time (for example, the cytoplasm, the landscape matrix).

References

Aicardi, M. (2004). *Teoria dei sistemi*. Dispense del corso. Dpt. di Ingegneria Elettronica, Università di Genova.

Behe M (2012) Irreducible complexity: Obstacle to Darwinian evolution. Cambridge University Press.

Fermi, E. (1972). *Termodinamica*. Bollati Boringhieri.

Forman, R. T. T., & Godron, M. (1986). *Landscape ecology* (p. XIX+619). Wiley.

Ingegnoli, V. (2002). *Landscape ecology: A widening foundation* (p. XXIII+357). Springer.

Ingegnoli, A. (2011a). Note sintetiche sui sistemi complessi. In V. Ingegnoli (Ed.), *Bionomia del Paesaggio*. Springer.
Ingegnoli, V. (2011b). *Bionomia del Paesaggio*. Springer.
Ingegnoli, V. (2015). *Landscape bionomics. Biological-integrated landscape ecology* (p. XXIV +431). Springer.
Ingegnoli, V., & Giglio, E. (2005). *Ecologia del Paesaggio: manuale per conservare, gestire e pianificare l'ambiente* (p. 685+XVI). Sistemi editoriali SE.
Lovelock, J., & Margulis, L. (1974). Atmospheric homeostatsis by and for the biosphere: the Gaia hypothesis. *Tellus, XXVI*, 2–10.
Monaco S (2011) Sistemi lineari, Elementi di analisi. Esculapio Ed
Odum, E. P. (1971). *Fundamentals of ecology*. Saunders.
Pignatti, S. (2015). Foreword. In V. Ingegnoli (Ed.), *Landscape bionomics*. Springer.
Prigogine, I. (1996). *La fin dès certitudes: temps, chaos et les lois de la nature*. Editions odile Jacob.
Prigogine, I., Nicolis, G., & Babloyatz, A. (1972). Thermodynamics of evolution. *Physics Today, 25*, 23–28, 38–44.
Rinaldi, S. (1973). *Teoria dei Sistemi*. Clup.
Ruelle, D. (1991). *Hasard et chaos*. Éditions Odile Jacob.
Samways, M. (1996). *Landscape ecology and biological conservation*. Biological Conservation.
Turchiano, B. (2000). *Dispensa di Teoria dei Sistemi*. Università di Roma, Dpt. di Informatica e Sistemistica.

Chapter 3
Structure, Functions, Transformations of a Landscape

Contents

Abstract The contents of Landscape Bionomics (LB) start from the integration of the separated fields of study of traditional ecology, developing the structure, functions, transformations, and evaluations of landscape units (LU) and of their natural and anthropic components, already deepened in the previous book.

Every biological system shows a very complex structure consisting of well-defined functional sets in a context of substrates varying in space-time. Remembering that the structure and the functions of a complex system are closely interrelated, two faces of the same evolving system, processes take shape creating the form of the landscape and of it subsystems.

V. Ingegnoli, *Landscape Bionomics and Planetary Health*, Landscape Series 21,
https://doi.org/10.1007/978-3-032-08908-3_3

The ecotissue represents the general structural model of a landscape unit (LU), that is, of a well-characterized part of a landscape, a territorial niche, the minor multidimensional subsystem of which can be called an ecotope. A system of LUs forms a landscape. Cybernetic processes are realized through interacting chains of organisms and communities which behave like information networks in the ecotissue, so much so that through them and their emergent properties, it is possible to maintain a certain level of metastability. Following Landscape Bionomic (LB) and the capacity of vegetation to control energy, matter, and information, we discovered a systemic function able to evaluate the energy flux available to maintain the order reached by a complex phytocoenosis: the biological territorial capacity of vegetation (BTC).

The main functions of a LU can be distinguished through actual ecological-functional apparatuses, which are analogous to the apparatuses of living organisms. Strictly related are the processes enhancing the main link between animal and human populations and their landscapes, which bring to their living space per capita: the processes can be expressed and quantified through the concept of "standard habitat per capita" (SH); SH is the inverse of the ecological density of the population, measurable in m^2/organism.

Concerning the reproductive landscape processes, it should be underlined that the reproductive processes of a landscape respond to the reproductive functions valid for all biological systems: the same for all the main other processes of living systems (Ingegnoli, Israel J Plant Sci 53:155–166, 2005).

Keywords Landscape structure · Landscape Unit (LU) · Ecotissue · Biological Territorial Capacity of vegetation (BTC) · Landscape Apparatuses (L.Ap) · Standard Habitat per capita (SH) · Cybernetic · Reproduction · Carrying capacity

3.1 Main Structural and Functional Characters of a Landscape

3.1.1 Landscape and Its Subdivisions in Traditional and Advanced Ecology

Concerning the levels of organization of life on Earth (see Table 2.1), it was noted that the *landscape* is a system that results from the integration of (a) a set of biotic communities (including the human ones), (b) a set of functional systems, (c) a spatial configuration at a territorial scale (*chora*) plus, more and more frequently, (d) the cultural and economic human district system. A landscape is born when the interaction and hierarchical integration of

- all the previous (a) + (b) + (c) + (d) elements and of the processes carried out by them, and of

- elements & processes and constraints & properties coming up by the upper and the lower organization levels,
- thanks to one or more emerging properties enhancing roles and meanings,

create a System which is unique, self-transcendent, identifiable and not repeatable and in a metastable state.

The *ecocoenotope* is a unitary entity that exists on a local scale, the lower life organization level respect to the landscape (territorial scale); it's the multifunctional—but not multidimensional- unit (recognizable only if an integrated point of view is adopted, for example, if one "looks" at the environment with an integrated perspective).

The substantial difference concerning the ecosystem is that 'ecosystem' is an abstract term (system of relationships): at the same time, ecocoenotope refers to a concrete reality, tangibly identifiable in the territory with a specific place (= *topos*) and describable as the integration of a specific set of communities of organisms (animals, plants, humans, and microorganisms) (= *coenosis*) in an environment (= *eco*, from *oikos* = home). Indeed, every biological system shows a very complex structure (Fig. 3.1), made up of well-defined functional sets (e.g., mitochondria at the cellular level or a wooded patch at the landscape level) in a context of substrates varying in space-time (e.g., the cytoplasm, the landscape matrix respectively).

As we have already said, this consideration is not easily understandable because, in traditional ecology, the reference unit is the ecosystem, which is an ambiguous concept. For many scholars, community, and ecosystem occupy the same vast range of scales between population and biome, but this contrasts with the principle of emergent properties. Table 3.1 (Ingegnoli, 2015) tries to give denominations and average dimensions of all the levels from the tessera to the landscape system.

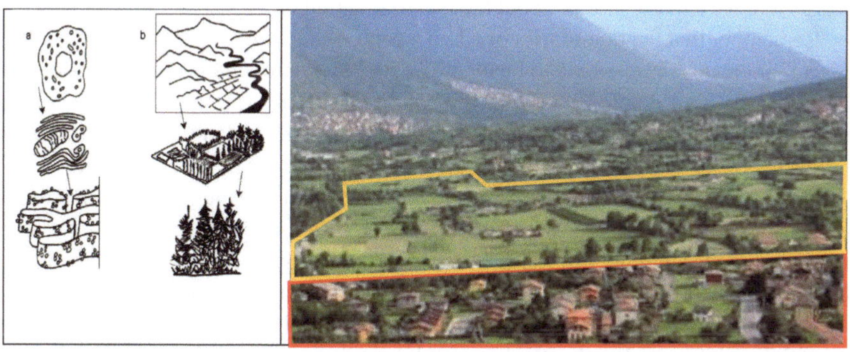

Fig. 3.1 Hierarchical levels of biological structures: small (**a**) and large (**b**) scale. Respectively: (**a**) from cell to ribosome and (**b**) from landscape to tile. The photo on the right shows two ecotopes within the territory of Capodiponte (ValCamonica, Lombardy), one agrarian (yellow), the other urbanized (red)

Table 3.1 Main ecological structures and ecological systems of life on Earth in traditional and advanced ecology and their common space-time dimension

Scale	Traditional ecology	Advanced ecology types of living systems	Landscape bionomics operative units	Normal surface (km^2)	Normal time (years)
Regional	Regional Ecosystem, Biome	Ecoregion (E-Reg)	Ecoregion, Land-System	5000–100,000	1000–5,000,000
Provincial/Territorial	Ecosystem	Landscape (L)	Landscape (L)	500–10,000	100–5000
District	Ecosystem	Sub-Landscape	Complex Landscape Unit (c-LU)	50–1000	50–2000
Municipality	Ecosystem	Landscape Unit (LU)	Landscape Unit (LU)	5–100	20–250
Meso-Zonal	Ecosystem	Ecotope (Ect)	Ecotope (Ect)	0.5–10.0	10–150
Micro-Zonal/Local	Ecosystem/Community	Ecocoenotope	Tessera (Ts, patch, tile, L. element)	0.005–1.0	5–100

3.1.2 *Frame of the Functional Processes*

Remember that the structure and the functions of a complex system are closely interrelated, complementary aspects of the same evolving whole, and even if educational reasons should have suggested a separation of the two topics into two different sub-chapters, they will be presented here in a sequence, guided by the words that we can use to talk about the materialization of a process: "it takes shape".

Remember that the boundaries of a Landscape Unit as well as that of a Landscape, even if not always immediate to be identified, occur where (that is in the precise sites and/or along well defined lines, i.e. the course of a river or a watershed line or an impressive human infrastructure or an abruptly change in land use…) the major processes of the environment change clearly making easier to be recognized.

Table 3.2 (Ingegnoli, 2015) can give an idea of the vastness of the functional processes concerning a landscape: they will be treated only briefly in a very close summary. Regarding the last topic, Landscape Pathology and Health, please refer to Chaps. 8, 9, 10, 11, Section II: Environmental Health and Human Health.

Among all these processes, here the attention will focus on the ones strictly essential to define and understand the landscape formation and behavior.

Table 3.2 Main articulations of the functional processes of a landscape

Root of processes	Sections
1- Geomorphical (Skeletal)	1.1 General processes 1.2 Geomorphic processes and soil formation 1.3 Signs of hydrological processes
2- Of connection and movement	2.1 Geo-climatic movements 2.2 Biological movements 2.3 Human movements
3- Autotrophic	3.1 Bio-energetic of vegetation (the BTC concept) 3.2 Ecological Efficiency of vegetation (the CBSt concept) 3.3 Vegetation dynamics (new theoretical principle) 3.4 Some vegetation processes in the landscape
4- Main configurative	4.1 Landscape biodiversity 4.2 General landscape metastability 4.3 Ecotone network and delimitations 4.4 Fragmentation and permeability 4.5 Other landscape processes
5- Protective and of recovery	5.1 Self-protection 5.2 Interface elements 5.3 Self-recovery
6- Reproductive and of colonisation	6.1 Ecotope reproduction 6.2 Colonisation processes
7- Heterotrophic	7.1 Animal/Human Relation to territory: the vital space per capita 7.2 Biologic and systemic processes within an Urban Region 7 .3 Signs of animal processes
8- Transformative	8.1 Transformation modalities 8.2 Control of the process
9- Cybernetic and of control	9.1 Ecological control of human culture 9.2 Buffer effect of a cultural change on the bionomic state of a LU
10- Evolutive	10.1 Limits of Neo-Darwinism 10.2 Evolutive modalities of a Landscape
11- Landscape syndromes	11.1 Landscape alteration 11.2 Landscape pathology
12- Landscape pathology and human health	12.1 New concepts of health damages 12.2 Environmental stress 12.3 Ecological damages and human health 12.4 Health and territorial planning

3.1.3 Processes Concerning the L. Skeleton

Geomorphic phenomena have three important functions in a landscape: (1) the generation of the load-bearing structure of the landscape; (2) the generation of fluxes of matter and the consequent modification of the regional surface, which signifies

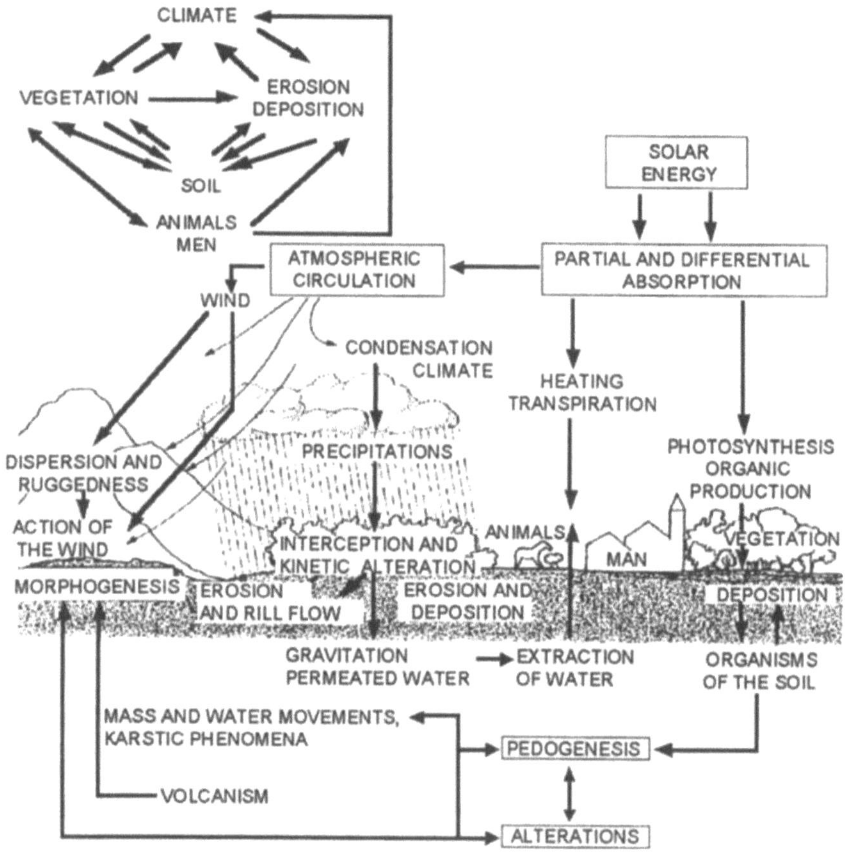

Fig. 3.2 Summary diagram of the main geomorphological processes that affect a landscape. (From: Ingegnoli (2002))

energetic ties for a landscape; (3) the derivation of characters from the transport of eroded matter (Fig. 3.2).

They can be distinguished into:

- *morphogenetic*, such as:
  - endogenous processes
  - littoral processes (marine and lacustrine)
  - fluvial processes
  - karst processes
  - wind processes
  - glacial processes
  - periglacial or cryo-nival processes

  - processes on slopes in a temperate climate
  - anthropic processes

- *pedogenetic*: the processes of formation and evolution of soils and mixed soils (Tricart & Kilian, 1979). Consequently, we can distinguish:

  1. *stable geodynamic environments* (characterised by the predominance of pedogenic processes), with slow evolution, with a tendency towards a stationary state, characteristic of regions with both internal geodynamic activity and intensity of external processes, both weak. They are environments favorable to the formation and transformation of soil, especially if maintained with a mature plant cover, and can be of recent or long-term stability.
  2. *unstable geodynamic environments* (characterized by the predominance of morphogenetic processes), with absence of forest cover, in which pedogenesis is extremely difficult to take place, so that mosaics of litho- and rego-sols are formed, the land is uncultivable and subject to landslides and mudslides, because even reforestation may not have sufficient effects; they are recognizable thanks to the presence of impervious reliefs, with strong inclinations, evident internal geodynamics, extreme bioclimatic conditions with strong variations that move and transmit large quantities of energy;
  3. *intermediate geodynamic environments* (in which there is the coexistence and interference of the two types of processes), with widespread pluvial erosion, surface movements of the land, with the only possibility of development of basic soils, or with widespread surface movements of land masses, sol flow and reptation and within which the presence of forest cover can also be non-positive.

In the study of the landscape, the pedological aspects must be studied dynamically and evolutionarily rather than chemically or physically (Duchaufour, 1983). The same pedogenic environment does not always produce the same soils; soils and plant cover originating on different substrates converge towards a uniform ecological system.

In extreme summary, we will also remember the hydrological processes, the movements of surface waters, which are important for their morphogenetic capacity and require a careful study of the floodable areas (for example, problems connected to the abuse of the floodplain areas of the river courses water, their low water levels, floods and floods), without forgetting the hydrogeological processes, with the dynamics of water in aquifers and underground aquifers, also important in the study of landscapes.

The search for levels of biological characterization of geomorphological landscapes today opens interdisciplinary fields of study of considerable interest. For example, the frequency of distributions of measurable geomorphological properties (altitude and slope of mountains, sinuosity of rivers, extent of land cover in LU, slope-drainage area relationships) seem to differ in relation to the degree of bioticity of geomorphological landscapes (Dietrich & Perron, 2006).

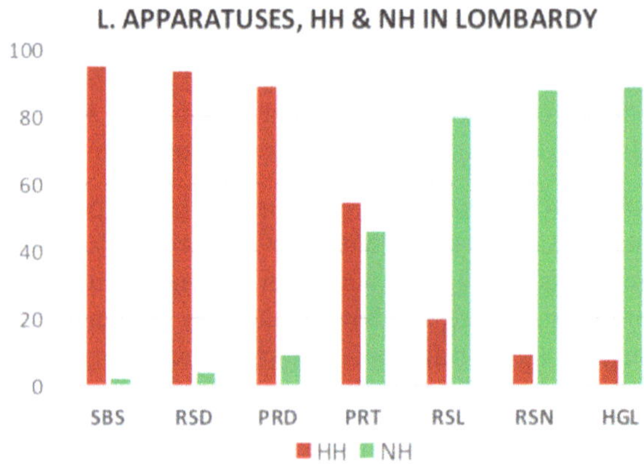

Fig. 3.3 Example of proportionality between the human (HH) and natural (NH) habitat components in tiles with functionality of residential (RSD), subsidiary (SBS), productive (PRD), protective (PRT), connective (CON), excretory (EXR), source landscape system (SOU). Typical values for Lombardy. (From: Ingegnoli (2002))

3.1.4 Processes Concerning the Self-Regulation Capacity of Landscapes: HH and NH

In relation to the processes of self-regulation and self-maintenance of each landscape unit, it is necessary to distinguish between the set of areas in which the self-regulation capacity of natural systems is limited and subsidiary energy is added, named human habitat (HH) (i.e., in which the human population lives or the ones they permanently manage), and the set of areas within which this capacity is not limited and the landscape is depending only on energy coming from the sun, called natural habitat (NH).

However, in man-dominated scrublands and corridors, it is possible to find natural components, just as in natural landscapes, it is possible to find anthropic elements (Fig. 3.3) (Ingegnoli, 2002). The mosaic of NH and HH is therefore different from the mosaic of land use, in coherence with the concept of ecotissue that we are going to see, and the percentage sum of its surface extension can be greater than 100%.

3.1.5 From Ecomosaic to Ecotissue: The Landscape Takes Shape

Following the previous paragraph, the traditional ecomosaic models describing the structure of a landscape need to be revised. To scientifically understand the shape of a landscape, it is better to speak of an "ecological tissue" formed by series of

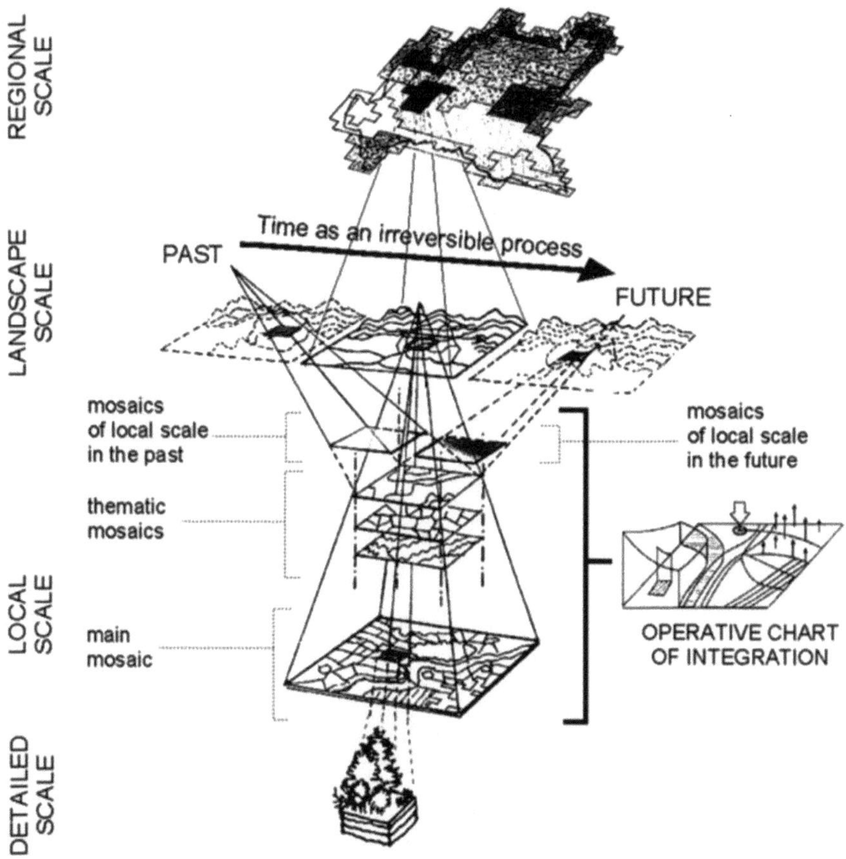

Fig. 3.4 Ecotissue model. The basic mosaic is the vegetational one. The complex structure of a landscape derives from the integration of different components: temporal, spatial, thematic, informative. An operational integration map may be necessary for planning. (From: Ingegnoli (2002))

interrelated and interactive information, elements and processes and of their redundancy, within which the vegetation component plays the key role of intercepting and making available the sun energy, and consequently life processes.

So, the general structural model of a landscape unit (LU), that is, of a well-characterized part of a landscape—the minor subsystem of which can be called "ecotope" (formed by sets of ecocoenotopes)—, is the eco-tissue (Fig. 3.4) (Ingegnoli, 2002): it is a multidimensional structure identified by an essential mosaic and a hierarchical series of related mosaics, as well as by a series of interrelated and interactive information that can be hierarchically integrated but cannot be represented in the form of a mosaic.

The identification of landscape units (note that a system of LUs forms a landscape) depends first on the geomorphological structure of the territory, such as

Fig. 3.5 The territory of the municipality of Tesero (Fiemme Valley, Italy) is divided into three landscape units (LU): a stretch of the Avisio valley and two valleys, one to the north (Stava) and one to the south (Lagorai). Failure to recognize these divisions leads to even serious planning errors. (From Ingegnoli (2015))

ridges, watersheds, alluvial fans, rock deposits, faults, bars, hydrological network, changes in slope, paleo-landslides, paleo-channels, terraces, karst). The example of Fig. 3.5 is noteworthy because, as it happens especially in the mountains, some municipal territories are made up of two or three LUs and, as we will see, failure to recognize these divisions leads to even serious planning errors: here it is shown that Tesero is made up of three Lus, that are (1) the northern valley (Stava), (2) a stretch of the Avisio River valley, and (3) the southern valley (Lagorai).

To the geomorphological structure of the territory however further characterizations must be added, due to:

1. distribution of vegetation (e.g., different formations with the same substrate, ecological state of forested patches).

2. historical-geographical characteristics (e.g., disturbed areas, areas influenced by neighboring events, natural barriers),
3. historical-anthropic characteristics (e.g., different infrastructures, different types of crops, artificial barriers, urbanization),
4. system of undisturbed areas of strategic importance and ecological networks (see Par. 7.5)

In summary, the ecotissue, which represents the multidimensionality of the structure of a complex system of ecocenotopes, is composed of:

1. simple elements, that can be traced back to tiles distributed in characteristic configurations about local geography and ecological processes of the system.
2. configurations of elements and their sets, which can also be seen as spots and corridors placed on a fundamental matrix (sensu Forman & Godron, 1986).
3. functional subsystems, that is those formed by ecotopes, habitats (human and natural), and "context role subsystems" (sensu Ingegnoli & Pignatti, 2007), which are highlighted in "geomorphological subsystems" and "landscape systems."

3.2 Landscape Elements Configuration and Sense

3.2.1 Simple Elements: Characters of Tesserae

The *tessera (tile or patch, from Latin Tessera-ae)* is defined as the smallest simplest element of the basic ecomosaic that can be found in the territory—with characteristics of homogeneity of the geopedological substrate and vegetal cover or actual use of the land, but also with characteristics of multifunctionality and integration of the biotic, functional and spatial components and concerning a specific context of a geographical surrounding: that is, the tessera corresponds to the ecocoenotope. Be careful, however: the tessera is the component but not the smallest element of a landscape (as it should have been multidimensional, while it is only multifunctional) because it belongs to the lower level of biological organization (local).

Considering that a landscape arises from the co-evolution of vegetal, animal, and anthropic components with a specific territory, it is important to underline that the *tesserae* making up the basic ecomosaic also include those of an anthropic type. Furthermore, it should be noted that, within the same vegetation formation, e.g. in a forest of a particular type, the tessera can also be distinguished by the stages of development of the phytocoenosis: juvenile, adult, mature, and senescent. A tessera may present a structural difference between its edge band and interior, given that the peripheral boundaries are usually affected by the influences of the surrounding environment. They may exchange some species or show plants with structural behavior different from that of the interior (e.g., for reasons even just of brightness and conspecific density).

To identify a tessera directly "in the field," it is necessary to observe the characteristics of the vegetation (not only the phytosociological associations but

Fig. 3.6 (Left) An example of distribution of tiles/ecocoenotopes (red) and ecotopes (blue) in the LU-2 landscape unit of Tesero (Trento). (Right) Vinyard landscape in the Oltre-Po Pavese, with well characterized tesserae, the matrix being evidently vineyards and viticultural infrastructures

also any ecotones or para-anthropic formations), the geomorphology (even the less conspicuous ones), the anthropic uses (even only partial), wildlife habitats (including potential ones) and the role of the area in question in the landscape context. In Fig. 3.6, we see three wooded ecocoenotopes within the southern half of LU number 2 of Tesero (Fiemme Valley, Italy) and the repeating type of ecocoenotopes in a LU of Oltre-Po Pavese, Italy.

Even the shape of the tesserae can have meanings in the structure of a landscape unit, certainly in its surroundings, especially in any crossing directions of flows of energy, materials, and species.

Figure 3.7 (Grassi & Ingegnoli, 2006) is reported to show the two landscape units (LU1 and LU2) of Robecco sul Naviglio, a municipality within the floodplain of the Ticino River in the Regional Park of the same name. All the tiles composing it were carefully marked. There are 42 types of tesserae and 11 types of corridors, i.e., rows, hedges, and canals. The tesserae can be grouped into sets of different territorial functions, which, as we will see in the next paragraph, are called landscape systems. To better understand the survey of the tesserae, we can anticipate that, in this case, the tesserae with anthropic functions are not as dominant as in other plain municipalities. Only 4.61% are urbanized, and 57.59% are of agricultural type. Beds and river waters cover another 5.40%, while all the rest are of natural or semi-natural vegetation (32.40%). Therefore, the forests in this LU2 are very extensive, reaching 25.18% of the surface, equal to 258 ha out of 1027. The forest formations are also very diversified, as can be seen from the high number of tesserae, which are divided into 14–15 typologies.

Another characteristic of the structure of an LU is shown in Fig. 3.8, where the number of essential components (different types of tesserae) is a function of the surface of the LUs examined (Lombardy and Trentino/South Tyrol regions). These are landscape Units (LUs) that include—at least in part—an anthropic component. The correlation line is a logarithmic one with $R^2 = 0.789$. The change in curvature indicates the transition from ecotopes to true LUs, which occurs for surfaces >6–9 km^2.

Fig. 3.7 Landscape units LU1 and LU2 of Robecco sul Naviglio, year 2006. 42 types of tessere and 11 types of corridors can be distinguished in an area of only 25 km^2. The forestry component is particularly high, 25.18% in LU2 (floodplain). (From Grassi and Ingegnoli (2006))

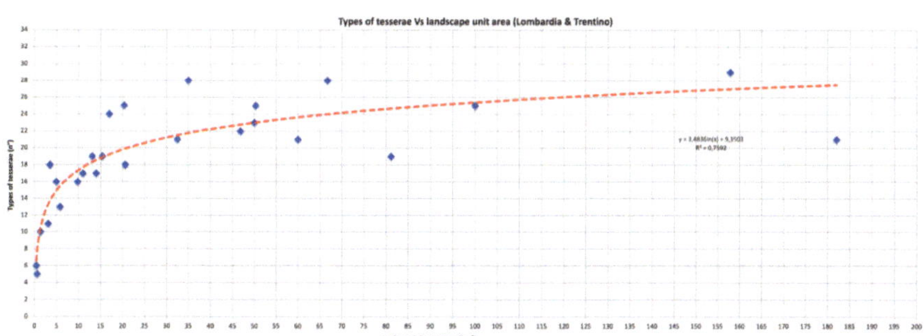

Fig. 3.8 The main types of tiles composing a landscape unit are relatively few, as is clear from the correlation exposed here between the surface area and the diversity of tiles in an LU

The number of types of elements (TS-type) composing a LU is generally somewhat limited: it does not exceed TS-type = 28 in the reported cases and TS-type = 35 for a district of approximately 5000 km^2 (e.g., Geneva-Vaud). Consequently, it must be remembered that the LUs (even in district sets) belong to an organized complexity, which is particularly difficult to study. In much rarer cases, 40–42 types of TS can be found in territories of only 25 km^2.

3.2.2 Configurations of Elements: The Landscape Patches

Generalizing the concept of *tessera (tile)*, Forman and Godron (1986) defined any area that differs from the surrounding area for some reason as a "landscape *patch*." A grouping of similar tesserae correctly forms a patch. In practice, it is legitimate to refer to patches of tesserae, especially when moving from a detail scale towards a larger synthesis scale. For example, a grouping of arable agricultural fields is identifiable as a patch of tesserae at a 1:25,000 scale, while at a 1:10,000 scale, each individual field is usually identifiable as a patch.

The patches can also be distinguished by their origin, due to:

- disorder (isolated, chronic).
- relict (whether associated with a disorder).
- resource (biotic or abiotic).
- anthropic activity (temporary or fixed).
- colonization (natural or anthropic).

The influence of the edges of a patch is proportionally inverse to its length. Larger patches usually have more species than small ones. A maquis surrounded by a vastness of elements of different types (for example, a forested maquis in the middle of agricultural fields) behaves like an island, and some concepts of the well-known island theory are often applicable to it (*sensu* Mac Arthur & Wilson, 1972). If a patch has a digitate shape, peninsula effects (species loss towards the tip) and interdigitation (gradient modification with the surrounding elements) occur. The density of patches on a matrix and the spatial distribution of their types characterizes a landscape.

3.2.3 Configurations of Elements: The Ecological Corridors

Landscape elements linked along a line form an ecological corridor. The connections of different corridors form an ecological network. Corridors can be mainly formed by the hydrological network (rivers and their tributaries, canals), by the road network (roads, motorways, and railways), by tree lines and hedges, but also by the edges of large patches with the very different exterior, for example, the edge of a forest relative to vast surrounding grasslands or a band of undeveloped areas around an

Fig. 3.9 Example of an ecological corridor formed by a canal with a country road next to it but with few trees remaining on the banks (Ticino Park in Morimondo). On the right, a semi-natural tree corridor in the hills of Tuscany, a remnant of the wood edge

urbanized environment. When studying the structure of the corridors, it is necessary to consider curvilinearity, nodes, open intervals, connectivity, the average height of the dominant plants, density, and stratification of the vegetation, and cross sections (Fig. 3.9). We must add the need to know the density of corridors per unit area in a LU, measurable in m/km^2, and the interference between natural and semi-natural corridors with artificial barriers.

It must be noted that, by the term 'corridor', we can never mean only the element generating linearities, such as a stream or a road: still, its marginal, intrinsically characterizing parts must also be included, such as a bank forest or a tree-lined row.

It is also interesting to note that, within the countryside, the tree-lined corridors are of particular importance and can be flanked by a canal or a farm road too and often have a double row with shrubs at the base. According to Forman and Godron (1986), the minimum width for the ecological efficiency of these rows is 12 m. This measure is a significant fact, as it must be remembered that agricultural rows and hedges play a connecting role and are a refuge for wild species. An unpublished study by Ingegnoli and Giglio revealed that in 100 m of *Prunus laurocerasus* hedge, as many as 46 species of plants are taking refuge there. The role is also strategic for insects, which have specific effects on the biological regulation of crops, with less need for the use of pesticides. Unfortunately, it is known that these structures have been in sharp decline for some decades in agricultural landscapes throughout Europe and the United States of America (Burel & Baudry, 1999).

3.2.4 *Configurations of Elements: The Landscape Matrix*

The study of the relationships between configurations of patches and corridors capable of characterizing a landscape can only be done by considering the so-called landscape matrix. This matrix is formed by the particular and the most extensive and the most connected functional component of a landscape unit. For example, the matrix of a landscape composed of a series of productive tiles, therefore with elements that have cultivation functions, detectable with the presence of >50% of the area of a LU, presents the essential characteristics of an agricultural landscape. In the case of a LU made up of a myriad of heterogeneous elements, the most functional apparatus may even be <50%.

Various types of matrices can be distinguished based on the distribution of the dominant element on the territory. Note, however, that the recognition of the type of the matrix is not equivalent to the definition of the type of landscape.

3.2.5 Functional Subsystems: The Landscape Apparatuses/Eco-countries Functional Systems

The main functions of LU form structures that can be distinguished through actual ecological-functional apparatuses (Ingegnoli, 2002) in analogy with the apparatuses of living organisms (i.e., digestive apparatus, respiratory apparatus, …: note that English language translate 'apparatus' with 'system', and 'landscape' with 'eco-countries' but we prefer to maintain the previous Italian term to avoid misunderstandings). Each type of apparatus comprises one or more functional systems of tesserae (tiles), each one characterized by a dominant function alongside multiple others: we can speak of the multifunctionality of landscapes (Brandt et al., 2000; Naveh, 2000). Usually, 4–5 apparatuses can be considered typical of the Human Habitat (HH) (for example, residential, subsidiary, productive, protective) and 8–10 of the Natural Habitat (NH) (for example, resistant, stabilizing, ecotonal, resilient, connective, excretory, hydrogeological). The HH functions include five types of devices:

1. *residential development* elements (RSD), for example, residential building lots, neighborhood service building areas, school areas, administrative office areas, urban streets and squares, parking areas, worship areas, and market areas.
2. elements of the *subsidiary apparatus* (SBS), such as industrial areas, commercial warehouses, railway and car stations, national transport networks, important technological networks, and large storage areas.
3. elements of the *productive apparatus*, i.e., agricultural ones (PRD), for example, arable land, vegetable gardens, vineyards, olive groves, orchards, poplar groves, agricultural woody crops, fodder crops, country roads, small farms.
4. elements of the *protective apparatus* (PRT), for example, hedges, rows of trees, urban parks, extensive gardens, residual groves, vegetated protection corridors in general, sports greenery.
5. HH elements *included in the natural habitat* (HH < NH), such as forest roads, timber storage areas, refuges, ski slopes, spring collection, corridors for the passage of technological pipelines, recreational and sports areas, sheep tracks, livestock pens, and areas of intensive forest exploitation.

The NH functions include at least 10 types of devices:

1. elements of the *resistant apparatus (RNT)*, for example typically the patches of mature forest, i.e. any element formed by vegetation with high territorial bio-potentiality, such as BTC (see… and 7.2.3) of standard class, VI or VII.
2. elements of the *resilient apparatus (RSL)*, for example low BTC elements such as meadows and shrubbery and pastures (for example BTC of standard classes I, II, III);
3. elements of the *ecotonal apparatus (ETN)*, for example the elements with a strong gradient such as coastal vegetation bands or semi-natural vegetation bands, i.e. elements that divide types of landscapes or units in a scalar way of landscape with strong contrast.
4. elements of the *stabilizing apparatus (STB)*, for example elements with vegetal formations capable of containing morphogenetic disorders, such as stains on unstable slopes, salt marshes in tidal lagoons, willow groves on riverbeds with frequent flooding.
5. elements of the *connective apparatus (CON)*, for example elements, often corridors or steppingstones, necessary to connect the main natural areas of a landscape and its units.
6. elements of the *excretory apparatus (EXR)*, for example the elements of the surface hydrological network (aquatic ecosystems), capable of purifying the catabolites coming from the basic mosaic within water bodies.
7. elements of the *hydrogeological apparatus (GEO)*, for example the elements dominated by hydrological and/or geological components, such as lakes, lagoons, areas of outcropping rock, beaches, screes, glaciers.
8. elements of the natural *productive apparatus (PRD)*, for example the vegetated elements in which production dominates, such as forest patches in the juvenile phase of recreation, or certain herbaceous formations in wet areas.
9. elements of the natural *protective apparatus* (PRT), for example the vegetated elements with a pre-eminent function of sheltering other elements, such as certain riparian woods or spots capable of slowing down the wind, salt water, gravel.
10. NH elements *included in the human habitat (NH < HH)*, for example parks and gardens with the role of refuge for native animal and plant species, tree-lined corridors and hedges in agricultural areas, protective ruderal areas, meadows, and semi-natural pastures.

Pay attention to the concept of productive apparatus: from an ecological/bionomic point of view, the only production is the one coming from the soil, thus from agriculture and growing stages of vegetation, not the one coming from secondary and tertiary economic sectors: the latter items must be ascribed to the subsidiary apparatus.

A survey map of the main types of natural anthropic systems in each LU is shown (Fig. 3.10), considering the intensities of the prevailing function.

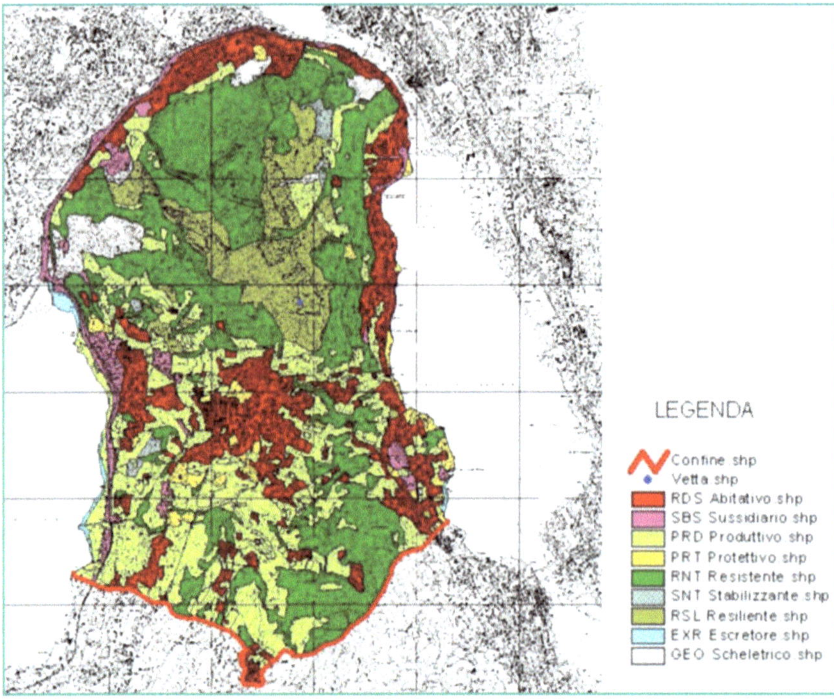

Fig. 3.10 Landscape Unit (LU) of the Monte Barro Regional Park (Lecco), of approximately 16 km^2. Distribution of the main landscape apparatuses. The productive (agricultural) apparatus was approximately halved from 1888 to 1995 in favor of the growth of the residential and subsidiary sectors

3.2.6 The Landscape Morpho-Functional Classification

Based on the co-presence of the landscape apparatuses/eco-countries systems, it is possible to arrive at a classification that is more suited to the bionomics of the territory and has a more excellent balance between the natural and anthropic forms. Even if you want to limit yourself to the possible relationships of the most common apparatuses (geological, excretory, stabilizing, resilient, connective, protective, productive, residential, subsidiary) Ingegnoli (Fig. 3.11) (Ingegnoli, 2011) proposed the following sequence of 16 types of landscapes, from the five to seven basic types shown by Naveh & Lieberman (1984/1994):

7 naturals = desert, semi-desert, meadow, grassland-bushy, shrubby, open forest, closed forest.
2 semi-natural = semi-natural high biomass, semi-natural low biomass.
7 anthropogenic = agricultural-protective, agricultural, rural, suburban-rural, suburban-industrial, open urban, closed urban.

Geo	Exr	Stb	Rsl	Con	Prt	Prd	Rsd	Sus	Tipi di paesaggio:
+ +	-	-	+ -	-	-	-	-	-	Deserto
+ +		-	+ +	-	+ -	-	-	-	Semi-deserto
-	+ -	-	+ +	+ +	+ -	+ -	-	-	Prateria
+ -	+ -	+ -	+ +	+ -	+ +	+ -	-	-	Prateria arbustiva
+ +	+ -	+ -	+ -	+ -	+ -	+ -	-	-	Steppico-arbustivo
+ -	+ +	+ +	+ +	+ -	+ +	+ +	-	-	Forestale aperto
-	+ -	+ +	+ -	+ +	+ +	+ -	-	-	Forestale chiuso
+ -	+ -	+ +	+ +	+ -	+ -	+ +	-	+ -	Semi-naturale ad alta biomassa
+ -	+ -	+ -	+ -	+ -	+ -	+ +	-	+ -	Semi-naturale a bassa biomassa
+ -	+ +	+ -	+ +	+ +	+ +	+ +	+ -	-	Agricolo protettivo
+ -	+ -	-	+ +	-	+ -	+ +	+ -	-	Agricolo produttivo
+ -	+ -	-	+ -	-	+ -	+ +	+ -	+ -	Rurale
+ -	+ -	-	+ -	-	+ -	+ -	+ -	+ -	Suburbano-rurale
+ -	+ -	-	-	-	+ -	-	+ -	+ +	Suburbano-industriale
+ -	+ +	-	-	+ -	+ +	-	+ +	+ -	Urbano aperto
-	+ -	-	-	-	+ -	-	+ +	+ +	Urbano chiuso (denso)

Con, connettivo; *Exr*, escretorio; *Geo*, geologico; *Prd*, produttivo; *Prt*, protettivo; *Rsd*, residenziale; *Rsl*, resiliente; *Stb*, stabilizzante; *Sus*, sussidiario; ++, presenza piena; + -, presenza parziale; - -, assenza.

Fig. 3.11 Classification of landscapes through the presence of different landscape apparatuses/eco-countries systems that characterize them. (From Ingegnoli (2011)). Natural Devices (green) and Anthropic Devices (pink)

Also, in this case, the geographical area specifications must be added to classify thousands of typologies.

On a larger scale, some biogeographers, such as Bailey (1996), have divided the emerged lands into 30 macroscale regions, grouped into the domains (I) polar, (II) humid-temperate, (III) arid, and (IV) humid-tropical, identifying the 30 regions according to the climatic zone and macro-vegetation zones.

3.3 Two Main Capacities of Landscape: Carrying and Bio-territorial

3.3.1 Vital Space per Capita: Landscape Resource Relationships Linked to Living Populations

The process of animal or human populations to relate to their landscapes is a function of their *per capita* living space and of the functional configuration of the landscape. To express the processes, it is necessary to define the concept of "*standard habitat per capita*" (SH). SH is the inverse of the ecological density of the population, measurable in m^2/organism.

$$SH_H = (GEO + EXR + PRT + PRD + RES + SBS)/(n^\circ \text{ of peoples}) \left[\text{m}^2/\text{inhabitant}\right].$$

At the basis it's necessary to define the set of portions of the landscape apparatuses (within the LU examined) indispensable for the survival of an organism, that is the minimum standard habitat per capita SH* (Table 3.3). For bionomics the ratio SH/SH* between detected SH and the minimum theoretical SH* is the *Carrying Capacity* of a territory: remember that it can be properly evaluated for both animals of different species and humans.

If we were to refer to altitudinal bands, it must be noted that there are some differences compared to what was said about climatic bands. In the Alpine zoning, as can be seen from Fig. 3.12, the change in SH* is greater than in the climatic zones of

Table. 3.3 Theoretical Minimum Standard Habitat per capita (SH*) related to the main climatic belts of the biosphere

Climatic belts	Kcal/inhab[a]	SH* m^2	Min. t °C[b]	PRD m^2
Arctic	3500	**2500**	−45.6 to −34.4	1650
Boreal	3100	**1850**	−34.4 to −23.3	1230
Cold-temper.	2850	**1475**	−23.3 to −12.2	1050
Warm-temper.	2750	**1360**	−12.2 to −1.1	950
Sub-tropical	2550	**1250**	−1.1 to +10	850
Tropical	2350	**1020**	+10 to +21.2	730

PRD = minimum field available to satisfy the edible energy per capita needed in 1 year. Note that these values were estimated in 2012–2015; today they tend to reduce by about 5%
[a]Minimum edible kcal/day per capita (very different from the 3500 kcal/day of USA population)
[b]Following USDA Plant Hardiness Zone Map for Cultivations (Average Annual Extreme Minimum Temperature (USDA Plant Hardiness Zone Map, 2012))

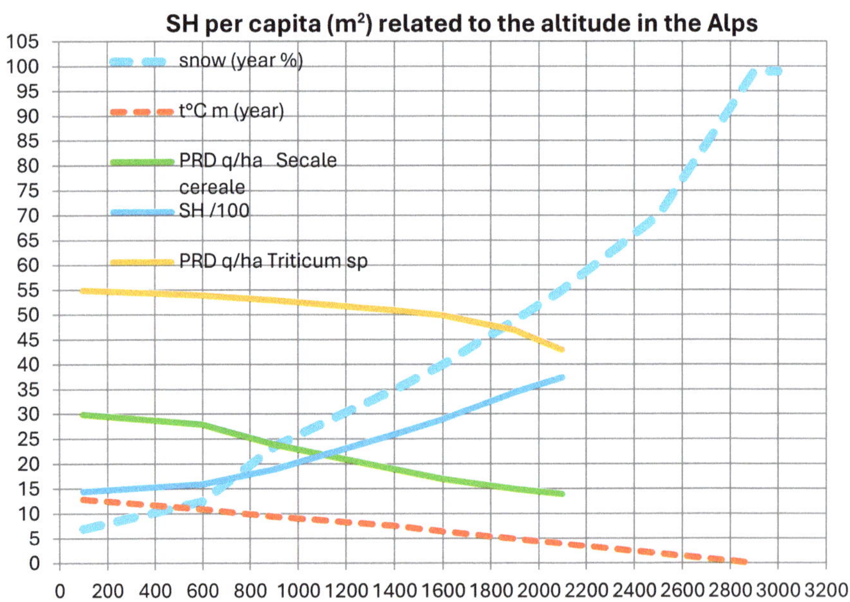

Fig. 3.12 Variations in the standard habitat (SH/100) by altitudinal range in the Alps (sky blue line) and variation in the yield of Secale and Triticum cereale (green and yellow)

the biosphere. The reasons are many, for example (i) a faster and more unpredictable modification of the climate, (ii) the presence of even notable and very diversified slopes, (iii) a much shallower depth of the soil, (iv) a lower yield of the cultivable species, (v) a lesser variety of cultivable species.

3.3.2 Biological Territorial Capacity: Discovery of a Systemic Vegetational Function

The role of the vegetation in shaping the landscapes and making life processes possible is so crucial that it need to be properly investigated and evaluated. Knowing that the physiology of plant communities is closely linked to their latent capacity for allostasis and that the possibility to estimate the flow of energy available to maintain the order achieved by a complex system could be strategic, a function has been studied based on dynamic resistance balance, vegetation type, and data relating to metabolism and organization of these communities. It was named: Biological Territorial Capacity of vegetation or Territorial Bio-potentiality (BTC).

The physiology of vegetation involves the concept of the latent capacity for allostasis of a formation, which can be studied based on:

(a) The concept of resistant stability.
(b) The type of vegetation community.
(c) The basic metabolic data of the considered system (biomass = B, Net Production = NP, Gross Production = GP, Respiration = R).
(d) Their metabolic R/GP relationships.
(e) Their order relations R/B = dS/S (antithermic maintenance).

For systemic reasons, we will have to link the most significant processes considering:

1. the set of *phytocoenosis* (Phc) of the Emerged Lands.
2. the metabolic and order (entropic) functions of the ith $Phc_N = i$

We need to work out two coefficients:

$$a_i = (R/GP)_i/(R/GP)_{max} \qquad b_i = (dS/S)_{min}/(dS/S)_i$$

a_i, measures the degree of relative metabolic capacity of the main plant communities.

b_i, measures the degree of relative antithermic (i.e. order) maintenance of the main plant communities themselves.

The degree of allostatic capacity of a phytocoenosis (plant community) is proportional to its respiration and can be expressed as the *flux of energy that the system must dissipate to maintain its condition of order and metastability*.

Consequently, the following equation can be written:

$$BTC_i = (a_i + b_i)R_i w \left(\text{Mcal/m}^2/\text{year}\right)$$

Where: w = 1.4–1.7 (safety coefficient that estimates the root and soil biomass).

It is observed that the BTC is also a state function and is systemic since it was found by applying systemic correlation principles. Furthermore, the complexity of a highly ordered system is expressed by a high level of energy, e.g. a *Querco-Carpinetum* (high resistance) usually has a BTC = 8.50–9.50 Mcal/m^2/year, while a cultivated meadow (low resistance) has a value of BTC = 0.45–0.55 Mcal/m^2/year.

3.3.3 First Applications of BTC

The BTC function is very important because it can evaluate the flow of energy available to maintain the order achieved by a complex system, as can be seen by comparing two agricultural landscapes near Milan (Fig. 3.13). Note how the HH values are not enough different, while those of BTC decrease drastically in the transition from traditional to industrialized agriculture.

Table 3.4 (Ingegnoli, 2002 [upgraded]) shows the esteem of the optimal BTC values for the most characteristic vegetation formations of the Emerged Lands, following what was written previously.

Fig. 3.13 Comparison between two agricultural landscapes near Milan. Note how the HH values are not too much different, while those of BTC decrease drastically in the transition from traditional to industrialized agriculture

Table 3.4 Plant biomass (at the ecosphere scale: PB_b, at the ecocoenotope scale: PB_e), net primary production (NP), respiration vs gross primary production rate (R/GP), respiration vs biomass rate (R/B, where $B = PB_{max}$), and calculated BTC values ($Mcal/m^2/year$)

Main types of vegetation	PB_{max}	R/GP	R/B	a_i	b_i	R_e	(a + b) R	BTC_{max}
Tropical rain forest	10.5	**0.75**	0.078	**1.00**	0.41	8.19	11.55	**18.00**
Tropical seasonal for	8.0	0.7	0.072	0.93	0.44	5.76	7.94	12.10
Mediterranean forest	8.5	0.71	0.063	0.95	0.51	5.36	7.87	**12.00**
Temper. deciduous for	8.0	0.69	0.061	0.92	0.52	4.88	7.05	**11.50**
Boreal/alpine forest	7.3	0.71	0.062	0.95	0.53	4.53	6.70	**10.50**
Temperate woodland	5.0	0.69	0.054	0.92	0.59	2.70	4.08	**6.50**
Mangrove forest	5.5	0.65	0.04	0.87	0.80	2.20	3.67	6.00
Boreal open forest	4.0	0.71	0.062	0.95	0.52	2.48	3.63	5.60
Temper. perennial cult.	1.5	0.6	0.227	0.80	0.14	3.41	3.20	**5.00**
Chaparral bush-land	2.0	0.65	0.13	0.87	0.25	2.60	2.89	4.50
Tropical marsh/swamp	3.6	0.39	0.083	0.52	0.39	2.99	2.71	4.10
Tropical perennial cult.	1.5	0.55	0.19	0.73	0.17	2.85	2.57	4.00
Dry forested savannah	3.8	0.42	**0.032**	0.56	**1.00**	1.22	1.90	**3.00**
Shrub savannah	1.8	0.42	0.111	0.56	0.29	2.00	1.69	2.70
Temperate marsh/swa	1.8	0.39	0.117	0.52	0.27	2.11	1.67	**2.60**
Estuaries	1.8	0.39	0.072	0.52	0.44	1.30	1.25	2.20
Graminoids savannah	4.5	0.4	0.377	0.53	0,08	1.70	1,05	1.80
Suburban green	6.0	0.55	0.167	0.73	0.19	1.00	0.93	**1.40**
Peat bog	1.2	0.4	0.067	0.53	0.48	0.80	0.81	**1.00**
Tropical yearly culture	2.8	0.41	0.464	0.55	0.07	1.30	0.80	1.20
Temper. yearly culture	3.0	0.41	0.333	0.55	0.10	1.00	0.64	**1.00**
Moist temperate grass	3.5	0.4	0.266	0.3	0.12	0.93	0.61	**1.00**
Dry temperate grass	2.8	0.4	0.179	0.53	0.18	0.50	0.36	0.60
Urban green	3.5	0.5	0.16	0.67	0.20	0.45	0.39	**0.70**
Tundra/alpine vegetat.	3.0	0.35	0.167	0.47	0.19	0.50	0.33	0.60
Shrub semi-desert	4.2	0.45	0.059	0.60	0.54	0.25	0.28	0.50
Continental shelf	2.0	0.42	0.14	0.56	0.23	0.28	0.22	0.40
Cold desert	1.3	0.38	0.061	0.51	0.52	0.08	0.08	0.15
Lake and stream	0.5	0.4	0.16	0.53	0.20	0.08	0.06	**0.10**
Sand warm desert	0.25	0.45	0.14	0.60	0.23	0.04	0.03	0.05

From Ingegnoli (2002), [upgraded]
NP Net primary production ($kg/m^2/year$) and PB_b (kg/m^2) normal ranges at ecosphere scale (Golley & Lieth, 1972; Golley & Vyas, 1975; Whittaker and Likens in Lieth & Whittaker, 1975; Kimmins modified in Piussi, 1994). PB_e ($kg/m^2/year$) near mature stages at ecocoenotope scale; Ingegnoli (2002)

Table 3.5 indicates in greater detail the seven classes of BTC, which correspond to different anthropic and natural vegetation sets, from thickets and sparse meadows to mature natural forests with high BTC. We will notice that there are globally nine classes of BTC.

Table 3.5 BTC classes and threshold values

Biological territorial capacity	BTC class	Mcal/m^2/yr	range
Semi-desert, scrubs	I	0.0–0.4	0.4
Grass, fields and urban green	II	0.4–1.2	0.8
Vines, shrubs, plantations, gardens	III	1.2–2.4	1.2
Young forest, temperate coppice, urban parks	IV	2.4–4.0	1.6
Near adult forest, young "macchia" woods	V	4.0–6.0	2.0
Adult forest, temperate woodland, adult "macchia"	VI	6.0–8.4	2.4
Mature boreal and temperate forests	VII	8.4–11.2	2.8
Dense sclerophyll forests	VIII	11.2–14.4	3.2
Dense tropica/pluviall forests	IX	14.4–18.0	3.6

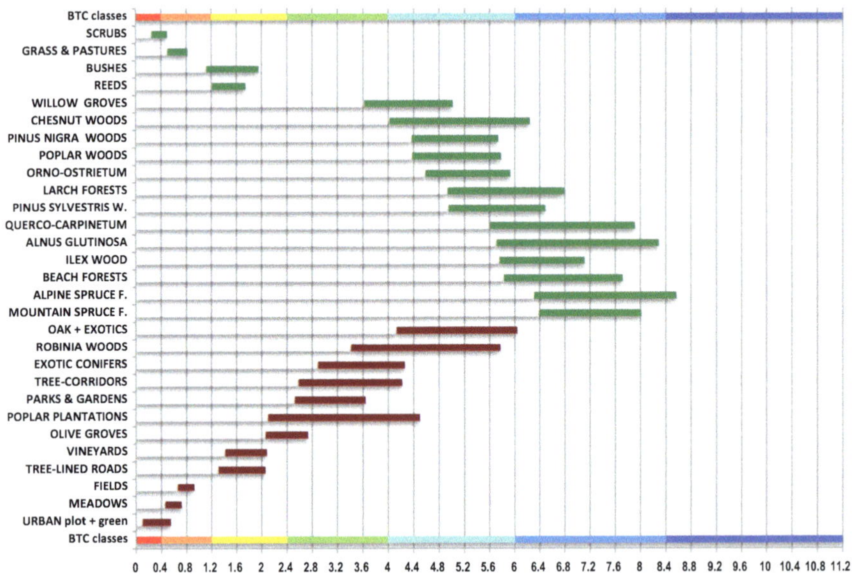

Fig. 3.14 BTC values (Mean ± St. Dev.) from 430 surveys in Lombardy and Trentino (2005–2015). The quantities (Mcal/m^2/year) can refer to seven standard classes as per the colored bars. (From Ingegnoli (2015))

Figure 3.14 (Ingegnoli, 2015) gives an idea of the average BTC values for around thirty natural (green) or anthropic (dark red) vegetation formations. The colored bars in the figure show the classes into which it is helpful to divide the BTC values.

Figure 3.15 shows that the BTC function allows us to build a model (blue curve) necessary to carry out vegetation surveys according to bionomic principles, following the LaBiSV (Landscape Bionomics Survey of Vegetation) method, which will be illustrated in the following chapter (see Chap. 5).

This method is based on phytomass measurements, which already systematically involve various vegetation parameters, with the addition of further parameters, 28 in

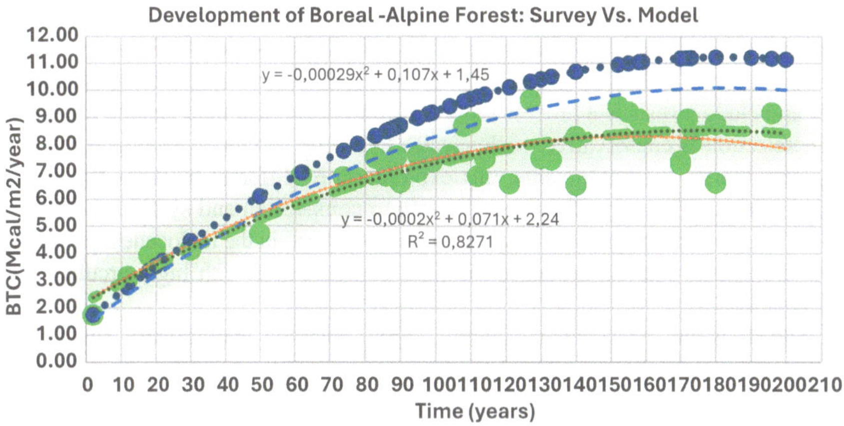

Fig. 3.15 Parametric model (blue) necessary for vegetation surveys according to the LaBiSV method. In green the real Time/BTC correlation regarding the Boreal-Alpine forests detected. The difference is due to natural and anthropogenic disturbances

total (for forests). The correlation of BTC levels as a function of time (green curve) is obtained from the study of 43 Boreal-Alpine forests: it is noted that, after the first 50 years, this curve differs significantly from that of the model (which is based on conditions of optimality) due to both natural and anthropogenic disturbances. The full maturity of these forests is between 125 (threshold) and 190 years. Then, with senescence, the BTC tends to decline but could also suffer more damage.

The key importance of Biological Territorial Capacity in quantitatively evaluating the behavior of a landscape or of its LUs will emerge in quite all the following chapters.

3.4 Processes of Information, Coevolution, and Interface Systems

3.4.1 Cybernetic Within and Around the Ecotissue

Interacting chains of organisms and communities behave like information networks in an ecotissue, so much so that through them and their emergent properties, it is possible to maintain a certain level of metastability. Studies on mycorrhiza networks in forests tell us that trees sense what grows around them: for example, the noise of an insect's jaws can generate information even among trees of different species and induce the production of chemical defenses (Simard et al., 2020) (see Chap. 5). The landscape is an essential information system for coevolution and processes of group selection because genetic characterization is linked to three levels of scale: cell, population, and landscape. Many animal populations in a new landscape tend to lose

Fig. 3.18 Example of an Interface Area in Milan. This angle is a Residual Interface among a canal (background), a road (see the bus stop), and the School Area (see the hedge). It is not managed by the park section of the municipality, so the vegetation is near chaotic but prevents the enlargement of a paved square

According to Ingegnoli (2015) the spatial "spots" applicable to the human population, which follow the forms of Euclidean geometry, frequently find themselves colliding with natural forms, which instead follow the forms of fractal geometry; consequently, this inevitably produces many interface elements, which only apparently appear not to matter. The interface elements form memory systems of the natural characteristics pre-existing in the anthropic transformations and the possibility for nature not to lose its functions completely; moreover, it often favors the entire landscape unit (LU) and the human population.

Thus, through similar processes, the emergence of rules of correspondence between the complex structure of natural habitats and the formation of new human habitats leads to a complex structuring of landscape units (LUs), which must be carefully re-evaluated (Fig. 3.18).

3.4.3 *Reproductive Landscape Processes*

Reproductive landscape processes manifest themselves differently than what happens at the population level. However, they respond to the reproductive functions valid for all biological systems (Ingegnoli, 2005) (Table 3.6). Given that the reproductive capacity of an ecotope is essentially dependent on the presence and characteristics of the vegetation component, it is necessary to introduce the concepts of "zero event" (Oldeman, 1990) and of "ecological memory" (Bengtsson et al., 2003).

The first has the meaning of lethal disturbance (for example, small and medium scale fire, gusts of wind, crashes, or clear cut), the second is made up of two parts: rules internal to the landscape element (for example, propagule bank) and context relationships (e.g., dispersion filters). Among the main factors capable of producing

Table 3.6 Reproductive processes at animal population and landscape ecotope level

Population renewal	Reproduction process	Ecotope renewal
Predisposed gonads	Reservoir of information	Ecological memory (e.g. propagule bank)
Chromosome crossing over, coping and coding	Mutation	Local disturbances
Nest or parental cares	Young structure protection	Nursery niches
Competition and predation, bio-semantics	Selection and correspondence	Competition and predation, bio-semantic
Death, often deferred	Old structure death	Crucial disturbance

Fig. 3.19 (Left) Spruce Nursery at the edge of a forest in Ziano (Fiemme Valley, Trentino, 2023). In this patch of about 50 m^2 we count about 24–25 small plant (height from 0.25 to 0.90 m) age from 30 to 45 years. (Right) Patches of a zero event, 2018, October, tornado “Vaia” in Gambìs Valley (Fiemme). Example of the destruction of hundreds of ha of spruce forests

zero events there are fire, water, wind, air humidity in some situations, earthquakes and earth movements, many human actions, the actions of animals, the actions of microorganisms or the combined effects of multiple components.

In a nutshell the process is activated after the “death” of one or more tiles, due to an above mentioned “zero event”, i.e. an out-of-scale disturbance capable of destroying the mature structures of an ecotope. The role of the gonads is assumed by the bank of propagules, the stabilizing mutations instead of “crossing over” occur due to local disturbances, the young are protected in special “nursery” niches (Fig. 3.19) and selection occurs mainly through competition and predation.

Colonization by plant (and sometimes animal) species plays an important part in the reproduction of the landscape too, as does anthropic management, if it does not prove too predatory. It will be noted that human populations are also colonizers through a method often repeated with few variations, which is like the *limitatio* Romana, not surprisingly reused in the United States and Brazil. It is reiterated that the guiding component capable of reproducing elements of a LU is the vegetation, which plays a crucial role in characterizing a landscape. Colonization will depend on

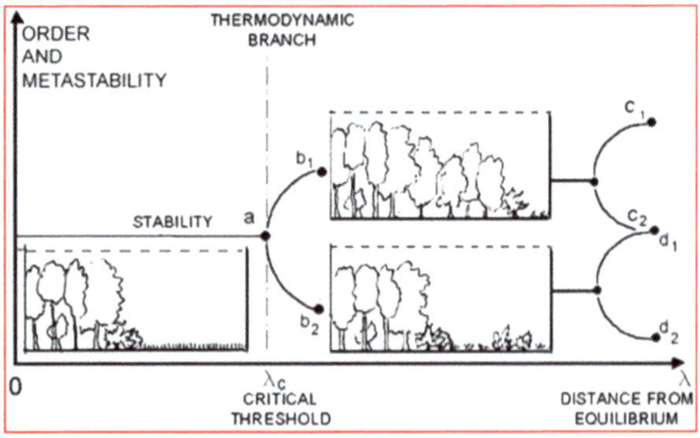

Fig. 3.20 A macro-fluctuation process (e.g. climate disturbance) leads to a bifurcation point with two possible ecological transformations: b1 and b2, which have different metastabilities. Note that after another macro-fluctuation the resulting transformation can coincide with d1 = c2 (reachable in different ways) or not (c1 or d2). Examples like these show the need to review the concept of succession. (From Ingegnoli (2015))

the size of the tiles and their number, as well as the intensity and breadth of the triggering event, compared to the total size of the ecotope. Thus, renewing the pre-existing ecotope or creating a totally new ecotope will be possible. In the first case (a) the reproduction process falls within the regenerative phase of vegetation that is in a situation of fluctuation (Falinski, 1998); in the second case (b) the reproduction process must be defined as ecological succession but in a new meaning. As it will be seen (Fig. 3.20), ecological succession can no longer be considered mechanistic, passing through predetermined evening stages and arriving at a climax phase. This process must be updated according to non-equilibrium thermodynamics.

References

Bailey, R. G. (1996). *Ecosystem geography*. Springer.

Bengtsson, J., Angelstam, P., Elmqvist, T., et al. (2003). Reserves, resilience, and dynamic landscapes. *Ambio, 6*, 389–396.

Brandt J, Tress B, Tress G (2000) Multifunctional landscapes: interdisciplinary approaches to landscape research and management. Centre for Landscape Research, .

Burel, F., & Baudry, J. (1999). *Écologie du paysage: concepts, méthodes et applications*. Technique & Documentation.

Dewaelheyns, V., Bomans, K., & Gulinck, H. (Eds.). (2013). *The powerful garden. Emerging views on the garden complex* (p. 237). Aantal Pagina's. Garant.

Dietrich, E., & Perron, T. (2006). The search of a topographic signature of life. *Nature, 439*, 411–418.

Duchaufour, P. H. (1983). *PÉdologie et classification (I) Constituants et propriÉtÉs du sol (II)*. Masson.

Falinski, J. B. (1998). *Dioecious woody pioneer species in the secondary succession and regeneration*. Supplementum Cartographiae Geobotanicae 8. Phytocoenosis, Vol. 10 (N.S.).

Forman, R. T. T., & Godron, M. (1986). *Landscape ecology* (p. XIX+619). Wiley.

Godron, M. (1984). *Ecologie de la vÉgÉtation terrestre*. Masson.
Golley, F. B., & Lieth, H. (1972). Summary. In P. M. Golloy & F. B. Golley (Eds.), *Tropical ecology with anemphasis on organic production*. University of Georgia.
Golley, F. B., & Vyas, A. B. (1975). Ecological principles. In *International Scientific Publications*.
Grassi, E., & Ingegnoli, V. (2006). *Caratterizzazione ecologico-paesaggistica del territorio dei comuni di Robecco sul Naviglio e Cassinetta di Lugagnano: Confronto tra l'unità di paesaggio di terrazzo*. Tesi di laurea in Scienze Naturali, Università di Milano.
Gulinck, H., Marcheggiani, E., Lerouge, F., & Dewaelheyns, V. (2013). The landscape of interfaces: Painting outside the lines. In *UNISCAPE conference landscape and imagination, Paris, 2–4 May 2013*.
Ingegnoli, V. (2002). *Landscape ecology: A widening foundation* (p. XXIII+357). Springer.
Ingegnoli, V. (2005). An innovative contribution of landscape ecology to vegetation science. *Israel Journal of Plant Sciences, 53*, 155–166.
Ingegnoli, V. (2011). *Bionomia del paesaggio. L'ecologia del paesaggio biologico-integrata per la formazione di un "medico" dei sistemi ecologici* (p. XX+340). Springer.
Ingegnoli, V. (2015). *Landscape bionomics. Biological-integrated landscape ecology* (p. XXIV +431). Springer.
Ingegnoli, V., & Pignatti, S. (2007). The impact of the widened landscape ecology on vegetation science: towards the new paradigm. *Springer, Rendiconti Lincei Scienze Fisiche e Naturali, s. IX, XVIII*, 89–122.
Kimmins, J. P. (1987). *Forest ecology*. Macmillan.
Lieth, H., & Whittaker, R. H. (1975). *The primary productivity of the biosphere*. Springer.
Mac Arthur, R. H., & Wilson, E. O. (1972). *Geographical ecology: patterns in the distribution of species*. Princeton University Press.
Naveh, Z. (2000). Introduction to the theoretical foundations of multifunctional landscapes and their application in transdisciplinary landscape ecology. In J. Brandt, B. Tress, & G. Tress (Eds.), *Multifunctional landscapes: interdisciplinary approaches to landscape research and management*. Centre for Landscape Research published.
Naveh, Z., & Lieberman, A. (1984/1994). *Landscape ecology: theory and application* (p. XXVII +360). Springer.
Oldeman, R. A. A. (1990). *Forests: elements of sylvology*. Springer.
Piussi P (1994), *Selvicoltura generale*. Utet, Torino.
Simard, S. W., Roach, W. J., Defrenne, C. E., Pickles, B. J., Snyder, E. N., Robinson, A., & Lavkulich, L. M. (2020). Harvest intensity effects on carbon stocks and biodiversity are dependent on regional climate in Douglas-Fir forests of British Columbia. *Frontiers in Forests and Global Change, Canada, 3*, 88.
Tricart J, Kilian J (1979) L'éco-géographie et l'aménagement du milieu naturel. Librairie François Maspero, .
USDA Plant Hardiness Zone Map. (2012). *Average annual extreme minimum temperature 1976–2012, USA*.
Whittaker, R. H. (1975). *Communities and ecosystems* (2nd ed.). Macmillan Publishers.

Chapter 4
Renewal of Vegetation Science: Theory

Contents

V. Ingegnoli, *Landscape Bionomics and Planetary Health*, Landscape Series 21,
https://doi.org/10.1007/978-3-032-08908-3_4

Abstract The studies on vegetation carried out by Braun-Blanquet were undoubtedly helpful, but they needed help to keep up with the upheavals of scientific paradigms. So, phytosociology studies present advantages regarding the description (syntaxonomy) of vegetation but also limitations, being based on a positivist epistemology and on ecological concepts that are now outdated.

The differences in the representation of the ecological space between phytosociology and the bionomic view is sharp, as we can see from the beginning of this chapter. We must consider the concept of "fittest vegetation for", i.e. a reinterpretation of potential vegetation indicates the most 'suitable' vegetation.

These limits of Phytosociology explain why Ingegnoli, Giglio, and Pignatti (2007) have proposed a new methodology called LaBiSV (Landscape Bionomic Survey of Vegetation), whose theoretical bases can be summarized as follows: (a) reference to the concepts of ecocoenotope and ecotissue, (b) use of the biological-territorial capacity of vegetation (BTC) as the primary integrative function, (c) drafting of development models of the different types of vegetation (time-BTC) based on the dynamic relationships between gross primary productivity, net primary productivity and the respiration of plant phytocoenosis summarized in an exponential and logarithmic function, (d) possibility to study the ecological state of natural and anthropic vegetated patches.

The Concise Bionomic State (CBSt) of the formations under examination is the indicator that considers the degree of maturity of the phytocoenosis too, and, therefore, estimates the ecological efficiency of the vegetation in percentage.

Keywords Fittest vegetation · LaBiSV (Landscape Bionomic Survey of Vegetation) · BTC · Summary parametric sheets · Forest damages and regrowth · Biodiversity · Vegetation science

4.1 Theoretical Motivations of Renewal

4.1.1 Main Limits of Phytosociology

As already underlined, the urgent need for a change in the scientific paradigm (Agazzi, 2014), from the reductionism still dominant today towards a more decidedly systemic study (Urbani-Ulivi, 2019), emerges from many underlying issues. Here, we will note that reductionism is linked to radical empiricism and, therefore, to determinism, which does not give importance to time and calls human freedom and the true knowledge of reality into question.

Like all ecology studies, the studies on vegetation carried out by Braun-Blanquet were undoubtedly valuable and important. However, they needed help to keep up with the upheavals of scientific paradigms. For these reasons, phytosociological studies (Braun-Blanquet, 1926) present obvious advantages regarding the description (syntaxonomy) of vegetation, but also certain limitations, being based on a positivist epistemology and on ecological concepts that are now outdated. Here a quick summary of the primary limits of phytosociology (Ingegnoli, 2005; Naveh & Lieberman, 1984/1994; Pignatti et al., 2002):

1. reference to a concept of naturalness that excludes man in any case
2. dynamics based on the concept of ecological succession often still understood as linear and deterministic
3. reference to an 'ecological space' that does not consider the principle of emergent properties (Fig. 4.1)
4. use of the concept of 'potential vegetation', which does not consider the role of disturbances in ecological systems
5. ignorance of complex and scale-dependent system functions, claiming to be able to study the landscape with the deterministic approach linked to the concept of 'sygmetum' and 'geo-sygmetum' (Rivas-Martinez, 1987; Tuxen, 1956). To this we must add:

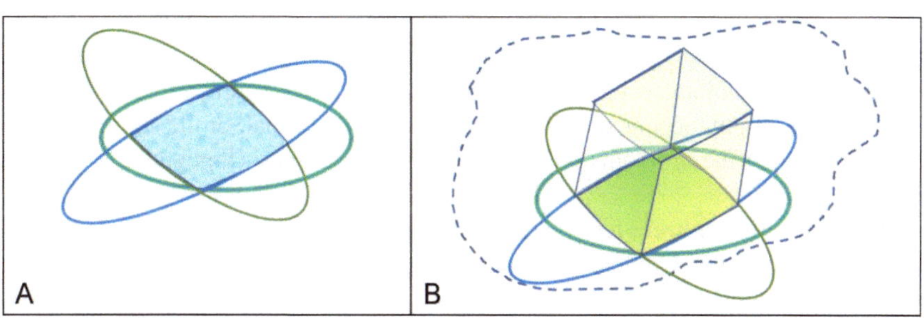

Fig. 4.1 Representation of an "ecological space" in the study of vegetation. (**a**) In the phytosociological model, (**b**) in the landscape bionomics model. According to the principle of emergent properties, the ecological space acquires characteristics that go beyond the sum of the individual species. Furthermore, the context is considered

6. the reference to a different level of organization, that is the community/ecosystem in one case, the landscape unit in the other
7. the existence of limits to the capacity of species to act as 'environmental indication'.

Furthermore, the method of estimating the ecological distance between potential and actual vegetation assumes that the optimal landscape is homogeneous, thus implying that the set of all components of a *series* of vegetation has reached the stage of *climax*, thus converging to the final dominant association. This single dominant phytocoenosis is contrary to all the principles of landscape bionomics and the real concept of biodiversity. It is a sort of virtual ecology!

4.1.2 The Concept of "Fittest Vegetation for..."

The coherent responses to the Bionomics Landscape theory begin with the proposal of the new concept of *the fittest vegetation for* (Ingegnoli, 2002; Ingegnoli & Giglio, 2005; Ingegnoli & Pignatti, 2007) to overcome that of potential vegetation. This reinterpretation of the concept of potential vegetation indicates "the most 'suitable' vegetation: in climatic and geomorphic conditions of a limited period in a certain defined place; depending on its history and with a certain set of disorders that can be incorporated (including human ones); in natural and non-natural conditions".

The implications are different: for example, the concepts of primary or secondary vegetation lose meaning; furthermore, in heavily anthropized regions, it no longer makes sense to compare actual vegetation with potential vegetation, given the ongoing environmental changes and the concept of allostaticity. The concept of *the fittest vegetation for*, combined with the principle that "...the behavior of a system depends not only on its component elements but also on how they are assembled and arranged..." (Principle of Emergent Properties), highlights a significant change in perspective and in applications too. Furthermore, it is no longer possible to talk about the recovery and restoration of ecological vegetation systems in the sense of a return to an unaltered state. The concept of *allostaticity* implies that the return to '*quo ante*' conditions is *unthinkable*, not only due to the shortness of time and with the mere removal of anthropic disturbances, but due to the complex dynamics of the development of phytocoenosis and differences, even minimal, of the starting conditions (remember the problems related to deterministic chaos (Ingegnoli, 2015, Par. 1.3.8)).

These last aspects were tested starting from the studies carried out by Ingegnoli for the CNR, National Council of Research (Ingegnoli et al., 1995) on the Gariglione forest (*Asyneumati-Fagetum Abietetosum*) in the Sila Piccola Park. It was demonstrated that while the phytosociological map showed only two types of tesserae, dominated by *Fagus sylvatica* or *Abies alba*, 12 types of tiles could be identified, including many derived from the ancient custom of charcoal pits, which, after centuries of use, have modified the soil to the point of creating conditions suitable almost exclusively for stands of poplars (*Populus tremula*).

4.1.3 *Other Systemic Limits of the Present Approach of Vegetation Science*

Moreover, we note:

(a) The concept of ecological succession is not valid in its traditional definition and there are limits to species as indicators.
(b) Species fluctuations (Sp) both in number and types do not follow a linear succession and lead to a critical threshold of rapid community change.
(c) Animal, and especially human, influences on vegetation emerge at this scale.
(d) Ecotissue transformations guided by geomorphological and anthropic conditions also emerge, whose disturbances, when not destructive, positively influence the vegetation too, given that the conditions of complexity depend on the entire LU.

And what about some other questions, like:

1. How to consider the contribution of a vegetated tessera to the general metastability of a landscape unit (LU)?
2. How to compare the ecological data of a forested patch with those of another type of vegetation?
3. How to use the ecological characteristics of the different types of vegetation to arrive at a diagnostic evaluation referring to a certain landscape unit (LU)?
4. How to integrate the other ecological parameters of a LU (HH, SH etc.) with those relating to vegetation?

With reference to Fig. 4.2, we focus on the different level of organization: single community Vs. landscape unit; therefore, they change genetic relationships because only at the LU level can polymorphism and metapopulation be interpreted.

4.2 The Bionomics Method of Studying Vegetation (LaBiSV)

4.2.1 *Landscape Bionomics Survey of Vegetation: Theoretical Basis*

The theoretical bases of Landscape Bionomics Survey of Vegetation (LaBiSV) can be summarized as follows:

A. Reference to the concepts of ecocoenotope and ecotissue as structural entities of the landscape.
B. Use of the biological-territorial capacity of vegetation (BTC) as the primary integrative function.

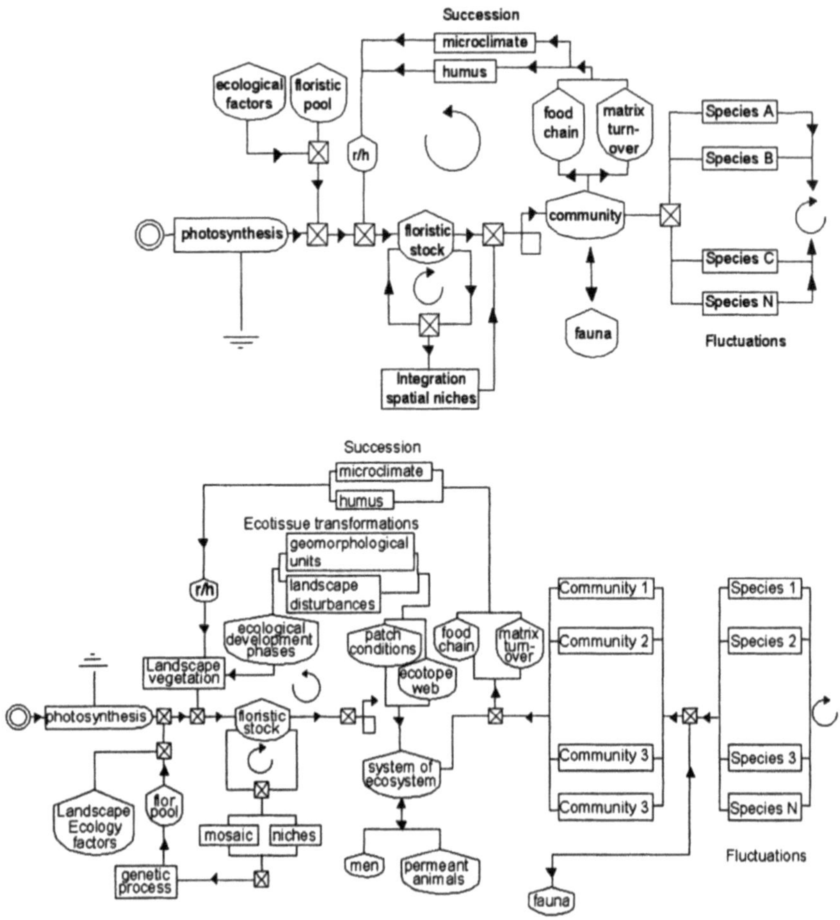

Fig. 4.2 Comparison between the flow diagrams of a single community (e.g. Vegetation Association) (above) and that of a community system (below). Note the increase in complexity in the transformation of the vegetation of a landscape unit. (From Ingegnoli (2002))

C. Possibility of arriving at a comparison between the ecological state of natural and anthropic vegetated patches, according to the principles of landscape bionomy.
D. Possibility of determining the state of normality of the ecological parameters of the different vegetation types.
E. Ability to measure the concept of biodiversity at landscape level (diversity of biological organization of context).
F. Possibility of drafting the development models of the different types of vegetation (time-BTC) based on the dynamic relationships between gross primary productivity, net primary productivity, and the respiration of plant

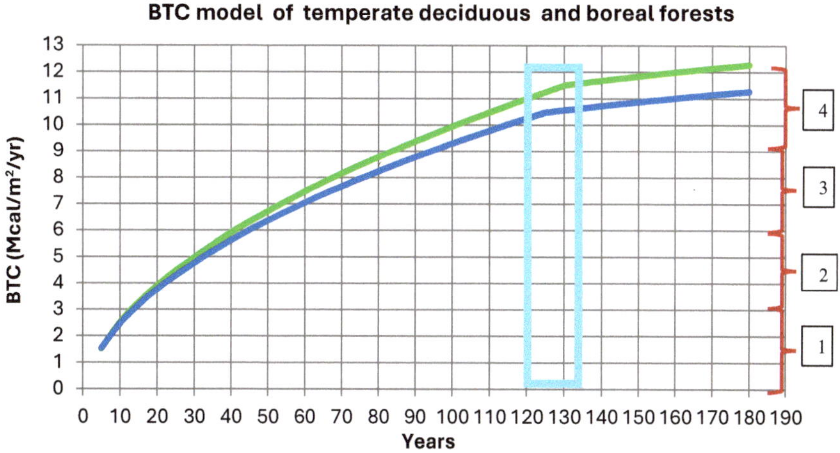

Fig. 4.3 Development model of temperate (green) and boreal-alpine (blue) forests as a function of the biological-territorial capacity of the vegetation (BTC). The light blue rectangle marked on the graph marks the transition to the maturity phase of the forest system. The 4 divisions on the right correspond to the 4 evaluation columns of the parametric form (see below Table 4.1)

phytocoenosis (Duvigneaud, 1977; Odum, 1971; Piussi, 1994) summarized in an exponential and logarithmic function (Fig. 4.3).

An exponential process expresses the phases from juvenile to adult up to the passage towards maturity; a logarithmic process expresses the phases from the beginning to full maturity. The behavior of this model is recorded by dividing the growth phases into four intervals of the same width, whose values are considered as weights for the four columns of the Survey Parametric sheets (see Table 4.1). The reference values of the BTC are to be considered at the maturity threshold. In the case of the forest tesserae, both temperate-deciduous forests and alpine-boreal coniferous forests were considered, with BTC reference maximum of 12 and 11 Mcal/m^2/year, respectively. The equations relating to the model are:

$$BTC_{(tp_1)} = t^{0.515}–0.77 + BTC_{(tp_2)} = 2.45\ ln \times t \tag{4.1}$$

$$BTC_{(br_1)} = t^{0.50}–0.70 + BTC_{(br_2)} = 2.25\ ln \times t \tag{4.2}$$

where in tp_1 and br_1 the BTC ranges are 0–12 and 0–11, while in tp_2 and br_2, they are reduced to 12–13 and 11–12.

Table 4.1 Summary of *LaBiSV* parametric sheet for the evaluation of Boreal-Alpine and Temperate-Deciduous Forests

Boreal forest	**1**	**5**	**14**	**25**	**Score**
Temperate forest	**1**	**5**	**12**	**22**	**Score**
T. Tessera characters (Ts)					
T1 - Vegetation height (m)	<9	9.1–18	18.1–29	>29.1	Canopy Hc = ...m
T2 - Cover of the canopy (%)	<30	>90	31–60	61–90	Ts surface
T3 - Structural differentiation	Low	Medium	Good	High	Age, space groups, etc.
T4 - Interior/edge (%)	None	<30	31–89	>90	(% Ts)
T5 - Management	Simple coppice	Coppice	Wood	Natural forest	Natural must be uneven-aged
T6 - Permanence (years)	<80	81–160	161–240	>240	Old trees (Mother Tree)
F. Vegetational biomass (Above ground)					
F1- Dead plant biomass	Near 0	>10	1–5	5–10	% of living biomass
F2- Litter depth	Near 0	<1.5	1.6–3.5	>3.5	cm
F3 b - Biomass volume (m^3/ha)	<200	201–500	501–950	>950	pB = ...m^3/ha
F3 t - Biomass volume (m^3/ha)	<150	150–350	350–600	>600	pB = ... m^3/ha
E. Ecocoenotope parameters					
E1- Dominant species (n°)	>3	3	2	1	As pB volume
E2- Species richness	<15	16–30	31–40	>40	n° sp./Tessera
E3- Ectomycorrhizal Fungi	<2, no web	3–6 - part.web	-12-(web)	>12 (web)	Species n° and web
E4- Allochthonous species (%)	>10	10–4	<4	0	From other ecoregions
E5- Infesting plants %	Near all	>25	<25	0	Covering area on Ts
E6- Diseased plants	Evident	Suspect	Risk	0	Even acid rain damage
E7- Biological forms (n°)	<3	4–5	6–7	>7	Cfr. Box (Naveh and Lieberman 1984/1994), mod.
E8- Vertical stratification	2	3	4	>4	Traditional
E9- Renewal capacity	None	Intense	Sporadic	Normal	Dominant species
E10- Dynamic state	Degradation	Recreation	Regeneration	Fuctuation	Cfr. Ingegnoli (2013)
U. Landscape Unit (LU) Parameters					
U1- Similar veg. contiguity	0	<25	26–75	>76	% of perimeter
U2- Source or sink	Sink	Neutral	Partial	Actual source	Species & resources
U3- Functional role in LU	Reduced	Minor	Evident	Important	Context & typology
U4- Disturbances incorporation	Insufficient	Scarce	Normal	High	Local disturbances
U5- Geophysical instabilities	Evident	Partial	Risk	None	On the physiotope
U6- Permeant fauna interest	Low	Medium	Good	Attraction	Key species
U7- Transformation modalities of the Ts	Strong disturbances	Gradual changes	Temporal instabilities	Fluctuation	Today + tendency
U8- Landscape pathology interference	Serious	Near chronicle	Easy to incorporate	None	Coming from surroundings ecotopes
U9- Permanence of analogous vegetation (years)	<100	100–300	300–1200	>1200	Historical presence
Results of the survey					
Total score Y (= H+J+K+W)	h = ...	J = ...	K = ...	w = ...	Y = ...
Bionomic quality of the Ts (bQ)	bQ_b = Y/700 (b) bQ_t = Y/ 616 (t)				bQ = ... [%]
Estimation of the BTC Boreal (conifer) Forest	BTC (b) = 0.01339 (y-28) + 0.12 (pB / 70)				BTC = ... [Mcal/m^2/year]
Estimation of the BTC Temperate (deciduous) Forest	BTC (t) = 0.01667 (y-28) + 0.13 (FM / 65)				BTC = ... [Mcal/m^2/year]
Ta = (apparent age)	Ta = $(BTC+0.77)^{1/0{,}515}$				Ta = ... [year]
Tb = balanced Age	(Dmd+Dold)/4/[0.4 to 0.7]**				Tb = ... [year]
Concise Bionomic State	CBSt-2 = (MtL x bQ) (%) where: MtL= Ta/125				CBSt = ... [%]

4.2.2 LaBiSV Main Phases and Summary Parametric Table

For the study of the vegetation of a landscape unit (LU), this methodology can be divided into at least six phases, which are summarized below:

- Phase I: study of geographical and historical characteristics of the LU: collection of geographical data (for example, paleoclimate, pedological substrate, morphology, etc.) of the LU.
- Phase II: Identification of landscape elements: once the ecological boundaries of the landscape unit in question have been established, the ecotopes and vegetated tiles (Ts) are highlighted. Please remember that the vegetation types do not always coincide with the phytosociological associations; reconstruction of the historical transformations, also of anthropic origin, through old maps and land uses in the past.
- Phase III: detection of sample tesserae for vegetation parameters; sample tiles are chosen for each type of vegetation present, and, for each tile, four groups of parameters are detected: (1) tile characteristics (T), (2) phytomass characters (aboveground) (F), (3) ecocoenotope characters (E), (4) landscape tile/unit relationships (U);
- Phase IV: ordering and evaluation of the main bionomic parameters; the above-mentioned parameters are ordered into 4 classes, evaluating them by columns with scores dependent on the different development models of the vegetation types. It is thus possible to estimate the bionomic and ecological qualities (bQ) of the parameters and/or estimate the BTC of the Ts with equations linked to the model of the type of vegetation under examination (Ingegnoli & Giglio, 2005);
- Phase V: development of the diagnostic aspects necessary to know the ecological state of the LU under examination, by referring to the principles of landscape ecology and bionomics comparing with normality intervals.
- Phase VI: intervention criteria and directions; an intervention therapy for the rehabilitation of the vegetation examined is identified, always considering the principles of landscape bionomics.

The proposed detection method (Ingegnoli, 2002; Ingegnoli & Giglio, 2005) makes use of a multiple response form (Table 4.1) to be used on the field, designed to evaluate the level of self-organization and to estimate metastability for phases III and IV of a vegetated tessera. The parametric detection tessera of the vegetated tiles is divided into four parts and records 27–32 parameters (rows) divided as in phase III.

For each parameter four evaluation classes (columns) have been considered: the resulting "weight" of each (e.g., 1, 5, 12, 22 for temperate deciduous forests) is proportionate to the curves in the previous model (Fig. 4.3) through the sections 1, 2, 3, 4 (marked with red brackets). Once all the items on the form have been filled in, the number of items marked for each column is counted and multiplied by the relevant score. The possible values of the sheet vary from 28 to 616 and from 28 to 700, respectively.

The sum of the four values thus obtained will give the total Y score of the sheet, which, compared to the comparison values and inserted into the appropriate equations, will complete phase IV. For alpine-boreal coniferous forests and temperate-deciduous broad-leaved forests the evaluation equations are:

$$\text{-temperate forest}\, BTC(tp) = 0.01667(y\text{–}28) + 0.13(pB/65)\ Mcal/m^2/year.$$

$$\text{-boreal forest}\, BTC(br) = 0.01339(y\text{–}28) + 0.12(pB/70)\ Mcal/m^2/year.$$

where the slope is the "BTC range/score range" ratio, for example 9.8/(616 − 28) = 0.01667, and where the second part depends on the ratio of biomass volume (pB) value to a standard minimum (65) in m^3/ha.

Oher results can be obtained and evaluated at the end of the table (Table 4.2).

Thus, for each type of vegetation, a form was studied (Ingegnoli, 2002; Ingegnoli & Giglio, 2005) (see also Chap. 5). A series of equations like the previous ones were expressed, as shown in Table 4.3.

In Figs. 4.4 and 4.5 two examples of application of the LaBiSV method are shown, comparing two forest tiles in the Groane Park, 20 km north of Milan, municipality of Misinto and two stretches of tree-lined corridor, municipality of Bollate. Bionomic qualities by parameter group (QT, QF, QE, QU) can be significant; the BTC of the oak forest is double that of the pine forest (8.37 vs. 4.05 Mcal/ m^2/year).

Even the size of the trees itself is very different, with an average height of 27.1 m vs. 21.3 (see photo). Regarding the rows of trees, we note that the differences in BTC are not huge (3.80 vs. 2.70), while the CBSt is very significant (10.0 vs. 3.0). However, the method proves to be very sensitive.

Table 4.2 Other results of the vegetation survey according to the summary sheets

1	Bionomic Quality (bQ), based on the Y/Ymax ratio (where: Y = score detected)
2	Territorial bio-potentiality of vegetation, BTC (Mcal/m^2/yr), with the specific estimation equation
3	Apparent age T_0 = (BTC + 0.70)1/50 for Boreal forests or T_0 = (BTC + 0.77)1/0.515 for temperate forests
4	Estimated age T = D_{md}/an-incr (mm)[a] for Boreal forests or T = 1/4 (D_{md} + D_{old})/[0.4–0.8][b] for Temperate forests
5	Concise bionomic state CBSt = (MtL × bQ) (%), where: MtL = T_0/(120–135)

Notes: [a]An-incr = annual growth (mm) of the average diameter (D_{md}) by altitudinal range (boreal-alpine conifers); D_{old} = average diameter of the oldest trees

[b]Ring growth [rings/year (mm)] in relation to the dominant species in the Ts (e.g., 4–6 mm × Quercus robur; 6–8 mm × Populus nigra). In any case, local conditions must be taken into account, e.g. with core drilling

Table 4.3 Evaluation equations of BTC ranked for the main vegetation types in Europe

Vegetation types	Max BTC value ($Mcal/m^2/yr$)	Maturity (age) thresholds (years)	BTC Evaluation equations ($Mcal/m^2/year$)
Sclerophyll Forest	12.5	120–150	0.01705 (y-28) + 0.13 (pB/60)
Temperate Forest	12.0	120–140	0.01667 (y-28) + 0.13 (pB/65)
Boreal alpine Forest	11.0	120–130	0.01339 (y-28) + 0.12 (pB/70)
Medit. Pine Forest	10.5	100–130	0.01510 (y-28) + 0.12 (pB/65)
Corridors with trees	9.8	90–120	0.0072 (y-33) + 0.10 (pB/75)
Urban Parks & Gardens	8.1	70–90	0.00526 (y-30) + 0.10 (pB/45)
Wooden Agrarian	5.0	25–45	0.00575 (y-29) + 0.15 (pB/35)
Tall Shrubs	4.2	30–40	0.00344 (y-30) + 0.10 (pB/17)
Low Shrubs	2.6	25–35	0.00247 (y-30) + 0.03 (pB/0.2)
Reed thicket	2.2	36–48	0.0023 (y-29) + 0.04 (pB/0.3)
Agricultural Fields	2.3	10–20	0.00192 (y-27) + 0.09 B3
Prairie and Pasture	1.35	20–30	0.001335 (y-29) + 0.02 (pB/0.14)
Salt marshes prairie	1.1	15–20	0.0026 (y-28) + 0.10 (pB/1.4)

Data from: Ingegnoli (2002) and Ingegnoli and Giglio (2005)
Note that some descriptors (as Corridors with trees or Tall shrubs), are physiognomic, because the differences among vegetation types in these cases do not change the BTC evaluation in a too sensible way. In case of landscape unit in which prevail these physiognomic types, they can be expressed in specific vegetation types, e.g. rhododendron shrubs, etc. after adjusting some ecological parameters

Example of reliefs of two forest tiles in the Groane Park (Milan)		SP.tot n°	Q.T %	Q.F %	Q.E %	Q.U %	Hc m	v FM m^3/ha	BTC $Mcal/m^2/yr$	bQ %
A – P. Groane, Misinto, Oak Forest	*Quercus robur 63, Q.petraea 9.5, Carpinus betulus 7, Pinus sylvestris 6, Q.rubra 3.5*	39	84,9	84,9	72,3	76,3	27,1	436	8,37	77,6
B – P. Groane, Misinto, Pine forest	*Pinus sylvestris 77, Robinia pseudoacacia 9.5, Betula pendula 6.5, Quercus robur 6.5*	24	51,3	46,7	34,8	44,9	21,3	233	4,05	42,8

Fig. 4.4 Example of survey of forest tiles in the Groane Park (north of Milan, Italy) following the LaBiSV method. Note the sensitivity of this method

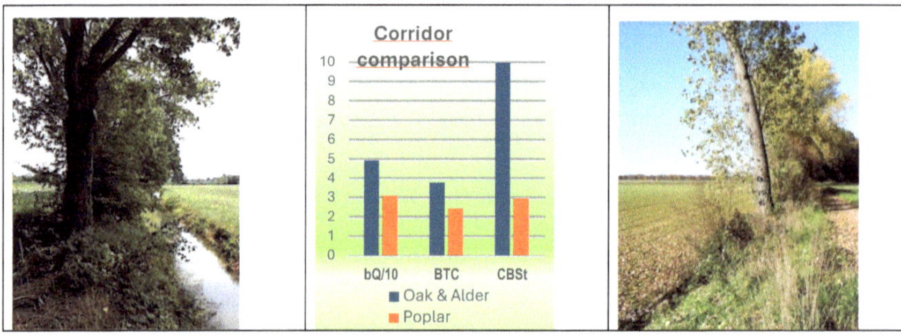

Fig. 4.5 Comparison of two tree-lined corridors in the agricultural landscapes of the Milan area. On the left a natural stretch of Quercus robur and Alnus glutinosa (blue bars in the graph); on the right an artificial stretch of Populus × canadensis and Robinia pseudoacacia, which shows a contraction of CBSt values equal to 75%

4.3 Analysis of the Characters of a Tessera

For simplicity, only the "standard" description of the entries relating to forest synthetic parametric form is reported here, referring to the manual for the others (Ingegnoli & Giglio, 2005).

4.3.1 Average Canopy Height (m)

The height of the vegetation must be considered as a weighted average of the highest heights of all the species making up the "canopy" of the tile in question: the height of at least 6–12 individuals is measured for each plant species present in the dominant layer (individuals that are already representative of the average height of the species in that specific tile, taking care to avoid the tallest and the shortest), we count how many individuals for each species are present in the tile in question and calculate the weighted average.

4.3.2 Canopy Coverage (%)

The coverage of the canopy must be considered with respect to the entire extension of the tessera in question (generally more than 1 ha) and not to individual plots (such as the one in which the phytosociological survey is carried out). It only considers the coverage of the tree highest component.

4.3.3 Structural Differentiation

Unevenness is measured through the distribution of diameters on the examined tessera. As regards to the spatial arrangement (or horizontal structure, or texture), the main typologies are considered. Please remember that both items must be verified by referring, firstly, to all the arboreal plant species, whether they belong to the dominant tree layer or to the dominated tree layers, and, secondly, to the species composing the upper layers shrubby. The evaluation will correspond to an overall consideration of the state of the card. Therefore, a tile in which the individuals are uneven-aged and with a normal aggregate arrangement is considered well-structured (high), and this condition is verified for both tree and shrub species; is attributed:

- the good value when there are not very accentuated different ages of trees combined with a spatial arrangement with signs of aggregation, or with certain species in an aggregate arrangement and others not, or even with a regular arrangement of aggregates.
- the average value when only one of the two conditions is verified, i.e. alternatively unevenness or spatial arrangement in groups.
- the low value when the tile in question is almost the same size and with a random arrangement of the plants or, on the contrary, identifiable according to a precise planting scheme, therefore regular, all for both components.

4.3.4 Inside/Margin (%)

The percentage ratio between the surface of the tessera identifiable as "internal" (depending on the type of plant and animal species that characterize it and the habit of the plant species) and the surface of the tile identifiable as "margin" is considered. In this regard, remember that the margin band is one to two and a half times deep than the average height of the species in the canopy. Please note that there are tiles that are all marginal, recognizable by a fairly sparse canopy cover, by the shape of the tree species with foliage development more similar to that of isolated trees (therefore with leafy branches also along the trunk) by a number of higher species which also includes species specific to the types of bordering tiles: it is underlined that the edge tiles are not always actually located in the marginal band of the forest, but can also be found within forest ecotopes.

Similarly, there are all internal tiles, in this case always located in the innermost parts of forest eco-topes (or in those that were internal areas, before recent deforestations or selective cutting of surrounding parcels) and recognizable for the specificity of the flora present and the animals found there, as well as for the shape of the trees, characterized by bare trunks and very reduced foliage concentrated in the upper part of the canopy. Remember that there are phytocoenosis which in themselves are "open formations" and others which are "closed formations".

4.3.5 Type of Management

Remember that here we are referring to the actual type of management by man and not to similarities of vertical or horizontal structuring with these stated types:

- all those patches, including new-growth, and high forest, which can be defined as "tree plantations" (for example, plantations of poplars or young fir trees), which will be subject to clear cutting, are considered similar to a simple and/or matrixed coppice after a certain number of years or of steering cutting (i.e. part of the shoots at regular intervals) and in which the trees are all of the same age (simple coppice) or with the presence of plants of a generation older, acting as matrices (coppice matriculated); furthermore, the soil between the rows of arboreal individuals is often weeded, there is no herbaceous layer, no shrub layer, or other arboreal layers except the planting one and propagation occurs mainly via vegetative means;
- all those pieces of arboreal vegetation which, following the rhythm of the cuts, take on an uneven-year pattern are considered similar to a compound coppice (basis of coppice with matrices of one shift, two shifts, three shifts, four shifts, of five shifts): the cuts are successive (i.e., the trees are all eliminated but in the course of 2–5 subsequent cuts) or occasional for the high forest component, clear for the coppice component; furthermore, vegetated patches in which the collection of wood or charcoal pits are carried out, or in which selective cutting of certain species is carried out, are considered as such;
- those patches governed by high forests are considered similar to a forest in which the propagation occurs mainly by sexual means, the management is carried out with the aim of producing wood or for purposes of a tourist-recreational or protective nature: therefore they are carried out periodically collecting dead wood that falls to the ground or felling dead trees still standing, or mature, senescent or diseased trees (renewal cuts), or intercalary or thinning cuts are carried out, or undergrowth species are eliminated or shrubby above a certain height. Furthermore, we recognize the presence, at the edges of the card, of forest roads or lay-bys for collecting piles of wood or sheds for collecting tools or for the overnight stay of lumberjacks, traces of passage of heavy vehicles, including tracked ones, for the removal of wood timber, fire lines, cableways.
- those patches of forest vegetation in which propagation occurs solely by sexual process are considered like natural forest management; dead wood on the ground is neither collected, nor dead trees still standing or mature, senescent or diseased nor have any cuts of any kind been made nor have any undergrowth species. We can add: possibly weeding species eliminated (whether herbaceous, shrubby or tree-like), are absent or no longer actively used (except in exceptional cases) or forest roads (or similar) and places of stable human presence. Therefore, this category can include both simple, matrixed or composite coppice tesserae and those of forest in which the management regime has changed for at least 4–5 decades.

4.3.6 Permanence (Years)

This item considers the average age of the card, which can be assessed based on the average age of the oldest individuals present on it. This assessment can be carried out quite precisely with core sampling and ring counting (without prejudice to observations of damage resulting from the plants themselves) or through an estimate of the age based on the diameter of the individuals, measured at a height of 130 cm from the base: however, it must be remembered that the data in this case may be incorrect, considering the possibility that the seedlings spend up to 50 or 70 years of their life in the form of young seedlings waiting for a crash to give them the opportunity to complete their development. This data should then be calibrated with an evaluation of the favorable or unfavorable climatic conditions that occurred during the period in question. In this regard, there are growth tables drawn up by foresters, to which a tolerance value of 10% can be applied. On the other hand, remember that in the form what is needed is not so much an exact value as the position within four quite distinct age ranges, and this allows us to circumvent some objective difficulties.

4.4 Analysis of Epigeal Phytomass

4.4.1 Amount of Dead Phytomass

All necro-mass is considered whether it consists of trunks on the ground or still standing, hollow or solid, branches or bark, including the bases of the cut trunks, excluding leaves which will be counted in the litter. The value as a percentage of the volume must be compared to the calculated live epigeal phytomass.

4.4.2 Litter Depth

Remembering that the term litter indicates the set of more or less decomposed organic residues (in particular leaves and needles, small twigs, etc.) in which the original structure is still recognizable and there are no obvious signs of decomposition, and which is directly superimposed on the mineral soil (which it supplies); it must be measured in three or four different points of the tile, always far from the tree trunks, and the final value will be a weighted average of the values detected.

4.4.3 Phytomass (Fm) by Volume (m^3/ha)

The entry refers to the epigean phytomass estimated per hectare, calculated not in weight but in volume, a value which is more explanatory of the occupation of space for the purposes of a landscape ecology discussion. For the estimate of the

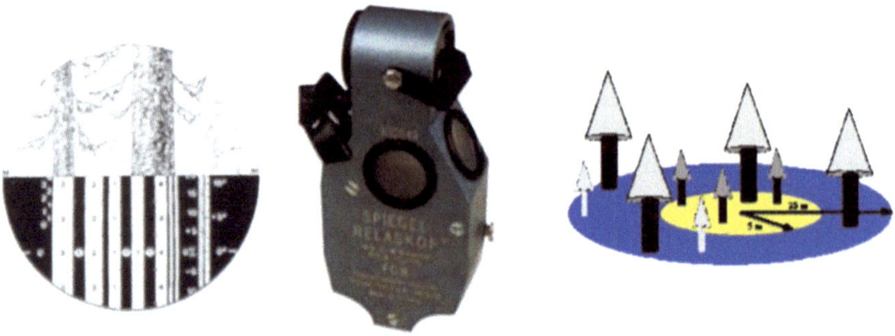

Fig. 4.6 Bitterlich mirror relascope, University of Vienna. In the centre, the instrument, on the left the optical targets, on the right an example of a tree phytomass survey station from the instrument's station point

phytomass, compare Ingegnoli and Giglio, 2005. In the cases of multi-stratified forests, i.e., in which in the survey of the basal area a high number of individuals with a diameter comparable to that of the individuals making up the canopy but of height significantly lower is found, it will be necessary to consider the weighted average height between the two layers in question when calculating the phytomass, instead of limiting it to the canopy alone. In the LaBiSV tab, this entry is twofold: it provides the phytomass intervals characteristic of boreal and temperate forests.

4.4.4 *Relascope*

The method most used today for measuring the volume of phytomass is based on an instrument called Bitterlich's "mirror relascope" (Bitterlich, 1970) (Fig. 4.6), which is an optical instrument. The most frequently used *survey angle (K)* is K = 2 (white in Fig. 4.6), even if the relascope can evaluate K = 1 (small trunks), or K = 3 (large) or K = 4 (extra-large).

It follows that the mirror relascope equation for determining the basal area in m^2/ha is:

$$G = K \times N^*$$

where: K = BAF (Basal Area Factor), N = [n. trees > (bK) + 1/2 n. trees = (bK)], being (bK) the broadband K in the instrument.*

4.4.5 *Allometric Coefficients*

The study of the shape of a tree or shrub, understood as an architectural model, is of considerable importance as the external appearance is evidence of a specific adaptive meaning to the changing conditions and stresses of the environment, that is, of the ".. result of the overall growth project of the individual codified by his genotype, common to all individuals of that species, but filtered through the action of the environment..." (Piussi, 1994).

In Landscape Bionomics, this knowledge serves two specific purposes: the first consists in helping to identify the characteristics of a vegetated patch with respect to the "edge" factor of the forest or shrubland (with a habit therefore more like that of an isolated tree) or "inside" factor of the forest or shrubbery; the second in the development of specific indices, called allometric indices (Al), useful as a corrective factor in the calculation of the overall phytomass of a vegetated patch. In fact, the volume of a tree calculated simply as the product of basal area (base surface of the trunk) by height, would provide the volume of a cylinder, regardless of the shape and consistency of the crown, which is of fundamental importance. The corrective factor takes into consideration the relationship between the crown and the trunk (Table 4.4; Figs. 4.7 and 4.8).

Table 4.4 Allometric coefficients of the main European tree species for which data were available

Plant species	Greenery/trunk (mass ratio)	Allometric coefficient (A)	(A) for an isolated tree	Specific weight
Abies alba Miller	∅ 20 = 0.167	0.47	0.51	0.44
	∅ 50 = 0.152	0.46		
	∅ 95 = 0.127	0.45		
Picea abies (L.) Karsten subsp. *abies*	∅ 20 = 0.500	0.55	0.6	0.44
	∅50 = 0.467	0.54		
	∅95 = 0.445	0.53		
Larix decidua Miller	∅ 50 = 0.450	0.57	0.62	0.6
Pinus sylvestris L.	∅ 50 = 0.300	0.46	0.5	0.53
Fagus sylvatica L.	∅ 20 = 0.267	0.71	0.95	0.74
	∅50 = 0.528	0.84		
	∅95 = 1.450	1.35		
Prunus avium L.	∅ 30 = 0.186	0.73	0.8	0.66
	∅50 = 0.410	0.7		
Robinia pseudoacacia L.	∅ 30 = 0.170	0.65	0.75	0.78
Populus alba L.	∅ 50 = 0.450	0.77	0.85	0.5
Alnus glutinosa (L.) Gaertner	∅ 30 = 0.180	0.6	0.7	0.56
Quercus robur L.	∅ 50 = 0.560	0.85	1.05	0.75
Quercus pubescens Willd.	∅ 30 = 0.190	0.7	0.8	0.96
Quercus ilex L.	∅ 30 = 0.250	0.75	0.88	0.9

From: Susmel (1980), in Ingegnoli (2002), integrated
These values are to be considered only indicative, because each local situation may present its own peculiarities

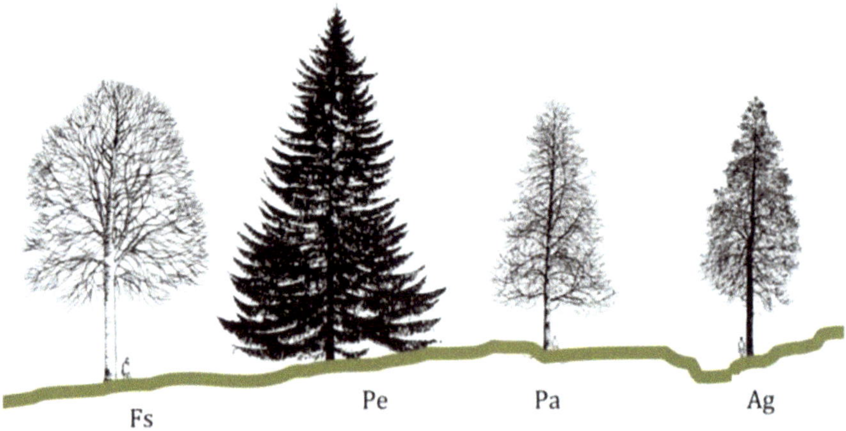

Fig. 4.7 Example of trees with isolated or semi-isolated growth, which have higher allometric coefficients (c.al.) than within forests. From left: Fs, Fagus sylvatica, c.al. = 0.80–0.88; Pe, Picea excelsa, c.al. = 0.57–0.60; Pa, Prunus avium, c.al. = 0.70–0.76; Ag, Alnus glutinosa, c.al. = 0.68–0.72

Fig. 4.8 Example of monumental beech on Mount Biaena, 1300 m altitude, in the municipality of Mori (Trient). The Author's children (3 and 7 years old) with a forest caliber of 1.2 m in the foreground, give an idea of the size and powerful ramifications of the plant, approximately 33 m high and with approximately 4 m in circumference. In these cases, the allometric coefficient can exceed unity: c.al. = 1.07

4.4.6 *Calculation of the Volume of Phytomass*

All these data and considerations give the possibility of establishing the equation that allows the calculation of the phytomass for each species under examination in m^3/ha:

$$vFM = G \times Hc \times Al \; \left[m^3/ha\right]$$

where: G = basal area, Hc = canopy height, Al = allometric coefficient.

Note that the Hc is not the same of Sect. 4.3.1; Hc is the weighted average of all the measured trees per station.

This result is usually reduced by a value between 5% and 10%, given that in surveys with the relascope there is a tendency to overestimate the trees of limit size (which are recorded as 1/2 n. trees).

4.5 Analysis of Ecocenotopic Characters

4.5.1 *Dominant Species (n°)*

The dominant species is established not in relation to the coverage value, as in phytosociological surveys, but as a function of the volume of aboveground phytomass. Therefore, where more than one species exists in the canopy, the procedure is as follows: once the total volume of phytomass has been calculated, it is divided by the number of species found in the canopy, to obtain the equitability value, i.e., the quantity of phytomass which would belong to each individual species if they all had the same importance. This value is then compared with the actual phytomass values measured for each individual species: if one or more species are found to have a phytomass clearly higher than the average equitability value, that species/species are considered dominant. For this purpose, a significance threshold of no less than the order of magnitude of 20% was established (i.e., the dominant class(es) must have a percentage of phytomass at least 20% greater than the others). Consider the example: given a canopy with 5 species, the equitability would be 20% of phytomass for each; if a species has 40% phytomass (and the other four values fluctuate between 10% and 20%, for an overall total of 100%), this species can be considered dominant as its phytomass value exceeds by at least the 20% the value of the other classes); if there were two species with approximately 30% each (and with the remaining attested at around 13%) the situation at the time of the survey would not be sufficiently characterized and we would have to talk about more than three dominant species.

4.5.2 Species Richness

By species richness we mean the overall number of plant species found on the total surface of the *tessera*/tile (and not just those of the individual phytosociological surveys) considering both the three arboreal, shrubby and herbaceous layers, and the mussy and lichen species, i.e. mosses and liverworts present on the soil and non-corticolous lichens (where possible, make at least an estimate of the possible species), and, in addition, the species of fungi (considering those with evident epigeal carpophores). Basically, the complete floristic list with the addition of mushrooms.

4.5.3 Alien Species (%)

Species that belong to a region or eco-region different from the one to which the examined *tessera* belongs are considered alien species from an ecological point of view. Thus, not only those reported as such in the Flora of Italy (Pignatti, 1982) will be alien species, but also, for example, the steno-Mediterranean species for the Middle European region, or vice versa, for example, the Central European species for the Mediterranean region. For the definition of regions and ecoregions to which they belong, refer to (Ingegnoli, 2015). Their percentage must be evaluated on the complete floristic list.

4.5.4 Fungi Ectomycorrhiza

It is now known that mature natural forests of the temperate and boreal zones have a cybernetic network (Simard, 2018) in the soil formed mainly by the hyphae of ectomycorrhiza fungi. Therefore, it is important to verify its presence, even if only partially. See Par. 5.4.

4.5.5 Weeds (% Coverage)

By weed plant we mean any species, even characteristic of the association to which the *tessera* in question refers, belonging to a layer other than the dominant one, which however has, in the tile, within the layer that belongs to it, a covering such as to suffocate or severely limit the development of the other species of that same layer or of other subordinates: remember that the percentages must be calculated with respect to the total surface area of the *tessera* and not to that of one or more phytosociological surveys or study plots. This happens for example with brambles

Fig. 4.9 Serious alteration of a willow forest (*Salix alba*) on the Po River right floodplain near Pancarana (Voghera, Lombardy, Italy) due to the invasion of Sicyos angulatus (Bur Cucumber) exotic cucurbitatcea from North America

(*Rubus* sp.*pl.*), which often make the undergrowth impenetrable, but also with *Pteridium aquilinum* (the aquiline fern), with a herbaceous species such as *Vinca minor* in an oak-hornbeam forest of the Po Valley or even with *Luzula nivea* in a spruce forest, or *Clematis vitalba* in a deciduous forest. The worst alteration of a forest is done by exotic invasive, like *Sicyos angulatus* as shown in Fig. 4.9.

4.5.6 Sick Individuals

It involves recognizing, within the examined *tessera*, of the presence of dead or sick individuals, or with signs of non-optimal health, considering the state of the bark, trunk, branches, and leaves and the so-called "new damage". In the risk item, it is important to consider any damage from acid rain, even if it has not yet manifested itself, or from other polluting sources, but also from instability which could create risks for the vegetation of the tile in question.

4.5.7 Biological Forms (n.)

Biological forms, from an ecological point of view, must be evaluated not based on Raunkier's division, but on that of Box (1987) as modified in Ingegnoli (2002). This is for a better correspondence between the shape of the plant verified in the *tessera* and its ecological role even in the different life stages (a beech plantlet or seedling does not have the same ecological role in the *tessera* that a young beech individual can have, or even that of an adult beech individual). It is recommended to pay careful attention not to confuse the recognition of biological forms (therefore the shape of plants) with that of the vertical structure of the *tessera* under examination, as per the following point. Remember that the term Krummholz, for which there is no equivalent either in English or Italian, indicates shrubs with the twisted and/or prostrate shape typical of the mountain pine.

The resulting forms (Table 4.5) are the following: thallophytes and bryophytes, epiphytes (including parasites), lianas, ferns, graminoids (i.e. species with a habit similar to that of grasses), other herbaceous plants (herbaceous species not graminoids, including hemicryptophytes and geophytes), reeds, rosette shrubs (with leaves at ground level, often succulent), semi-shrubs or suffruticose (chamaephytes, renewals, seedlings, seedlings and seedlings of woody species), succulents (perennials with or without woody skeleton), Krummholz (twisted shrubs, with one or more stems less than 1 m in height—except for the main stem—prostrate, branched; for example, mountain pine in mountain pine forests), shrubs arborescent (with one or more stems higher than 1 m in height, but indeterminate form of growth), "tuft trees" (tree ferns and similar, with a trunk but without bark and with the leaves arranged above in a tuft), trees (even young ones or saplings).

4.5.8 Vertical Stratification

From the point of view of landscape bionomy, a layer is considered present when it is represented by several individuals such as to give coverage equal to at least 25–35% of the surface of the tessera. Otherwise, note in the marginal notes the presence of a certain number of individuals attributable to a certain height, but do not consider it an actual layer. The vertical stratification adopted for these survey sheets is schematized as follows (Fig. 4.10) (Ingegnoli, 2002):

- layer of large trees = height greater than 30 m.
- second size tree layer = height between 20 and 30 m.
- layer of third-size trees and tall shrubs = height between 5 and 20 m
- shrub layer and arborescent trees = height between 1 and 5 m.
- herbaceous layer or low shrubs = height between 0.45 and 1 m.
- medium grass layer = height between 0.15 and 0.45 m.
- moss and low grass layer = height between 0.1 and 0.15 m.

Remember that what matters in this case is the height real value of the species at the time of the survey and in the tile considered.

Table 4.5 Dichotomous table for the recognition of plant forms

1.	Vascular plants (kormophyte)	2
	Non-vascular plant (cryptogam)	Thallophytes (incl. mosses, algae, liverworts, lichens)
2.	Plant rooted in ground	3
	Plant not rooted in ground, sitting on other plants	Epiphytes (incl. parasitic)
3.	Plant self-supporting (at least at maturity)	4
	Plant permanently sprawling, climbing, or otherwise not self-supporting	Vines/Lianas
4.	Plant with stems	5
	Stems essentially absent above ground (ab.gr.); leaves in terminal rosette at ground level (often succulent)	Rosette-shrubs
5.	Stems not permanently woody or succulent (ab. gr.)	6
	Stems permanently woody (ab. gr.) or succulent	10
6.	Plant totally herbaceous	7
	Plant not totally herbaceous	9
7.	Spermatophyte (especially angiosperm)	8
	Pteridophyte	Ferns
8.	Plant grass-like	Graminoids
	Plant not grass-like	Forbs (incl. geophytes)
9.	Plant perennial with progressively sclerotic culm, >0.8 m tall (or similar)	Reeds
	Plant perennial from woody *xylopodium* (or similar)	Semi-shrubs (incl. trees *plantulae* or seedlings)
10.	Stems succulent, with or without woody skeleton; plant perennial	Stem-succulents
	Stems permanently woody (ab. ground)	11
11.	Plant with multiple main stems, often with none dominant; plant generally <5 m tall	12
	Plant with a single main stem (trunk); plant generally >5 m tall	13
12.	Plant with 1 or more main stems, <1 m (except for emergent trunks), prostrate, usually highly branched	Krummholz (e.g. twisted shrubs, incl. cushion-shrubs, dwarf-shrubs)
	Plant with 1 or more main stems, >1 m; growth form indeterminate (e.g. overgrown bushes, scandents)	Arborescent shrubs (incl. cushion-shrubs, dwarf-shrubs)
13.	Plant growing from terminal bud, with wood produced secondarily, no bark; rarely branching; leaves in rosette(s), usually terminal	Tuft-trees (incl. tuft-treelets, tree-ferns)
	Plant with true wood growing outside; usually with bark, plant usually branching	Trees (incl. treelets, small trees)

Box (1987), modified and integrated

Fig. 4.10 Example of vertical structure of vegetation in mixed forest and in tall prairies with bushes. Low shrubs and grasses below 1 m layer, tall shrubs in the second layer (5 m), small trees in the third layer (5-20), medium-high trees in the fourth layer (20-30 m), and the higher trees in the last layer (>30m).

4.5.9 Renewal Capacity

The renewal capacity is measured on the dominant species(s) in the tile at the time of the study in the form of sexual propagation, i.e., from seed. If the dominant species (s) do not present renewal, but there is very abundant renewal of another species, even arboreal, the leftmost column must be marked. The abundant renewal of this other species will instead be indicative of the dynamic state of the *tessera*, referred to in the next point.

4.5.10 Dynamic State

The presence of a strong renewal of a species other than the dominant one(s) is indicative of the possibility of a change. If this renewal is accompanied by the presence of the same species in the different vertical layers of the tile, this indicates that the change is underway; otherwise, that there are some factors, or complex of factors, which are contrasting the shift, i.e., can incorporate such change, independently of any evaluation on the positivity or less than such "incorporation". The recognition of these phases in the field is not simple and is favored by knowledge, or by some inspections also of tiles surrounding the one directly under examination. In a very schematic way, it can be stated that:

- The characteristics detectable in the field survey are indicative of a phase of 'degradation', in progress, in conclusion or in continuous fluctuation (i.e. in dynamic equilibrium), when disturbances that cannot be incorporated are present: for example, when the advance or penetration of inclusion in the *tessera* of invasive or alien plant species that take on a consistent cover or biomass or are characterized by abundant renewal, capable of winning the competition with native species; or when the *tessera* is reaching a state of widespread senescence, due to the lack of embeddable disturbances, or in the case of destructive events, or even when there is a progressive destructuring (in a vertical or horizontal sense) both due to natural events and anthropic or to the strong reduction of the biological forms present or even to a state of widespread plant disease. As regards the species richness aspect, remember that even a high biodiversity can indicate degradation, while a low value of specific biodiversity in a forest context does not necessarily indicate degradation (always refer to the average number of species per survey, to then be compared to the tile, indicated by type of forest in Pignatti (1998)).
- The characteristics detectable in the *tessera* are indicative of an ongoing or concluding phase of 'recreation' when the *tessera* is changing in floristic composition or in structure and organization under the pressure of forces coming from the surrounding tiles or ecotopes. Generally this phase is the one that follows a zero event or a strong natural disturbance (for example, floods, landslides, volcanic eruptions), or even a sudden, clear modification of multi-annual balances and therefore of the range of disturbances to which the *tessera* appeared to have adapted (for example, consider the case of the sudden cessation of a thousand-year-old forestry-pastoral activity managed by man following the establishment of a protected area, or of a certain type and method of cultivation, or even the sudden abandonment of certain hill or mountain areas for urbanization) following which even the seed bank contained in the soil or the own forces of the *tessera* may not be able to trigger a recovery process on their own. The tile must thus be completely reorganized. In practice it is recognizable by the presence of contingents of characteristic species attributable to three or more different phytosociological associations or phytocoenosis, usually coming from surrounding areas, so that at the time of survey it is not possible to predict where the path taken will lead, or by the infesting behavior assumed by one or more of these, by strong characterizations of the margin. In practice, it is the situation closest to the concept of succession typical of traditional ecology.
- The characteristics detectable in the *tessera* are indicative of a 'regeneration' phase in progress or nearing completion when the *tessera* is recovering from an embeddable disorder, or from a phase of degeneration that is not particularly strong, or, again, was in a phase of fluctuation which was followed by only a

slight change in the range of disturbances: the dynamics of the tile is moved only by the internal forces of the tile itself, by its potential (think of the abandonment of a coppice of a species belonging to the floristic procession of vegetation reference for that specific area or when starting it with tall trees). In this case, the seedlings and seedlings belong to the dominant species, the vertical structure denotes the reconstitution of a missing or discontinuous layer. A fundamental distinctive character is the absence or very low incidence of species coming from plant associations different from that characterizing the reference vegetation (fittest vegetation) of that area. In practice, only a change in the order of the system occurs. This point also includes the case of a tile undergoing normal growth.

- Characteristics detectable in the tessera are indicative of an ongoing maturity 'fluctuation' phase when the tile oscillates around a metastable position, without entering the instability field. It is recognized by the change of only marginal characters, as happens in the case of a normal alternation of dominance between the two dominant species in that type of vegetation (for example, *Fagus sylvatica* and *Abies alba* in a beech forest) or by the entrance, or the exit of some species, even gradual and recurrent, without this altering the relationships of alternation and dominance.

4.6 Analysis of Vegetation in Relation to Landscape Unit

4.6.1 Contiguity with Similar Vegetation

It measures the percentage of the perimeter, out of the total, that is in close and direct contact with vegetation of a similar type. For this purpose, it must first be clarified which vegetations can be considered like those present in the tessera in question: all, and only those, referable to the plant association(s) or phytocoenosis.

- Other woodland tiles referable to different associations cannot be considered as analogous (a larch forest compared to a spruce forest, or an *Orno-ostrietum* forest compared to a chestnut forest).
- Other shrubby, meadow, pasture or agricultural vegetation of any kind cannot be considered similar, nor urban, suburban, or industrial.
- The roads that can be found delimiting the tile examined constitute an impediment to contiguity if, whether forestry or asphalted, their width prevents contact between the canopy crowns of the two tiles whose contiguity is examined, or if they are widely trafficked.
- Any fences impede contiguity if they are continuous and closed; discontinuous and/or open ones (like networks) are not.

4.6.2 *"Source" or "Sink" Characteristics*

This entry refers to Pulliam's (1989) source-sink theory. The tessera is considered as a sink if it is in an ecological state such as to act mainly as a receptor for species or resources: in general (but each case must be evaluated directly) these are very young or very degraded tile, parcels or coppices recovering after a recent cut, or patches affected by fire or survivors of some epidemic or attack by insects, fungi or parasites, or patches of woody plants when kept without undergrowth (for example, poplars), or with a high number of alien species, few biological forms and few key species etc.

- The tessera is considered *neutral* if it is in an ecological state such as not to be able to export either species or other resources to the surrounding tiles, but not even to need to receive them, or in a condition of substantial parity between import and export. These are generally partially degraded tiles in recovery, or tiles coming from clear cuts or older coppice, or even tiles from woody plants managed in a more "ecological" way, or even tiles in the transition phase from juvenile to adult in good condition, or of adult but degraded tiles, or with a high number of cosmopolitan or non-characteristic species in relation to the low number of characteristic species, or of tiles with individuals suspected of disease.
- The tessera is considered as a *partial source* if it is in an ecological state such as to be able to partially export both species and resources towards the surrounding tiles: this happens for adult tiles in good condition or mature but partially degraded, or tiles with an average number of key species, biological forms and chorologically autochthonous species, not a high internal/margin percentage ratio (item T4) or higher but internal with signs of degradation.
- The tessera is considered as an *effective source* if it is in ecological conditions such as to be a source of species and other resources towards the surrounding tiles: these are mature or partially senescent tiles in good conditions, in a dynamic phase of fluctuation or regeneration, with good vertical stratification, a high number of key species present, scarcity or absence of alien species, high number of biological forms, high percentage of interior surface in good ecological conditions compared to the margin.

4.6.3 *Functional Role in the LU*

The role of the *tessera* in question must be directly evaluated based on the LU to which it belongs (and not the ecotope to which it belongs) but should not be confused with the source function referred to in point U2. It is rather to be evaluated considering the landscape functions that the *tessera* can perform (or, for simplicity, the various landscape systems to which it could be ascribed). This role is primarily a function of the position of the tile (or the ecotope to which it belongs) in the ecomosaic: the role of a tile of arboreal vegetation will be the stronger the more the LU matrix is of a different type (in an increasing sense silvo-pastoral, agricultural, rural, suburban, urban matrix); in a subordinate place it is a function of the overall ecological state of the tessera as can be assessed from the previous items: the better this ecological state, the greater its role in the context. It is necessary to underline that, when dealing with the role of the tessera in the context of the LU, several tesserae of the same forest ecotope with similar characteristics could have the same important or evident role in relation to the LU, even if they would instead have a smaller or different role between them, if considered instead with respect to the ecotope itself; furthermore, even a tile of low ecological quality can take on an important, or even evident, role if inserted in a small wooded ecotope but belonging to a suburban or urban matrix.

4.6.4 Incorporation of Disturbances

The ability to incorporate disturbances at the *tessera* level considers local disturbances typical of the environment in which the *tessera* itself is inserted, in particular meteorological and climatic events (for example, thunderstorms, tornadoes, hailstorms, lightning, rains of different frequencies and intensity) and recurring geophysical events (for example, soil creeping, small landslides, flooding), even if over multi-annual cycles, but of an order of magnitude comparable to that of the development of a forest tile; or periodic or chronic anthropic activities (for example, management interventions, collection of wood, charcoal, passage or stationing of people for recreational or sporting activities or study); or, again, disturbances linked to the activities of wild, livestock and domestic fauna (for example, the holes and removal of the sward carried out by wild boars or the elimination of all young shrubs by goats or roe deer or fawns). If these disorders are localized, not too frequent and affect only limited surfaces, they can be incorporated. Otherwise, or in the case of events outside the temporal or spatial scale, the incorporation occurs at a higher scale than that of the tile.

4.6.5 Geophysical Instabilities

Possible factors of instability are highlighted, linked to climatic, geophysical events and activities in the broad sense, and/or hydrogeological ones with even very long recurrence cycles, to which the tessera is potentially subject, and which can be evaluated based on information obtainable from reports and/or maps relating to geological, hydrological, hydrogeological "Risks of instability" (c/o geologists) or "Risks of meteorological events" (c/o agronomists).

4.6.6 *Interest on the Part of the Permeating Fauna*

The interest that the *tessera* has for the permeating fauna is generally a function directly proportional to the structural and functional characteristics of the tile itself (when inserted in a forest ecotope), but also of the distribution of *tesserae* of the same type in the matrix (in particular, the interest may increase when the tile constitutes almost a unique example in the matrix, as in the case of one or a few cultivated poplar grove tiles in an agricultural matrix). For its correct evaluation it is necessary to understand whether it presents elements of attraction or interest for the fauna characteristic of that certain environment (for example, diversified ecological niches, place of food supply, refuge against possible predators, nesting or shelter for the night, nursery etc.).

4.6.7 *Transformation of the Ts as an Element of the Landscape*

The dynamics of a *tessera* as an element of the landscape is influenced by the surrounding conditions, that is, by what happens in the ecotissue. They concern both the reasons that determined the current state of the *tessera* and those that will guide its transformations soon. It is identified with:

- a strong disturbance when a clear intervention is carried out on the tessera as such or on the directly bordering tile (for example, clear cutting, drastic selective cutting, fire, attack by various diseases or parasites or other) and a strong transformation is occurring at the margins even not directly neighboring, but whose influence can affect the tessera in question (for example, construction of a large road, railway, industrial or commercial infrastructure, sudden change of use of neighboring land with the construction of residential buildings or of various types of recreational area or services, or reservoirs for hydroelectric purposes…);
- a gradual change when dealing with the gradual and slow evolution of the ecotissue as a whole, within which even the transformation of the tessera is guided by the internal reorganization of the system in the search for a new threshold of metastability, which however is no longer guaranteed to be higher than the previous one; in practice, a gradual change is implemented when the system changes step by step following the change in the range of disturbances to which it was previously subjected, slowly adapting its organization to the new parameters. But this change can be directed both towards higher and lower levels of metastability and organization (for example, the same forest tile can belong both to a silvo-pastoral landscape gradually evolving towards an open forest landscape, then a closed forest one for abandonment of pastoralism and forest management activities, both to a silvopastoral landscape gradually changing towards seminatural agriculture, then industrialized agriculture);

- a temporary instability, when dealing with the presence of an event that has created a disturbance in the LU as a whole, lowering its level of metastability, organization and order, a disturbance however necessary for a broader reorganization of the landscape unit in a higher level of metastability which has not yet been achieved and whose effects are also felt on the tile under examination, despite not having directly involved it (because for this tile the disturbance would probably have had a destructive effect). An example could be an accidental fire in a prairie, which does not substantially alter its floristic or faunal composition but can modify the dominance relationships between both the plant and animal species that inhabit it.
- a fluctuation, when dealing with the situation of an ecotissue that has reached a threshold of high metastability and which, therefore, oscillates around its attractor, allowing the ecotope tiles that compose it to evolve according to their intrinsic potential.

In case of discrepancy between the event that produced the currently present situation and the event that guides its transformations in the nearest future (as far as can be estimated based on the knowledge possessed at the time of drafting the form), mark the column corresponding to the more negative event of the two.

4.6.8 Interference with Pathologies Typical of the Surrounding Landscape

This item instead highlights the actual pathologies of the surrounding LUs which may have an influence on the tessera in question. They are distinguished according to their severity, both intrinsic to the LU itself, and in relation to the effects they can have on the surrounding LU, and therefore on the tiles (where the distance of influence is different from case to case depending on the type of pathology). Concerning pollution, a very important aspect to verify this item is the presence or absence of corticolous lichens on trees: according to the parameters of the ANPA protocol, the absence (lichen desert) or scarcity (under 15 species per phorophyte) of the number of different species indicates a strong alteration in air quality, perhaps coming from surrounding LUs, even in forests in apparently good condition (check which tree species are most favorable to the installation of lichens). For the list of all the other possible pathologies, refer to Table 7.1.

4.6.9 Permanence of the Same Current Vegetation Type (Years)

Here we consider the number of years, possibly reconstructed on a cartographic basis, for which that tile has been continuously being covered by arboreal vegetation, even if referable to different phytosociological associations or phytocoenosis. The ecological significance of this evaluation is to be found in the importance it plays for the quality and health of a forest tile, the availability of a time interval suitable for correct forest pedogenesis. Therefore this value can be totally different from that reported in point T6: in fact, you can have a very young tile (less than 80 years old) on land been being covered for millennia by forest vegetation (older than 1200 years), or on the contrary a very mature forest, even over 300 years old, but in a position where until a few years before the creation of the forest itself there was arable land or perhaps a suburban area.

4.7 Chorological Analyzes and Bioindicators

4.7.1 Chorology

The floristic list of phytosociological surveys carried out is accompanied, for each species, by the relative attribution of the chorology. Based solely on the presence/absence of the species (therefore regardless of their coverage in the survey or presence class), the percentage presence in the survey itself is calculated for each chorological type.

However, the chorological typologies must be reduced in number, grouping them by similar area of origin. A typical sequence for the pre-Alpine environment can be: (1) Eurosiberian, (2) circumboreal, (3) endemic alpine, (4) orophyte, (5) steppic, (6) Eurocaucasian, (7) Eurasian and paleotemperate, (8) Atlantic, (9) eurimediterranean, (10) stenomediterranean, (11) cosmopolitan and subcosmopolitan, (12) exotic.

As suggested by Pignatti (1998) and Biondi and Baldoni (1994), the comparison of the chorological data of a LU with the chorological spectra at the syntaxonomic alliance level can be useful in verifying the discordances of local phytocoenosis with the Italian statistical averages by type of alliance. An example of the results is graphed in Fig. 4.11 (Ingegnoli, 2015), which compare the surveyed values of two real LU to the typical distribution of the association or alliance of reference.

In northern Italy, the calculation of the value of the Mediterranean/Circumboreal ratio can be very useful. This ratio is higher the more the species detected are better adapted, among other aspects, to high annual temperatures.

4.7.2 Ellenberg Bioindicators

The completion work carried out by Pignatti (2005) to adapt the Ellenberg bioindicators (1991) to the Italian flora allows them to be used for interesting comparisons between the floristic lists of different Lus too. Please remember that these indicators are: (1) light, (2) temperature, (3) continentality, (4) humidity, (5) soil reactions, (6) soil nutrients, (7) salinity (usually negligible).

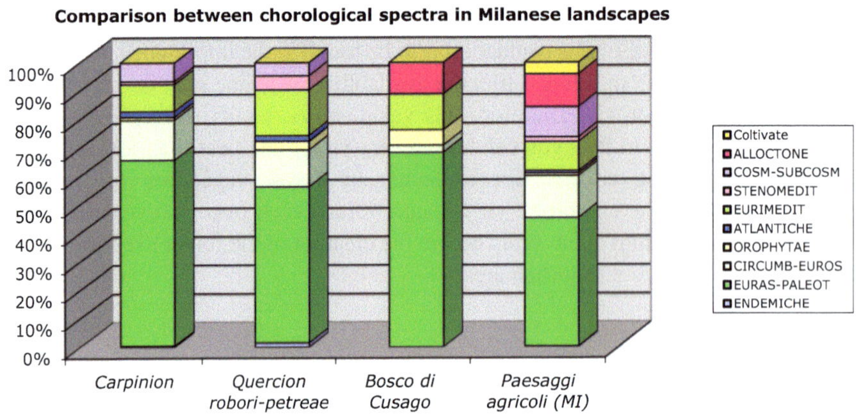

Fig. 4.11 Comparison between the chorological spectra of a relict forest near Milan (Cusago) and the average of 10 agricultural landscape units in the Milan area. The spectra of the two phytosociological alliances refer to the theoretical formations reported by Biondi and Baldoni (1994). (Figure by Ingegnoli (2015))

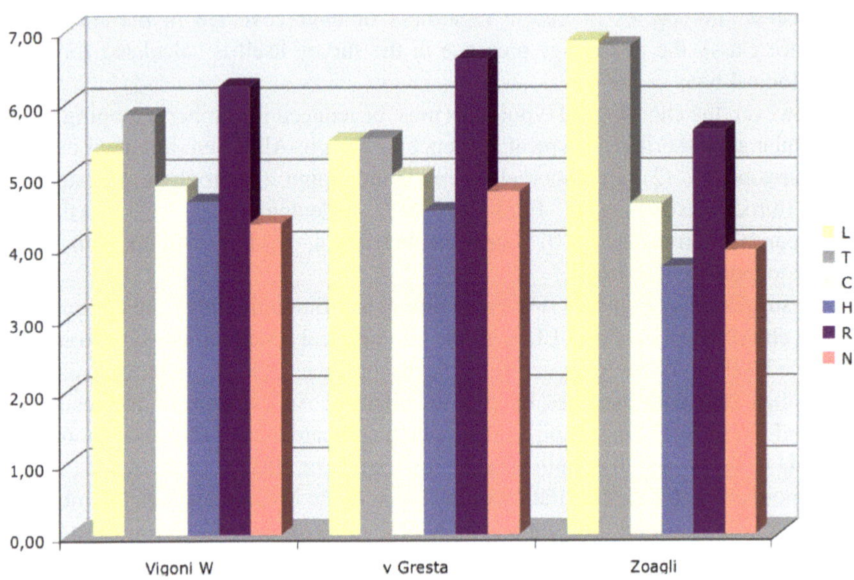

Fig. 4.12 Ellenberg bioindicators concerning the forest vegetation of the lower Val di Gresta, compared with a similar pre-Alpine area (Boschi di Villa Vigoni, Menaggio) and with a Mediterranean LU (Zoagli). Despite the influences of Garda and the tendency towards climate warming, the LU in question is not sub-Mediterranean. (From Ingegnoli (2015))

From Fig. 4.12, we see the comparisons between the Ellenberg bioindicators measured in the woods of the lower Val di Gresta (Garda Lake, TN, Italy) and in the woods of Menaggio (Como Lake, CO, Italy) and Zoagli (Genoa, Italy). We always notice the good correspondence with the pre-Alpine woods of Menaggio (similar altitudes and latitudes) and the clear distance with the Gallo-Mediterranean belt of Zoagli. With this we can dispel the assertion that the surroundings of the large pre-Alpine lakes are of a sub-Mediterranean environment: they are thermophilic, but of a Central European type.

4.8 CBSt or Evaluating the Efficiency of Vegetation

4.8.1 Meaning of the Concise Bionomic State, CBSt

The CBSt (Concise Bionomic State) of a formation under examination is the indicator (Ingegnoli, 2013)—already mentioned and evaluated in Tables 4.1 and 4.2 (reportable for each sheet)—that considers the degree of maturity of the phytocoenosis too, and therefore estimates the ecological efficiency of the vegetation, in percentage. To apply the formula expressed in the table we must evaluate MtL (degree of maturity):

$$MtL = (BTC + b)^{1/a} / Mod.flex.\%,$$

where: b depends on Eq. (4.1), and *Mod. flex.* = maturity threshold, see Fig. 4.3.

Figure 4.13 (Ingegnoli, 2015) shows a series of correlations between the biological-territorial capacity of vegetation (BTC) and its degrees of bionomic efficiency (CBSt). Remembering the 7 classes of BTC found in the temperate zones, we immediately notice that in class IV (BTC = 2.4–4.0) there are almost exclusively types of anthropic vegetation (ochre color), while in classes VI and VII (BTC > 6.0) only natural (green) vegetation types appear. Anthropic vegetation develops in areas not affected by natural vegetation, significantly completing the spectrum of efficiency levels of vegetation formation. This allows the presence of a high territorial biodiversity in landscape units, which leads to a greater gain of information and an increase in the capacity for self-rebalancing (greater incorporation of disturbances). This observation is important, as it allows us to also give a high value to agricultural crops, if they do not follow excessive criteria of agricultural industrialization or GMO modifications.

In add, within Fig. 4.13 the colored lines mark the value ranges. From below, they are:

1. Precarious CBSt = 0–6, below the gray line.
2. Stunted CBSt = 6–18, gray-yellow range.
3. Mediocre CBSt = 18–36, yellow-blue range.
4. Valid CBSt = 36–60, light blue-green range.
5. Optimal CBSt = > 60, above the green line.

In Fig. 4.14, showing the distribution of CBSt in the regions of Lombardy and Trentino Alto Adige, it is also possible to notice a more alarming aspect due to the downward concentration of the tesserae and ecotopes detected compared to the five

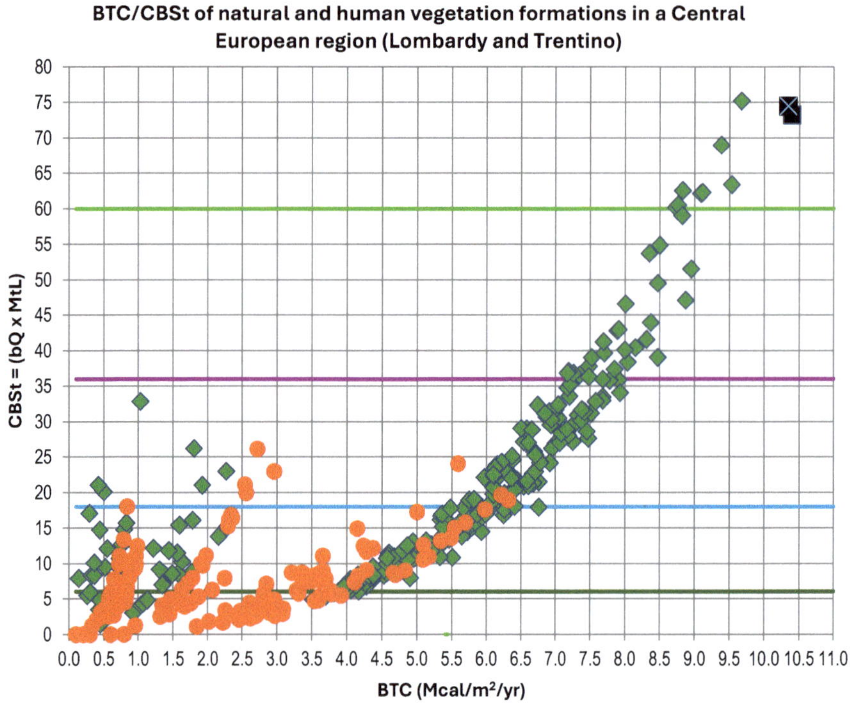

Fig. 4.13 BTC/CBSt correlations, from 427 case studies in Lombardy and Trentino (2005–2015). Natural formations in green, anthropogenic in ocher. The two blue elements top right were detected in Bialowieza Park in 2001. The 4 horizontal lines define 5 areas of bionomic efficiency

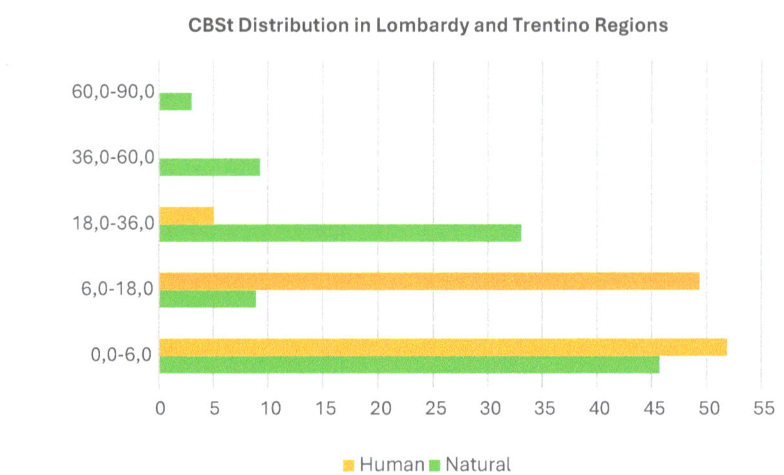

Fig. 4.14 Distribution of CBSt values in the vegetation formations of the Lombardy and Trentino regions. Note that only 12.3% of the natural formations are in the high efficiency classes (>36) and that only 5% of the anthropic ones are normal

bionomic efficiency classes marked by the horizontal lines. Since there are 269 surveys of natural formations and 158 anthropic ones, and considering the relative percentages, it will be noted: (a) the natural formations present valid to optimal conditions only in 12.26%, (b) the anthropic formations do not present good conditions, and those judged to be just within the norm/mediocre are 5.06%. This small value means there are too many alterations for both vegetation formation types.

References

Agazzi, E. (2014). *Science, metaphysics, religion*. Franco Angeli Ed.

Biondi, E., & Baldoni, M. (1994). *The climate and vegetation of peninsular Italy* (pp. 675–721). Colloques Phytosociologiques XXIII.

Bitterlich, W. (1970). *Spiegel-Relaskop, metric standard and metric CP*. Relaskop-Technik Vertriebsges. m.b.H.

Box, E. O. (1987). Plant life form and mediterranean environments. *Annali di Botanica, XLV*, 7–42.

Braun-Blanquet, J. (1926). *Pflanzensoziologie: Grundzüge der Vegetationskunde*. Springer.

Duvigneaud, P. (1977). Ecologia. In *Enciclopedia del Novecento* (Vol. 2). Enciclopedia Italiana Treccani.

Ellenberg, H. (1991). Zeigerwerte von pflanzen in Mitteleuropa. *Scripta Geobotanica*, 1–248.

Ingegnoli, V. (2002). *Landscape ecology: A widening foundation* (p. XXIII+357). Springer.

Ingegnoli, V. (2005). An innovative contribution of landscape ecology to vegetation science. *Israel Journal of Plant Sciences, 53*, 155–166.

Ingegnoli, V. (2013). Concise evaluation of the bionomic state of natural and human vegetation elements in a landscape. *Rendiconti Lincei, Scienze Fisiche e Naturali, 24*, 329–340.

Ingegnoli, V. (2015). *Landscape bionomics. Biological-integrated landscape ecology* (p. XXIV +431). Springer.

Ingegnoli, V., & Giglio, E. (2005). *Ecologia del Paesaggio: manuale per conservare, gestire e pianificare l'ambiente* (p. 685+XVI). Sistemi editoriali SE.

Ingegnoli, V., & Pignatti, S. (2007). The impact of the widened landscape ecology on vegetation science: towards the new paradigm. *Springer, Rendiconti Lincei Scienze Fisiche e Naturali, s. IX, XVIII*, 89–122.

Ingegnoli, V., Aquila, C., & Padoa-Schioppa, E. (1995). Rapporto preliminare sullo studio dell'ecomosaico forestale del Gariglone. In *Atti 5° Workshop "Progetto Strategico Clima Ambiente e Territorio nel Mezzogiorno" Amalfi 1993* (pp. 397–419). Tomo I, Guerrini editore, C.N.R., Roma.

Naveh, Z., & Lieberman, A. (1984/1994). *Landscape ecology: theory and application* (p. XXVII +360). Springer.

Odum, E. P. (1971). *Fundamentals of ecology*. Saunders.

Pignatti, S. (1982). *la Flora d'Italia*. Edagricole.

Pignatti, S. (1998). *I boschi d'Italia. Sinecologia e biodiversità*. UTET.

Pignatti, S. (2005). Valori di bioindicazione delle piante vascolari della flora italiana. *Braun-Blanquetia, 39*, 1–97.

Pignatti, S., Box, E. O., & Fujiwara, K. (2002). A new paradigm for the XXIth century. *Annali di Botanica, II*, 31–58.

Piussi, P. (1994). *Selvicoltura generale*. Utet.

Pulliam, R. (1989). Sources and sink complicate ecology. *Science*, 477–478.

Rivas-Martinez, S. (1987). Nociones sobre Fitosociologıa, biogeografıa y bio-climatologıa. In *La vegetacion de Espana* (pp. 19–45). Universidad de Alcala de Henares.

Simard, S. (2018). Mycorrhizal networks facilitate tree communication, learning and memory. In F. Baluska, M. Gagliano, & G. Witzany (Eds.), *Memory and learning in plants*. Springer.
Tuxen, R. (1956). Die heutige potentielle naturliche. Roma Vegetation als Gegenstand der Vegetationkartierung. *Angew. Pflanzensoziologie Stolzenau/Weser, 13*, 5–42.
Urbani-Ulivi, L. (Ed.). (2019). *The systemic turn in human and natural sciences. A rock in the pond.* Springer-Nature.

Chapter 5
Renewal of Vegetation Science: Applications

Contents

Abstract The study of forest patterns according to the LaBiSV methodology was reported in Chap. 4, concerning boreal or temperate forest phytocoenosis with a list of notes on the involved main parameters. Here some summary parametric sheets for the LaBiSV survey of other types of vegetation formations are presented: (a) Mediterranean sclerophylls, (b) meadows-pastures, (c) tree-lined corridors, (d) urban parks and gardens, (e) arable fields. Research on forest damages and their regrowth in the Fiemme Valley (Trento, Italy) shows the application of the LaBiSV methodology.

In add, the importance of forest soil is underlined and the functions of mycorrhizae, whose systems can be compared to the microbiota of the human body and, at least in part, to the neuronal system is briefly summarized. Other important applications of LaBiSV concern (a) the demonstration that the increase in species biodiversity does not indicate an increase in the level of organization of the phytocoenosis considered, (b) the bionomic limits of the Lombard plain or low hill forests around the metropolis of Milan. The modest values of BTC and CBSt and the high percentage of alien species are unfortunately very alarming.

V. Ingegnoli, *Landscape Bionomics and Planetary Health*, Landscape Series 21,
https://doi.org/10.1007/978-3-032-08908-3_5

Keywords Fittest vegetation · LaBiSV (Landscape Bionomic Survey of Vegetation) · BTC · Summary parametric sheets · Forest damages and regrowth · Forest soil · Mycorrhizae · Biodiversity

5.1 Examples of Summary Parametric Forms

We now present some Summary Parametric Sheets for the LaBiSV survey of vegetation formations, following the cited studies of Ingegnoli and Pignatti (Ingegnoli, 2002, 2005; Ingegnoli & Pignatti, 2007; Pignatti, 1998), to be added to the one already presented concerning boreal and temperate forest phytocoenosis (Table 4.1, where one can observe how the two types of forest are very similar, but not the same, as expressed in Table 4.3):

- Table 5.1, Mediterranean sclerophylls,
- Table 5.2, Meadows-pastures,
- Table 5.3, Tree-lined corridors,
- Table 5.4, Urban parks and gardens,
- Table 5.5, Arable fields.

For other formations, it is possible to see the manual by Ingegnoli and Giglio and the volume of Ingegnoli (Ingegnoli, 2011, 2015; Ingegnoli & Giglio, 2005) which also contains notes for the use of the sheets. Plant life forms followed Box (1987), modified by Ingegnoli & Giglio (2005).

All this has been studied with a 10-year parallel field control in vegetation at different stages of maturity, in different climatic conditions and in different areas of

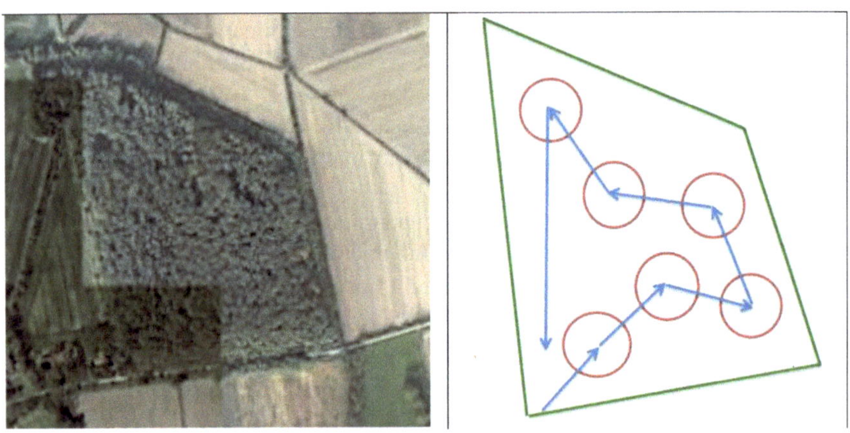

Fig. 5.1 Case study in Cusago (Milan, Italy). This remnant Ts of Querco-Carpinetum, surrounded by crop fields, has been analysed with 6 Relascope Stations (red circles of about 50 m). Its vegetation was detected in these 6 stations and in their 7 linkage strips. Thus, the detailing investigated area resulted about 25–30% of the total Ts. Here the edges are very disturbed by the presence of allocthonous species: so, if we remove its margins (DE ≈ 30 m) this Ts is reduced of about 1/3, from 12 to 7.5 ha

Table 5.1 LaBiSV parametric form for Mediterranean sclerophylls and Pines Foreste

Mediterranean pine forest	1	5	13	23	Mod.Inflex. =120 a
Medit. sclerophyll forest	1	5	11	22	Mod.Inflex. = 140 a
T. Tessera (Ts) Characters					
T1p -Vegetation height (m)	<9	9.1–17	17.1–25	>25	Canopy Hc=….m
T1s -Vegetation height (m)	<8	8.1–15	15.1–21	>21	s. canopy trees
T2 -Canopy cover (%)	<30	>90	31–60	61–90	Ts surface
T3- Structural differentiation	Low	Medium	Good	High	Age, groups
T4- Interior ratio (%)	Absent	<30	31–89	>90	vs ecotope
T5- Management	Simple coppice	Complex copp.	Wood	Natural forest	or similar
T6- Permanence (years)	<80	81–160	161–240	>240	Ts real age
F. Vegetational biomass (Above ground)					
F1- Dead plant biomass in Ts	Near 0	>10	1–5	5–10	% of pB
F2- Litter depth of the Ts	Near 0	<1.2	1.2–3.0	>3.0	cm
F3p – pB volume (m^3/ha)	<160	161–450	450–750	>750	pB = ...m^3/ha
F3s – pB volume (m^3/ha)	<120	120–300	300–500	>500	pB = ...m^3/ha
E. Ecocoenotope parameters					
E1- Dominant species	>3	3	2	1	pB of trees
E2- Sp Diversity	>65	<13	14–35	>35	n° sp./Ts
E3- Key species presence %	<10	11–40	41–75	>75	Botanical
E4- Allocthonous sp. (%)	>10	10–4	<4	0	Not regional
E5- Infesting plants	Near all	>25	<25	None	% of covering
E6- Diseased plants	Evident	Suspect	Risk	0	Even acid r.
E7- Plant forms (n°)	<3	4–5	6–7	>7	Cfr. Box 1987
E8- Vertical stratification	2	3	4	>4	>25-35% Ts
E9- Renewal capacity	None	Intense	Sporadic	Normal	Dominant sp.
E10- Dynamic state	Degrading	Recreation	Regeneration	Steady attractor	Transformation process
U. Landscape unit (LU) Parameters					
U1- Boundary connections	0	<25	26–75	>76	% perimeter
U2- Source vs surroundings	Sink	Neutral	Partial	Actual source	Sp & resources
U3- Role in the LU	Reduced	Minor	Evident	Important	Contest etc.
U4- Disturb incorporation	None	Scarce	Normal	High	Local disturb.
U5- Geo-physical instability	Evident	Partial	Risk	None	Physiotope
U6- Permeant fauna interest	None	Medium	Near good	Attractive	Ts/key sp.
U7- Transformation reason of the Ts as landscape. Element	Strong disturbance	Temporary instability	Gradual change	Steady attraction	LU trend
U8- Landscape pathology interference	Extremely serious	Near chronic	Easy to recover	None	Coming from surroundings
U9- Permanence (years)	<100	100–300	300–1200	>1200	Age of LU
R. Results of the survey					
R1- Total score Y (= a+b+c+d)	a = ...	b =.....	c =.....	d =.....	Y = ...
R2- Quality of the Ts (%)		bQ_s = Y/616	bQ_p = Y/644		Q = ...[%]
R3- BTC estimation (Mcal/m^2/year) Mediterranean Pine Forest	BTC(p) = 0.0151 (y-28) + 0.12 (pB/65)				BTC = ... [Mcal/m^2/year]
R3- BTC estimation (Mcal/m^2/year) Schlerophyll Forest	BTC(s) = 0.01705 (y-28) + 0.13 (pB/60)				BTC = ... [Mcal/m^2/year]
Maturity level [a,b from mod eq] MtL = $(BTC + b)^{1/a}$/Mod.Inflex. %	CBSt = (MtL x bQ)/100				CBSt = ... [%]

Italy and Europe, to give the model the maximum guarantee. The evaluations carried out are about 500, of which 270 in person by Ingegnoli and Giglio and about 240 by students in thesis of the Landscape Ecology courses in Natural Sciences.

Figure 5.1 remembers the vegetation survey method to plan the relascope stations (see Par. 4.4.4) reporting an opportune example: a small deciduous forest ecotope (13.5 ha).

Table 5.2 LaBiSV parametric form for grass/prairies and reed thickets

Grass & pastures (p)	1	4	13	32	Score
T. Tessera characters (Ts)					
T1p- Vegetation height (m)	<0.2	0.2–0.6	0.6–1.0	>1.0	Dominant layer
T2-Vegetation cover(%)	<30	>90	31-60	61–90	Ts surface
T3-Shrub presence (%)	<5	5–15	15-25	>25	<34% of TS
T4-Structural differentiation	Low	Medium	Good	High	Age, space, Sp. Groups, ...
T5-Edges	None	>30	30-10	<10	(% Ts)
T6-Management	Artificial	Near natural	Nat. grazed	Natural	Or similar
T7-Permanence (years)	<20	21–60	61–120	>120	See also old shrubs
F. Vegetational biomass (Aboveground)					
F1- Dead plant biomass	>60	60–21	<20	Quasi 0	% on living biomass
F2p- Litter depth	Quasi 0	<1.5	1.2–3.5	>3.5	cm
F3p – Plant biomass (kg/m^2)	<0.6	0.6–1.2	1.2–2.0	>2.0	Dried pB
E. Ecocoenotope parameters					
E1p- Dominant species (n°)	Not clear	1	2-3	>3	cover
E2p- Species richness	>15	16–25	26–40	>40	n° sp./Tessera
E3- Key species (%)	<5	6–20	21–75	>75	Phytosociological
E4- Allocthonous species (%)	>10	10–2	<2	0	From other ecoregions
E5- Infesting plants (% Ts)	All	>25	<25	0	Covering area on Ts
E6- Diseased plants	Evident	Suspect	Risk.	0	All pathologies
E7- Biological forms (n°)	<2	3	4	>4	See Box (1987), mod.
E8- Vertical stratification	1	2	3	4	From musk to shrub
E9- Renewal capacity	None	Sporadic	Normal	Intense	Dominant species
E10- Dynamic state	Degradation	Recreation	Regeneration	Fluctuation	See Ingegnoli, (2002)
U. Landscape unit (LU) Parameters					
U1- Similar vegetation contiguity	0	<20	21–80	>80	% of perimeter
U2- Source & sink	Sink	Neutral	Partial	Actual source	Species & resources
U3- Functional role in LU	Reduced	Minor	Pvident	Important	Context & typology
U4- Disturbances incorporation	Insufficient	Scars	Pormal	High	Local disturbances
U5p- Geophysical instabilities	Evident	Partial	Pisk	0	On the physiotope
U6- Permeant fauna interest	Low	Medium	Good	Attraction	Key species
U7- Transformation modalities of the	Strong	gradual	Temporal	Fluctuation	Today + tendency
U8- Landscape pathology	Serious	Near chronicle	Easy to	0	From landscape
U9- Permanence of analogous	<70	70–150	150–300	>300	Historical presence
Results of the survey					
Total score Y (= h + j + k + w)	h = ...	j =......	K = ...	W = ...	Y= ....
Bionomic quality of Ts		bQ$_c$ = Y/1073	bQ$_p$ = Y/928		bQ = ...[%]
Estimation of the BTC	BTC (p) = 0.001335 (Y-29) + 0.02 (Fm/0,14)				BTC = ...[Mcal/m^2/a]
Maturity lev.[a,b from mod eq] MtL = (BTC + b)$^{1/a}$ / Mod.flex.	CBSt = (MtL x bQ)/100				CBSt = ... [%]

Table 5.3 LaBiSV parametric form for corridors with trees

Corridor with trees	1	7	17	31	Score
Cd. Corridor characters (Cd)					
Cd1- Wide of the corridor (Wc)	<2.5	2.6–10	10.1–20	>20	m ...
Cd2- Canopy height (h)	<3	3.0–12	12.0–24	>24	m ...
Cd3- Tree cover	<30	30–60	60–90	>90	(%) of the tract
Cd4- Presence of stream (Str)	0	Canal	Near natural	Natural	Str wide < Wc
Cd5- Road presence (Rd)	Traffic	paved	rural	0	Rd wide < Wc
Cd6- Interruptions (> h)	>4	3.0–4.0	2.0–3.0	0	N° in the tract
Cd7- Linearity	Rectilinear	Near-rectilinear	Mixed	Irregular	Tract
Cd8- Management	Strong cut	Pruning	Light pruning	None	Human control
Cd9- Permanence (years)	<60	60–120	120–180	>180	Old trees
4. Plant biomass Aboveground					
F1 Dead plant biomass	Near 0	Too much	Medium	Normal	% on living biomass
F2- Litter depth	quasi 0	<1.5	1.2–3.5	>3.5	cm
F3- Plant biomass in volume	<100	100–300	300–500	>500	pB = ...m^3/ha
E. Ecocoenotope parameters					
E1- Dominant species (n°)	1	1.0–2	2.0–3	>3	On pB value
E2- Species richness (n°)	>15	16–29	30–44	>45	N° .... Sp
E3- Key species (%)	<5	5.0–20	21–75	>75	K-SP = ...%
E4- Allocthonous species (%)	>10	10–4.0	<4	0	All.SP = ...%
E5- Infesting plants (% Ts)	all	>25	<25	0	Covering area on Ts
E6- Diseased plants	Evident	Suspect	Risk.	0	All pathologies
E7- Biological forms (n°)	<3	4.0–5	6.0–7	>7	See Box 1987, modified
E8- Vertical stratification	1	2	3	4	
E9- Renewal capacity	None	Sporadic	Normal	Intense	Dominant species
E10- Dynamic state	Degradation	Recreation	Regeneration	Fluctuation	Transform. Process
U. Landscape unit (LU) Parameters					
U1- Internal species presence	0	Sporadic	Few	Many	e.g. forest Species
U2- Lichen presence	0	1.0–15	16–35	>35	N° of bark species
U3- Source & sink	Sink	Neutral	Partial	Actual source	Species & resources
U4- Connections	0	1	2	>2	With other corridors
U5- Network participation	None	Potential	Partial	Effective	Today situation
U6- Bird fauna exchanges	None	Reduced	Normal	High	With the landscape
U7- Mammal fauna presence	None	Reduced	Normal	High	Around the corridor
U8- Apparatus functions	Not clear	1	2	>2	Landscape apparatuses
U9- Corridor presence in LU	<80	80–160	160–240	>240	Years
U10- Disturb. Incorporation	Insufficient	Scars	Normal	High	Local disturbances
U11- Landscape type of LU	Urban	Suburban	Rural	Near-natural	Landscape unit (LU)
Results of the survey					
Total score Y (= h + j + k + w)	h = 3	j = 7	k = 18	w = 5	Y= 513
Bionomic quality of the Ts	bQ = Y / 1023				bQ = ...[%]
Estimation of BTC	BTC (cd) = 0.0072 (Y-33) + 0.10 (Fm / 75)				BTC= ..... [Mcal/m^2/a]
Concise Bionomic State	(MtL x bQ) (%)				CBSt = ...(%)

Table 5.4 LaBiSV parametric form for urban parks and gardens: ecological evaluation of the tessera quality (bQ) and estimation of biologic territorial capacity of vegetation (BTC)

Urban Parks and Gardens	1	7	20	39	Score
T. Garden characters					
T1- Canopy height (m)	<6	6–12	12–24	>24	Weight average
T2- Canopy cover (%)	<25	25–50	50–75	>5	Ts
T3- Underground	Built	Terraced	Mixture	Natural soil	See also percolation
T4- Garden design	Technical	Formal	Mixture	Natural	General composition
T5- Peculiar species	Exotic	Rare exotic	Local	Monumental	Characterizing garden
T6- Management	Strong pruning	Light pruning	Marginal	None	Or similar
T7- Permanence (years)	<50	50–100	100–250	>250	Oldest trees
F. Plant biomass aboveground					
F1- Litter depth	quasi 0	<1.5	1.5–3.5	>3.5	cm
F2- Plant Biomass volume	<100	100–200	200–300	>300	pB = (m^3/ha)
E. Ecocoenotope parameters					
E1- Dominant species (n°)	Not clear	3	2	1	On pB value
E2- Species richness	>10	11–25	26–40	>40	n° sp./Garden
E3- Key species (%)	<5	6–20	21–50	>50	Phytosociologic
E4- Allocthonous species (%)	>30	11–30	5–10	<4	Outside the ecoregion
E5- Diseased plants	Evident	Suspect	Risk.	0	All pathologies
E6- Biological forms (n°)	<3	4–5	6–7	>7	See Box (1987), modified
E7- Vertical stratification	2	3	4	>4	From musk to high tree
E8- Structural differentiation	Low	Medium	Good	High	Age, groups
E9- Renewal capacity	None	Sporadic	Normal	Intense	Dominant species
E10- Green proportion	<0.3	0.31–0.50	0.51–0.75	>0.75	
U. Landscape Unit (LU) parameters					
U1- Similar vegetation contiguity	0	<10	11–50	>50	% of perimeter
U2- Internal species presence	0	Sporadic	Few	Many	e.g. forest Species
U3- Lichen presence	0	1-15	16-35	>35	N° of bark species
U4- Ecological network	None	Neutral	Potential	Effective	Garden participation
U5- Apparatus functions	Not clear	1	2	>2	Landscape apparatuses
U6- Functional role in LU	Reduced	Minor	Evident	Important	Context & typology
U7- Disturbances incorporation	Insufficient	Scars	Normal	High	Local disturbances
U8- Geophysical instabilities	Evident	Partial	Risk	0	On the physiotope
U9- Permeant fauna interest	Low	Medium	Good	Attraction	Key species
U10- Transformation modalities of the Ts	Strong disturbances	Gradual changes	Temporal instabilities	Fluctuation	Today + tendency
U11- Landscape pathology interference	Serious	Near chronicle	Easy to incorporate	0	From landscape
Results of the survey					
Total score Y (= h + j + k + w)	h = ...	j =.......	k = ...	w = ...	Y= ....
Bionomic quality of the Ts	bQ = Y / 1170				bQ = ... [%]
Estimation of the BTC	BTC (vu) = 0.00526 (y-30) + 0.1 (Fm/45)				BTC = ... [Mcal/m^2/a]

Table 5.5 LaBiSV parametric form for agricultural fields: ecological evaluation of the tessera quality (bQ) and the estimation of biologic territorial capacity of vegetation (BTC)

Agricultural Fields	1	5	15	31	Score
T. Tessera characters (Ts)					
T1- Vegetation height (m)	<0.5	0.5–1.0	1–2	>2	Weight average
T2- Field shape	Geometrical	Polygonal	Irregular	Natural	Ts
T3- Tree presence	0	1	Few	Many	Even at the edge
T4- Management	Industrial	Near industrial	Traditional	Biological	Cultivation
T5- Irrigation	Technical	Near technical	Canal	Near natural	Main method
F. Plant biomass aboveground					
F1- Dead plant biomass	0	Low	Medium	High	On soil
F2- Litter depth	quasi 0	<1.5	1.5–3.5	>3.5	cm
F3p – Plant biomass in volume	<1	1–2	2–3	>3	Dried pB = Kg/m^2
E. Ecocoenotope parameters					
E1- Species richness	>10	11–20	21–30	>30	n° sp./Tessera
E2- Allocthonous species (%)	>10	10–2	<2	0	Outside the ecoregion
E3- Natural phytocoenosis	0	Sporadic	Marginal	Patches	Species presence
E4- Diseased plants	Evident	Few plants	Risk	0	All pathologies
E5- Genetic characters	Transgenic	Allocthonous	Current	Traditional	Cultivars
E6- Chemicals	>4	3	1–2	0	All types
E7- Soil alterations	Patches	Spots	Marginal	0	Ts humus, ...
E8- Soil fertility classes	IV	III	II	I	Cfr. Land Capability
E9 Earthworms	Absence	Only few	Frequent	Abundant	
E10- Soil preparation	Technical	Mixed	Marginal	0	e.g. tilling
E11- Presence of margins	0	<50	50–80	80–100	% perimeter
U. Landscape Unit (LU) parameters					
U1- Contagion with natural elements	0	<10	11–50	>50	% del perimeter
U2- Ecological network	0	Marginal	Partial	Effective	Presence
U3- Functional role in LU	Reduced	Minor	Evident	Important	Context & typology
U4- Permeant fauna interest	None	Medium	Near good	Attractive	Ts/key sp.
U5- Geophysical instabilities	Evident	Partial	Risk	0	On the physiotope
U6- Landscape pathology interference	Serious	Near chronicle	Easy to incorporate	0	From landscape
U7- structural & functional congruence with LU	Contrast	Poor	Near adequate	Good	Main parameters
U8- Permanence of analogous field (years)	<25	26–100	101–200	>200	Even diverse cultivation
Results of the survey					
Total score Y (= h + j + k + w)	h = ...	j =.......	K = ...	W = ...	Y= ....
Bionomic quality of the Ts	Q = Y/806				bQ = ... [%]
Estimation of the BTC	BTC (ca) = 0.00192 (Y-27) + 0.09 pB				BTC = ...[Mcal/m^2/a]
Maturity lev. [a,b from mod eq] MtL = (BTC + b)$^{1/a}$ / Mod.flex.	CBSt = (MtL x bQ)/100				CBSt = ...[%]

5.2 Design with Summary Parametric Forms

There are 13 Summary Parametric Sheets tested in the field, as shown in Table 4.3. They are helpful not only in field surveys to obtain the values of bQ, BTC, and CBSt, which characterize a landscape unit (LU) under examination but also in

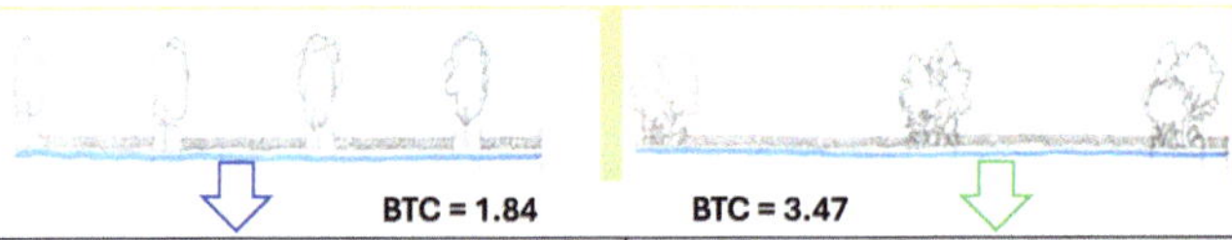

Tab.5.3. LaBiSV Parametric Form for Corridor with Trees: Ecological Evaluation of the Tessera Quality (bQ) and Estimation of Biologic Territorial Capacity of Vegetation (BTC).

Corridor with trees	1	7	17	31	Mod.inflex. = 105 a
Cd. CORRIDOR CHARACTERS (Cd)					
Cd1- Wide of the corridor (Wc)	< 2,5	2,6-10	10,1-20	> 20	Canopy expansion (m)
Cd2- Canopy height (h)	< 3	3-12	12-24	> 24	Weight average (m)
Cd3– Tree cover	< 30	30-60	60-90	> 90	(%) of the tract
Cd4- Presence of stream (Str)	0	canal	near natural	natural	Str wide < Wc
Cd5 – Road presence (Rd)	traffic	paved	rural	0	Rd wide < Wc
Cd6- Interruptions (> h)	> 4	3-4	1-2	0	N° in the tract
Cd7- Linearity	rectilinear	near-rectilinear	mixed	irregular	Tract
Cd8– Management	strong cut	pruning	light pruning	none	Human control
Cd9- Permanence (years)	< 60	60-120	120-180	> 180	Old trees
F. PLANT BIOMASS Aboveground					
F1 Dead plant biomass	near 0	too much	medium	normal	% on living biomass
F2- Litter depth	quasi 0	< 1,5	1,2-3,5	>3,5	cm
F3 – Plant biomass in volume	< 100	100-300	300-500	> 500	pB =30 m³/ha
E. ECOCOENOTOPE PARAMETERS					
E1- Dominant species (n°)	1	1-2	2-3	> 3	On pB value
E2- Species richness (n°)	> 15	16-29	30-44	> 45	Tract
E3- Key species (%)	< 5	6-20	21-75	> 75	Phytosociological
E4- Allochthonous species (%)	> 10	10-4	< 4	0	From other ecoregions
E5- Infesting plants (% Ts)	all	> 25	< 25	0	Covering area on Ts
E6- Diseased plants	evident	suspect	risk.	0	All pathologies
E7- Biological forms (n°)	< 3	4-5	6-7	> 7	See Box 1987, modified
E8- Vertical stratification	1	2	3	4	Musk to tall Trees
E9- Renewal capacity	none	sporadic	normal	intense	Dominant species
E10- Dynamic state	degradation	recreation	regeneration	fluctuation	transform. process
U. LANDSCAPE UNIT (LU) PARAMETERS					
U1- Internal species presence	0	sporadic	few	many	e.g. forest Species
U2- Lichen presence	0	1-15	16-35	> 35	N° of bark species
U3- Source & sink	sink	neutral	partial	actual source	Species & resources
U4- Connections	0	1	2	> 2	With other corridors
U5- Network participation	none	potential	partial	effective	Today situation
U6- Bird fauna exchanges	none	reduced	normal	high	With the landscape
U7- Mammal fauna presence	none	reduced	normal	high	Around the corridor
U8- Apparatus functions	not clear	1	2	> 2	Landscape apparatuses
U9- Corridor presence in LU	< 80	80-160	160-240	> 240	years
U10- Disturbances incorporation	insufficient	scars	normal	high	Local disturbances
U11- Landscape type of LU	urban	suburban	rural	near-natural	Landscape uunit (LU)
RESULTS OF THE SURVEY					
Total score Y (= h+j+k+w)	h = 10	j = 16	k = 4	w = 3	Y= 283
Bionomic quality of the Ts		bQ = Y / 1023			bQ = 28.22 [%]
Estimation of BTC		BTC (cd) = 0,0072 (Y-33) + 0,10 (Fm / 75)			BTC = 1.84 [Mcal/m²/a]

Tab.5.3. LaBiSV Parametric Form for Corridor with Trees: Ecological Evaluation of the Tessera Quality (bQ) and Estimation of Biologic Territorial Capacity of Vegetation (BTC).

Corridor with trees	1	7	17	31	Mod.inflex. = 105 a
Cd. CORRIDOR CHARACTERS (Cd)					
Cd1- Wide of the corridor (Wc)	< 2,5	2,6-10	10,1-20	> 20	Canopy expansion (m)
Cd2- Canopy height (h)	< 3	3-12	12-24	> 24	Weight average (m)
Cd3– Tree cover	< 30	30-60	60-90	> 90	(%) of the tract
Cd4- Presence of stream (Str)	0	canal	near natural	natural	Str wide < Wc
Cd5 – Road presence (Rd)	traffic	paved	rural	0	Rd wide < Wc
Cd6- Interruptions (> h)	> 4	3-4	1-2	0	N° in the tract
Cd7- Linearity	rectilinear	near-rectilinear	mixed	irregular	Tract
Cd8– Management	strong cut	pruning	light pruning	none	Human control
Cd9- Permanence (years)	< 60	60-120	120-180	> 180	Old trees
F. PLANT BIOMASS Aboveground					
F1 Dead plant biomass	near 0	too much	medium	normal	% on living biomass
F2- Litter depth	quasi 0	< 1,5	1,2-3,5	>3,5	cm
F3 – Plant biomass in volume	< 100	100-300	300-500	> 500	pB = 45 m³/ha
E. ECOCOENOTOPE PARAMETERS					
E1- Dominant species (n°)	1	1-2	2-3	> 3	On pB value
E2- Species richness (n°)	> 15	16-29	30-44	> 45	Tract
E3- Key species (%)	< 5	6-20	21-75	> 75	Phytosociological
E4- Allochthonous species (%)	> 10	10-4	< 4	0	From other ecoregions
E5- Infesting plants (% Ts)	all	> 25	< 25	0	Covering area on Ts
E6- Diseased plants	evident	suspect	risk.	0	All pathologies
E7- Biological forms (n°)	< 3	4-5	6-7	> 7	See Box 1987, modified
E8- Vertical stratification	1	2	3	4	Musk to tall Trees
E9- Renewal capacity	none	sporadic	normal	intense	Dominant species
E10- Dynamic state	degradation	recreation	regeneration	fluctuation	transform. process
U. LANDSCAPE UNIT (LU) PARAMETERS					
U1- Internal species presence	0	sporadic	few	many	e.g. forest Species
U2- Lichen presence	0	1-15	16-35	> 35	N° of bark species
U3- Source & sink	sink	neutral	partial	actual source	Species & resources
U4- Connections	0	1	2	> 2	With other corridors
U5- Network participation	none	potential	partial	effective	Today situation
U6- Bird fauna exchanges	none	reduced	normal	high	With the landscape
U7- Mammal fauna presence	none	reduced	normal	high	Around the corridor
U8- Apparatus functions	not clear	1	2	> 2	Landscape apparatuses
U9- Corridor presence in LU	< 80	80-160	160-240	> 240	years
U10- Disturbances incorporation	insufficient	scars	normal	high	Local disturbances
U11- Landscape type of LU	urban	suburban	rural	near-natural	Landscape uunit (LU)
RESULTS OF THE SURVEY					
Total score Y (= h+j+k+w)	h = 4	j = 6	k = 18	w = 5	Y= 507
Bionomic quality of the Ts		bQ = Y / 1023			bQ = 50.55 [%]
Estimation of BTC		BTC (cd) = 0,0072 (Y-33) + 0,10 (Fm / 75)			BTC = 3.47 [Mcal/m²/a]

Fig. 5.2 Example of a study for an environmental restoration project of an agricultural landscape on the outskirts of Milan. Comparison between two types of tree-lined rows with different structures and estimate of their potential BTC using the parametric sheets of tree-lined corridors (Table 5.3)

environmental design studies. In fact, after having hypothesized some restoration interventions, e.g., of an agricultural landscape, it is necessary to have an estimate of the changes in the bionomic parameters of what is proposed: thus, to check that it is possible to achieve the restoration objectives with a given project proposal. An example is shown in Fig. 5.2, where two structurally different design choices are compared based on the BTC of the tree-lined fields. The example is based on Table 5.3 (tree-lined corridors) and compares two alternatives that can be proposed in the project to improve an area of South Milan Park without any more rows of trees. On the left, a system of single rows of Italian black poplar spaced about 40 m apart; on the right, double rows with oaks and alders spaced about 60 m apart. The difference in *corridor BTC* goes from 1.84 to 3.47 Mcal/m^2/year, equal to 189%, and the difference in *arable field with corridors BTC* goes from 0.95 to 1.60, equal to 168%.

5.3 Monitoring After Strong Disturbances

5.3.1 *The Example of Val Gambìs Before and After Vaia gale*

Related or not to present climate change, storms of consistent energy occur even in temperate regions, producing terrible destructions, especially in forest landscapes.

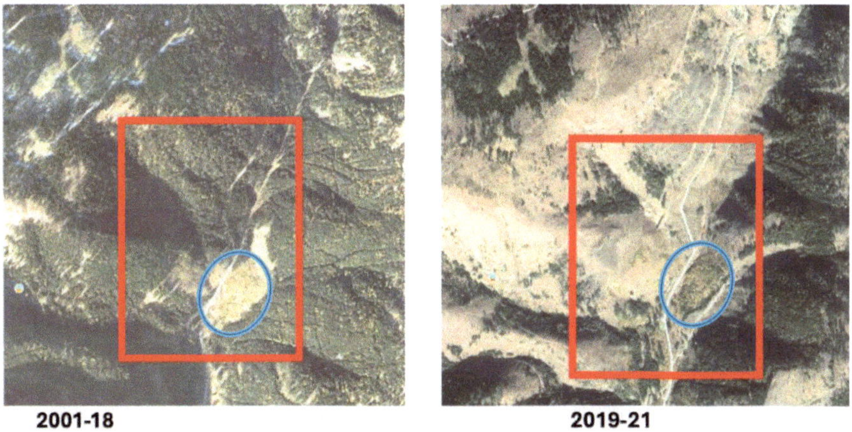

Fig. 5.3 Val Gambìs (Fiemme), tornado destructions. Forest damage in 2000 (left red square) vs 2018 (right red sq.) Note the pale green regrowing in S-E corner of the red squares (blue perimeter)

Conventional ecology can assess these damages mostly as loss of photosynthetic area, of defense against avalanches and of wood volume, predicting the regrowth of secondary forest. But after the present scientific paradigm shift and under sustainability pression, it is necessary to approach the evaluation of the effects within a systemic vision, so we must follow the principles and methods of Landscape Bionomics.

An example is reported in Fig. 5.3. This study, by Ingegnoli and Giglio (2019), regards the surroundings of Fiemme Valley, Northern of Italy, and concerns the drastic change in the spruce forest landscape unit of Gambìs valley after the tornado of September 20th, 2000, which repeated with greater force and extension on October 30th, 2018. The blue perimeter in this figure indicates the regrowing forest tessera after the year 2000, which has been monitored with repeated surveys up to 2024. Figure 5.4 resume the last survey.

Certain that sustainable forestry needs bionomics functions to widening the knowledge of ecotope alterations in a systemic approach, the Bionomic Territorial Capacity of vegetation (BTC, Mcal/m^2/year) following LaBiSV (Landscape Bionomic Survey of Vegetation) method was evaluated. Moreover, with the help of the model of BTC dynamics it was possible to find: (a) the loss of BTC due to forest destruction, (b) the change in forest recreation, which has nothing to do with secondary forest, (c) a credible hypothesis of the time needed to recover the examined ecotope, arriving to the same BTC level ante perturbations, (d) the resistance to perturbation according to forest bionomic parameters. Finally, on the recurrence of October 30th, 2018, on the same surfaces, some preliminary considerations are given.

Figure 5.5 shows the sequence of the mentioned surveys. The four parametric columns for each of the seven sheets gathered (boreal forest phytocoenosis, see Table 4.1) summarizing the detailed state of the forest surveyed in 2000

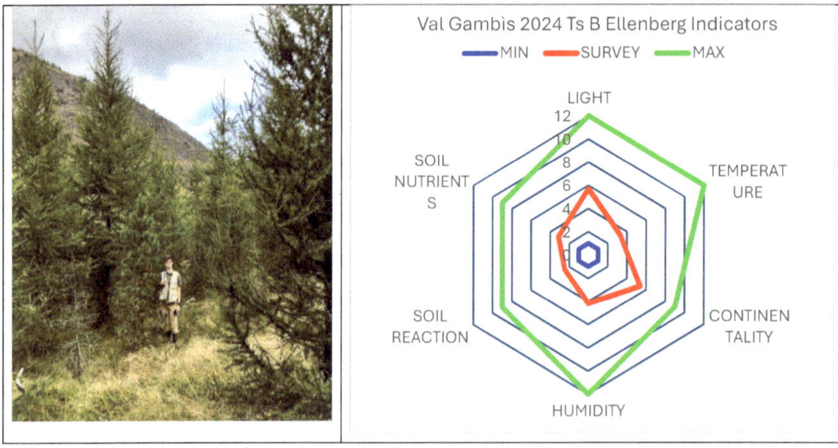

Fig. 5.4 23–26-year-old *Larix decidua* 8–9 m height in August 2024 on the studied remnant patch of forest. Surveyed values of Ellenberg bio-indicators (Ellenberg, 1991) (red) on the right

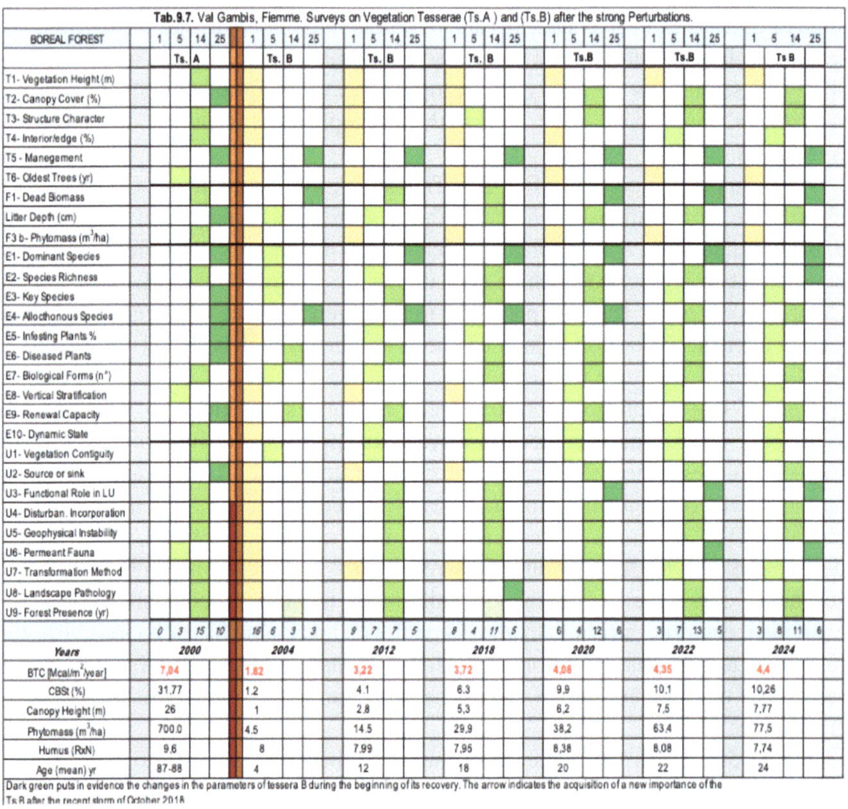

Tab.9.7. Val Gambis, Fiemme. Surveys on Vegetation Tesserae (Ts.A) and (Ts.B) after the strong Perturbations.

BOREAL FOREST	1	5	14	25	1	5	14	25	1	5	14	25	1	5	14	25	1	5	14	25	1	5	14	25	1	5	14	25
		Ts.	A			Ts.	B			Ts.	B			Ts.	B			Ts.B				Ts.B				Ts B		
T1- Vegetation Height (m)																												
T2- Canopy Cover (%)																												
T3- Structure Character																												
T4- Interior/edge (%)																												
T5 - Manegement																												
T6- Oldest Trees (yr)																												
F1- Dead Biomass																												
Litter Depth (cm)																												
F3 b- Phytomass (m³/ha)																												
E1- Dominant Species																												
E2- Species Richness																												
E3- Key Species																												
E4- Allochthonous Species																												
E5- Infesting Plants %																												
E6- Diseased Plants																												
E7- Biological Forms (n°)																												
E8- Vertical Stratification																												
E9- Renewal Capacity																												
E10- Dynamic State																												
U1- Vegetation Contiguity																												
U2- Source or sink																												
U3- Functional Role in LU																												
U4- Disturban. Incorporation																												
U5- Geophysical Instability																												
U6- Permeant Fauna																												
U7- Transformation Method																												
U8- Landscape Pathology																												
U9- Forest Presence (yr)																												
	0	3	15	10	16	6	3	3	9	7	7	5	8	4	11	5	6	4	12	6	3	7	13	5	3	8	11	6
Years	**2000**				**2004**				**2012**				**2018**				**2020**				**2022**				**2024**			
BTC [Mcal/m²/year]	7,04				1.82				3,22				3,72				4,08				4,35				4,4			
CBSt (%)	31.77				1.2				4.1				6.3				9,9				10,1				10,26			
Canopy Height (m)	26				1				2.8				5,3				6,2				7,5				7,77			
Phytomass (m³/ha)	700.0				4.5				14.5				29,9				38,2				63,4				77,5			
Humus (RxN)	9.6				8				7,99				7,95				8,38				8,08				7,74			
Age (mean) yr	87-88				4				12				18				20				22				24			

Dark green puts in evidence the changes in the parameters of tessera B during the beginning of its recovery. The arrow indicates the acquisition of a new importance of the Ts B after the recent storm of October 2018

Fig. 5.5 Dynamic of re-growth after a zero event on a boreal-alpine forest due to a strong tornado (2000, October) in Val Gambìs, Trento province. Dominant trees changed from *Picea abies* to *Larix decidua*

(pre-hurricane), 2004, 2012, 2018, 2020, 2022, 2024 can be displayed in sequence to better understand the dynamics of destruction and regrowth of the forest in question. We can highlight the points of most significant interest with different colors. At the base of these four matrices, the most significant bionomic and ecological data are listed. It is thus possible to monitor the different processes that most influence the regrowth in question, always considering the overall picture of the situation. For example, $CBSt_{2000} = 31.77$ (in September, reduced to 0.9 in a few hours); $CBSt_{2004} = 1.18$; $CBSt_{2012} = 4.08$; $CBSt_{2018} = 6.30$; $CBSt_{2020} = 9.92$; $CBSt_{2022} = 10.10$; $CBSt_{2024} = 10.26$.

5.3.2 *Processes of Regrowing: Val Gambìs and Paneveggio*

Starting from the data of Fig. 5.5, it is possible to evaluate the regrowth of the tessera (Ts) under examination (blue-green) compared with the development according to the model presented in Fig. 3.15. Since the layers of humus, litter, grasses, seedlings, and shrubs were not destroyed by the hurricane, the new growth (BTC/years) in the monitored tessera of Val Gambis is more significant than the model predictions (Fig. 5.6). Note the importance of the first phase of regrowth, in which 36.5% of the

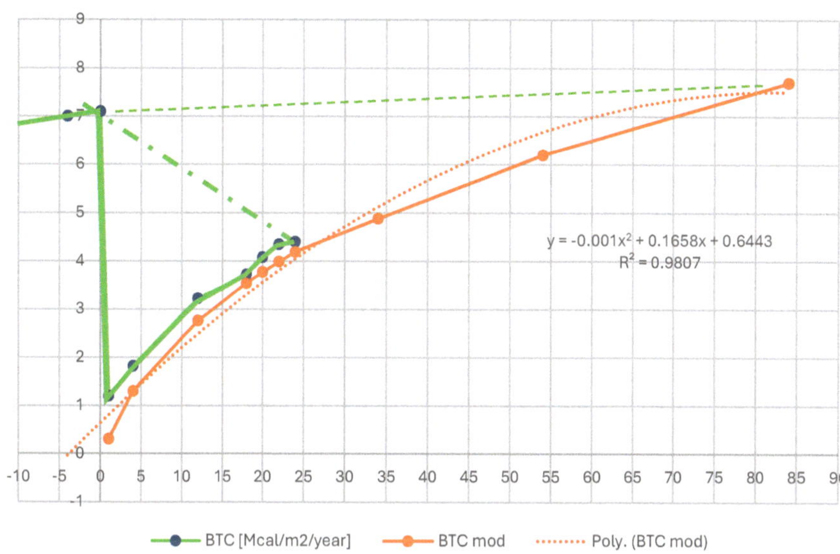

Fig. 5.6 The regrowth of the tessera (Ts) under examination (blue-green) compared with the development according to the model shown in Fig. 3.15. The hurricane did not destroy the litter layers, grasses, seedlings, and shrubs; therefore, the new development is greater than the model's predictions. The total Transformation Deficit is represented by the "triangle" formed by the orange curve (up to 85 years), the green line of the fallen of BTC and the green light dotted line representing the theoretical development of the BTC of the forest without perturbations

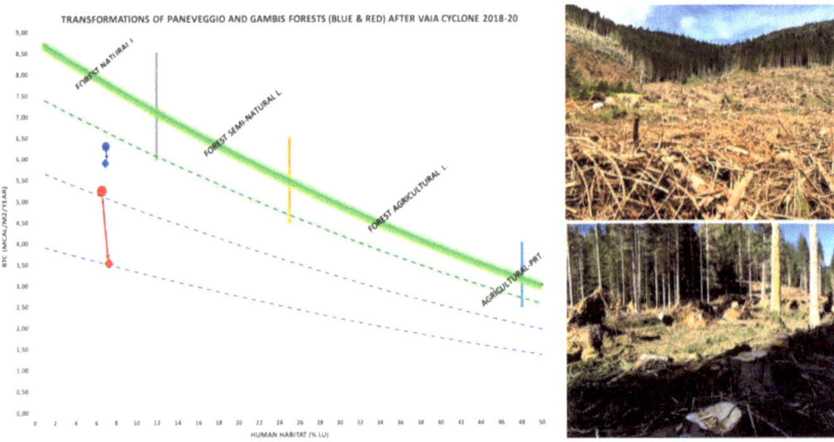

Fig. 5.7 Right: a stretch of the Val Gambìs (above) and Paneveggio forest (below). Left: the Ecological/Bionomic status in Val Gambìs (red arrow) and in the Paneveggio Park (blue arrow). After the most recent forest destruction (2018), the Gambìs valley, already partly altered by the predominantly coeval forest management, presented five times more damage than the Paneveggio Park

Transformation Deficit TD (which will be filled in about 84–85 years) is reached in only 23 years (note that 23/84 = 27.5%).

Let me compare now in Fig. 5.7 the damage assessment of Vaia tempest (October 2018) of Val Gambìs to that of the forests of Paneveggio (both related to Fiemme, TN, Italy): the principles of bionomics allow us to compare the differences between the conditions of Val Gambìs (photo at top) and those of Paneveggio (photo below). The forests of the Gambìs valley, mainly of the same age and therefore poorly structured, have suffered damage in the order of 40–45% of their total surface area and have a rather low average BTC (ante quam): they went from a $BTC_{ANTE} = 5.20$ Mcal/m^2/yr to a $BTC_{POST} = 3.50$ Mcal/m^2/yr (−32.7%). On the contrary, the forests of Paneveggio Park, mainly uneven-aged and therefore well-structured, suffered damage of 15–20%: with a better average BTC (*ante quam*) (6.40 Mcal/m^2/a), they went down from a $BTC_{ANTE} = 6.30$ Mcal/m^2/yr to a $BTC_{POST} = 5.85$ Mcal/m^2/yr (−7.15%).

It can be deduced that the semi-natural condition of forest management can also have a decisive impact on their protection in the event of catastrophic disturbances.

5.4 Importance of Forest Soil

5.4.1 Main Soil Characters

Figure 5.8 compares the first and last case studies detected in the Milan hinterland: the forest of Cusago (Parco Sud Milano) and the wood of Casorezzo (PLIS Roccolo).

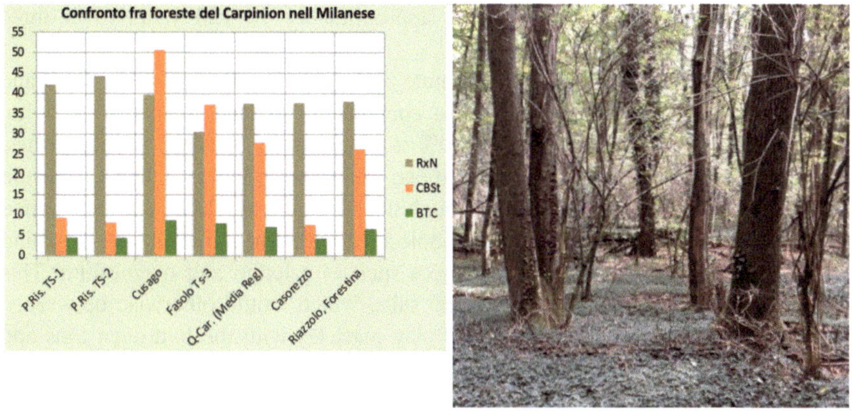

Fig. 5.8 It is possible to demonstrate that in the examined ecotope of the wood of Casorezzo, the ancient soil (in the forest residue and adjoining corridor) shows the humic value R × N = 37–38 (Ellenberg coefficient; Pignatti, 2005). The destruction of 2.5 hectares of such soil would require at least 2 centuries to reconstitute it. To the right, a patch of the forest of Cusago, dominated by *Quercus robur* and *Carpinus betulus* (South-Milan Park) BTC 8.89

The Cusago forest (*Quercus robur* 66% of the canopy, *Carpinus betulus* 28%, *Quercus rubra* 4%, *Castanea sativa* 2%) with oaks of about 40 m, is in decent conditions, even if not optimal, given the percentage of non-native species pair to 10.6% and the limited number of Species (27 in summer). On the other side, the example of Casorezzo represents a case of clear degradation (*Prunus serotina 59%* of the canopy, *Robinia pseudoacacia 39%*). However, a good parametric survey such as the LaBiSV one allows us to notice aspects sufficient to warn the quarrymen in the nearby extraction area from destroying this forest relic. Cusago Vs. Casorezzo show the following values respectively: BTC 8.89 Vs. 4.18 Mcal/m^2/a; phytomass 801 vs. 235 m^3/ha; CBSt 52.79 Vs. 7.59 (Ingegnoli, 2013).

Typical species of the *Querco-Carpinetum* have also been highlighted in Casorezzo, such as *Vinca minor* and *Poligonatum multiflorum*, and the presence of seedlings of *Quercus robur* and *Carpinus betulus*. It can, therefore, be stated that the "ecological memory" of a tessera of this forest is still maintained, despite the non-native adventitia reaching 34%, due to post-war exploitations and the last 50 years of disturbances. This is confirmed by the soil humification value R × N = 37.6 sensu Ellenberg (1991) and Pignatti (1995, 2005) higher than the Lombard average (R × N = 34.1) and not far from the R × N = 39.8 value of Cusago (Fig. 5.8).

It must be noted that the pedogenetic processes, from the source rock to the superficial layers present five horizons R, C, B, A, O, where: R = Rocks; C = Mineral slightly altered; B = Differentiated mineral layer; A = Evolved mineral layer; O = organic). Soil of total thickness (1.1–1.5 m) can be very slow in its evolution (Duchaufour, 1984) and require almost 4000 years, on average, to express 80% of the soil properties. Meanwhile, the humic horizons, which affect the presence of life

(O and A), reach the same value in approximately 90–120 years (over three generations!).

We must remember that in the temperate zones, each hectare of soil in the O and A horizons (average thickness 40 cm) contains on average 8–9 tons of living biomass per hectare (ha), of which 80% is made up of bacteria, actinomycetes, fungi and algae and 20–40 tons/ha of roots. The roots in layers A and B transmit hydrogen ions (positive) to the soil through the root hairs to release nutrients bound to the clay components. Since the clay molecules have a negative charge, they bind to the positive ions of mineral substances such as calcium and magnesium. This binding allows the soil to retain mineral salts, which would otherwise be washed away by rain. The root hairs flood the clay particles with the hydrogen ions and release some mineral dies bound to the clay surface.

And we must not forget the Wood-wide web of mycorrhizae (Simard, 2018; Simard et al., 2020), already cited in paragraph 2.7.

5.4.2 *Cybernetic Mycorrhizae Networks*

The fungal hyphae that spread like a network in the soil are much thinner than root hairs and can, therefore, penetrate the roots (Figs. 5.9 and 5.10) as mycorrhizae. This fundamental symbiosis integrates the ability of plants to produce sugars through photosynthesis and that of fungi to extract mineral salts from tiny soil particles.

You should note that row E4 of the parametric sheet of Table 4.1 has been updated compared to what has already been published (2015) to give the possibility of detecting the status of mycorrhizae in terms of number of species and presence of fungal networks. Given the importance of these parameters (Simard et al., 2020), some insights are reported below.

Fig. 5.9 Fungal hyphae of the mycelium of Lactarius sp. in a Picea forest in Ziano, and (Dx) etcomycorrhizae on small Picea roots (white windings)

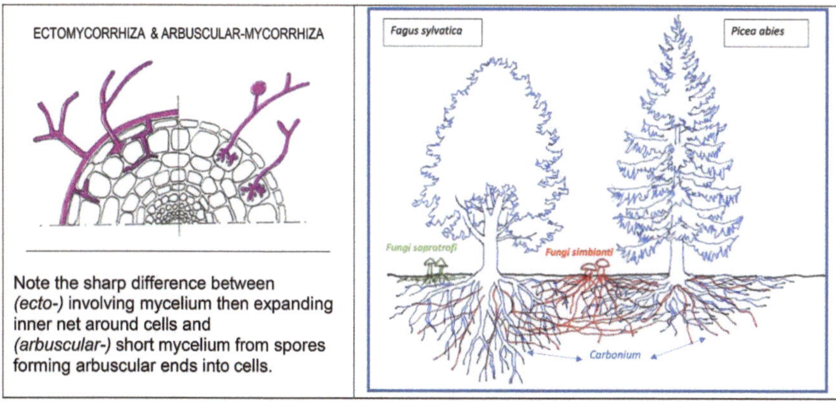

Fig. 5.10 Main types of mycorrhizae (Ectomycorrhiza & Arbuscular mycorrhiza) and differences between saprotrophic and symbiotic fungi, capable of exchanging carbonaceous products even between different species, as shown in the figure

Please remember that there are mainly two forms of mycorrhizae:

(a) *ectomycorrhizae*, whose mycelium envelops the rootlets from the outside (Fig. 5.9), with network-like introflections behind them,
(b) *arbuscular mycorrhizae*, whose hyphae penetrate the rootlets, expanding into internal arbuscules (Fig. 5.10, left).

Furthermore, remember that fungi can be (Nonis, 2001/2020): (i) parasites, often pathogenic for roots (e.g., *Armillaria* sp), (ii) saprophytic, capable of destroying dead organic residues, e.g., *Pluteus* sp. (iii) symbionts, such as mycorrhizae, e.g., *Boletus*, sp.

We must observe that forests form living systems in which not only the plant species have coevolved with each other and adapted to the environment to the point of composing very complex structures both by stratification (moss, herbaceous, shrubby, liana, submissive, and dominant arboreal), both for spatial and uneven-aged distribution in diversified groups with nurseries and mature mother trees. This process leads to the formation of different habitats of animals that can be considered macro-symbionts (e.g., herds of ungulates, with multiple functions, such as fertilization of the soil) and micro-symbionts, such as soil fauna, ants and pollinating insects, both for the presence of networks of mycorrhizae, which perform functions of the utmost importance. We try to summarize these functions of mycorrhizae, whose systems can be compared to the *microbiota* of the human body and, at least in part, to the *neuronal system*.

1. System of exchange of resources (e.g., sugars, water, enzymes), even between different species (e.g., spruces and birches), alternating according to the seasons but in balance over the year.

Fig. 5.11 Example of MotherTrees in Ziano (Fiemme, Trentino) at 1010–1020 m of altitude. This tessera presented BTC = 8.26 Mcal/m^2/years. Some Picea abies present a diameter D = 1.08–1.15 m, height H = 50–51.0 m

2. System of perception of the environment, with information molecules exchanged in the hyphae-radicle synapses and consequent behavior modification (e.g., stomata closure, speed of photosynthesis, growth). Some hyphae transmit information over a long distance (*Rhizopogon*), others at a short range and high speed (*Wilcoxina*).
3. The defense system has specific enzymes. For example, if a Spruce receives damage, it transmits a warning enzyme to the Pine, which it is linked to by the hyphae, with a stress warning within 24 hours.
4. Filial recognition system, where the mother plant demonstrates that it recognizes its seedlings and nourishes them better than other plants.

Simard observed (Simard, 2018; Simard et al., 2020) that mother trees (Fig. 5.11) constitute important hubs in the communication networks formed by fungal hyphae. When these older trees (Mothertrees) die, they release CO_2 and defense compounds into the underground network providing nourishment and information for future generation.

Concluding, we expose a table with some of the most frequent species of ectomycorrhizal fungi (Table 5.6).

Table 5.6 Main Ectomycorrhizal fungi in alpine forests. Species in red color are poisonous. Many species are the same or very similar in North America (USA and Canada)

Tuber	*Tuber aestivum V.*	**Amanitopsis**	*Amanitopsis crocea*
	Tuber magnatum		
		Lactarius	*Lactarius deliciosus*
Cantharellus	*Cantharellus foriesii*		*Lactarius piperatus*
	Cantharellus cybarius		
		Boletus	*Boletus aereus*
Russula	*Russula cyanoxantha*		*Boletus satanas*
	Russula rubroalba		*Boletus edulis*
	Russula virescens		*Boletus impolitus*
	Russula emetica		
		Suillus	*Suillus granulatus*
Cortinarius	*Cortinarius praestans*		*Suillus grevillei (elegans)*
Lepiota	*Lepiota brunneoincarnata*	**Morchella**	*Morchella conica*
	Macrolepiota procera		*Morchella esculenta*
Amanita	*Amanita cesarea*	**Inocybe**	*Inocybe godeyi*
	Amanita phalloides		
	Amanita muscaria	**Hygrophorus**	*Hygrophorus arbustivus*
	Amanita verna		

From Nonis (2001/2020) elaborated and modified
In red poisonous/mortal species

5.5 Validity of the BTC Function

5.5.1 LaBiSV Model and Disturbances

As expressed in paragraph 3.4, the biological-territorial capacity of vegetation (BTC) is a state function of a complex system capable of relating both to the metabolism and the degree of organization of the phytocoenosis in question. As we have seen, the model used for the evaluation of the vegetation surveys (Fig. 4.3) is of the exponential/ logarithmic type, as all biological systems in the transition between the juvenile/adult phases and those of maturity/senescence change their metabolic regime and level of organization following said curves, with asymptotic development up to senescence. We can render the representation of this model (BTC/Time) for greater convenience with a polynomial equation (Fig. 4.3), where T is the time:

$$BTC = 0.00029\ T^2 + 0.106\ T + 1.55\ (\text{Boreal} - \text{Alpine Forest}),$$
$$BTC = 0.00029\ T^2 + 0.109\ T + 2.1\ (\text{Temperate Forest})$$

We did the compatibility check of the model with the ecological conditions of the forests detected in the field after studying enough cases (67 forests) (Ingegnoli, 2015; Ingegnoli & Giglio, 2005) to express the correlation with various parameters (e.g., phytomass) of phytocoenosis related to the ages, to consider the form of development. As we have seen, the summary parametric sheets modulated on said curves made it possible to evaluate quantities capable of being used by the estimation equations. These quantities made it possible to assess the extent of disturbances that vegetation systems cannot incorporate and trends that lead to a decline in growth curves.

Figure 5.12 compares the theoretical BTC model and the detected BTC of Boreal-Alpine (blue) and Temperate Deciduous (green) forests. The correlation curves of the field surveys are lower than those of the theoretical model by approximately 20–23% at the maturity threshold (130–140 years), even if at this threshold approximately 40% does not fall below 10–12% from the expected values of the model. It is immediately clear that the two curves are very similar and that, despite the dispersion in maturity, the correlation significance (Pearson) is good: 2.9 times the minimum necessary. The model is therefore compatible with the major real parameters.

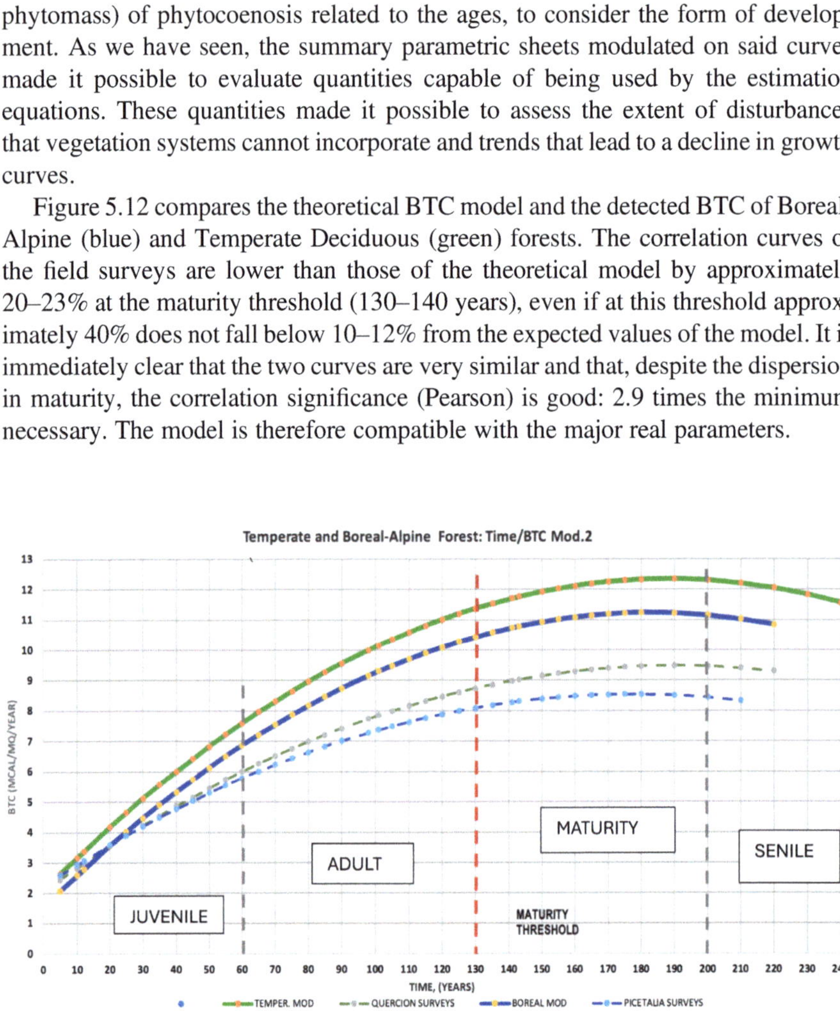

Fig. 5.12 Comparison between the theoretical BTC model and the BTC detected in Boreal-Alpine (blue) and Temperate Deciduous Forests (green). Due to the disturbances, the correlation curves of the field surveys are decidedly lower than those of the theoretical model, even if, at the maturity threshold (130–140 years), approximately 40% do not fall below 10–12% from the expected values of the model

5.5.2 Biodiversity and BTC Relationship

Figure 5.13 simultaneously shows an essential and delicate problem concerning biodiversity. This ecological concept is among the most cited today, even if often without reason. In our case, 211 forests in Lombardy and Trentino (mostly belonging to protected territories) were studied. It was possible to demonstrate that the increase in species biodiversity does not indicate an increase in the level of organization of the phytocoenosis considered (Ingegnoli, 2011, 2015).

Note that the regression curve has an $R^2 = 0.11$ (low) and a maximum of BTC = 7.00 for Sp.N° = 50: this is a misleading condition. It is possible to observe that the highest BTC (Mcal/m^2/yr >8.00) corresponds to Sp.N° = 36.5 ± 11 (ellipse).

Figure 5.14 present the results of a study to understand the forest state/altitude relationship as proof of the inadequacy of specific biodiversity values for measuring the ecological efficiency of vegetation formation: on the left, using specific biodiversity, the study of the ecological state of Boreal-Alpine forests concerning the altitudinal bands on a sample of 211 forests in the Lombardy, Trentino, and Val d'Ayas Alps, did not give the expected results: $R^2 = 0.13$ and Pearson coeff. = 0.15 (since the significance = 0.14 for 200 samples): the Sp.N° remains approximately

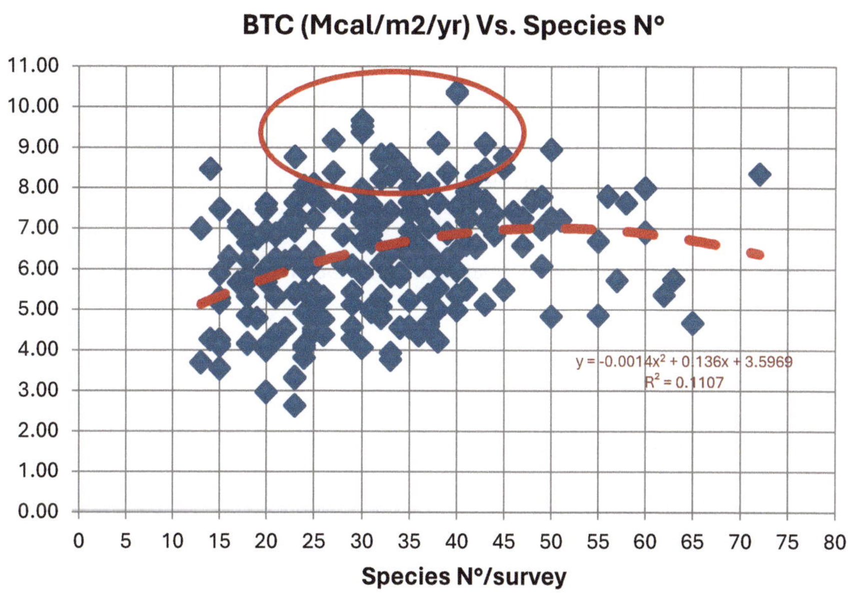

Fig. 5.13 The increase in species biodiversity does not indicate an increase in the level of organization of the vegetation formations considered. Note that the regression curve has a low R2 and that the highest BTC (Mcal/m^2/a > 8.0) corresponds to Sp.N° = 36.5 ± 11 (ellipse)

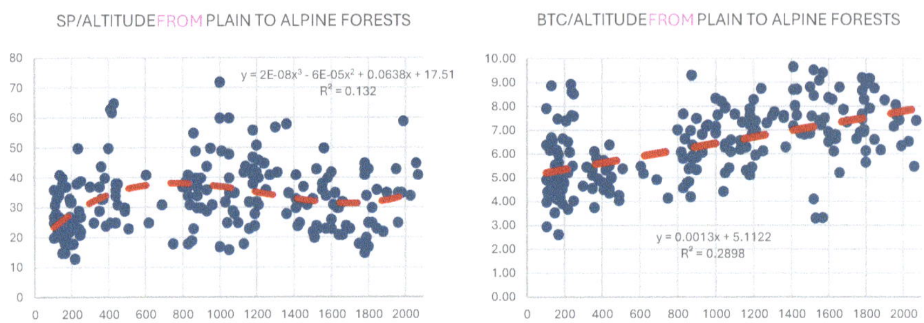

Fig. 5.14 Study of the ecological efficiency of alpine forests in relation to altitudinal ranges. Referring to the biodiversity of Species, the correlation is insufficient (Pearson = 0.15), while referring to the BTC an excellent correlation is found (Pearson = 0.54)

constant; on the right, we note that only using the BTC can we arrive at a credible correlation: $R^2 = 0.32$ and Pearson coeff. = 0.54 and consequent excellent correlation significance, equal to 3.8. Only in this way can it be stated that altitude influences the ecological state of forests (decreases anthropogenic disturbance).

5.5.3 BTC Belts for Euro-Siberian Riparian Vegetation

An item of great importance for all the different types of landscapes is the considerations on the riparian vegetation and on the importance of the elevations in the floodplain profile. *Querco-Carpinetum boreo-italicum* forests (Pignatti, 1982, 1998) can be found within river floodplains, of the Euro-Siberian Riparian Vegetation in bumps sufficiently distant from the water altitude, such as along the Ticino River to Robecco.

These Association can be summarized, from the river to the terraced plain, as follows:

1. *Salicetum albae,*
2. *Salicetum triandra,*
3. *Salici-Viburnetum opuli,*
4. *Carici brizoidis-Alnetum glutinosae,*
5. *Populetum albae,*
6. *Aro italici-Ulmetum minoris,*
7. *Querco-Ulmetum minoris,*
8. *Polygonatum multifloris-Quercus roboris,*
9. *Carici remotae-Fraxinetum oxycarpae,*

(from Pedrotti et al. modified) (Pedrotti et al., 1996).

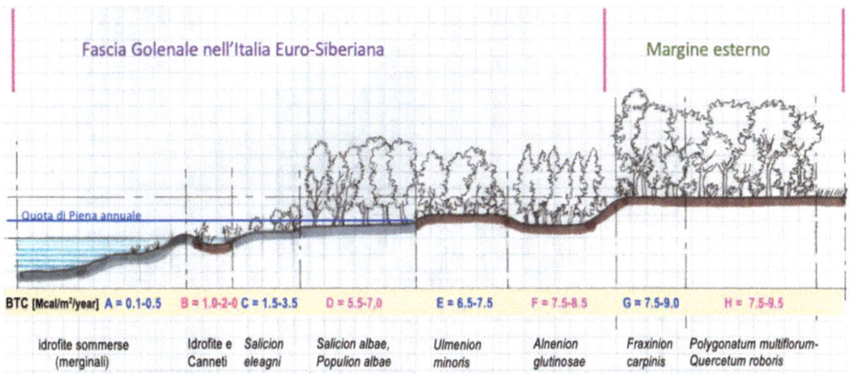

Fig. 5.15 Optimal theoretical structure of riparian vegetation deduced from field observations and processing, and partly deduced from studies by Pedrotti et al. (1996) and Pignatti (1995). Transect by Ingegnoli

Furthermore, it may be significant to know that a hemi-floodplain of approximately 400–430 m, covered by 80% with forest formations, can have on average a value of BTC = 6.0–7.0 Mcal/m^2/year, values that are decidedly higher than the current averages (BTC = 3.0–5.0). Figure 5.15 (Pedrotti et al., 1996; Pignatti, 1995) shows the average values of BTC, to which reference should be made, which are significantly higher than those recorded.

5.5.4 State of the Forests of the Milanese Hinterland

Table 5.7 reports the survey of 20 forests in protected areas in the Milan hinterland, listed according to decreasing CBSt (Ingegnoli, 2013) values (blue). Only the first 5 cases present CBSt >36.0%, and in any case not higher than 53.8%. So, most of these forests present alterations, sometimes even serious ones, given that eight have CBSt <18.0%. The average values are disappointing: BTC = 6.45 Mcal/m^2/year, the presence of exotic species of almost 15%, and phytomass of 433.7 m^3/ha.

So, the bionomic limits of the forests of the Lombard Plain or low hills around the metropolis of Milan are evident: the modest values of BTC and CBSt and the high percentage of alien species are very alarming.

Table 5.7 Survey of lowland forests in protected areas of the Milan Hinterland, LaBiSV method (years 2010–2015)

Surveyed protected areas	Tree species (% on all canopy trees)	N°SP	HC	FM	BTC	CBSt	RxN	Exotic
P. Sud-MI, Cusago	Quercus robur 66, Carpinus betulus 28, Quercus rubra 4	27	29.9	801	8.89	53.8	39.77	10.56
P. Lambro, Tregasio	Quercus robur 33, Q. petraea 19 (52), Q. rubra 23, Populus Nigra 8, Carpinus betulus 6.5	40	29.1	557	8.85	51.4	35.83	13.25
P. Groane, Misinto, S. A.	Quercus robur 63, Q. petraea 9.5, Carpinus betulus 7, Pinus sylvestris 6, Q. rubra 3.5	32	27.1	460	8.54	46.1	33.13	10.71
P. Groane, Misinto, S. A.	Fraxinus excelsior. 62, Robinia pseudoacacia 33, Betula pendula 3, Q. robur 2	30	32.8	632	8.63	45.5	37.69	10.42
P. Ticino, B. Fasolo	Quercus robur 70, Carpinus betulus 16, Robinia pseudoacacia 9	28	24.0	477	7.94	37.3	30.61	12.24
P. Groane, Misinto, S. A.	Q. robur 48 Q. petraea 6 (54), Pinus sylvestris 18.5, Acer pseudoplatanus 15, Robinia psa 7,	27	25.8	348	7.62	34.3	27.83	12.27
P. Colli Briantei Ts 2	Quercus robur (+ Q. petrea) 95+ Robinia pseudoacacia 2 + Platananus x Hyb. 2	18	28.7	691	7.42	28.9	37.75	14.29
P. Sud-MI, Forestina	Quercus robur 54.3, Rob. psa. 20.4, Carpinus betulus 4, Platan. x hyb. 10	47	23.9	258	6.57	26.3	38.04	14.70
P. Monza, Ts 4, old-Q	Quercus robur (petraea) 82, Q. rubra 8.5, Carpinus betulus 6	32	26.4	633	6.67	21.6	36.12	16.92
P. Colli Briantei Ts 1	Quercus robur (+ Q. petrea) 87+ Castanea sativa 3+ Robinia pseudoacacia. 7	27	25.9	526	6.54	21.2	31.03	12.50
P. Sud-MI, Forestina	Quercus robur 12.8, Carp. bet. 83.4, Robinia pseudoacacia 2.3	25	20.7	506	6.19	20.5	37.21	18.80
P. Groane, Bollate	Q. robur 41, Carpinus betulus 27, Q. rubra 14, Pinus sylvestris 7, Robinia pseudoacacia 5	24	22.0	437	6.19	18.8	25.23	15.22
Oasi WWF, Carugate	Ulmus minor 47, Robinia pseudoacacia 26, Populus nigra 14, Q. robur 11	40	22.3	458	5.56	15.6	41.10	18.85
P. Groane, Barlassina	Pinus sylvestris 37, Betula pendula 30, Q.rubra 23, Q. robur 9	15	20.2	305	5.53	14.7	19.64	19.17
P. Groane, Bollate,	Quercus robur (+ Q. petraea) 51, Carpinus betulus 18, Robinia pseudoacacia 21	24	17.0	322	5.29	12.7	34.00	13.64
P. Monza, Ts 3, juv-Q	Quercus robur 79, Carpinus betulus 10, Prunus avium 5, Tilia cordata 4.5	29	22.2	321	5.04	11.1	36.64	16.37
Osio Sotto, Brembo	Ulmus minor 36, Q. robur 19, Robinia pseudoacacia 15	20	10.5	145	4.62	9.5	48.67	8.11
P. Monza, Ts 2, Qr.	Quercus robur 65, Carpinus betulus 19, Robinia pseudoacacia 16	25	24.4	321	4.73	9.3	32.48	17.46
P. Groane, Misinto, S. A	Pinus sylvestris 77, Robinia pseudoacacia 9.5, Betula pendula 6.5, Q. robur 6.5	20	21.3	241	4.05	7.7	22.45	6.25
Plis Roccolo, Casorezzo	Prunus serotina 65, Robinia pseudoacacia 34, Quercus robur 1	24	15.8	235	4.18	7.6	37.63	34.00
	Media	27.70	23.50	433.70	6.45	24.70	34.14	14.79
	Standard deviation	7.78	5.23	172.43	1.59	15.14	6.72	5.75

HC = height of canopy; FM = m^3/ha plant biomass (root included); BTC = Mcal/m^2/yr; CBSt = concise bionomic state (%); RxN = soil humification (Ellenberg/Pignatti); Exotic = allocthonous species (%)

References

Box, E. O. (1987). Plant life form and mediterranean environments. *Annali di Botanica, XLV*, 7–42.

Duchaufour, P. (1984). *Pedologie: Sol, vegetation, environment*. Masson.

Ellenberg, H. (1991). Zeigerwerte von pflanzen in Mitteleuropa. *Scripta Geobotanica*, 1–248.

Ingegnoli, V. (2002). *Landscape ecology: A widening foundation* (p. XXIII+357). Springer.

Ingegnoli, V. (2005). An innovative contribution of landscape ecology to vegetation science. *Israel Journal of Plant Sciences, 53*, 155–166.

Ingegnoli, V. (2011). *Bionomia del paesaggio. L'ecologia del paesaggio biologico-integrata per la formazione di un "medico" dei sistemi ecologici* (p. XX+340). Springer.

Ingegnoli, V. (2013). Concise evaluation of the bionomic state of natural and human vegetation elements in a landscape. *Rendiconti Lincei, Scienze Fisiche e Naturali, 24*, 329–340.

Ingegnoli, V. (2015). *Landscape bionomics. Biological-integrated landscape ecology* (p. XXIV +431). Springer.

Ingegnoli, V., & Giglio, E. (2005). *Ecologia del Paesaggio: manuale per conservare, gestire e pianificare l'ambiente* (p. 685+XVI). Sistemi editoriali SE.

Ingegnoli, V., & Giglio, E. (2019, February 19). *Val Gambìs Storms*. Preprint in Researchgate.

Ingegnoli, V. & Pignatti, S. (2007). The impact of the widened landscape ecology on vegetation science: towards the new paradigm. *Springer, Rendiconti Lincei Scienze Fisiche e Naturali, s. IX, XVIII*, 89–122.

Nonis, U. (2001/2020). *500 Funghi*. Priuli & Verlucca.

Pedrotti, F., Gafta, D., & Grossoni, P. (1996). *Ecologia delle foreste ripariali e paludose dell'Italia*. Ed. Università degli Studi di Camerino.

Pignatti, S. (1995). *Ecologia Vegetale*. UTET.

Pignatti, S. (2005). Valori di bioindicazione delle piante vascolari della flora italiana. *Braun-Blanquetia, 39*, 1–97.

Pignatti, S. (1982). *Flora d'Italia*. Edagricole.

Pignatti, S. (1998). *I boschi d'Italia: sinecologia e biodiversità*. Utet.

Simard, S. (2018). Mycorrhizal networks facilitate tree communication, learning and memory. In F. Baluska, M. Gagliano, & G. Witzany (Eds.), *Memory and learning in plants*. Springer.

Simard, S., Roach, W. J., Defrenne, C. E., Pickles, B. J., Snyder, E. N., Robinson, A., & Lavkulich, L. M. (2020). Harvest intensity effects on carbon stocks and biodiversity are dependent on regional climate in Douglas-Fir forests of British Columbia. *Frontiers in Forests and Global Change, 3*, 88.

Chapter 6
Outline of Bionomical Evaluation of Fauna

Contents

Abstract Even with regards to the faunal component, bionomic studies on the landscape bring both conceptual and methodological innovations, even if there is, at present, a more limited number of examples on this approach.

Animals play a vital role in structuring the landscape too. Those who have had the opportunity to observe the behavior of the most critical animal populations in the large African parks have realized that animals act on the structure of the landscape in two main ways: (a) as users of resources, (b) as an engineer species. On the other side, the significant effects of fragmentation on fauna are, in decreasing order: isolation, increase in generalists and multi-habitat species, edge species, nest predation and the rate of extinction. A study of European Environment Agency demonstrates a very significant law: up to 9 fragmentation elements per 1000 km^2, the survival of large mammal populations is possible, but if the density increases from 9 to 14, survival drops from 80 to 20%.

V. Ingegnoli, *Landscape Bionomics and Planetary Health*, Landscape Series 21,
https://doi.org/10.1007/978-3-032-08908-3_6

In accordance with what was stated for vegetation, a new method of studying the animal component is being developed, which will be called "LaBiSf." This method currently studies faunal sensitivity and the Standard Habitat (SH) per individual in the Natural habitat (NH). But Urbanized landscapes may contain hundreds of wild species, too. In the Western World, the most common animal species living in towns are found in Europe and North America,

Detecting the presence of invertebrates, through pit traps, the level of ecological efficiency of a restored green tessera (Ts), after 2–3 years of green planting, can be compared to the natural ones: the results of an investigation to measure the validity of a LU restoration in a mining area are shown.

Then, in summary, the effects of the bark beetle epidemic on *Picea* forests are presented. The epidemic broke out after the destruction of the Vaia storm (October 2018), which felled from 15 to 35% of the trees in a few hours (wind with gusts of 200–220 km/h). The abnormal quantity of dead wood has multiplied the populations of *Ips typographus*, which are now causing perhaps more damage than the storm itself, even if their numbers are starting to decrease.

Keywords Engineers species · Generalist species · Fragmentation · Faunal sensitivity · Ecological efficiency · Bark beetles · Animal standard habitat per individual

6.1 Interaction Among Faunal Component and the Landscape: An Overview

6.1.1 Animal Populations Shaping the Landscape

Even with regards to the faunal component, bionomic studies on the landscape bring both conceptual and methodological innovations, even if there is, at present, a more limited number of examples on this approach.

One of the most important criteria to evaluate faunal aspects in a landscape is to study the *engineer species* (Sanderson & Harris, 2000). The structure of the landscape is due in the first instance, as we know, to the relationship between geomorphological processes and the distribution of vegetation. Yet, animals also play a very important role in structuring the landscape. Those who have had the opportunity to observe the behavior of the most important animal populations in the large African parks have realized that animals act on the structure of the landscape in two main ways:

(a) as *users of resources*, being capable in complex systems ecocenotopes of exploiting the greater resources of plant components (for example, prairies), but also of contributing to their maintenance (preventing the expansion of forests, distributing fertilizing manure, transporting seeds, etc.);

Fig. 6.1 Animals and landscape structure. Transformations in a LU due to engineer species such as African elephants. Author's study in the Serengeti (Tanzania), where two ramps can be seen to go down to the river banks

(b) as an *engineer species*, being capable of modifying the environment where they live to their advantage, forming peculiar structures or maintaining others, creating paths or tracks, connections with water bodies (for example, *elephant watering tracks*; Fig. 6.1), by building homes (ants, termites) or dams and lakes (beavers), or niches and paths on steep high mountain slopes (for example, ibexes; Fig. 6.2).

The effects on the structure of the landscape are usually enhanced by a series of correlations with the vegetation, so even if the alterations of certain populations or communities of animals seem modest or marginal, on the contrary it can be great.

The case of beavers (for example, *Castor fiber, Castor canadensis*) is the most obvious: a dam on a small stream is not spatially relevant compared to a valley, but this involves regulating the water and continuously using wood, and in short the consequences spread to the entire landscape unit, whose constituent elements change within a few tens of years, to the point of forming a basic ecomosaic completely different from the previous one, so much so as to give the name to entire valleys (for example, "val Bevera", Lombardy and Liguria).

Furthermore, the symbiosis between trees and birds sometimes becomes important, as can be observed in the Alps with the well-known links between nutcrackers (*Nucifraga caryocatactes*) and stone pine trees (*Pinus cembra*), as the birds bury the

Fig. 6.2 Female ibex (*Capra ibex*) with their young in the moraines of the Verra glacier, altitude 2350, in the upper Ayas valley (AO)

Fig. 6.3 Example of Cembreta (*Pinus cembra*) in Val de Cornon (above Panchià di Fiemme), altitude 2100. They are formations in symbiosis with the populations of nutcrackers

seeds as a food reserve, but not using them all, they allow some of them to germinate, expanding stone pine forests in areas of conflict with pasture (Fig. 6.3): an example of which is reported in the upper Val del Cornon (Fiemme Valley, TN, Italy).

Similarly important is the case of red ants (*Formica rufa*) in a subalpine spruce forest (Fig. 6.4). Please remember that the nest, an acervus that can exceed 1.5–2.0 m, is also dug underground, therefore it is very vast, being able to contain hundred thousand individuals with more than 100 queens. An acervus like the two ones shown in the figure can destroy around five million insects in one season. Note that these ant acervus can help also some birds, first the nutcrackers, land on the nests of red ants to be hit by their sprays of formic acid and thus be freed from mites and lice.

Fig. 6.4 The author is photographed next to a *Formica rufa* acervus more than 1 m tall, in the forest of Passo di Lavazé (TN). The nest is also dug underground and can contain a few hundred thousand individuals with more than 100 queens. An acerve like this, which is also very rich in herbaceous cover, can destroy around five million insects in one season

6.1.2 Animal Populations Structuring the Ecotissue

The presence of populations of large herbivorous mammals is also closely linked to the structuring of the ecotissues of the landscape, especially because of grazing and fertilization. Here too, however, there are correlations with vegetation which increase their influence on the structure of the landscape: for example, animals can induce a felting of pastures, which selects the reproduction of certain plant species of trees (including bushes and herbs) acting on biodiversity in a positive but also negative sense. Herbivores, however, manage to enrich meadows that live on incomplete soils with low humus content with nitrogenous substances, as happens in the mountains.

Often the faunal studies required by landscape ecology can concentrate on a series of careful observations on the presence or absence of species that have a strategic relationship with the structure of the landscape unit (LU) in question; or on observations on the presence of impacts of animal origin in the natural habitat or in areas with little or no anthropization. It is always necessary, however, to try to understand the relationship of animal species with the structure of the landscape, starting from their distribution in relation to the distribution of ecotopes and LUs (at least in principle). In fact, the structure of the landscape also acts on animal populations in different ways:

- through the heterogeneity of the landscape
- through its load bearing capacity
- through the conformation of spots and corridors
- through the metapopulation model
- through movements in the landscape
- through the presence of anthropic structures.

For example, a landscape ecology study on the Cornino nature reserve in Friuli, set up for the protection of vultures (*Gyps fulvus*), showed a non-compliance of the protection perimeter with the structural pattern of its LU, because it did not include the entire range of overhanging rocks suitable for nesting and did not consider the plateau behind, where various species of birds of prey usually go to find suitable prey.

Furthermore, from the data of Opdam (Opdam et al., 1992), we note that the probability of the presence of avifauna for an isolated grove goes from 76% at 1 km from the forest, to 64% at 18 km.

It should also be noted that some animals (for example, wolves, foxes, coyotes, lions, etc.) also use artificial corridors (Odum, 1971), such as roads (grassy and/or tree-lined banks or dirt roads), obviously if there is little traffic.

6.1.3 Animal Populations and Landscape Dynamics

The heterogeneity of the vegetated tesserae and ecotopes produced by fires has an influence on the fauna, which finds resources changed both in quality and in spatial arrangement. From studies carried out after the great fire in Yellowstone Park (United States), it was noted that the main factor for the survival of ungulates (*Cervus elaphus* and *Bison bison*) is the winter following the fire, when nourishment is lacking. If the winter is mild, an area greater than 60% of a LU must burn for the ungulates to be affected. Furthermore, as Turner et al. recall (Turner et al., 1992), the spatial configuration of the unfired residual patches, the semi-fired ones and the burned ones is important: if heterogeneous enough, sufficient compensation can be created to spare the animal populations; if, however, the fire is limited to a single tile with good resilience (for example, a meadow-pasture) inserted in an ecotissue in good condition, it has almost no influence on the invertebrate community of the tile itself, as noted by Filippo Bernini (personal communication) in a study on the Monte Barro Park, especially for arachnids (the first recolonizers).

In Mediterranean areas, spontaneous fires are usually more limited than arson and are notoriously part of recurrent disturbances (Naveh & Lieberman, 1984). However, these events also create a particular dynamic towards grazing, given that the strong regrowth of geophytes (for example, *Asphodelus* spp.) prevents sheep grazing in forest clearings, as can be seen from Fig. 6.5, where the flock remains at the outside of a half-burnt pine forest.

Fig. 6.5 Semi-burnt Mediterranean pine forest, with sheep grazing on its edges. Before the fire, the flock used to graze in the internal clearings, but the overwhelming dominance of the asphodels after the fire drove the animals away from the forest

Even slow movements, such as bradyseism, can affect fauna, as it has been found in the Venice lagoon. A subsidence (combined with slight eustatism) of about 2 mm/year is sufficient to modify part of the shape of the salt marshes and to vary the halophilous vegetation, so in a few decades changes can be noted in the mosaic of salt marshes, with consequences both for nesting and for the resources of some types of birds, such as Ardeidae (for example, *Egretta alba*), Sternidae (for example, *Sterna caspia*), Shelduck (for example, *Tadorna tadorna*), Oystercatchers (for example, *Haematopus ostralegus*).

Climate change can be felt by modifying the mosaics of humid areas in once temperate-cold areas (such as Germany) and allowing the introduction of species typical of Mediterranean countries, as reported by Ott (2001). In a short time, in addition to species such as *Crocothemis erytraea*, which had already been moving northwards for 10 years, another 8 species of dragonflies from temperate-warm zones have become frequent in Germany. The movement of many species in a short time can cause effects on the landscape, through alterations in the floristic composition and behavior of animals towards the various biotopes.

Four general reactions are foreseeable according to Graves and Reavey (Graves & Reavey, 1996): (1) change in the inhabited area, (2) a priori tolerance towards new conditions, (3) adaptation to new conditions through microevolution, (4) increased mortality.

Urbanization (see further on) has led to a change in the behavior of some groups of seagulls (for example, *Larus ridibundus*) which have even appeared in cities like Milan, in areas far from the sea or lakes, and which compete with the crows (*Corvus corone cornix*) for the waste in urban landfills.

Particularly interesting are extinction delays, due to destructive changes in the agricultural landscape (Burel & Baudry, 1994, 1999). Burel and Baudry have demonstrated in Brittany that the spatial distribution of *Abax parallelepipedus* (a carabid) is more correlated to the connectivity of a network of "bocager" trees from 1952 than to the current network, whose connectivity has decreased by 35%. Den Boer (1985) observed that carabids from stable habitats (with little dispersal power) remain at least 40 years after their hedge has been destroyed, while highly dispersing species remain for only 10 years.

Table 6.1 Comparison of different forest management in two theoretical LU in the Alps, with forests of boreal type and feeding areas of ungulates. Notice of the diversity of landscape ecology characters in relation to feeding areas

Forest management	Age (years)[a]	Height (m)[a]	BTC (Mcal/m²/year)	Feeding areas (%)[a]	Damages[b]
(a) Flush cut and reforestation	2–16	<8	2.0–3.5	30–100	3–5
	16–70	8–24	3.5–6.0	1–9	
	70–100	24–30	6.0–7.5	9–16	
(b) Natural renewal, patch cut	10–50	6–21	3.0–5.0	6–14	1–3
	50–80	21–27	5.0–7.0	14–30	
	80–120	27–36	7.0–8.5	30–100	

Order of magnitude, as a reference; [a]from Reimoser and Gossow (Forman, 2019), revised; [b]from Mustoni et al. (Opdam et al., 1992), damage by stripping and barking

Forest management dynamics (such as deforestation, reforestation, etc.) continually modify local landscape units and affect animals in various ways. As can be seen in Table 6.1, comparing two different managements in theoretical LUs of boreal-alpine forest (large clear cuttings with reforestation and small cutting with natural renewal), it emerges that the feeding areas of the ungulates are almost opposite (Reimoser & Gossow, 1996), as shown by the different damage from debarking and browsing (Mustoni et al., 2002). Also note the different level of BTC reachable in the two cases.

The dynamics of the landscape lead to spatial transformations which can be reduced, as we have seen, to six typologies. The relationship between the fauna and the typology in question still needs to be explored further, however it must be said that part of these studies can fall within the more strictly structural previous paragraph. Dissection and fragmentation (and their inverses) are in any case the most important spatial processes in landscape changes.

As it is found almost everywhere, even in Italy, dissection (Forman, 1995) is the first and most subtle of processes, which occurs (within certain limits) even without and against any planning. It involves the opening of a road in an agricultural or wooded landscape, followed by micro-interventions, often illegal, of construction and human activities resulting from this traffic line, some of which enlarging and bringing a disturbance towards the interior, which it recalls other minor road connections, in a very harmful positive feedback process. The disturbance to the fauna is evident from every aspect, and often, these are disturbances that cannot be incorporated at the LU scale. In studies on foxes (*Canis vulpis*), it is stated (Storm et al., 1976) that their dens remain at a minimum distance of 200 m from anthropized areas, even if foxes, as we know, come to look for food on the outskirts of towns as happens in Val di Fiemme (Fig. 6.6) or in the Abruzzo Park (where the Author also encountered a fox in the urban centre of Civitella Alfedena). Consequently, residual stains with pleasing potential naturalness remain very small.

Fig. 6.6 Fox in a freshly mown meadow 15 m from the houses of the village of Panchià, and at the margin of the road to Rolle Dolomite Pass, Val di Fiemme, Italy. Deers often also pass in the same places

Fragmentation is the best-known process of landscape dynamics, so we will discuss on it further on (Sect. 6.3.1).

6.2 LaBiSF: Landscape Bionomics Survey of Fauna

In accordance with what was previously stated for vegetation, a new method of studying the animal component is being developed which will be called "LaBiSV", which currently sees the study of faunal sensitivity and the study of SH per individual in the NH (see Sect. 6.3.2), in analogy with what has already been explored for human populations.

For the microbenthic component relating to inland waters, please refer to the River Functionality Index. Adequate studies have not yet been conducted for the fish component.

6.2.1 Analysis of Faunal Sensitivity (Sf)

The faunal sensitivity method (Ingegnoli & Giglio, 2005; Ingegnoli, 2002; Massa & Ingegnoli, 1999) reported here was developed with the collaboration of colleagues Massa and Fornasari (2002) to measure faunal sensitivity (Sf) for a landscape unit (LU). This sensitivity is assessed in two steps. The first consists of the evaluation that can be expressed with the following equation:

$$Sf = (Q1 + Q2 + Q3) \times (V1 + V2 + V3)$$

Table 6.2 Evaluation of faunal sensitivity in a Landscape Unit (LU)

Faunal attributes	0	1	2	3	Score
Q. Faunal quality (general characters)					
Q1- General rarity	Not Determin.	Rare, risk	Vulnerab. (general)	Threaten.	Species rarity
Q2- Chorology	Cosmopolites	Continent	Regional	Endemic	Areal
Q3- Critical state: local presence vs other areas	Other	Rare in district	Rare in region	Doubt in region	Presence in other areas
V. Faunal vulnerability (local characters)					
V1- Abundance	Abundant	Common	Scarce	Rare	*i/p°*
	>0.5	<0.5	<0.25	<0.125	
V2- Habitat extension	Not selective	Rural	Ecotonal	Forest	Habitat sensitivity
		Habitat	Habitat	Habitat	
V3- Fragility: living conditions	Any habitats	Not degraded	Colonial	Optimal habitat	Species needs vs. habitat
R. Results					
R1- Total scores	a	b	c	d	S(f)
Evaluation of faunal sensitivity				Total scores	
I	Critical S (f)			S(f) >72	
II	High S(f)			S(f) <72	
III	Relevant S (f)			S(f) <54	
IV	Mean S(f)			S(f) <36	
Level of attention					
V	Low S(f)			S(f) <18	
VI	Negligible S(f)			S(f) <9	

where: Sf is the faunal sensitivity of a species, Q1 the rarity, Q2 the chorology, Q3 the criticality, V1 the abundance, V2 the extension of the habitat, V3 the fragility (always of the same species)

To quantify the equation, it is necessary to refer to the sheet shown in Table 6.2, which is conceptually not too different from the sheets proposed for the evaluation of vegetation. However, the score values used in this case are limited to a linear arithmetic sequence, not a composite curve. The final evaluation of the sheet referred to in the upper part of the sheet must be compared with the faunal sensitivity classes of the lower part of the same sheet in relation to the total score recorded per species.

Obviously, if you do not have the time and means to carry out a total investigation, it is essential to carry out this evaluation on key species, to be chosen previously depending on the type of landscape unit and the study objective. In this regard, we recall the definitions of "umbrella" species and "focal" species: umbrella species are those that present the greatest ecological needs, therefore the first to disappear after out-of-scale destruction of the landscape; focal species are those capable of representing a set of structural and functional requirements of the landscape also necessary for the other species that characterize the landscape itself. In detail, the items of the proposed index (Sf) are presented:

Q1. Rarity (R.)

- Critical or High R. (3 points): the red lists for species threatened with extinction both on a global scale and on a European or Italian scale become important. Rare and vulnerable species for Italy or undergoing large declines in Europe or species included in Annex II of EEC Dir. 43/92. Bird species with fewer than 25,000 breeding pairs in Europe.
- Medium R. (2 points): species with indeterminate status for Italy or included in Annex IV of EEC Dir. 43/92 or in the lists of the Bern Convention with similar criteria, species with a contracted range and scarce populations. Bird species with fewer than 50,000 breeding pairs in Europe.
- R. low (1 point): Species not meeting the above criteria. Species of birds with less than 100,000 breeding pairs in Europe. Disperse species, typical of fragmented landscapes and colonial species; species excluded from the list of protected species in a Country.
- R. none (0 points): Species not meeting the above criteria, Species of birds with more than 100,000 breeding pairs in Europe. Common and widespread species.

Q2. Chorology (Cor.)

- Cor. with reduced range (3 points): species or subspecies endemic in Italy or in some Italian regions, particularly vulnerable to landscape modifications.
- Cor. of intermediate range (2 points): species with a reduced range (for example, Mediterranean or infra-European).
- Cor. with large range (1 point): species with continental:(e.g., European).
- Cor. with very vast range (0 points): species with range intercontinental (e.g., Eurasian, Palearctic) or cosmopolitan.

Q3. Criticality (Cr.)

- Cr. high regional (3 points): species not usually present in the region (doubtful in the literature), or species present in less than 50% of the regional territory, but as an elective area at national level.
- Cr. regional low (2 points): species present in less than 20% of the regional territory and present in the study area with any density, or present in the area with a density higher than the regional average.
- Cr. local (1 point): rare species at a regional level but relatively abundant at a local level.
- Cr. undefined (0 points): species not meeting the previous criteria.

V1. Abundance (A.) or Frequency

- rare (3 points): rare species. For example: for avifauna <0.125 individuals/point; for mammals and herpetofauna, the frequency in the survey units can be considered (% of occupied area), because species are not very detectable: rare species <12.5%.
- scarce (2 points): scarce species. For example: for avifauna <0.25 individuals/point; for mammals and herbal fauna: scarce species <25%.
- common (1 point): common species. For example, for avifauna <0.50 individuals/point; for mammals and herpetofauna: common species <50%.
- abundant (0 points): abundant species. For example: for avifauna >0.50 individuals/point; for mammals and herpetofauna: abundant species >50%.

V2. Habitat Extention (hab.)

- forestal hab. (3 points): species with a habitat restricted to forests or wooded patches with substantial tree cover.
- ecotonal hab. (2 points): ecotonal species (forest-countryside) less selective than the previous ones, or aquatic species from little altered areas.
- rural hab. (1 point): species adapted to agricultural landscapes, including open ones.
- non-selective hab. (0 points): species indifferent to the habitat, capable of living even in urban and suburban landscapes.

V3. Fragility (F.)

- High F. (3 points): species tightly linked to optimal habitat characteristics, good BTC, without pollution or otherwise, and species with concentrated and isolated populations.
- Medium F. (2 points): species negatively affected by degradation, although not requiring optimal habitats; colonial species not as above, sedentary species or with low reproduction.
- Normal F. (1 point): species that are affected by environmental disturbances only if they cannot be incorporated and are out of scale (i.e. destructive).
- Low F. (0 points): species with high resilience and mobility, capable of living even in altered environments.

6.2.2 Faunal Sensitivity by Ecotope Group or by Landscape Unit

To obtain eco-landscape indices on faunal quality in the typical subdivision compartments of a landscape, one can start from the previous method, summing the *Sf* values per compartment, then evaluating a faunal sensitivity per area (*Sa*). However, a species can be present with different densities in different environments, therefore a weighting must be carried out before the sum (Malcevschi, 1985) as follows:

$$Sa = \sum S_i \times K_i a$$

where: S_i is the sensitivity of the i-th species and K_i is the weight due to the relative density of the i-th species within *a*.

The following weights (K_i) can be assigned:

- high density = 1.25;
- normal density = 1;
- reduced density = 0.75;
- sporadic presence = 0.5.

It is noted that this index (*Sa*) effectively replaces other usual indices for the landscape field, such as the Shannon and Weaver diversity, as it does not measure the information content, which is uncritical (Fornasari, 1997), but also evaluates its importance for conservation.

We therefore move on with relative ease to the drafting of a faunal map suitable for the purposes of landscape ecology. This map can also contain the detections of species that exceed the minimum threshold of attention (*Sf* = 18) and possibly point emergencies. Within the same areas, parcels can also be highlighted, corresponding to clearly identifiable landscape elements, with high faunal sensitivity, possibly marked by an alphanumeric index.

6.3 Fauna and Alteration of the Landscape

6.3.1 Large Mammals and Landscape Fragmentation

Fragmentation (i.e. breaking a large habitat into smaller parts) is undoubtedly the best-known process, although often poorly defined: it tends to also include part of the other spatial processes already mentioned. The major effects of fragmentation on fauna are, in decreasing order: isolation, the increase in generalists and multi-habitat species, edge species, nest predation and the rate of extinction. Besides, the dispersion of inland species and those with large home ranges and the richness of inland species of scrubland decrease.

Fragmentation tends to influence almost all major ecological and even genetic processes. For example, increased metapopulation dynamics and backcrossing are common. On the other hand, the area of disturbing spots decreases. We must observe, however, that fragmentation in certain cases can also produce positive effects: in fact, it can be demonstrated (Ingegnoli & Giglio, 2005) that the general metastability of a natural landscape (LM) increases if a limited percentage of human habitat is introduced (never greater than the 8–12%); they also increases complex biodiversity (up to HH = 40%; see Sect. 3.2.10). This is also demonstrated by the habitat choices of many animal species: it is known that the creation of ecotonal bands and new heterogeneous tiles (including cultivation, if not intensive) are sought by roe deer and wild boars. Even deer can be attracted to it, as long as the anthropic SH remains very high, i.e., the population density remains low. Moreover, any naturalist can see for himself that the LUs near the small towns of the Alps or the Apennines present a particularly high variety of bird species.

With the domination of man over the ecosphere, there are many alterations in the relationships, including structural ones, between vegetated tesserae or ecotopes and animals, which are not always replaced (or replaceable) by the presence of herds of domestic herbivores. This too often makes us forget the role of animals also as a disturbance in the landscape, a disturbance that we know is of primary importance for the structuring of complex ecocoenotope systems. A comparison with African parks is a must to understand that, within natural landscapes, the environmental impact is not absent at all, but is due to the populations of animals, especially large mammals.

It is always necessary, however, to try to understand the relationship of animal species with the structure of the landscape, starting from their distribution in relation to the mosaic of ecotopes and Landscape Units. In fact, the structure of the landscape also acts on animal populations in different ways, such as: (a) landscape heterogeneity, (b) carrying capacity, (c) conformation of patches and corridors (d) metapopulation model, (e) movements in the landscape, (f) presence of anthropic structures.

The abandonment of agricultural activities causes the birth of new vegetated patches, therefore new habitats available to fauna. There is, however, a delay between the appearance of new habitats and the colonization of animal species, except in the case of arachnids. Burel and Baudry (1994) in a study on carabids in Normandy reported that only new vegetated patches less than 20 m from a source (for example, a tree-lined corridor) are colonized relatively early. Connectivity and circuitry are therefore essential for colonization, as is the area of the tiles. Furthermore, it should be noted that the dynamics of plant succession cannot always be correlated with animal colonization, both because this succession does not behave according to traditional rules (Ingegnoli & Giglio, 2005) and because polyphage species are independent of this process.

It should also be noted that the ecological consequences of the technical intensification of agricultural landscapes cannot be assessed by remaining at the cultivated tile scale. It is necessary to refer to the scale diversity in the exploitation of resources by stenotope and eurytope animal groups, to the structural complexity of the landscape, and to its heterogeneity of tiles (Ingegnoli, 2002).

The effects of opening artificial corridors (roads, railways, canals) on fauna are multiple: barrier, conduit, filter, source, drain. For example, for large mammals in forest landscapes, the effect of a busy road on the surrounding tiles can reach 90 to 200 m, and for birds from the forest interior, it remains around 100 m, while in a prairie landscape such disturbance distance can reach 1.5–2 km (Forman, 1995).

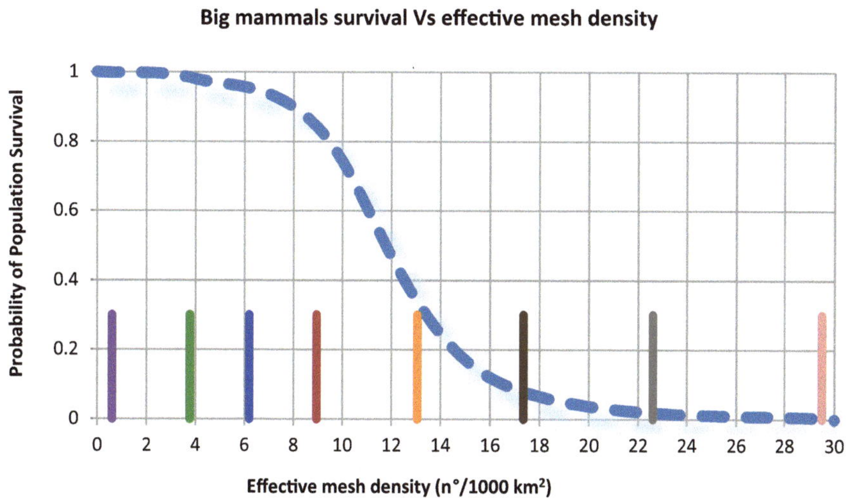

Fig. 6.7 Model proposed by EUEA to indicate the survival probabilities of large mammals in European landscapes in relation to fragmentation (mesh density). The colored bars represent the averages in: Sweden, UK, Austria, Italy, Switzerland, Poland, Czech Republic, France (redrawn and completed by Ingegnoli)

Figure 6.7 summarizes a study by the European Environment Agency (EUEA and FOE, 2011) to demonstrate a very significant law: up to 9 fragmentation elements per 1000 km^2, the survival of large mammal populations is possible, but if the density increases from 9 to 14 survival drops from 80 to 20%.

6.3.2 Ungulates and Natural Habitat in Alpine Landscapes

Figure 6.8 reports the minimum standard habitats SH'for ungulate populations in Italy and their relationships with the main landscape types and the relative NH value.

Note the large difference in living space per capita between the SH* (see Sect. 3.4) relating to the human population—which in Italy has a particularly vast range, from the sub-tropical of Pantelleria (SH* = 0.10 ha/inhabitant) to the Boreal-Alpine of Val d'Aosta (altitude of 1800 m a.s.l., SH* = 0.35 ha/inhabitant)—compared to 4.0–16.0 ha/animal for ungulates.

NH %	Landscape types	BOAR	ROE	DEER	CHAM	STEINB
>90	Natural, Alpine	–	–	–	+++	+++
>90	Natural, forested	++	+++	+++	+++	++
90–75	Semi-natural, forested	+++	+++	+++	++	+
75–60	Agricultural and Forestry	+++	+++	++	+	–
60–35	Agricultural, protective	+++	++	+	–	–
35–10	Agricultural, productive	+	+	–	–	–
15–5	Suburban	+	–	–	–	–
<5	Urban	–	–	–	–	–
	SH (ha/animal)	3–4	4–5	15–16	5–6	5–7
	Survival number[a]	10	15	70	60	60

CHAM Chamois; *STEINB* Steinbock
[a]Data from Perco [7], but intended as MVP (minimum viable population) of sub-populations

Fig. 6.8 Ungulates and natural habitat in an alpine landscape unit. Average standard habitat size per capita (SH) ha/individual. Standard habitat (SH) referred to Central European ungulates and their relations with the main types of landscapes. In the first column, an average value of their natural habitat (NH)

6.3.3 Evaluating Ecological Quality Through the Help of Fauna Component

In many cases a traditional and simple method of animal detection can also be used in landscape bionomics research. For example, in Table 6.3 we summarize the results of an investigation to measure the validity of a LU restoration, in a mining area of Tavernola Bergamasca (Iseo Lake, Northern Italy) (Fig. 6.9). To verify the level of ecological efficiency of the restored green tesserae (Ts), such as Ts A in the figure, after 2–3 years of green planting, we compared these tiles with the natural ones (Ts N in the figure), detecting the presence of invertebrates, through pit traps as exposed in Ingegnoli and Giglio (Ingegnoli, 2002).

The most significant results can be found by evaluating the ecological indices shown in the table. The ratio of insects to invertebrates in Ts A was only 82.5% of Ts N, while the invertebrate/plant biomass ratio in Ts A is double that of Ts N. Note that the BTC was exactly the inverse of the invertebrate ratio/plant biomass, confirming the good significance of the experiment and denoting a still unbalanced state phase in the artificial green Ts.

It is worth underlining with particular attention the perfect convergence of the evaluation of a tile (Ts) through animal parameters or through vegetation parameters. In fact, the best condition of a vegetative phytocoenosis means the best condition of the fauna habitat and, therefore, of the insect community.

The study of the presence or the absence of dragonfly species *(Odonata)* in a landscape unit helps to evaluate its ecological conditions. For example, in the central Chiaravalle district of the South Milan Agricultural Park 13 species of dragonflies were found (Ingegnoli, 1998), among which only two were quite uncommon: but compared with a potential 20–25 species in this humid field landscape (Ott, 2001) that value was very meaningful.

Table 6.3 Comparison between two vegetated tesserae (A = restored; N = natural) within the area of environmental remediation of the cement marl mining in Tavernola Bergamasca (Lake of Iseo, BG, Italy)

Surveys	Measures	Tessera A	Tessera N
Vegetation characters of the tessera	Phanerophytae shrubs (%)	22	13
	Alien and cosmopolite (%)	14	8
	Above ground plant biomass (g/m^2 d. m.)	800	1.600
Invertebrate species (n°)	Detected	68	66
	Species in common	27	27
	Species insects	40	47
Captured individuals (n°)	Total	302	312
	Insects	164	236
	Coleopteran	32	64
Biomass of invertebrates (g. d. m.)	Total (g)	36.77	35.68
	Coleopteran (g)	6.46	4.29
Ecological indexes	Sp. insect/Sp. invert. tot (%)	58.8	71.2
	Detected invert./plant biomass (n°/kg)	85.00	41.25
	BTC of the Ts Mcal/m^2/year	0.74	1.44

Fig. 6.9 The area of the Tavernola Bergamasca mine (Lombardy) and the location of the two sites (tiles A and N) of the experiment shown

Table 6.4 Biotopes, habitat characters and key species of Odonata

Biotope type	Landscape elements (habitat)	Key species of Odonata
Lakes	Colonisation zone, riparian reeds, semi-submersed vegetation, open water	*Coenagrion puella, Enallagma cyathigerum, Erythromma najas, Anax imperator, Aeshna grandis, Somatochlora metallica, Orthetrum cancellatum*
Small temporary water basins	Partially shaded by vegetation	*Lestes dryas, Sympetrum sanguineum*
Small perennial water basins	Partially shaded by vegetation	*Laestes sponsa, Platycnemis pennipes, Aeshna canea, Cordulia aenea*
Marshy waters and peats	Acid waters, partially shaded, presence of *Sphagnum*	*Coenagrion hastulatum, Aeshna juncea, Somatochlora arctica, Leucorrhinia dubia*
Secondary brooks and canals	Rapid change of vegetation types, with also hydrophites	*Erythromma viridulum, Libellula depressa, Crocothemis erytraea, Gomphus pulchellus*
Rivers and current waters	With meanders, mainly shaded, few sand banks, hydrophytes	*Calopteryx virgo, Calopterys splendens, Orthetrum coerulescens, Cordulegaster boltoni, Gomphus* sp.

The *Odonata* expert Juergen Ott published (Ott, 2001) a table of human impacts on dragonflies and a key species table (Table 6.4) for different biotopes, applicable to central European landscapes (data from Rheinland-Palatinate).

6.4 Landscape Urbanization and Wildlife

6.4.1 Common Urbanized Wildlife and the Problem of Zoonosis

It is known that in recent years various species have become accustomed to living in suburban landscapes: just think of the foxes that walk looking for food among domestic waste, or the mass appearance of starlings (*Sturnus vulgaris*) in Rome, who feel more protected by spending the night in the trees of urban avenues: as a result, kestrels (*Falco tinnunculus*) have also arrived. In Milan, in the Public Garden of Palazzo Dugnani, you often see a night heron (*Nycticorax nycticorax*) which has learned to exploit the presence of fish in the garden ponds. Also in that garden, animals never seen in large cities often appear, such as a kingfisher (*Alcedo atthis*). Some birds, such as the redbreast (*Erithacus rubecula*), have learned to sing at night due to the lighting and less traffic (Petretti, 2003). The most negative aspect of urbanized structures is that linked to the introduction of exotic species: from the gray squirrel to the black-cheeked turtles (water areas), to nutrias, to lake shrimps (all mainly American) or to parrots (for example, some African species are frequent in the gardens of Genoa).

Urbanized landscapes may contain hundreds of wild species. In the Western World the most frequent animal species living in towns are common both in Europe and North America, as we can see in Table 6.5.

Note that (a) the 21 most common species in towns of EU and USA count seven species from America Vs 14 from Europe; (b) many EU & USA additional species are replacements (e.g., Larus/Corvus, or Capreolus/Odocoileus).

Table 6.5 Common urbanized fauna

Towns	Birds	Mammals
EU	1. *Apus apus*	(i) *Capreolus capreolus*
	2. *Ardea cinerea*	(ii) *Erinaceus europaeus*
	3. *Asio otus*	(iii) *Males males*
	4. *Bubo bubo*	(iv) *Martes foina*
	5. *Coloeus monedola*	(v) *Sus scrofa*
	6. *Columba palumbus*	(vi) *Sorex araneus*
	7. *Corvus cornix*	
	8. *Delicon urbicum*	
	9. *Garrulus ghiandarius*	
	10. *Larus michehellis*	
EU & USA	11. *Anas platyrhynchos*	(vii) *Mus musculus*
	12. *Columbia livia*	(viii) *Myocastor coypus*
	13. *Falco peregrinus*	(ix) *Oryctolagus cuniculus*
	14. *Passer domesticus*	(x) *Procyon lotor*
	15. *Psittacula krameri*	(xi) *Riattus rattus*
	16. *Sturnus vulgaris*	(xii) *Sciurus vulgaris*
	17. *Titus alba*	(xiii) *Sciurus carolinae*
	18. *Turdus merula*	(xiv) *Vulpes vulpes*
USA	19. *Brantha canadensis*	
	20. *Bubo virginianus*	
	21. *Corvus brachyrhynchos*	
	22. *Cyanocetta cristata*	
	23. *Eudocimus albus*	(xv) *Canis latrans*
	24. *Larus marinus*	(xvi) *Lepus americanus*
	25. *Mimus polyglottos*	(xvii) *Mephitis mephitis*
	26. *Picoides pubescens*	(xviii) *Myrcomis minutus*
	27. *Spinus tristis*	(xix) *Odocoileus virginianus*
	28. *Zenaida macroura*	(xx) *Blarina brevicauda*

Note that 80–85% is linked with urban green

1. Swift; 2. Grey heron; 3. Common owl; 4. Eagl owl; 5. Jackdaw; 6. Wood pidgeon; 7. Crow; 8. Martin; 9. Jay; 10. Herring gull; 11. Mallard duck; 12. Pigeon; 13. Peregine falcon; 14. House sparrow; 15. Green parakeet; 16. Starling; 17. Barn owl; 18. Blackbird; 19. Canadian Goose; 20. American owl; 21. American crow; 22. Crested jay; 23. White ibis; 24. Great black backed gull; 25. Northern mockingbird; 26. Downey woodpecker: 27. American goldfinch. 28. Mourning dove. (i) Roe deer; (ii) Common hedgehog; (iii) Badger; (iv) Marten; (v) Boar; (vi) Shrew; (vii) House mouse; (viii) Nutria; (ix) Wild rabbit; (x) Raccoon; (xi) Rat; (xii) European squirrel; (xiii) American squirrel; (xiv) Red fox; (xv) Coyote; (xvi) American hare; (xvii) Striped skunk; (xviii) Harvest mouse; (xix) White tailed deer; (xx) Short tailed shrew

Table 6.6 Main Zoonosis in Italy (2022–23)

Sicks (IT)	%	Disease	Transmission	Mortality	MRw
3.302	50,25	*Salmonellosis*	Food, water, pets	0,08	0,04
1.539	23,42	*Campylobacteriosis*	food, water, chickens, pets	0,05	0,01
723	11,00	*West Nile Virus*	Birds and mosquitos(Culex)	4,5	0,50
345	5,25	*Listeriosis*	Food	0,2	0,01
220	3,35	*Dengue*	Mosquitos	3,6	0,12
125	1,90	*Lyme Borreliosis*	Ticks, from wild in humid/cold environment	0,4	0,01
118	1,80	*Escherichia*	Food	0,05	0,00
89	1,35	*Toscana virus*	Phlebotomus	0,4	0,01
50	0,76	*Neuro-invasiva TBE*	Ticks	1,8	0,01
36	0,55	*Yersinia*	Domestic and wild animals	2,5	0,01
20	0,30	*Brucellosis*	Rural animals	1,2	0,00
15	0,23	*Bovine Tuberculosis*	Aerosols from animals (sneezing, coughing), milk	1,8	0,00
6.582	**100,00**	**ITALY (2022–23)**		%	**0,73**

Many of these species can carry zoonoses that are dangerous for humans, as Raccoons in the Po plain may bring rabies or Rats may bring leptospirosis. In the 2021 EU Zoonosis Report, the first and second most reported human zoonoses were campylobacteriosis and salmonellosis, respectively. Cases of campylobacteriosis and salmonellosis increased in comparison with 2020 but decreased compared with previous years (Table 6.6). Seven human factors are driving the rise of zoonotic diseases:

1. increased human demand for animal protein.
2. unsustainable intensive agriculture.
3. increased use and exploitation of wildlife.
4. unsustainable use of natural resources accelerated by urbanization, land use changes and extractive industries.
5. increased travel and transportation.
6. changes in food supply; and
7. climate change.

Moreover, if we want to limit the zoonosis dangers, we must avoid the banalization of Urban parks and gardens, because the habitat alterations increase the zoonosis (see Sect. 13.4).

6.4.2 Road or Rail Effects on Animal Species in Near-Natural Landscape

The effects of the opening of artificial corridors (roads, railways, canals, etc.) on fauna are multiple: barrier, conduit, filter, source, drain. For example, for large mammals in forest landscapes the effect of a busy road on the surrounding tiles can reach 90 to 200 m and for birds in the forest interior around 100 m, while in a prairie landscape this disturbance distance it can reach 1.5–2 km (Forman, 1995). Roads are rarely crossed, even by animals that could do so easily in periods of little traffic, such as large mammals as the caribou (*Rangifer tarandus*) or the mountain goat (*Oreamnus americanus*) in America or the deer (*Cervus elaphus*) and the roe deer in Europe. Small forest mammals, such as red squirrels, generally do not crossroads wider than 15 m (including shoulders). It must be added that the killing of animals caused by the opening of roads (even with little traffic) are sometimes high, as happens with black bears and pumas in Florida, roe deer in Val Pusteria or badgers in England.

A well-known study made by Torres et al. (2016) in a Sub-sclerophyll Sparse Wood landscape in central Spain, mentioned by Forman (1995, 2019), allowed Ingegnoli to underline what expressed in Fig. 6.10.

The average of human activities results in the first km from the road, and this has direct interference (orange line) with many animal species, as shown in this figure.

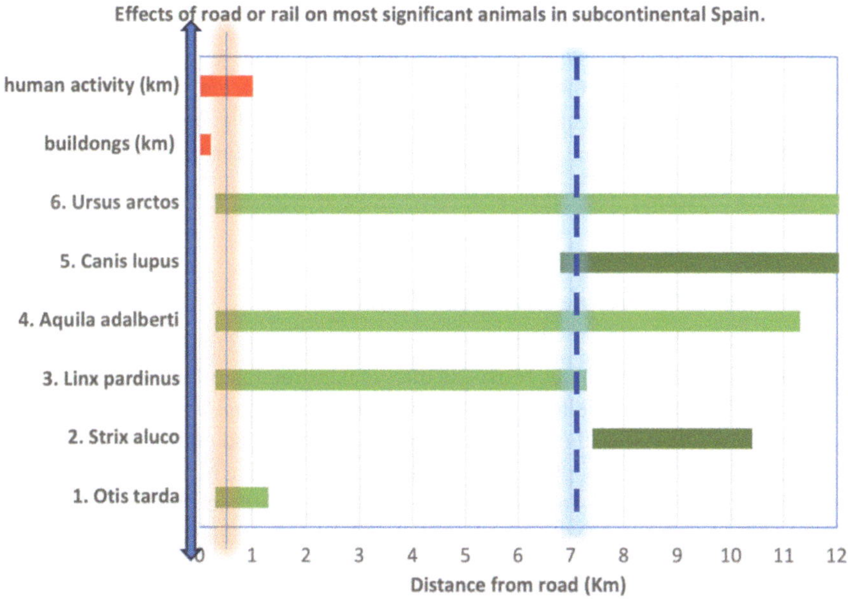

Fig. 6.10 Effects of roads or railways on main animal species in a near-natural landscape in central Spain (Sub-sclerophyll Sparse Wood). Note the direct interference animal/human activities (orange line) for 4 species and the habitat restriction for 2 species (blue dotted line)

The main consequence is, for instance, the decrease of *Linx* populations. However, other road damage is fragmentation (see Fig. 6.7) and habitat restriction (blue dotted line) of the most cautious species, such as *Canis lupus*.

6.5 The Forest Destruction Due to the Spruce Boyster

6.5.1 *The* Ips Thypographus *Beetle*

The *Ips thypographus*, popularly known as the spruce bark beetle, is a small beetle of the *Scolytidae* group, cylindrical in shape and brown in colour, approximately 4–5 mm long. It is endemic to the woods of the Alps and mainly attacks red spruce (*Picea abies*), where it develops under the bark, digging intricate tunnels which interrupt the flow of sap; thus, it inevitably causes the plants to die in a short time.

In spring, the males who survived the winter penetrate the plants and build a nuptial chamber, in which they mate with two to three females. These then dig tunnels up to 12–16 cm long and parallel to the axis of the trunk, where each female lays an average of 60–80 eggs.

The larvae, white with a dark head, dig 5–6 cm tunnels orthogonal to the axis of the stem, always subcortical, as they feed. At the end of development, they transform into adults, and the new generation will be able to settle on other plants in the same year or the following year, depending on the climate. In fact, after a couple of weeks with temperatures from −10 to −15 °C, the insects die.

However, *Ips typographus* is often linked to (or at least favors) some fungi, which, by slowing down the circulation of water in the phloem, make the situation much worse. In a few weeks a tree over 30 m high can die from the bark beetle epidemic. We report some photographs (Fig. 6.11) that illustrate what has been written above.

6.5.2 *Forest Horrible and Wide Damage*

We must say that the epidemic broke out after the destruction of the Vaia storm (end of October 2018), which felled from 15 to 40% of the trees in a few hours (wind with gusts of 200–220 km/h): the abnormal quantity of dead wood on the slopes favored the multiplication of the populations of bark beetles which are now causing the same or more damage than the storm itself, even if their numbers are starting to decrease. In Fiemme Valley, this beetle destroyed about 10–15% more than the Vaia tempest!

But even in this case, it must be remembered that man is not entirely free from blame: in fact, cultivation for wood production has over time eliminated various species of broad-leaved trees, like *Acer pseudoplatanus, Fraxinus excelsior, Prunus avium, Fagus sylvatica, Populus tremula,* etc. which would have formed a network of local barriers to the expansion of the beetles, while even-aged and very dense forest cultivations facilitated the spread (Fig. 6.12).

Fig. 6.11 An Ips typographus (left) enters the bark of a fir tree; in the centre, a larva digs a tunnel; on the right, a fir tree measuring 21 m and 80 cm in diameter will die 2 weeks after the photograph. All photos, taken by Ingegnoli in Val di Fiemme, August 2023

Fig. 6.12 Epidemic of Ips typographus in Val di Fiemme, at the entrance to the Cavelonte Valley. To the left, the dense forest in 2017–18. Dx: Note the severe thinning of trees in the Spruce Forest (August 2024)

6.6 Notes on Cattle Breeding

Following the principles of landscape bionomics, it is frequently suggested that grazing on agricultural estates and planting patches of forest and rows of trees should be encouraged, especially if there are livestock farms. Please remember, for example, that if cattle are raised, leaving them in open pastures as much as possible, at least (a) meat acquires antioxidant factors, beta-carotene, etc. and improves the flavour; (b) cheeses contain more CLA (conjugated linolenic acid). On the contrary, when cattle are fed with high doses of starch (feed), the pH of the rumen drops to 4: it acidifies.

Cattle conditions	Near-Natural	Artificial
Cattle Diet	Natural grasses	Grained mixture (Corn)
Veterinary Drugs	No antibiotics or hormones	++ antibiotics & hormones
Environment	Open pasture	Confined in Stable
Maturation (years)	1.5 to 2.0	1.0
Meat characters	2-4 x omega-3	1 omega-3

Fig. 6.13 Comparison between near-natural vs. artificial cattle conditions

The acidic rumen environment develops other lines of bacteria than usual, which prevent the formation of healthy substances and make the animal's digestive system ill. Above all, we must note that animals in stables cannot live as a population, which leads to ethological and psychological alterations to metabolism, which have repercussions on human health. We therefore report a comparison between the two breeding conditions: (a) herbaceous nutrition outdoors vs. (b) artificial feed in the stable (Fig. 6.13).

Could Be Useful to Remember

1. *Grass-fed beef contains less total fat than grain-fed beef.*
2. *Grass-fed beef contains an optimal ratio of omega-6 to omega-3.*

   Both omega-3 and omega-6 *polyunsaturated fatty acids (PUFAs)* are essential fatty acids, meaning we must consume them through our diet since we cannot produce them ourselves. Omega-3 s are associated with protective mechanisms (anti-inflammatory, neuroprotective, etc.) and, although essential, excess omega-6 s are generally associated with promoting inflammation and modern chronic diseases.
3. *3-Grass-fed beef contains two to three times more conjugated linoleic acid (CLA)*

   Conjugated linoleic acid (CLA) is a type of PUFA, which preliminary research suggests may offer great health benefits, such as improved glucose regulation, insulin sensitivity, and protection against cardiovascular disease due to its anti-hypertensive effects.
4. *Grass-fed beef contains more beta-carotene.*

   Beta-carotene is the precursor to retinol (vitamin A), a fat-soluble vitamin that is very important for good vision, bone health, reproduction, immunity, and various cellular functions. Additionally, beta-carotene has protective properties primarily through its antioxidant activity.

5. *Grass-fed beef contains more vitamin E.*
   Grass-fed beef contains significantly more vitamin E than grain-fed beef, in the form of α-tocopherol.
6. *Grass-fed beef contains higher antioxidant enzymes*
   Glutathione and the two enzymes, superoxide dismutase and catalase, were all found in higher concentrations in grass-fed beef than in grain-fed beef.

References

Burel, F., & Baudry, J. (1994). Reaction of ground beetles to vegetation changes following grassland derelictions. *Acta Oecologica, 15*, 401–415.

Burel, F., & Baudry, J. (1999). *Ecologie du paysage: concepts, methodes et applications*. Technique & Documentation Ed.

Den Boer, P. J. (1985). Fluctuation of density and survival of carabid populations. *Ecologia, 67*, 322–330.

EUEA and FOE (2011) *Study on landscape fragmentation*. EU edition.

Forman, R. T. T. (1995). *Land Mosaics: The ecology of landscapes and regions*. Cambridge University Press.

Forman, R. T. T. (2019). *Towns, ecology and the land*. Cambridge University Press.

Fornasari, L. (1997). *I rapporti tra i vertebrati e il paesaggio: teoria ed esempi*. In V. Ingegnoli (Ed.), *Esercizi di ecologia del paesaggio* (pp. 131–168). CittàStudi-UTET.

Graves, J., & Reavey, D. (1996). *Global environmental change. Plants, animals and communities*. Addison Wesley.

Ingegnoli, V. (1998). *Landscape ecological criteria as a basis for the planning of a Suburban Park in Milan*. In J. Breuste, H. Feldmann, & O. Uhlmann (Eds.), *Urban ecology* (pp. 657–662). Springer-Verlag.

Ingegnoli, V. (2002). *Landscape ecology: a widening foundation*. Springer.

Ingegnoli, V., & Giglio, E. (2005). *Ecologia del Paesaggio: manuale per conservare, gestire e pianificare l'ambiente*. Sistemi editoriali SE.

Malcevschi, S. (1985). *Approccio sintetico agli ecosistemi e valutazione di qualità ambientale. Considerazioni introduttive*. In M. Fasola (Ed.), *Atti III Convegno Italiano di Ornitologia*. Salice Terme (PV) 3–6 ottobre 1985 (pp. 191–194). Pavia.

Massa, R., & Fornasari, L. (2002). *Faunal sensitivity in landscape unit*. In V. Ingegnoli (Ed.), *Landscape ecology: A widening foundation*. Springer Verlag.

Massa, R., & Ingegnoli, V. (1999). Conservazione… .Utet.

Mustoni, A., Pedrotti, L., Zanon, E., & Tosi, G. (2002). *Ungulati delle Alpi*. Nitida Immagine Ed.

Naveh, Z., & Lieberman, A. (1984/1994). *Landscape ecology: Theory and application*. Springer-Verlag XXVII+360.

Odum, E. P. (1971). *Fundamentals of ecology*. Saunders.

Opdam, P., Van Apeldoorn, R., Shotman, A., & Kalhoven, J. (1992). *Population responses to landscape fragmentation*. In C. C. Vos & P. Opdam (Eds.), *Landscape ecology of a stressed environment* (pp. 147–171). Chapman & Hall.

Ott, J. (2001). *Expansion of Mediterranean Odonata in Germany and Europe- consequences of climatic changes*. In G. R. Walter, C. A. Burga, & P. J. Edwards (Eds.), *Fingerprints of climate change, adapted behaviour and shifting species ranges* (pp. 89–111). Kluwer.

Petretti, F. (2003). *Gestione della fauna. Il management delle popolazioni animali negli ambienti naturali, agricoli e urbanizzati*. Edagricole.

Reimoser, F., & Gossow, H. (1996). Impact of ungulates on forest vegetation and its dependence on the silvicultural system. *Forest Ecology Management, 88*, 107–119.

Sanderson, J., & Harris, L. D. (Eds.). (2000). *Landscape ecology: A top-down approach*. Lewis Publishers.

Storm, G. L., Andrews, R. D., Philips, R. L., Bishop, R. A., Sniff, D. B., & Tester, J. R. (1976). Morphology, reproduction, dispersal and mortality of mid-western red fox populations. *Wildlife Monographs, 49*(5–82), 1.

Torres et al. (2016). Sub-sclerophyll Sparse Wood Landscape in Central Spain.

Turner, M. G., Wu, Y., Pearson, S. M., Romme, W. H., & Wallace, L. L. (1992). *Landscape-level interactions among ungulates, vegetation and large-scale fires in Northern Yellowstone National Park*. In G. E. Plumb & H. J. Harlow (Eds.), *16th annual report* (pp. 206–211). University of Wyoming national Park Service Researche Center.

Chapter 7
Diagnosis of the Health of a Landscape

Contents

Abstract When some landscape unit alters and shows pathological signs, we can speak of real landscape syndromes. A syndromic classification of landscape alterations or, better, of landscape units (LU) is an ordering regardless of their causes, and we will summarize it in this chapter. The study of pathology becomes essential, a discipline that is not easy to apply, as the meaning of pathology is inextricable from that of physiology. As Konrad Lorenz had underlined, in physiology, to understand a

V. Ingegnoli, *Landscape Bionomics and Planetary Health*, Landscape Series 21,
https://doi.org/10.1007/978-3-032-08908-3_7

function, it is necessary to consider its alteration and vice versa in a continuous postponement of answers.

We can present the equation of the general metastability of an LU; this is a very significant indicator capable of considering the information content of the LU about the flow of energy necessary for maintaining the state of order of the LU itself. An exceptionally high correlation between BTC and human habitat (HH) brings us to measure the capacity for self-regulation of a natural system (HH). This diagnostic model is essential to arrive at a preliminary overall evaluation of a landscape unit with accuracy and consequently to check both a state of fact and a land planning project. Moreover, the local BTC can be balanced by an element of pleasing naturalness: so, the distance of influence from it is fundamental. It is called ER (eco-bionomic range) or radius of ecological influence.

Continued human fragmentation and alteration, due to the transformation of agricultural landscapes, seem to be the 'criteria' of the structuring of suburban landscapes, but leaves a chaotic mosaic of remnant patches or residual spots of different magnitudes and importance: sew these rags into fabric that performs an ecological sense is possible by the bionomic study of the remnant patches.

To understand the relationship between the state of health and diseases of a land unit, checking the compatibility with the needs and the human health, and to develop criteria of therapeutic rehabilitation with regenerative function must become the new concept of sustainability, too; so, it can no longer be to manage the resources to postpone as much as possible their depletion! Therefore, the ecological services of anthropic plant formations are of crucial importance, while they are, at present, seriously insufficient in terms of prevention and health care. We can indicate that without the presence of natural forests, human habitats are intrinsically prevented from achieving adequate health protection.

Keywords Landscape alteration · Pathological signs · Information content · HH/BTC model · Distance of influence · Transformation · Diagnostic index DI · Regenerative function · Therapeutic rehabilitation · Ecological services · Bionomic Functionality BF

7.1 Alteration and Pathology of a Landscape

Every biological system, as well as a landscape, can undergo alterations to its state in addition to transforming itself. This alteration can become pathological in different conditions, often connected, for example, when: (a) the level of metastability of the upper scale of the system is no longer able to incorporate the disturbance; (b) a large portion of information processes are blocked or seriously damaged; (c) the biological potential of the lower scale components is seriously damaged; (d) permanent alterations are caused to the main structures and functions of the system itself.

The process of incorporating large-scale disturbances can be found, e.g., when the average biological-territorial capacity of the landscape system remains almost constant for long periods, even in the presence of significant local changes. An example comes from the study of the transformations of Lombardy (Ingegnoli, 2002), which has been having an almost constant average BTC (1.9–2.1 Mcal/m^2/yr) for around two centuries.

However, even serious problems can hide at lower scales, especially when there are discontinuities in incorporating disturbances from the local to the regional scale. An example is that of Fig. 7.1, which concerns the dramatic transformation of the metropolitan area of Milan from the post-war period to the beginning of the new millennium. The component ecotopes are summarized in the following types: URB, urban; SURB, suburban; AGR, agricultural; S-NAT, semi-natural. The urbanization of the territory went from 45.2% to 78% in just 32 years (1969–2001), with an increase of 72.6% which produced a profound alteration of the landscape, as in such a short period the change could not be appropriate for every ecological parameter.

It should be noted that environmental diagnoses depend on the comparison among the conditions of the ecological system under examination and those of a state considered 'normal' for this type of ecological system. In other words, it is the relationship between 'pathology' and 'physiology' of the systems that allows a clinical diagnosis of the landscape under examination. It is necessary to understand how much the system shifts from the normal state due to pathogenic stimuli, and with a projection of information evaluate where the damage to the structure and functions could occur, in a congruent time.

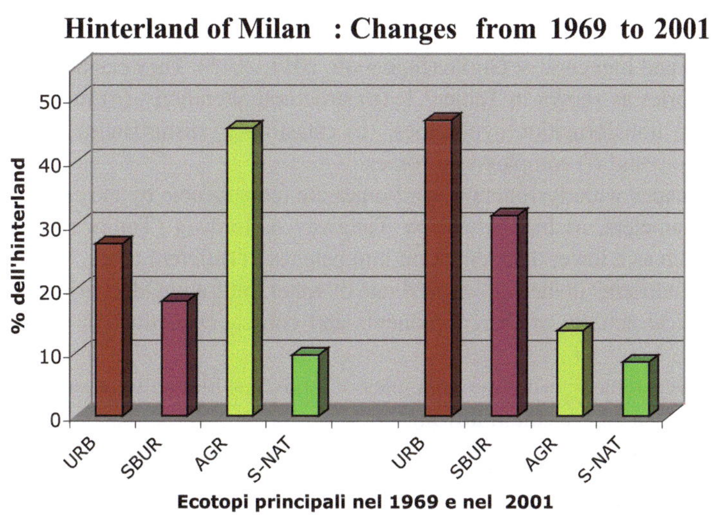

Fig. 7.1 Transformation of the Milan Metropolitan Area from 1969 to 2001. The component ecotopes are summarized in the following types: *URB* urban, *SURB* suburban, *AGR* agricultural, *S-NAT* semi-natural. The urbanization of the territory went from 45.2% to 78% in just 32 years, with an increase of 72.6% which produced a real alteration of the landscape

By 'normality of a landscape', or of one of its sub-units, we mean the quantified set of ecological characteristics typical of a recognizable landscape specificity (suburban, rural, agricultural, agro-forestry, lagoon, etc.) in its own metastable equilibrium state.

7.2 Landscape Syndromes

7.2.1 Landscape Can Get Sick

Dysfunctions in a landscape unit (LU) can be studied as in any level of biological organization by following iterative phases capable of considering the following:

(a) Recording of symptoms,
(b) Identification of causes,
(c) Analysis of reactions to pathogenic alterations,
(d) Control of risks of further deterioration,
(e) Choice of therapeutic criteria,
(f) Control of interventions.

When some landscape unit alters and shows pathological signs, we can speak of real landscape syndromes. While a disease can be defined as a set of symptoms traced back to a specific cause, a syndrome is a set of similar symptoms that may depend on different causes. A syndromic classification of landscape and landscape units (LU) alterations is an ordering regardless of their causes.

We can and must, therefore, talk about landscape syndromes, as indicated by Ingegnoli and Ingegnoli & Giglio (Ingegnoli, 2011, 2015). They can be grouped into six categories as shown in Table 7.1: (a) structural alterations, (b) functional alterations, (c) transformation syndromes, (d) catastrophic disturbances, (e) pollution degradation, and (f) complex syndromes.

Landscapes with dysfunction syndromes are recognizable by the examination of some parameters, as highlighted by Tongway & Ludwig (Tongway & Ludwig, 2004), such as a lower differentiation into patches of different types, lower concentration of nutrients in the soil, slowed rate of water infiltration, decrease in the levels of biological activity of the components and consequent reduction in production cycles.

However, remembering the principles of landscape bionomy, other observations must be added such as the following:

1. the appearance of an evident "transformation deficit",
2. a sharp decrease or strong variation in the average BTC level,
3. the loss or significant reduction of natural habitat (NH),
4. an anomalous ratio or a clear variation in the HH/NH proportion for that certain type of landscape under consideration,
5. a loss of correlation between heterogeneity and information,

Table 7.1 Example of intrinsic landscape syndrome

Categories	
Main landscape syndrome categories	Sub-categories of syndromes
A-Structural Landscape Alterations	A1-Landscape element anomalies
	A2-Spatial configuration problems
	A3-Functional configuration problems
	A4-Multiple structural degradation
B-Functional Landscape Alterations	B1-Geobiological alterations
	B2-Structurally dependent dysfunctions
	B3-Delimitation problems
	B4-Movement and flux dysfunctions
	B5-Information anomalies
	B6-Reproduction problems
	B7-Multiple dysfunctions
C-Transformation Landscape Syndromes	C1-Stability problems
	C2-Changing process dysfunctions
	C3-Anomalies in transformation modalities
	C4-Complex transformation syndrome
D-Catastrophic perturbations	D1-Natural climatic disasters
	D2-Natural geophysical disasters
	D3-Human-made destruction
E-Pollution degradations	E1-Dust pollution
	E2-Chemical pollution
	E3-Endocrine disruptor chemicals
	E4-Indirect pollution
F-Complex technol. Syndromes	F1-Acute
	F2-Chronic

6. the formation, at least in certain subsystems, of high fragmentation,
7. the loss of connections (connectivity and circuitry) with an increase in barriers,
8. a clear dysfunction between the landscape systems in relation to the type of landscape.

Identifying the causes that produce the found syndromes requires a good knowledge of (a) the 'anatomy' and physiology of the landscape and (b) the set of pathological disorders. It also requires a good anamnesis, which renews the importance of *history*, and a *semiotic* analysis, in which perceptive studies can contribute positively. Bear in mind, as doctors well know, that the etiopathogenesis of a syndrome is interpretative in most cases but is nevertheless of great importance, as seen in the case of agrarian industrialization syndrome (Fig. 7.2) in Central European regions.

In this figure, the reference landscape is Agricultural-Rural of Central-Southern Europe with BTC = 1,2–2,0 Mcal/m^2/yr.; SH/SH* = 3–5; HH = 50–75%, RSD = 4.0–9.0%, with discrete connectivity and circuitry. From these conditions, now stable for over a millennium (sometimes since the Roman Empire, as around

PATHOGENETIC SCHEME OF THE SYNDROME OF AGRICULTURAL INDUSTRIALIZATION OF TEMPERATE LOWLAND AGRICULTURAL LANDSCAPES.

Original state *permanence:* *3-17 centuries*	**TRADITIONAL AGRICULTURAL LANDSCAPE WITH "BOCAGE"** (BTC = 1,2-2,0 Mcal/m²/yr; SH/SH*= 3-5; HH = 50-75%) (heterogeneity, connectivity and circulation: good)
Main human cause	*socio-economic pressures for increased agricultural production* — Crop specialization *increase and facilitation for agricultural technologies* — Stream channelling
self-aggrandizing process	increase in arable land, increase in chemical fertilizers cutting tree rows, mechanical irrigation destruction of forest patches increase of parasites increase of chemical pesticides decrease of fauna soil impoverishment enlargement of field tiles, increase of road network independence from geomorphological constraints
Altered state *Permanence:* *2-8 decades*	**OPEN AND MONOCULTURAL AGRICULTURAL LANDSCAPE** (BTC = 0,9-1,3 Mcal/m²/yr; SH/SH*= 4-8; HH = 70-85%) (decreased heterogeneity, eliminated connectivity and circuitry)
	structural weakening increased fragility of the system recall of large road infrastructures recall of industrial areas in the countryside increased fragmentation
State upset *permanence:?*	**SUBURBAN RURAL LANDSCAPE** (BTC = 0,8-1,1 Mcal/m²/yr; SH/SH*= 0,4-1; HH = 80-90%) (increasing heterogeneity, increasing pollution)
outcome opposite to the cause	Decrease of agrarian production

BTC, SH/SH*, HH data are indicative, generally valid for the Po Valley

Fig. 7.2 Pathogenetic scheme of the urbanization syndrome of rural areas in Europe. Note that the three types of agricultural landscape are characterized by applying Landscape Bionomics dimensions

Padua), in just a few decades, human actions, such as the destruction of woodland patches and rows, the abandonment of agricultural fields, increased fragmentation, the construction of arable land, etc. have led to an altered rural-suburban state. In a few years (even just a decade), there was a weakening of the functional heterogeneity of the LUs and the development of road infrastructures, with a reduction of agricultural areas by up to 30% and metastability by at least 20%. This disrupted state has become a suburban tech landscape, with BTC=0.8–1.1; SH/SH* = 0.4–1.0 and HH = 80–90%.

With a few variations these pathological transformations can be found in much of Europe and similarly also in certain US states.

7.2.2 *Examples of Sorting of Landscape Syndromes*

We can now briefly summarize, by way of example, some of the most common structural and functional anomalies found in landscapes (European, but not only) with ongoing pathologies.

A1—Structural Anomalies in the Landscape Elements

1. Biological alterations to the structure of tiles/ecocoenotopes (for example vertical layers of vegetation).
2. Hyper-simplification of the shape of the elements (for example rectification of agricultural fields and their channels).
3. Anomalous grain of tiles or ecotopes (for example too small in relation to the internal structure: wooded patches as wide as they are tall).
4. Lack of form-function correspondence (for example fields that do not follow the direction of water drainage).
5. Geomorphological damage in tiles or ecotopes (for example forms of erosion).

A2—Spatial Configuration Problems

1. Too low heterogeneity of component elements (for example type and shape).
2. Excess of density of roads (for example >1000 m/km^2).
3. Labile differentiation into spots (for example all borders are blurred).
4. Excess or lack of contrast between elements (for example industrial zoning in a wooded area or in the middle of the city).
5. Strong difference (without reason) in the porosity of a matrix (for example distribution of patches or tree-lined rows).

A3—Functional Configuration Problems

1. Insufficient or excessive density of ecological networks in a LU (for example drastic decrease in the corridor network in a few decades).
2. Presence of some type of anomalous apparatus (for example the appearance of anthropic housing apparatus in a natural forest landscape).
3. Lack of natural habitat (e.g., HN deficiency in agricultural landscapes).
4. Lack or excess of subsystems in an ecotissue (for example lack protective apparatus in an urban landscape).

A4—Multiple Structural Degradation

1. Loss of congruence between the regime of geomorphological disturbances and the structure of the ecotissue (for example deepening, rectification of channels, change of regime at tidal inlets and destruction of the salt marshes of a lagoon landscape).
2. Chaotic structure, therefore, difficulties in localizing components at various scales (for example inability to distinguish both ecotopes and LUs).
3. Structural subsystems not compatible with the functional role of the LU in its landscape (for example presence of industrial areas and technological networks in a LU of an agricultural park).

B1—Geobiological Alterations

- Excess of morphogenesis (for example consequent slope instability).
- Soil degradation processes (for example loss of humified layer).
- Insufficient or excessive drainage (for example areas of dryness or stagnation).
- Erosion and alteration from extractive activities (for example formation of quarries).
- Heavy damming of watercourses (for example with barriers to the rising salmonids).

B2—Structure-Dependent Dysfunctions

- Excessive fragmentation of components (for example residual spots without core area).
- Excess landscape resistance to the movement of key species (e.g. too many barriers to the movement of badgers in agricultural landscapes).
- Connection disturbances (for example *stretch marks* in ecological networks).
- Negative variation to the fields of influence of elements in the matrix (for example restriction of these fields within the boundaries of a protected area).
- Problems at strategic points of a matrix (for example alterations of a nodal area between different ecotopes).
- Abrupt changes in the quantity or distribution of NH and/or HH (for example due to uncontrolled expansion of subsidiary settlements).

B3—Delimitation Problems

- Alterations in the formation of the boundaries of an element (for example rectification of the boundaries of a residual forest patch).
- Alterations in the ecotonal network or in the margins (for example formation of barriers instead of ecotonal bands).
- Loss of functional characteristics of a boundary (for example decrease in the porosity of an edge of a patch).
- Too high frequency of delimitations (for example fences in an agricultural landscape).
- Lack of margins in a patch (for example residual element with only internal species).

B4—Flow and Movement Dysfunctions

- Disturbances to the main inter-ecosystem flows (for example impediments to the transport of catabolites towards the excretory network, i.e. waterways).
- Disturbances in the seed advection process (for example disturbances due to emissions into the atmosphere).
- Obstructions and anomalies in the flow of flood waters (for example constructions in floodplains or regulation of rivers).
- Loss of purification capacity of the excretory system (for example destruction of riparian reeds).
- Decrease in exchanges of species and matter from ecotopes (for example destruction of patches suitable for metapopulations).

- Invasion of exotic species (for example introduction of alien vegetation in LUs close to suburban landscapes).
- Limitation to the seasonal movement of species between regions (for example interruption of a large passage corridor between different and complementary natural landscapes).
- Deviation of the migratory flow of species (for example modification of attractor habitats in suburban landscapes).
- High interference with the anthropic traffic system (for example passage of a motorway in an agricultural landscape).

7.3 Diagnostic Assessment of the Landscape

As we have already demonstrated, the ecological systems that constitute our environment, from ecocoenotopes to the entire landscape, must be considered as living systems, and consequently, they *can become ill.* We also underlined that in a complex system, knowledge of the components, their dynamics and their interactions are insufficient to arrive at a correct diagnosis. Therefore, we need to know the structure and functions acquired by the components in the 'system' condition; that is, we need to see the system's behavior as an 'organic whole'. For these reasons, system state functions, such as BTC or carrying capacity SH/SH*, become essential.

Furthermore, the study of *pathology* becomes important. This discipline is not easy to apply, as the meaning of pathology is inextricable from that of physiology. As Konrad Lorenz (Lorenz, 1978) underlined, in physiology, to understand a function, it is necessary to consider its alteration and vice versa in a continuous postponement of answers.

System Theory says that the scale capable of maximizing the importance and quantity of the relationships between the components of a system is the scale that allows the different relational forms and configurations to be optimally discriminated. Precisely for this reason, Ingegnoli (Ingegnoli, 2002, 2011, 2015) and Ingegnoli & Giglio (2005) have exalted the crucial importance of the landscape, as this territorial scale allows the optimization of the relationships between the anthropic and natural elements. Furthermore, as in Medicine, environmental assessment requires comparing the normal parameters of both ecotopes and landscape units. Consequently, the main problem of diagnostics becomes how to *know the state of normality* and, at the same time, the *levels of alteration of the system* under examination. We can proceed in that direction by using the main state functions of a landscape or of one of its units (LU). Four primary methodologies can lead to a correct diagnosis of the landscape as a complex system, and they can be applied in sequence. They are:

(a) The congruence of the environmental state of the landscape unit under examination with the normality function expressed by the correlation of the state functions of the bionomic-territorial capacity of the vegetation (BTC) with the percentage of human habitat (HH) detected, which leads to the measurement of the bionomic functionality (BF).

(b) The evaluation of the distance from the normal ranges of n^* ecological and bionomic parameters detectable in the LU under examination, which leads to highlighting a Parametric Spectrum quantifiable with a Diagnostic Index (DI).
(c) The control of the transformations of the LU in question, obtained by evaluating the relationship between the BTC of the forestry and agricultural components compared to the value of the regional BTC, calculated over a period of optimal conditions, which leads to the estimate of the Transformation Deficit (TD).
(d) Control of the dynamics of undisturbed areas of strategic importance called potential core areas (PCA).

7.4 Diagnostic Principles and Methods

7.4.1 Metastability and Information Content of a Landscape Unit

The level of *allostatic* capacity of a Landscape Unit (LU) is proportional to its respiration (dysmetabolic process). It can be expressed as an energy flow that the LU must dissipate to maintain its conditions of order and metastability. As we underlined, the estimate of this energy flow is possible using the function of the biological-territorial capacity of the vegetation or BTC [Mcal/m^2/year] (see Sect. 3.4.2). The BTC is therefore linked to the concept of metastability, and Bionomics allows the BTC of an entire LU to be estimated. This value is significant, especially if measured from a historical perspective. Still, it does not directly express the level of metastability reached by the landscape unit in question because it does not consider the distribution of the diversity of elements in the LU nor, therefore, the distribution of this flow between said elements nor, finally, the amount of information linked to this composition. To this end, please refer to Table 3.5, which specifies the values of the BTC classes.

Note that information is *more* than the physical dimension expressible with the well-known *H* (*heterogeneity*) formula of Shannon & Weaver (Shannon & Weaver, 1949). In fact, in the distribution of elements that make up a LU, H_{max} is the equiprobable one, but, for the biological organization of a LU, *H* is of interest if compared to the dominance D too, which is its opposite (Fig. 7.3). We must,

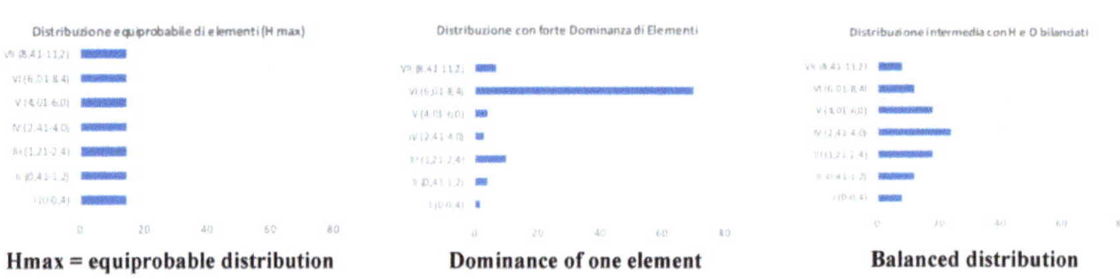

Fig. 7.3 Main types of BTC class distribution. Note that information is more than a physical dimension expressible with the well-known H formula (Shannon & Weaver). In fact, in the distribution of elements that make up a LU, H_{max} is the equiprobable one, but for the biological organization of a LU not only H is of interest if it is not also related to the dominance D

therefore, use a parameter capable of measuring both H and D, such as the synthetic landscape diversity τ (Ingegnoli, 2011):

$$\tau = H(3 + D),$$

evaluating H and D on the distribution of BTC classes in which the LU elements are included.

Recall that the concept of *negentropy* allowed Shannon and Weaver to use the Boltzmann entropy function as a measure of information. The information content of a LU can thus be expressed by the heterogeneity index H (average information) measured in a set n of habitats (Naveh & Lieberman, 1984):

$$H = -\sum_{k=1}^{n} Pk \ \log_2 Pk$$

where P is the distribution *probability*, i.e. the probability of each class of BTC detectable in the LU under examination (i.e. its real presence):

$$Pk = \sum BTC_k \ (\text{where} : BTC_k = BTC \text{ class } k)$$

If we define $\tau x \ BTC = gLM$, for negentropy the *information content* can be expressed by the heterogeneity index H, but also considering the diversity D:

$$D = log_2 n + \sum_{k=1}^{n} Pk \ log_2 Pk$$

We can then write the equation of the general metastability of an LU; this is a very significant indicator, see Fig. 7.4, capable of considering the information content of the LU in relation to the flow of energy necessary for maintaining the state of order of the LU itself:

$$gLM = [H(3 + D)] \times BTC \ \ [bit \times Mcal/m^2/yr]$$

The thin curves in the figure represent correlations with a partially impervious/barren territory within the LU. The vertical bars distinguish the main types of landscapes, from natural forest (left) to dense urban (right). The curves are polynomial, and the red one has the equation:

$$g\text{-}LM = 0,0001x^3 – 0,0158x^2 + 0,0738x + 53,5 \ \ [bit \times Mcal/m^2/yr]$$

It is interesting to note that the maximum value of g-LM does not correspond to an absence of HH, but to a $HH = 3.0\%$ and that, at the limit between the natural forest landscape and the semi-natural one (HH between 5% and 15%), g-LM varies slightly (from 54 to 52, with a decrease of <2%).

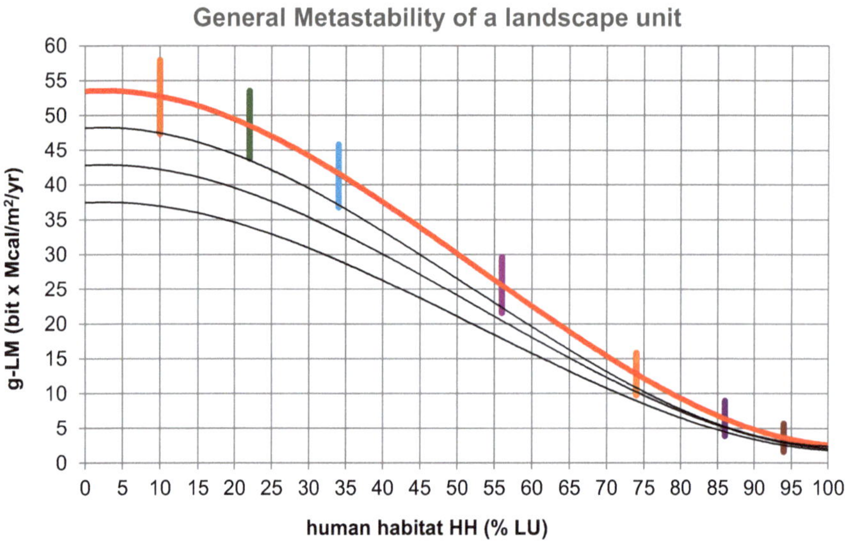

Fig. 7.4 General metastability correlation vs. HH, based on a set of studied examples of LUs and their main bionomic parameters. The vertical bars divide the landscape types. From the left: Natural forestal landscape, Seminatural forestal L., Agricultural forestal l., Agricultural Protective L., Agricultural productive L., Suburban L., Urban L., Dense Urban L

It is necessary to summarize an example of the application of the bionomic indicator g-LM evaluated on the Staffora-Foce LU (32.5 Km^2) in 2020 (see Chap.14). As it can be seen in Fig. 7.4, for a HH = 80%, you should have a g-LM = 9.00; therefore, the value of g-LM = 3.03 bit × Mcal/m^2/year (Fig. 7.5) is decidedly low!

In Fig. 7.6, a comparison is made between heterogeneity and information in 1780, 2020, and 2050 in the LU, already mentioned, north of Voghera. The target to be achieved in 2050 should be at least about 4.0 (3.95 in this Fig.). Note that today, certain classes of BTC, therefore certain types of landscape elements, are reduced so much as to be almost negligible. However, today (2020), g-LM = 3.03 vs. 6.75 in 1780: the alteration is impressive.

7.4.2 The Bionomic Functionality (BF): HH/BTC Diagnostic Model, Local to Province Scale

This method is derived from the study of 45 landscape units (LU), according to the principles of Bionomy, in Northern Italy, a study that showed an exceptionally high correlation between BTC and human habitat (HH). This correlation is very significant, as it indicates a relationship between the flow of energy necessary for a living system to reach and maintain its level of organization and structure (BTC) and the

LU Staffora Foce 2020		Sup. ha	%	HH	HH	BTC	BTC
Urbano		135,70	4,18	98	4,097	0,15	0,006
Cascine		14,10	0,43	97	0,421	0,20	0,001
	RSD	149,80	4,61				
Strade		20,80	0,64	99	0,634	0,02	0,000
Industrie		29,20	0,90	99	0,891	0,05	0,000
	SBS	50,00	1,54				
Giardini (15% URB)		24,00	0,74	90	0,665	1,70	0,013
Filari Staffora e campi		9,90	0,30	50	0,152	2,40	0,007
Filari golena PO		6,60	0,20	40	0,081	3,80	0,008
	PRT	40,50	1,25				
Seminativi		1.908,40	58,79	89	52,322	0,75	0,441
Prati e foraggi		828,00	25,51	75	19,130	0,55	0,140
Vigneti		6,00	0,18	80	0,148	1,45	0,003
Pioppeti		14,00	0,43	65	0,280	2,70	0,012
	PRD	2.756,40	84,91				
Arbusteti golena		12,40	0,38	20	0,076	2,50	0,010
Prati permanenti		25,00	0,77	50	0,385	0,60	0,005
	RSL	37,40	1,15				
Salici/Pioppi Pancarana		79,10	2,44	12	0,292	6,40	0,156
Boschi Staffora-Po		22,00	0,68	8	0,054	5,50	0,037
Boschi nella campagna		18,60	0,57	15	0,086	4,00	0,023
	RNT	119,70	3,69				
Po		82,30	2,54	10	0,254	0,05	0,001
Staffora		7,60	0,23	15	0,035	0,05	0,000
Altre acque		2,50	0,08	30	0,023	0,02	0,000
	HGL	92,40	2,85				
Totale		3.246,20	100		80,03		0,86

Tab.6.7. g.LM: esempio di applicazione		
LU Staffora-Foce 2020		
BTC class	P (elem)	P*log₂ P
I	0.0900	-0.313
II	0.8507	-0.198
III	0.0122	-0.078
IV	0.0158	-0.095
V	0.0068	-0.049
VI	0.0244	-0.131
VII	0.0001	-0.001
H	Heterogeneity	-0.864
D	Dominance	1.943
tau	Functional Dive	**4.272**
BTC		0.86
g-LM	g.LM [bit x Mca	3.67

Fig. 7.5 Application of the g-LM eco-bionomic indicator evaluated on the Staffora-Foce LU in 2020. The amount of information on the state of the complex system under examination (see photo) is really very limited: g-LM = 3.67 bit × Mcal/m^2/year

measure of anthropic control, i.e. the limit to the capacity for self-regulation of a natural system (NH). Note that this correlation has an $R^2 = 0.95$ and a Pearson coefficient = 0.910 (i.e. 3.14 times the significance level for this statistical population).

Figure 7.7 shows an HH/BTC model based on the above. It is a bionomic model of normality capable of evaluating any dysfunctions of a landscape or part of it (LU). The green line represents the normal condition of $BTC = f(HH)$ and is described by the polynomial equation:

$$BTC = 0,0007 \times HH^2 – 0,1518 \times HH, + 8,85 \quad \left[Mcal/m^2/year\right]$$

The bundle of dashed curves indicates four intervals of

$$BF = BTC_{RIL}/BTC_{NORM}$$

where: RIL = surveyed, and NORM = normal.

(detected concerning normality) which can be summarized as follows: the first (BF > 0.85) indicates the maximum tolerance concerning normality, the second (BF = 0.85–0.65) represents a first degree of alteration, the third (BF = 0.65–0.45) concerns a greater level of dysfunction, the fourth (BF < 0.45) indicates a complete degradation. The vertical bars distinguish the major landscape types in temperate

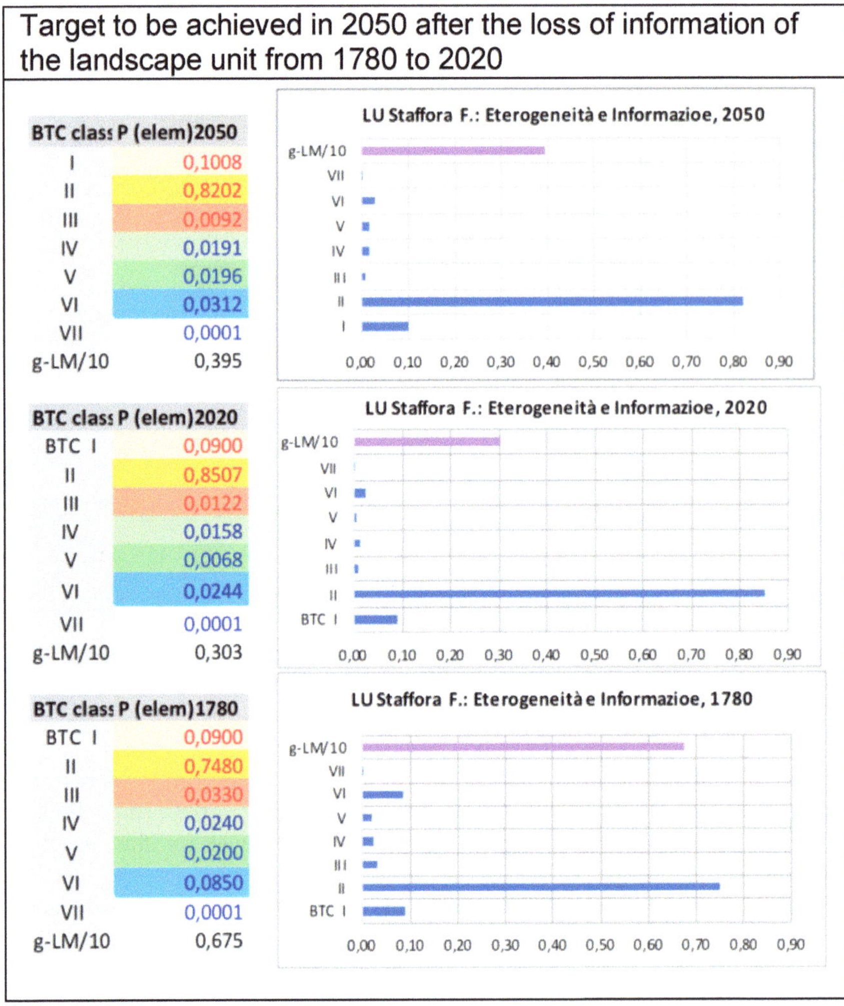

Fig. 7.6 Comparison between heterogeneity and information in 1780, 2020 and 2050. Note that today certain classes of BTC, therefore certain types of landscape elements, are reduced so much as to be almost negligible

zones through the presence of HH: from the natural forest landscape to the dense urbanized one.

This diagnostic model is essential to arrive at a preliminary overall evaluation of a landscape unit with accuracy and consequently to check both a state of fact and a land planning project. If it is possible to evaluate even the most significant past states of a LU, the movements of the system under examination can be marked on this model since they were traced on the system's state plane.

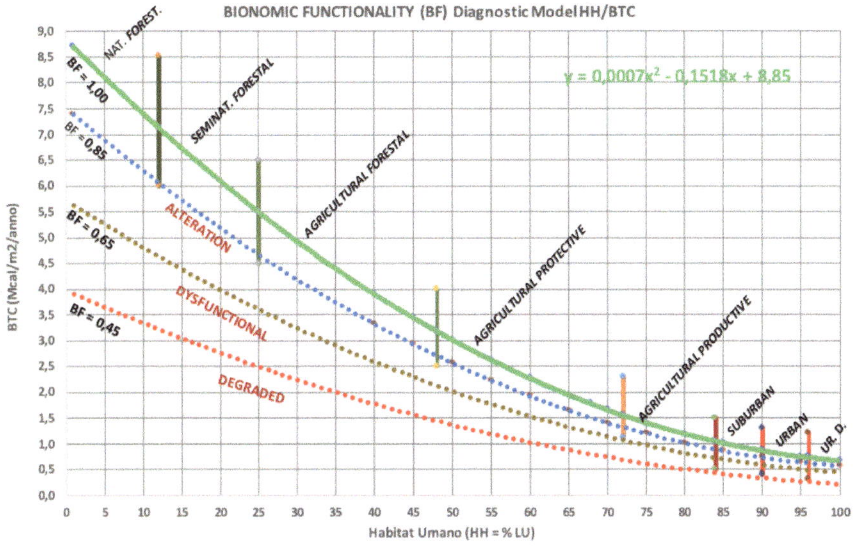

Fig. 7.7 The HH/BTC diagnostic model, capable of measuring the bionomic state of a landscape unit. The dotted lines separate the main levels of bionomic functionality (BF). The green curve represents the normal state of the function $BTC_{NORM} = f$ (HH): so, the value of BTC_{NORM} is gathered as the 'y' value corresponding to the HH value on the 'x' axis

Finally, the HH/BTC diagnostic model is not linear, as both variables are systemic.

7.4.3 The Bionomic Functionality (BF): HH/BTC Diagnostic Model, Regional to National Scale

As it is known, for bionomics it is important to follow the processes for space-time scale intervals. This leads us to remember that the previous HH/BTC diagnostic model (Fig. 7.7) is valid for landscape units (simple or complex) measuring from 5 to 2500 Km^2. Both in theory and in applications it may be necessary to refer to the regional (or regional-national) scale, i.e. for scales from 10^4 to 10^6 km^2. The experimental data processed on European regions/nations (from 1.2 to $54.0 \times 10^4\,km^2$), then improved by 5% to discard the worst cases, lead to a trend line with the equation:

$$BTC = -0,0362 \times HH + 4,50 \quad \left[Mcal/m^2/year\right]$$

As can be seen from Fig. 7.8, the correlation line is a straight line divisible into six intervals of which the most significant are nations characterized by:

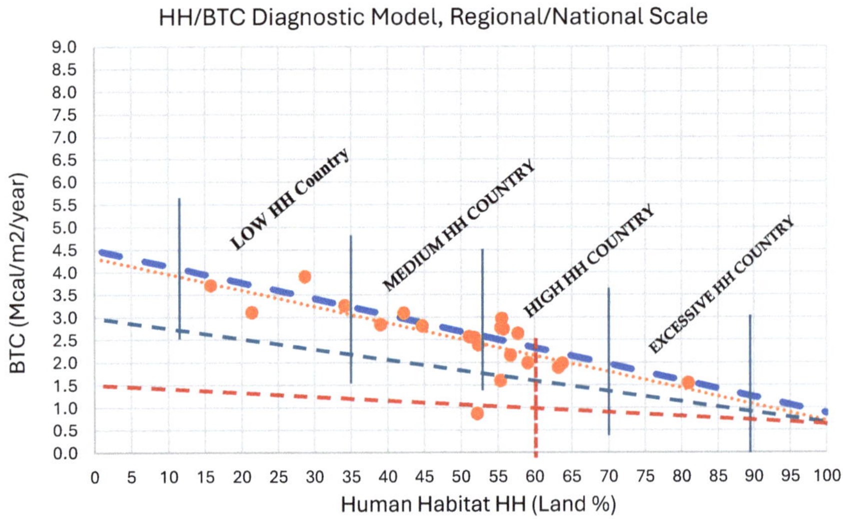

Fig. 7.8 HH/BTC correlation line at a regional scale. Orange dots represent the estimated values of some european States. Note that the hyper-anthropized range of HH is to be considered predatory for a nation's natural resources. An alteration is present for values inferior to grey dashed line; dysfunction is (**a**) inferior to red dashed line. (**b**) Note that dysfunctional conditions arise also when HH > 60%, independently from the medium BTC value of the country (see vertical dashed red line)

1. Semi-natural HH (HH = 16–35), e.g., Finland, Austria
2. Slightly high HH (HH = 35–55), e.g., Spain, Italy
3. High HH (HH = 55–70), e.g., Belgium, Hungary (altered)
4. Hyper-elevated HH (HH = 70–90), e.g., Denmark (dysfunctional)

7.4.4 Diagnostic Index (DI) of a Landscape Unit (LU)

This diagnostic method can achieve greater depth than the previous one because it uses more eco-bionomic parameters, up to 15–20. In the case of a full-blown pathology of the LU under examination, we often find a slight worsening of the state of the environmental health of a landscape unit (LU) compared to the previous method: this worsening is around 10–15%.

It should be noted that this method requires to establish precisely the type of landscape which the LU belongs to, since the calculation of the distance between the values of the parameters detected and the respective normal intervals depends on the type of landscape in question. In fact, the normal intervals can be quantified correctly only for this type. For this reason, a specific indicator called LtpE can be used (Ingegnoli, 2015).

The evaluation of the *Landscape Type Evaluation* (LtpE), an indicator capable of classifying landscape types in relation to HH as an independent variable, is based on two parts:

(a) *NHd/HH*, where: *NHd = NH—[NH × (SH/SH*deficit)]* and SH/SH*deficit is the amount of heterotrophy of a landscape when SH/SH* < 1.
(b) *BTC × (A + F)* = to the product of BTC due to the presence of the most relevant components: *agricultural (A) and forestry (F) in %* of productive (PRD) and resistant (RNT) elements /apparatuses. It results:

$$LtpE = 10[(NHd/HH) + (BTC \times (A + F)]$$

Table 7.2 shows the scale of LtpE index values, in relation to the main types of landscapes in temperate regions. These values are derived from an exponential curve, from 1.68 to 335.

After having identified the type of landscape under examination, reference is made to Table 7.3, which specifies the *normal ranges* of the most important parameters by type of landscape, derived by field studies. These intervals can also undergo local variations, so it is a good idea to check for any adaptations, usually partial.

After having quantified the distance from the norm of each variable, a score varying from 0 to 2 is added depending on the percentage of the deviation:

$$[0 - 10]\% = \mathbf{2}; \quad [10 - 30]\% = \mathbf{1}; \quad [30 - 60]\% = \mathbf{0,5}; \quad [>60]\% = \mathbf{0}$$

The sum of the deviations compared to the optimal total (in Fig. 7.9, the maximum score is 16 × 2 = 32) gives the value of the diagnostic index (DI), which in the example is equal to 0.76: this value indicates an apparent alteration, given that the tolerance must not be less than 0.85. We can also evaluate this structure for three different groups of parameters: (a) of structure, (b) of vegetation, and (c) of other functions; the average is, as we have already seen, equal to 0.76, but the variables structural are better (0.85) while the 'other functions' are worse (0.67). Regarding

Table 7.2 Range of LtpE related to the main landscape types

Landscape type	Min	Max	*mean value*
Dense Urban	0,30	3,00	1,68
Rare Urban	2,70	6,30	4,50
SB-Urban Rural	5,70	11,00	8,35
Agricultural	10,20	24,20	17,20
AGR-Protective	22,00	40,00	31,00
AGR-Forestal-Tour	34,00	55,00	44,50
AGR-Forestal	45,00	88,00	66,50
FOR-Tour	65,00	140,00	102,50
Forestal Semi.Natural	100,00	245,00	172,50
Forestal Natural	170,00	500,00	335,00

Table 7.3 Normal ranges by landscape type in the temperate belts of the Northern Hemisphere

Bionomic parameters	URB	SUB.T	SUB.R	AGR	AGR. PRT	Agr-Syl-Tour	AGR. FOR	For-Tour	FOR.SN	FOR. NAT
RNT, forest area (%)	4.0–8.0	5.0–12.0	8.0–25	10.0–30	20–55	25–65	40–75	45–80	60–85	65–95
CBSt of forest	16.5–22.5	18.5–25.5	22.5–30.5	23.5–32.0	27.5–37.5	28–38	29.5–40.5	30–41	33.5–45.5	> 39.5
Allocthonous forest formations (%)	2–5.0	1–4.0	1–2.0	1–2.0	0–1.5	0–1	0–1	< 0.5	0	0
pCA. potential core area (%)	5–10.0	10.0–20	20–35	35–50	50–65	45–70	60–75	65–80	75–85	> 85
CON, connection ($\alpha + \gamma$)	0.50–0.65	0.55–0.70	0.60–0.75	0.65–0.85	0.70–0.90	0.65–0.95	0.75–1.00	0.7–1.0	0.8–1.1	0.8–1.1
Network efficiency	0.6–0.8	0.6–0.8	0.7–0.9	0.8–1.2	1.0–1.8	1.1–2.0	1.5–2.0	1.2–2.1	1.6–2.2	> 1.8
PRD, agricultural areas (%)	5.0–15	10.0–35	30–55	45–75	35–60	15–40	20–50	10.0–25.0	5.0–20	3.0–6.0
RSD, urbanized areas (%)	25–50	10.0–25	8.0–15	4–9.0	2–6.0	2–7.0	1–3.0	0.5–2.0	< 1.5	< 1
SBS, industrial & transports (%)	10.0–20	15–50	7.0–12	4–7.0	2–4.0	1.5–3.0	1–2.0	< 1	< 1	< 0.5
HH, human habitat (%)	85–95	80–92	70–86	60–82	42–68	20–40	22–48	10.0–25	5.0–20	0–12
HS/HS* carrying capacity	0.1–0.5	0.4–1.0	0.6–3.0	1.5–9.0	3.0–12.0	1.2–12.0	3.5–15	1.5–15	4.0–18	> 4.5
BTC, LU ($Mcal/m^2/a$)	0.40–0.90	0.60–1.20	0.80–1.60	1.00–2.00	1.70–3.50	1.8–4.5	3.00–5.00	3.5–6.0	4.50–7.00	6.0–8.5
CBSt, of entire territory	4.0–12.0	6.0–16	8.0–20	10.0–24	12.0–30	14–34	16–38	18–42	20–46	> 45
g-LM, gen. Landscape metastability	3.5–6.0	5.5–8.0	7.5–15	14–24	20–30	27–37	32–43	40–47	48–52	52–55
LtpE = 10 (HNd/HU) + BTC_{A+F}	0.35–3.0	2.7–6.3	5.7–11.0	10.2–24	22–40	34–55	45–88	65–140	100–245	> 170
IFF, river functionality	120–180	120–180	180–200	180–200	200–250	200–270	250–270	250–300	261–300	261–300
Low noise areas (< 60 dB) %	40–50	35—45	45–55	50–60	55–70	70–80	65–75	75–85	80–90	> 90

Landscape types: URB = urban, SUB-T = suburban technologic, SUB-R = suburban rural, AGR = agricultural, AGR-PRT = protective agricultural

DI – BOLLATE	Bionomic probe spectrum	Suburban-RUR	Survey 2012	gap	score	
	POP/1000 (inhabitants)		35,75			
structure	LtpE (L. Type)	5,7-11	5,71	ok	2	
structure	Urbanized area	15-27	39,84	47,4	0,5	
structure	Gardens & Sport	3,5-7	8,84	ok	2	
structure	Agricultural components	30-55	38,2	ok	2	
structure	Potential core areas	20-35	21,4	ok	2	
vegetation-funct.	Forest & Heath	8,0-25	12,27	ok	2	
vegetation-funct.	Vegetation capacity (BTC)	0,8-1,6	1,03	ok	2	
vegetation-funct.	Photosyntetic area	60-70	57,2	4	2	
vegetation-funct.	Territorial CBSt	8,0-20	5,8	27,5	1	
vegetation-funct.	Forest & shrub CBSt	22,5-30,5	14,86	34	0,5	
Other functions	Human habitat HH (%)	70-86	79,84	ok	2	
Other functions	Standard habitat SH	800-1650	293	63,4	0	
Other functions	Carrying Capacity (σ)	0,6-3	0,2	65	0	
Other functions	Bionomics Functionality (BF)	0.85-1.15	0.86	ok	2	
Other functions	Rehabilitation potential	15-25	20,66	ok	2	
Other functions	L. Metastability (g-LM)	7,5-15	7,52	ok	2	
	Diagnostic Index (DI)	0.85-1.00			0,76	
	Bionomic state evaluation	normal			altered	
	Score ratio		0,85	0,75	0,67	0,76
			structure	veg-function	oth-function	mean

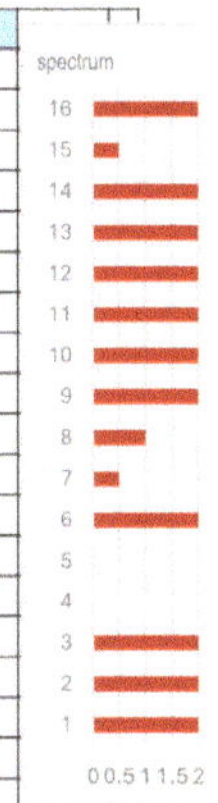

Fig. 7.9 Example of parametric diagnostic evaluation of a suburban municipality north of Milan. It is noted that every parameter outside the norm is not very significant but acquires a value in the whole spectrum. The Suburban-Rural column shows the normality range of values for each parameter

assessment in a park area, we can reduce the reference scores proportionally, by multiplying the results by a *safety coefficient* of 0.85–0.90.

In the case study of an LU, this parametric evaluation depends on the comparison of the data collected with a spectrum of typical values (by landscape type): the presence of gaps (distances from normality in (%) is estimated as expressed above for the deviations (%).The sum of these weights per parameter related to the same sum in case of full normality gives the value of DI. It should be remembered that the entire *profile of the spectrum* of parameters (diagnostic picture) acquires a clinical significance because it reveals a correlation structure between different variables (Table 7.3). In Table 7.4, the levels of alteration of the DI, with physio-pathological notes.

7.4.5 Diagnostic Index (DI) of a Set of Regions or Countries

In a larger scale case study, comparing a group of regions or countries with 10–15 bionomic-ecological landscape parameters (natural and social), the evaluation can be carried out by dividing each parameter into 5 classes (of the same dimension) with the following weight (colors as for traffic light): 10 (green), 8 (yellow), 6 (orange), 4 (brown), 2 (purple), see Fig. 7.10.

The sum of these weights (DI_{SUM}) must be related to the optimum, intended as 80% of the maximum parametric sum; if the parameters are 12, DI = DI_{SUM} / 96 (%).

Table 7.4 Levels of diagnostic evaluation of the state of a landscape unit

Pathology levels	Diagnostic index (DI)	Diagnostic evaluation	Physiological-pathological notes	Ecological health & interventions
I	0.85–1.00	Normal	Homeostatic plateau	Quite good health, only prevention
II	0.65–0.85	Alteration	Compensation needed	Instable health, some therapies & intervention
III	0.40–0.65	Dysfunction	Some physiological damages	Dysfunction, wide intervention needed
IV	0.10–0.40	Severe dysfunction	Harmful effects	High dysfunctions, difficult intervention
V	<0.10	Extinction	Irreversible damages	Degenerative transformations

weight	2	4	6	8	10									
Nation	BTC	RNT (%)	HH	BF	Org. Farm	Urb.Pop	GDP €/cap	RSL(%)	PRD(%)	Red Meat	Pest	Glyph	DI	ancer ons
AUSTRIA	3,45	32,82	39,8	1,08	25,30	58,30	38.170	32,50	23,70	19,70	3,30	0,16	89,58	4,88
ROMANIA	2,65	28,99	53,68	1,08	2,90	54,00	9.110	27,60	35,50	8	0,8	0,14	87,50	4,91
SWEDEN	4,51	66,82	18,74	1,04	20,40	87,4	43.900	10,70	9,00	22	0,6	0,19	85,42	5,57
POLAND	2,98	31,70	57,98	1,34	3,50	60,10	13.000	8,00	52,00	4,5	2,2	0,43	83,33	5,36
FINLAND	4,13	65,80	16,37	0,92	13,50	85,40	37.170	10,00	9,00	17	0,6	0,31	81,25	5,58
SPAIN	2,99	17,47	43,36	0,99	9,70	80,30	25.200	40,90	36,50	8,6	3,6	0,15	81,25	5,40
PORTUGAL	3,12	35,60	44,31	1,05	8,20	65,20	18.590	26,00	31,00	16	5,4	0,35	79,17	5,10
GREECE	2,92	29,84	43,29	0,99	10,30	79,60	17.750	28,50	34,50	16	2,9	0,25	77,08	5,26
ITALY	2,93	30,07	53,34	1,19	15,70	70,40	26.920	14,23	46,60	15,7	6,1	0,30	77,08	5,49
GERMANY	2,81	30,80	56,36	1,22	9,00	77,30	35.840	15,30	41,20	9	4	0,27	77,08	5,70
HUNGARY	2,05	21,60	64,55	1,13	5,70	71,35	13.260	13,50	56,00	2,4	2,2	0,31	75,00	6,23
FRANCE	2,59	26,35	53,93	1,07	7,70	80,40	33.270	24,50	38,50	23	3,6	0,32	66,67	6,21
UK	1,52	12,70	60,45	0,74	2,60	83,40	32.910	41,00	30,50	17,9	3,2	0,11	64,58	5,90
DENMARK	1,52	14,50	72,17	1,05	10,90	87,90	49.720	10,00	63,00	18,4	1,1	0,51	58,33	6,88
BELGIUM	1,61	22,75	67,28	0,97	6,90	98,00	35.940	15,00	50,00	15	6,7	0,30	56,25	6,40
IRELAND	0,98	10,50	53,33	0,42	1,60	63,20	60.170	63,00	8,00	35,8	6,5	0,14	50,00	7,18
NEDERLAND	1,17	8,60	67,26	0,69	3,70	91,50	41.870	34,00	37,00	14	7,9	0,39	45,83	6,55
w-average	2,97	33,43	45,91	1,05	10,01	76,40	30,00	27,16	33,02	15,00	3,00	0,25	72,67	5,65

Fig. 7.10 A set of 17 EU countries are evaluated with 12 parameters, dividing each column in five classes which acquire the weighted values (first row) DI_{SUM}. It must be related to the optimum condition established as 80% of the max weight of each parameter, here 0.8 (12 × 10) = 96. So, $DI = DI_{SUM} / 96$ (e.g., Austria = 86 / 96 = 89.58). DI is inversely related to Cancer onset (last two columns). See also Fig. 10.15. Org.Farm = organic farming; Urb.Pop. = Urban Population; Pest = use of pesticides; Glyph = use of glyphosate

DI can be compared with important data related with the set of regions or countries to measure the correlation with the evaluated environment. In Fig. 7.10, a set of 17 EU countries are evaluated with 12 environmental parameters dividing each column in five classes which acquire the weighted values (first row) DI_{SUM}. So, $DI = DI_{SUM} / 96$ (e.g., Austria = 86 / 96 = 89.58). DI is inversely related to Cancer onset (last two columns). See also Fig. 10.15.

7.4.6 Transformation Deficit (TD)

It should be noted that the basic elements of the human habitat (HH), i.e. agricultural and urbanized areas, require a transformation of the natural or semi-natural environment such as to produce a transformation deficit (TD), i.e. *a loss of the degree of natural organization and related self-rebalancing capacity of the concerned landscape or LU*. This loss must necessarily be compensated by a congruent part of the protective apparatus, therefore by high BTC PRT systems, under penalty of environmental degradation and consequences on human health.

The most significant evaluation of the TD is performed by considering the BTC of the two main autotrophic subsystems of the Natural Habitat and the Human Habitat, respectively, i.e., the forestry and the agricultural ones. As an example, the evaluation of the TD for the Lombardy Region (Fig. 7.11) is reported. The ratio between the BTC of forests (BTC_{FOR}) and the BTC of the Region (BTC_{REG}) and that between the agricultural landscapes (BTC_{AGR}) and the same BTC_{REG} in the most favorable period compared to the current one, which appears to be 1878, is therefore evaluated.

In 1878, the total energy flow of forest and agricultural landscapes was 97% of the region's total BTC; the light blue dotted line, sum of the two green and yellow dotted lines, represents the BTC of the two types of forest and agricultural landscapes. From a $BTC_{FOR} + BTC_{AGR} = BTC_{TRASF}$, which in 1878 reached 97% of the regional BTC, today it only reaches 85%. Consequently, the area between BTC_{TRASF} and BTC_{878} (flux tolerance) represents the TD.

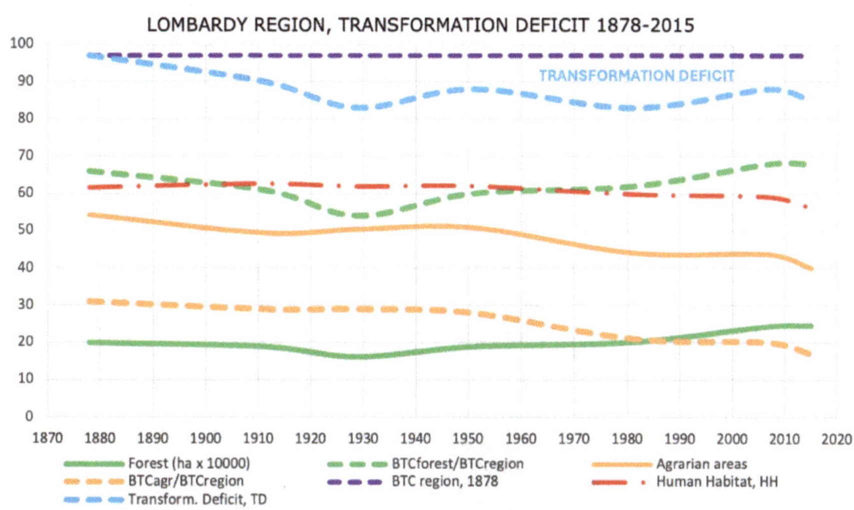

Fig. 7.11 The evaluation of the transformation deficit of the Lombardy Region in the period 1878–2015. Note that despite recent forest regrowth as 'land cover' (not as bionomic efficiency!), agricultural landscapes have decreased and degraded, so the TD is evident (light blue), today 85% of the BTC1878, sum of dotted lines (BTCfor/BTCregion + BTCagr/BTCregion)

TD = − 9.04% of the optimal total (rectangle 97% × 137 years), therefore the energy loss, being the BTC_{878} = 2.08 Mcal/m^2/year, is 6.133 × 10^{14} kcal. For a Region with a protected territory equal to almost 23%, this is a significant loss of protective apparatus PRT, not as a surface extension (as it could be considered by the reductionist approach), but as meaning and capacity.

This value of Transformation Deficit could correspond to an alteration of 0.6–0.7% per year of the natural and semi-natural components (natural infrastructures) of agricultural landscapes. This means that approximately 89% of agricultural landscapes have altered significantly in the considered 137 years.

7.4.7 *Estimation of the Ecological Radius of Re-balancing Influence*

The local detected BTC can be balanced by an element of pleasing naturalness: so, the distance of influence from it is fundamental, both in a diagnostic and therapeutic sense. It is called ER (eco-bionomic range) or radius of ecological influence (Fig. 7.12) (Ingegnoli, 2011): mathematically, it is equal to

$$ER = [BTC(+) \times Wc/2]/BTC(-)$$

where Wc is the width of the corridor or section of the patch, and BTC (+) indicates the BTC surplus while BTC (−) indicates the deficit of BTC.

The distance of microclimatic influence is equal to 6–7 times the average height of the corridor or patch, which must therefore be measured in the field, and is also a function of the porosity of the corridor or the density of the patch, as reported in Forman (Forman & Godron, 1986). If the presence of disturbing elements (country roads, power lines, buildings, etc.) that cross the matrix in proximity to the corridor or the scrub occurs, a reduction in influence can be estimated by counting the number

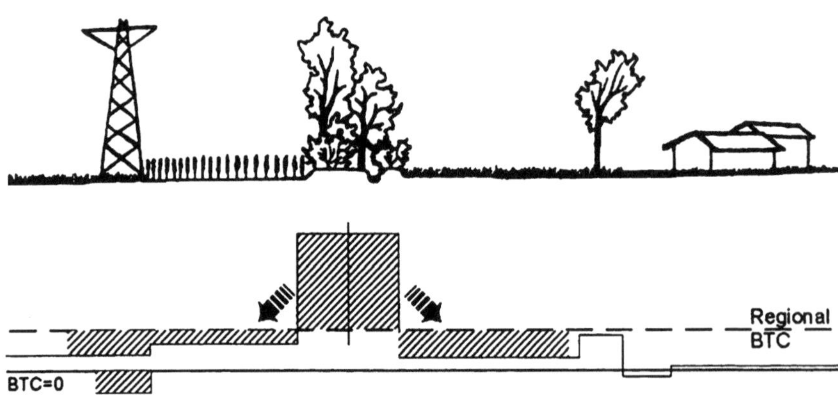

Fig. 7.12 Example of the sizing of the field of ecological influence of a tree-lined corridor in a rural landscape (from: Ingegnoli, 2002). The arrows indicate how far the surplus of BTC pertaining to the tree corridor can reach to fill the deficit (lateral shaded areas) compared to the regional average level

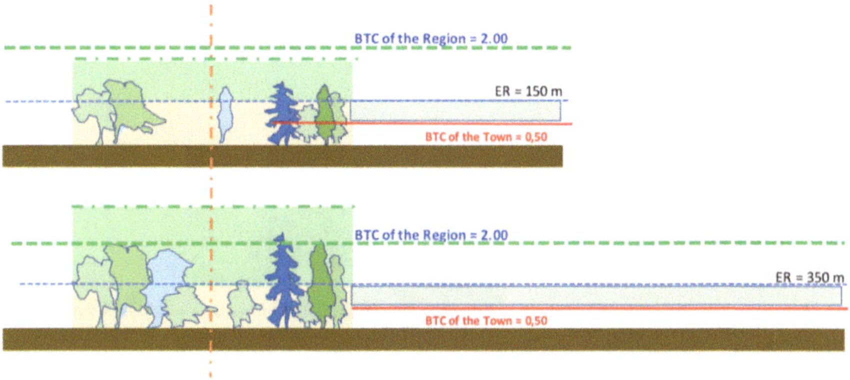

Fig. 7.13 Example of ecological influence: two 200 m wide urban parks with respectively BTC = 1.75 and 2.75 Mcal/m^2/year. Note that, dealing with urban parks, the ecological influence radius (ER) must guarantee the doubling of the surrounding urban BTC. Ecological rays have a lower ratio than that of BTC (0.428 Vs. 0.636)

of such disturbances if they affect at least half of the element itself, i.e. they are not just punctual.

In the example of Fig. 7.13, referring to a theoretical study for the city of Milan, CityLife park, it can be noted that given two urban parks of equal size (200 m), the first with BTC = 1.75 Mcal/m^2/year, the second with BTC = 2.75 Mcal/m^2/a, they produce a doubling effect on the surrounding urban BTC (= 0.50) respectively with ER = 150 Vs. ER = 350. Note that the BTC ratios are not proportional to those of ER (0.636 vs. 0.428).

7.5 The Anamnestic Role of the Drawing

7.5.1 Natural Harmony Surveys

It is said that art is a subjective expression, therefore it would be incompatible with a scientific preparation. Even if we wanted to forget the inseparable union of these two aspects in figures such as Leonardo da Vinci (Da Vinci, 1500) or Alexander von Humboldt (Von Humboldt, 1840) or J.W. von Goethe (Von Goethe, 1787) (Fig. 7.14) however, it is still Lorenz who states that the process of knowledge and the object of knowledge cannot be legitimately separated from each other; furthermore, it is simply false that subjective experience concerns exclusively the private sphere of the individual. Max Hartmann (Hartmann, 1964) defined the relationship between physiological processes and internal processes as illogical, but the correlation between them is so certain that a subjective phenomenon, for example the perception of a complementary color, can be used, in the phenomenon of contrast, as a reliable indicator of the occurrence or otherwise of a parallel physiological pathology.

Il Vesuvio, J.W. von Goethe, A.D. 1787

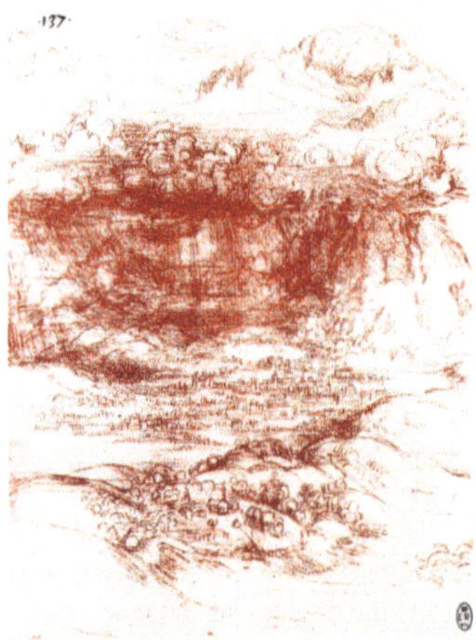

Storm on a pre-alpine landscape, Leonardo, about A.D. 1500.

Fig. 7.14 Drawings by Leonardo da Vinci and J. Wolfgang von Goethe. A landscape drawn by Leonardo da Vinci (about A.D. 1500) representing a storm in a pre-alpine landscape. While drawing, Leonardo began asking why the elements of the landscape had been shaped in that way and what had been the reasons for their development. Not dissimilar the vision of Goethe: the interest of painters was therefore not only aesthetic

Indeed, the overcoming of scientism (see Chap. 1) allows us to understand that the more rigorously human knowledge is defined as expressible on a verbal level, the clearer it becomes that 'many primary phenomena cannot be expressed at all through words', or, if you prefer, by means of rational logic. The good, the beautiful, the sensitivity for harmonies, the meaning of life or the ultimate truth, certainly exist, but they are unspeakable. Yet we can understand and express even the unspeakable, as every true artist, every poet knows. This is also important in nature, as in the case of beauty.

Beauty exists in nature, it manifests itself in many aspects and is also used by living systems as an attraction towards less immediate integration purposes or, better yet, more complex ones. For example, it is known that flowers are beautiful and colorful when they must attract pollinators, but if the pollen is spread by the wind this beauty is missing. The hyper complexity of large-scale living systems, such as the landscape, is striking for the obvious and hidden harmonies underlying it (Fig. 7.15), and it affects not only humans, as Jane Goodall (Goodall, 2000) demonstrated by studying the emotions of chimpanzees when faced with perception of landscapes of exceptional beauty.

It is Lorenz, again, who writes that science is today able to show that matter and energy, corpuscular radiation and electromagnetic waves, physiological events, and emotional events, are part of the same real process, of which we undoubtedly have experience, but through two immeasurable ways of knowing. Yet there is a principle accessible to both ways of knowing because it is the basis of knowledge itself: the truth (see also Chap. 15). Overcoming the contrasts between rationality and internal experience is possible through truth, which is one, unique, invariant: it provided that,

Fig. 7.15 Examples of an Author's drawings. Recent colonization of larches at Lago Bleu (Val d'Ayas, Aosta Valley) altitude 2200: comparison between a watercolor and a photo, highlighting how drawing allows us to put in evidence the aspects on which the author wants to focus based on his purpose, including scientific ones

as Wolfgang Pauli (Pauli, 1959) said, the state of an object is not considered independent of the way in which the possible sources of a certain information relating to the object are altered by observation.

7.5.2 The Clinical Eye of Both Art and Science

Speaking of epistemology, there is a precise path to approach the truth and this path is a correspondence of harmonies that allows the understanding of the multiple in unity (think of a musical orchestra). It is yet another aspect of the indistinguishable combination of truth/non-violence as it will be see (Chap. 15). Art and science have in common the same search for unity in variety. The beauty and emotion inherent in a creative act reveal, in a poem as in a theorem or in a symphony, that essential correspondence that allows the generation of thought itself and is identified with it. "The most beautiful sensation is the mysterious side of life—writes Albert Einstein (Einstein, 1965)—it is the profound feeling that is always found in the cradle of art and pure science".

Art and science are not opposed but are complementary activities and as such should not be radically separated by those who dedicate themselves to research. Indeed, both must be cultivated, because only in this way can knowledge be deepened.

Drawing can overcome the obligation of logical sequence of writing, and this fact is important to understand what Lorenz said about the painters' method in studying nature. This reminder is not only important for the method, but also for the development of sensitivity to differences in value among non-rational processes. The sensitivity for "harmonies", i.e. for the processes of integration of hypercomplex systems, depends largely on the ability to 'perceive shapes', a particular type of organization of our sense organs and our brain structures, also linked to the intuition (Fig. 7.16).

Fig. 7.16 (Left) The town of Glorenza at the foot of its alluvial fan in Val Venosta (Sud Tirol); (right) a view of the imponent fortified hill in the heart of Edinburgh like a modern acropolis, pen drawings of the Author (2018)

It has long been known that one of the most important functions of shape perception is to allow us to distinguish what is healthy from what is sick. "Which is indispensable—Lorenz wrote—for doctors, animal researchers and landscape ecologists" (Lorenz, 1978). Indispensable, because the 'clinical eye' can never be replaced by an accumulation of quantitative data, even if computerized, due to the non-expressibility, in a rational sense, of many essential functions of a biological system.

7.6 Mend the Fragmentation: The Strategic Role of Undisturbed Areas

7.6.1 Undisturbed Areas of Strategic Importance

Crucial is the following observation: the structuring of suburban landscapes, due to the transformation of agricultural landscapes because of continued fragmentation and alteration by humans, usually leaves a mosaic of remnant patches or residual spots (Forman & Godron, 1986) of different magnitude and importance. Pay attention to the fact that the remnant patches are not residues of wood scattered here and there in the countryside. On the contrary, all those areas are (remnant) patches of the current pre-existing landscape type/matrix located at a suitable distance from the primary sources of disturbance (chronic)—even in the case of some old agricultural fallow—distance sufficient to avoid the disturbances.

Furthermore, the term patch can also indicate a band with a direction of the prevailing development. Therefore, they are perhaps best defined as 'undisturbed areas'. The mosaic of remnant patches/undisturbed areas should be carefully evaluated using the principles of landscape bionomics if you want to prevent an ever-more-chaotic landscape and take action to bring greater order, going towards a good level of environmental remediation: for example in design checks, as shown in Fig. 7.17. The two routes are possible diversions for the final stretch of a stream, but they could also be alternative routes (see also Chap. 14).

The complete and accurate assessment of the ecological mosaic can turn the patches into *undisturbed areas of strategic importance* (or *undisturbed areas of strategic rehabilitation*). The structural pattern of remnant patches is composed of many elements, characterized by their size and shape (Forman, 1995), a role in the ecotissue and biological territorial capacity of vegetation (Ingegnoli, 2015), the ability to disturbances incorporation, etc. Consequently, the residual patches cannot be considered as indifferent to human activities, but on the contrary, their possible compatibility profile and their suitability to the planning and "biotope-und Artenshutz" should be evaluated (Ermer et al., 1996).

This can be done by mapping a defined set of belts, the width of which depends on the characters of the landscape unit; for example, there will be a difference

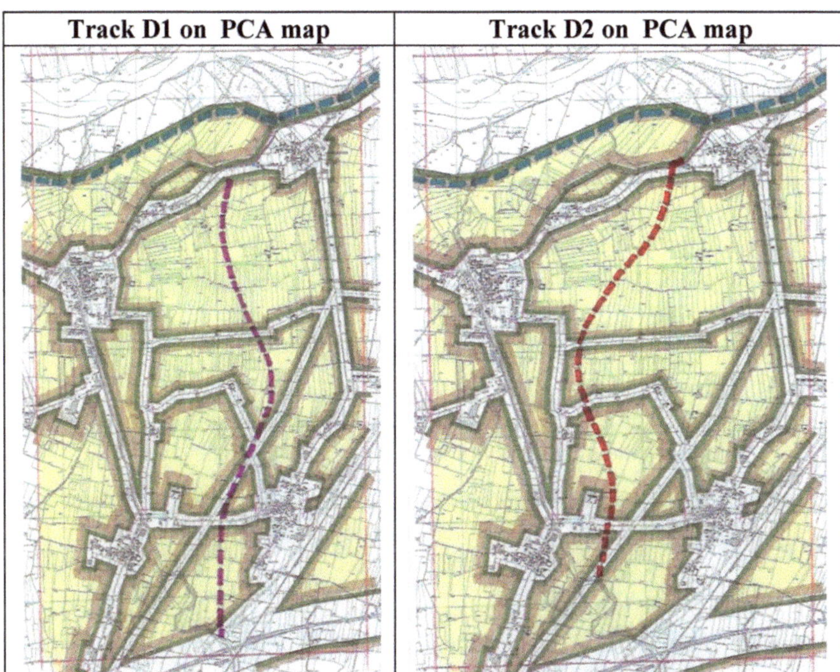

Fig. 7.17 Comparison between the traces of the diversions to be evaluated reported on the mapping of the PCA (Potential Core Areas) system, colored in yellow, in the LU under examination. Note the greater impact on fragmentation due to the D1 diversion (see also Chap. 14)

between an agricultural landscape unit and a suburban or open-forest one. In fact, one of the principal functions of a landscape element is its capacity to incorporate a disturbance. So, the width of a considered belt will be minor in the presence of a landscape element with a high BTC value and larger when the landscape element, on which the disturbance is exerted, has a lower BTC value (thus enhancing the extreme difficulty to incorporate the disturbance).

In the case of the study for the municipality of Capannori (Lucca) (Fig. 7.18; Table 7.5) (Ingegnoli, 2011), a set of belts, moving from the outside to the inside, has been defined as:

- Disturbance belt (Db), 10–100 m in width
- First buffer belt (B1), 60–100 m in width
- Second buffer belt (B2), 80–100 m in width
- Potential core area (PCa), 100–120 m in width
- Potential inner core area (PCi), >25 m in width

The resulting bounded area inside the third belt can be called the *potential core area*. Thus, a remnant undisturbed patch, even a large one, may lack of a potential core area due to its shape or too severe disturbances. Inside the potential core area, it is sometimes possible to define a *potential inner core area* with better restoration possibilities.

The second step will be to give each remnant undisturbed patch a specific quality value (Table 7.6): the presence of a potential natural or semi-natural core area, a strategic linking capacity to connect other remnant patches, a good level of BTC of its vegetation, will lead to a high value of the patch. For example, two close *core areas* can have different values of BTC, needing different evaluation types. Analogously, a patch without intrinsic value could be a critical steppingstone for an ecological network, so assuming a strategic functional role (SR).

Obviously, the strategic roles of remnant undisturbed patches must be clearly expressed after a specific study of their distribution in a landscape unit. For example:

- Stepping-stone function for connection
- Meso-climatic mitigation of an intensively urbanized area
- Island for birds' temporary refuge (e.g. along a flyway)
- Protection for hydrologic processes (e.g. a spring or resurgence)
- Necessity for maintaining good porosity of the landscape matrix.

Depending on their evaluation, remnant undisturbed patches present different constraints which are reflected in the different planning capacities. To measure these capacities, the main transformation processes need to be checked. For instance,

- N: Each transformation improving the naturalness of the patch (e.g. reforestation)
- M: Each transformation bringing the patch back to the original landscape matrix condition (e.g. from abandoned fields to cultivation)
- RG: The creation of recreational green areas (e.g. public gardens)
- HS1: The building of light human structures (e.g. rural roads or canals)

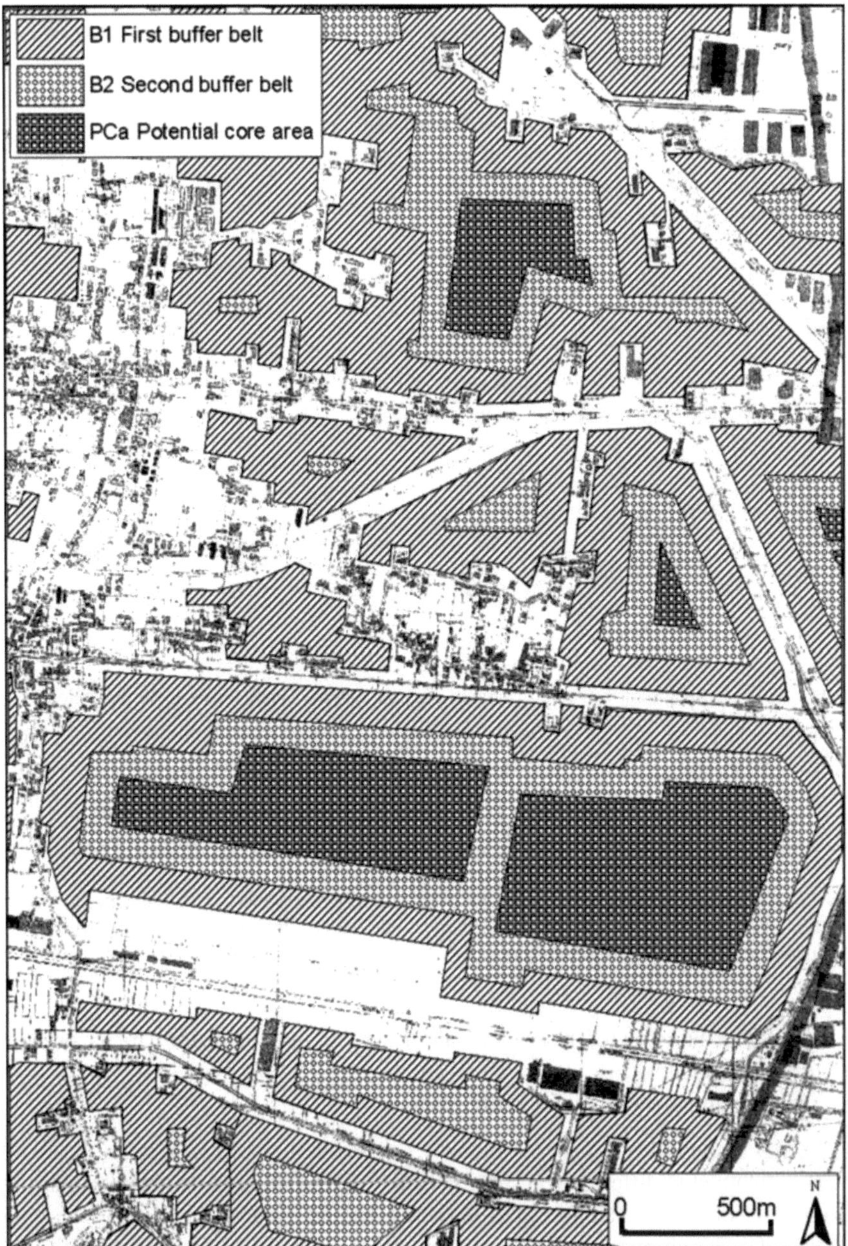

Fig. 7.18 An example of the application of a set of belts to measure the gradient of disturbances exerted on the territory of Capannori (Lucca, Italy) and the resulting remnant undisturbed patches. The potential core area is darker

Table 7.5 Example of a set of belts for the evaluation of remnant patches studied for the municipality of Capannori (Lucca). Note that the width of the belts depends on the BTC value of the landscape element through which the belt passes

Type of disturbance	BTC value of the landscape element close to the disturbance source	Belt type	Distance from the source of disturbance (m)
Residence buildings, small railroads, small roads, small industries, etc.	Low BTC	Db	20
	High BTC	Db	10
State roads, large industrial plants, main railroads, quarries, etc.	Low BTC	Db	40
	High BTC	Db	25
Airports, highways, dangerous industrial plants, etc.	Low BTC	Db	100
	High BTC	Db	60
Buffer zone perimeters.	Low BTC	B1 and B2	100
	High BTC	B1	60
		B2	80

Table 7.6 Evaluation levels of a mosaic of remnant patches characterised by a five belt set

Number of belts	i < 20%	i > 20%	BTCp > BTCm	SR
2 (Db + B1)	A	A	C	D
3 (Db + B1 + B2)	A	B	C	D
4 (Db + B1 + B2 + PCa)	B	C	D	D
5 (Db + B1 + B2 + PCa + Pci)	C	C	D	D

i is the most inner belt mapped, *BTCp* the BTC value of the remnant patch, *BTCm* the mean value of the BTC of the landscape unit, *SR* the strategic role of the patch within the landscape unit; *A, B, C, D* evaluation levels from worst to best

- HS2: The building of heavy human structures (e.g. industrial roads or new technical plants)
- TU: The urbanization of the entire patch

Referring to the last two Tables 7.6 and 7.7 note that the patch values (A, B, C, D) can be obtained in different ways: for instance, a 3D value can be obtained starting from a 3A or a 3C; analogously, a 4D value can derive from a 4B or it can be a simple 4D. Therefore, the value degrees are generally more than four and they could be indicated, respectively, as 3D(A), 3D (C), 4D(B) or 4D(D).

These evaluation degrees allow one to map the system of patches of a landscape with different colors (Ingegnoli, 2015): a map like this can help to study an

Table 7.7 Different planning capacities of the remnant patches

Patch values	N	M	RG	HS1	HS2	TU
A	+	+	+	+	+	+
B	+	+	+	+	(+)	[+/−]
C	+	+	(+)	(+)	[+/−]	−
D	+	(+)	(+)	[+/−]	−	−

Where: + freedom of transformation; (+) transformation possibility only marginal and only with mitigation projects; [+/−] transformation possibility only marginal and only with mitigation and compensation projects; (−) no transformation possibility

ecological network and compare some case studies of transformation proposed by the municipality's urban planners. Moreover, it can be used to restore the landscape structure and divide the landscape into different planning subunits.

In conclusion, we should note that the preservation of nature needs to protect the most critical parts of the system of residual patches of the previous, often agricultural landscape to sew up the fragmentation and arrive at an environmental rehabilitation of the territory. The methodology presented here is insufficient to reach this purpose, but it certainly can be of considerable utility, together with the principles of landscape bionomics already known in the literature. Note that the system of residual undisturbed patches should be considered in conjunction with:

(a) The distribution of cumulative impacts, e.g. the dangerous pollution patches around chemical industries or toxic waste dumps.
(b) The system of the river network, to which you usually give the appropriate values for the bands of protection.

In the study of an ecological network for a municipal or provincial territory to carry out an evaluation of the system of residual undisturbed patches remains essential.

7.6.2 Mapping and Evaluation of Ecological Networks at Different Scales

We are now able to relate these transformation processes to the evaluations received from the residual patches in the previous tables.

To identify key gaps and to apply the connectivity and circuitry indexes at a territorial sclae, the conversion of the basic mosaic of a LU into a network is needed (Fig. 7.19). In general, it is necessary to put in evidence the remnant undisturbed patches as nodes and/or linkages and the Key ganglia. However, even without a core area, residual spots can be significant for connection reasons at territorial scale. The connection of the analysed landscape structure is always monitored by comparison with its past configurations.

Remember that the *connection analysis* acquires a significative importance in ecological network design (Fig. 7.20). We can derive two useful indexes: for a network *connectivity* (γ) and for a network *circuitry* (α):

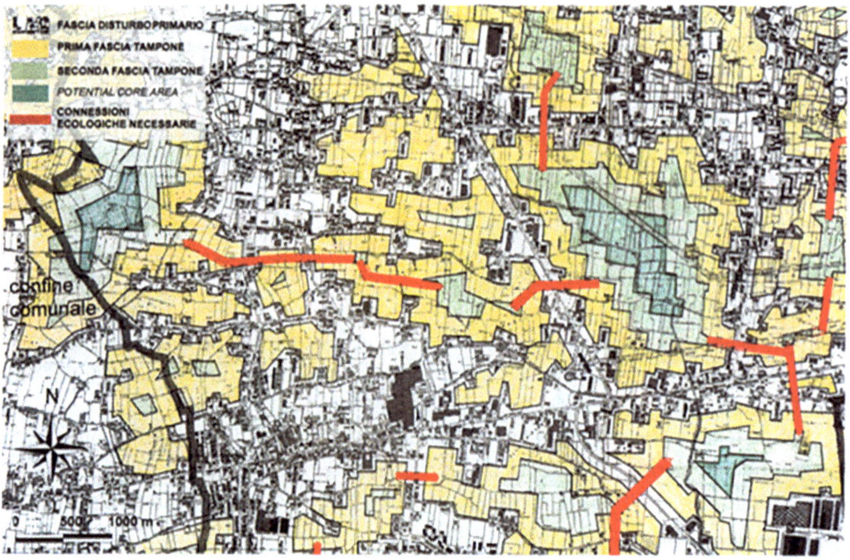

Fig. 7.19 Example of application of the method of the strategic undisturbed spots in the municipality of Capannori to identify also the key ganglia to mend the territory. The potential core areas are in dark green; see the very evident differences in the presence and extension of the B1 and B2 bands and the core areas (from: Ingegnoli and Giglio, 2005)

Fig. 7.20 Example of connection graph useful to study the territorial networks, mainly composed by vegetation patch and corridors and road/transport maps. Red arrows indicate an example of the potential connections to be created. Remember that connection needs the analysis of connectivity and circuitation

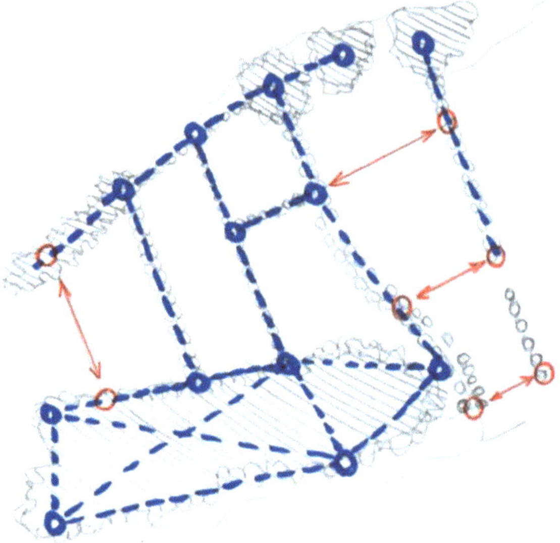

$$\gamma = L/3(V-2); \quad \alpha = (L-V+1)/2V-5$$

In general, it is necessary to put in evidence the vegetation patches and the main species sources as nodes and the existing corridors and sporadic routes as linkages. The connection of the analysed landscape structure is always monitored by comparison with its past configurations.

7.7 Care Functions/Services of Natural and Anthropic Vegetation

7.7.1 Ecological Services of Anthropic Plant Formations

We should note that the four canonical categories of Ecological Services (incorrectly called "Ecosystem Services") must be increased. In fact (Table 7.8):

1. The "Ecosystem Services" should be called "Eco-Bionomics Services", given the errors inherent in the concept of ecosystem which limit the ability to recognize the set of the same services, given the lack of integration already mentioned (see Sect. 1.2.1).
2. The "Eco-Bionomic Services" must refer not only to the human population, but to all the components of a Landscape Unit (LU);
3. The ecological functions of Health and Protection must appear explicitly among the guiding concepts of the Services. Today health is mentioned, but subordinate to the 4 canonical categories and only for the human population, while:

   (a) the concept of protection is not directly expressed and
   (b) health must be placed as a further primary category, and
   (c) it must be understood as dual, concerning landscape pathologies and human pathologies.

Table 7.8 The main sectors of ecological services

Ecologic Services (ES) following L. Bionomics (Bocchi & Ingegnoli 2019)	
ES types	ES (ES = Eco-bionomic Services),
1.Supporting	Primary production, soli formation, nutrient cycling, etc.
2.Regulating	Water & air purification, pollination, etc.
3.Provisioning	Food & fiber production, biodiversity, etc.
4.Cultural	Arch. Heritages & Recreations, sense of place, etc.
5.Health preserving	Geo-health/human-health. Risk factor control. LU diagnostic evaluation
6.Protecting	Transformation deficit; equivalence: Bionomics & Economics costs

7.7.2 *Proportionality of Eco-bionomic Services to the Bionomic Vegetational Value*

Here, we present a complex table, Fig. 7.21, made up of two parts, separated by the line showing the BTC intervals, which identify 10 columns from the minimum value (0.0–0.5) to the maximum value (9.6–11.2 Mcal/m^2/year). The upper part indicates the ecological services for the six categories of Table 7.8, and the lower part indicates the leading natural and anthropogenic phytocoenosis of the temperate eco-regions.

Within the figure, above, the main 23 Vegetation Services (and/or functions) are listed and divided into six categories. Their effects vary in relation to the BTC intervals, showing at least four intensity levels: 0, 1, 2, and 3 (from white to darker green). Pink colors underlined the direct or indirect link with human health, detected by 13 services/functions (56.5%). Functional scores and service effect values per BTC interval are summarized in color beige.

MAIN ECO-BIONOMIC FUNTIONS/SERVICES PROPER OF NATURAL & HUMAN VEGETATION											
Supporting	Trigger of rain	0	0	1	1	2	2	2	3	3	3
	Source of floristic biodiversity	1	2	2	1	3	3	3	3	2	2
	Mother-tree network	0	0	0	0	1	1	2	2	3	3
	Soil organic components	1	1	2	2	2	3	3	3	3	3
	Soli anti-erosion defense	0	1	2		3	3	3	3	3	3
Provision.	Supply of organic hay	0	1	3	1	0	0	0	0	0	0
	Wood prduction	0	0	0	1	2	2	3	3	2	2
	Biological Control of pests	1	1	2	2	3	3	3	3	3	3
Protective	Wild animal shelter	0	1	1	1	2	2	2	3	3	3
	Resilience (recovery after disturbance)	2	2	3	3	3	2	2	2	1	1
	Resistence (disturbance incoporation)	0	0	1	1	2	2	2	3	3	3
	Slope stabilization	1	1	1	2	2	2	3	3	3	2
Health	Psychoneuro treatment	0	1	1	1	2	2	2	3	3	3
	Immune system improving	0	0	1	1	1	2	2	2	3	3
	GM (Gut Microbiote) regulation	0	1	1	1	2	2	2	3	3	3
Regulating	Water regulation	0	1	2	2	3	3	3	3	3	3
	Air purification	0	1	1	2	2	2	3	3	3	3
	Climate regulation	0	0	1	1	2	2	2	3	3	3
	CO2 absorbance	0	1	1	1	2	2	2	3	3	3
	PM-10 absorbance	0	1	1	2	2	2	3	3	3	3
	TD (transfprm.deficit) compensation	0	0	1	1	1	2	2	3	3	3
Cultur.	Break of UHI (urban heath isle)	0	1	1	2	2	2	3	3	3	3
	Social & aesthetic recreation	1	1	2	2	2	3	3	3	2	2
	BTC (Mcal/m^2/year)	0.0-0.5	0.5-1.2	1.2-2,4	2.4-3.6	3.6-4.8	4.8-6.0	6.0-7.2	7.2-8.4	8.4-9.6	9.6-11.2
	FUNCTIONAL score (0,1,2,3)	7	18	31	33	46	49	55	63	61	60
	Function importance/ BTC set: score/69*100	10,14	26,09	44,93	47,83	66,67	71,01	79,71	91,30	88,41	86,96
Natural Formations	Dry pasture & Steppes										
	Alpine Pastures										
	Wet grasslands & Tall Grasses										
	Halophylous communiities										
	Swamps & Peat Bogs										
	Moors & Shrubs										
	Sclerophyll scrub & Maquis										
	Mediterranean Pine Forest										
	Temperate deciduous Forest										
	Mixed boreal/temperate Forest										
	Boreal-Alpine Forest										
	Humid & Riparian Forest										
Human Formations	a. URBAN PARKS										
	b. URBAN FORESTS & HISTORICAL PARKS										
	c. PRIVATE GARDENS										
	d. SPORT GREEN										
	e. ROAD GREEN										
	f. ARBORICULTURE										
	g. COPPICE WOOD										
	h. AGRARIAN HEDGEROWS										
	i. VINEYARDS & ORCHARDS										
	j. PERMANENT GRASS										
	k. MEADOWS										
	l. SEEDS WITH CEREALS										

Fig. 7.21 The Ecological/Eco-bionomic Services of anthropic plant formations are seriously insufficient with respect to prevention and health care (pink). The red square indicates that without the presence of natural forests, human habitats are intrinsically prevented from achieving effective health protection

In the same figure, below, are listed the major plant formations of Natural Habitats (NH) and Human Habitats (HH), and the extension of the normal range of BTC classes/intervals of each of them. It is evident that Human Habitat formations remain below BTC = 6.0, and only 2/12 = 16.7% reach the BTC = 6.0–7.2 range.

Pay attention now that approximately 70% of the most intense and effective ecological services for health protection (dark green) are in the red square, i.e. in the last four columns with high BTC. Therefore, human habitats must be restructured to broadly include green components with high BTC if we want to protect health. In the most anthropized territories (e.g., Po Valley), it is essential to change the relationship with the vegetation (HH/BTC), giving greater importance to the elements with high BTC.

Otherwise, the benefits for human health would not be activated and the deterioration of the environment would increase the onset of the most serious diseases, primarily oncological ones. It should be noted that epigenetics confirms the importance of the exchange of environmental information between different scales, as already mentioned. Therefore, by modifying the environment, men also act towards themselves.

7.8 Shifting from Sustainability Towards Regenerative Rehabilitation Concepts

The Systemic paradigm may help also in the logical interpretation of the items of the UN Agenda 2030 (UNESCO, 2015) for Sustainable Development. The key concept is n°4 Education (Fig. 7.21)(see also Chap. 15): *sine qua non* to affirm the other, as two crucial but different ethical-epistemological pathways may be identified and underlined (Fig. 7.22).

1. Human Rights and Democracy (HRD), most relevant to politics and administration.
2. Ecological and Medical Principles (EMP), most pertinent to Life Sciences.

Concerning the second one (see also Sect. 1.4), it goes around the new concept of life (see Sect. 15.3.1), not centered on the organism but on the entire 'biological spectrum' (see Sect. 1.2.1): Life cannot exist without its environment. The violence against nature damages human health, so the regeneration of our environment affect all, but need scientists able to follow the System Theory.

From the point of view of Landscape Bionomics, significant changes are occurring in the study of sustainability, too. Remember that James Lovelock (Lovelock, 2006) wrote that we must move from Sustainable development to Sustainable retreat. Pope Francis (Pope Francis I, 2015) wrote: "We also need to think about slowing down a bit, about setting some reasonable limits and also about going back before it is late." (Laudato Sì, 193). Let's realize that it is too late to act differently!

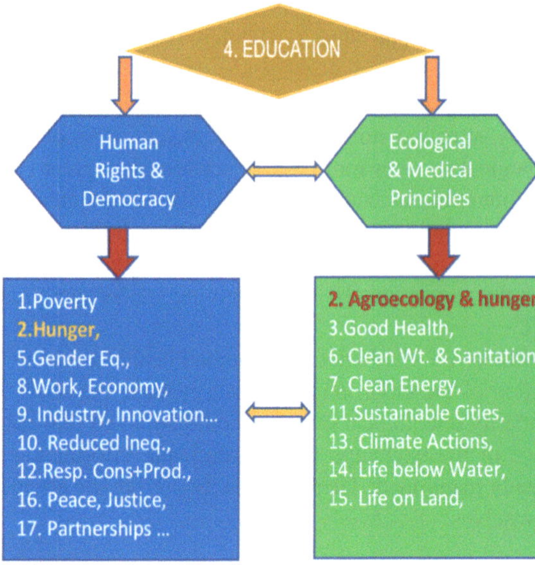

Fig. 7.22 Education as a priority component in the goals of Agenda 2030 and the two pathways

The first of these changes is the overcoming of the concept itself of Sustainable development by the principle of "rehabilitation strategy," linked to the redefinition of the landscape as a living complex adaptive system (see Sect. 3.1.1).

Sustainability is no more to manage the resources to postpone their depletion as much as it is possible, but to *understand the relationship between the state of health and diseases of a land unit, to check its compatibility with the needs of human health and develop criteria for therapeutic rehabilitation with regenerative function* (see also Chap. 15).

References

Da Vinci, L. (1500). *Windsor codex*. Royal Academy.

Einstein, A. (1965). *Pensieri degli anni difficili*. Bollati Boringhieri.

Ermer, K., Hoff, R., & Mohrmann, R. (1996). *Landschaftsplanung in der Stadt*. Ulmer.

Forman, R. T. T. (1995). *Land mosaics: the ecology of landscapes and regions*. Cambridge University Press.

Forman, R. T. T., & Godron, M. (1986). *Landscape ecology*. Wiley.

Goodall, J. (2000). *40 years at Gombe*. Stewart, Tabori and Chang.

Hartmann, N. (1964). *Der Aufbau der Realen Welt*. De Gruyter.

Ingegnoli, V. (2002). *Landscape ecology: A widening foundation* (p. XXIII+357). Springer.

Ingegnoli, V. (2011). *Bionomia del paesaggio. L'ecologia del paesaggio biologico-integrata per la formazione di un "medico" dei sistemi ecologici* (p. XX+340). Springer.

Ingegnoli, V. (2015). *Landscape bionomics. Biological-integrated landscape ecology* (p. XXIV + 431). Springer.

Ingegnoli, V., & Giglio, E. (2005). *Ecologia del Paesaggio: manuale per conservare, gestire e pianificare l'ambiente* (p. 685+XVI). Sistemi editoriali SE.

Lorenz, K. (1978). *Vergleichende Verhaltensforschung: Grundlagen der Ethologie*. Springer.

Lovelock, J. (2006). *The revenge of Gaia: Why the earth is fighting Back and how we can still save humanity*. Penguin Books.

Naveh, Z., & Lieberman, A. (1984/1994). *Landscape ecology: theory and application*. Springer pp. XXVII+360.

Pauli, W. (1959). *Aufsätze Vortäge über Physik und Erkenntnistheorie* (tr. it. Fisica e conoscenza. Boringhieri, 1964).

Pope Francis I. (2015). *Laudato Sì*. Encyclical.

Shannon, C. E., & Weaver, W. (1949). *The mathematical theory of communications*. University of Illinoiss Press.

Tongway & Ludwig (2004), Procedures for monitoring and assessing landscapes with special reference to Minesites and rangelands. CSIRO

UNESCO (2015) Moving forward the 2030 agenda for sustainable development. .

Von Goethe, J. W. (1787). *Viaggio in Itala*. New publ. by Vallecchi, Firenze. 1955.

Von Humboldt, A. (1840). *Cosmos: A sketch of the physical description of the universe*. Re published in 1997, by Rarewaves.

Part II
Alteration of Human Landscapes and Planetary Health

Qui tempus valetudinis non habet,
citius aut serius tempus infirmitatis suae inveniet.
Lucius Anneus Seneca

"He who has no time for his health, sooner or later
will have to find time for his illness".
Lucius Anneus Seneca

Chapter 8
Alterations of Agro-Forest and Urbanized Landscapes

Contents

Abstract To understand the significance of the current ecological crisis from a systemic point of view and to provide a qualitative-quantitative framework on the global health state of the Earth, studies were carried out using Landscape Bionomy

V. Ingegnoli, *Landscape Bionomics and Planetary Health*, Landscape Series 21,
https://doi.org/10.1007/978-3-032-08908-3_8

principles and methods. Thus, data on the Transformation deficit at Global scale are provided: note that from 1880 to 2020, the environmental potentiality of self-recovering of our Planet decreased from 88% to 64.3%, with a 4% tolerance: a difference of −23.7%! The compensation capacity for the increase in CO_2 disappeared. A synthetic framework on the health state of Spain, France, Germany, Italy and UK is also provided.

At the Regional scale (Lombardy and Trentino), in the Alpine Parks, whose forests have an ecological efficiency CBSt 1/3 higher than other Parks, it doesn't exceed the average value of 33%, but in other forest CBSt = 23.75. This alteration is primarily due to the change in the types of forest that developed in Lombardy during this period. Starting from 1890, the growth of coppice forests and the consequent decrease in temperate-deciduous forests have been robust: today we have 67% of coppiced, 23% of Boreal-Alpine, and only 8.5% of high-stem Temperate!

Concerning the agricultural landscapes, if we compare the three types—Protective, Productive, and Industrial- following the bionomic parameters, we can measure sharp differences. An example of the Alpine agrarian landscapes, the best at BTC level, is essential to clarify that the Alps agricultural landscapes have substantially altered in the last century. Lowland corridors, which have characterized the agricultural landscapes of Lombardy for over two centuries, have significantly been affected by agricultural industrialization. In the example of Albairate, there was a decrease in surface area and BTC. A reference to how the significant eco-bionomic parameters change in relation to the altitude ranges in the Alps is shown.

Focusing on urban landscapes, the city is the most typical of the ecological systems formed by man and one of the most recent in the natural evolution of the biosphere. The town was born from the integration of neighborhoods with complementary functions. Now, however, the disorganized growth has led to the creation of urban landscapes, the structure of which, according to Richard Forman, is very complex. Thus, what can be considered as urban landscape rather than urban region must be detected. Also, the delimitation of the metropolitan area of Milan compared with the current Metropolitan City shows how political-administrative choices can be in contrast with scientific ones. The bionomic-ecological state of the Metropolitan Area and Historical Center of Rome and Milan are compared and the incidence of cancers and cardiovascular diseases in the two cities is enlightened.

But the functionality of an ecological system is also revealed by the presence and characteristics of the Landscape apparatuses or, at least, by the % presence of the major categories of urbanization, agriculture, forest. Deepening some other integrated indicators are useful such as carrying capacity. The shifting between the simple LaBiSHH (Landscape Bionomic Survey of Human Habitat) to the LaBiDHH (Landscape Bionomic Diagnosis of Human Habitat) is presented through different case studies.

Keywords CityLife · Urban region · Metropolitan area · Green urban plan · Traditional planning · Bionomic planning · BTC national distribution

8.1 Agro-Forest Landscapes: An Overview on the Last 140 Years

8.1.1 Transformation Deficit at Global Scale

The previous chapter ended focusing on the assertion that "Life cannot exist without its environment. The violence against nature damages human health, so the regeneration of our environment affect all, but need scientists able to follow the System Theory". Therefore, studies were carried out using Landscape Bionomy principles and methods and, thus, the concept of Hierarchic Levels of Biological Organisation on the Earth (see Sect. 2.4) and the ontological importance of history (see Sects. 2.3 and 8.8), to provide a qualitative-quantitative framework on the global state of the health of the Earth.

The considered period goes from 1880 to 2020: the major changes have concerned the increase in human habitat (HH) decidedly high, going from 20% to 30% of the Emerged Lands, and that today reaches 50%, not counting extreme deserts, waters, ice, and high forests naturalness (land cover data from (Buringh & Dudal, 1987; FAO, 1962, 1990, 2010, 2015; Goldsmith & Allen, 1972; Marklund & Batello, 2008)). Starting from 1970, the destruction of forests (and the increase in cultivated land) slowed just as population growth skyrocketed, going from 3000 to 7800 million people in just 50 years. The contribution of the so called "Green Revolution" to this phenomenon seems evident: this change has strengthened trust in *industrialized agriculture*, which currently no longer yields as much as in previous decades. At the same time, its contribution to the profound alteration of the environment requires recovery actions that are very difficult to manage due to their vastness and urgency and to the need of quantifications. Furthermore, we should note that these studies significantly undermine the so-called 'Borlaug principle' (Borlaug, 1974), which links agricultural industrialization to the protection of forests due to the increase in production/ha. So, in the last 50 years, forests have slowed down their destruction but also their bionomic efficiency, while the energy flow to maintain an optimal bionomic organization of agricultural landscapes remains wholly insufficient.

Below is a figure specifically designed to evaluate the transformation deficit (TD) (see Sect. 7.4.6) of a large-scale ecological system. Please remember that the TD represents the loss of the energy flow necessary for maintaining the order achieved in the development of the system under consideration or for self-recovering: we will have to take into consideration the global scale, that of the Emerged Lands of our planet. In Fig. 8.1 one can see:

1. The decrease of forest cover from 36.6% in 1882 to 27.4% in 2020 (−25.8%), but (worst) the loose of bionomics efficiency (−34.1%)!

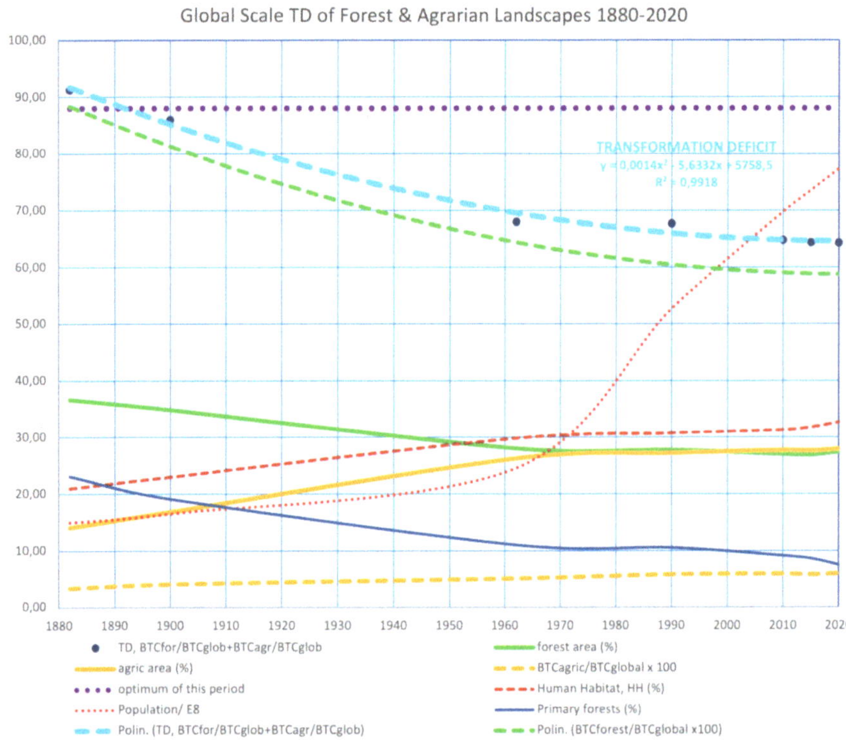

Fig. 8.1 Transformation deficit (TD) of emerged lands 1880–2020. In green, forest cover (%) (continuous line) and its ecological efficiency (dashed l.). In yellow, crop and grassland cover and its ecological efficiency (dashed). Blue dots represent the sum of the values of Agricultural and Forest BTC related to the global BTC. The space in between this line and the violet one, measures the transformation deficit (TD). In dotted red the increase of human habitat (HH, %). Primary forest (i.e., mature natural) is today reduced to 8% of Land cover

2. The increase of agrarian areas (green and crop) from 14.1% to 27.9% (+197.8%), but (worst) followed by a poor increase of bionomics efficiency (+178.7%), due to the degradation of agricultural landscapes (−19.1%).
3. The environmental potentiality of self-recovering passed from 88% (reference value on 1890) to 64.3%, with a 4% tolerance: a difference of −23.7%, which leads to a total Transformation Deficit of 8.29×10^{18} kcal.

It is noted that curbing forest destruction and increasing reforestation is necessary but will only produce serious effects in about a century; therefore, it is inevitable and urgent to curb the destruction of agricultural landscapes and increase their ecological efficiency, which requires less time (about half), provided that there is a political and professional will and that ecology can give space to bionomic principles.

8.1.2 Agro-Forest Landscapes and the Theory of the "Biotic Pump"

The large transformation deficit (TD) reported above has certainly weakened the self-rebalancing capacity of the ecosphere and, consequently, has contributed to the ongoing climate change.

To better understand this crucial aspect, it is considered appropriate to recall the recent bio-geophysical theory of the "Biotic Pump" (Makarieva & Gorshkov, 2012): this theory is based on the observation that forest landscapes create and control ocean winds that, passing over them, are further enriched with humidity that is condensed into clouds, then into rain and subsequently distributed to benefit all living systems on Earth. It is observed that atmospheric pressure does not depend on the weight of the air column on a given surface, but on the number of gas molecules, among which water vapor dominates. When the vapor condenses, the number of gas molecules greatly decreases and the pressure drops, forming a relatively persistent depression. Water vapor in the atmosphere is unstable to condensation and, once enough has been reached in the lower atmosphere, condensation begins. This phenomenon is much faster and more consistent in the forest than on the sea, because evapotranspiration occurs on a three-dimensional scale and not just superficially. Forest landscapes therefore become depression zones that attract ocean winds and form clouds. The hydrological cycle is completed with precipitation on land areas, then returning to the sea by way of the river network.

Like all vital processes, the Biotic Pump must be highly organized: the evapotranspiration of forests must not exhaust the soil moisture but, at the same time, it must be intense enough for the moisture drawn from the ocean to compensate for this loss. Meanwhile, the temperature gradient (due to the shade and height of the trees) and the distribution of evapotranspiration by layers, often with the addition of volatile chemical components (VOCs, capable of providing condensation nuclei) triggers the formation of clouds and the consequent precipitation. Some of those compounds are deposited on tiny airborne particles such as dust, bacteria, pollen, and fungal spores. As the particles grow with the deposition of VOCs, they promote condensation and gather the resulting moisture, hastening cloud formation.

Attention: this is possible (a) for forest systems, with mature natural phytocoenosis, evolved over thousands (sometimes millions) of years to reach optimal bionomic state (evaluable through the CBSt, see Sect. 4.8) and (b) for agricultural systems, maintaining the integrity of ecological systems from which they are obtained, thus enriching the BTC of current landscapes. Interestingly, a study on Borneo (Hance, 2019) shows that deforested areas appear to have an average temperature 1.5–1.7 °C higher than forests, and agricultural areas an even higher temperature, equal to 2.8–6.5 °C, even when related to oil palm plantations, i.e. arboriculture, because the local landscapes do not comply with the bionomic requirements mentioned above.

Note that considering the Biotic Pump theory, the process of climate change should also be reviewed. We know that the greenhouse effect is mainly determined by water vapor and cloud cover: the absorption of CO_2 molecules covers less than 20% of the thermal radiation spectrum, while atmospheric humidity uniformly absorbs the entire spectrum. Consequently, *the impact of the increase in* CO_2 (which must in any case be reduced) on the greenhouse effect *could be compensated* by a minor change in the hydrological cycle, i.e. *by significantly reducing the destruction of forests* (and changing their management) *and rehabilitating agricultural landscapes*.

If the transformation deficit seen in the previous paragraph has reached an abnormal level, it is obvious that the compensation capacity for the increase in CO_2 has disappeared. Please remember that the quantities of energy that have been lost in a very short time (135 years) are truly enormous: $TD = 7.974 \times 10^{18}$ kcal.

8.2 The Framework: Health State of Landscapes in Western Europe

8.2.1 BTC Distribution in Most Important European Nations

The hierarchical organization of life on the Earth and the concept of ecotissue as well as the diagnosis of a LU, and so even ecological territorial planning, requires a serious geographical framework on the bionomic state at a synthetic scale, i.e., at the scale of the major European nations. Figure 8.2 is therefore shown, which considers

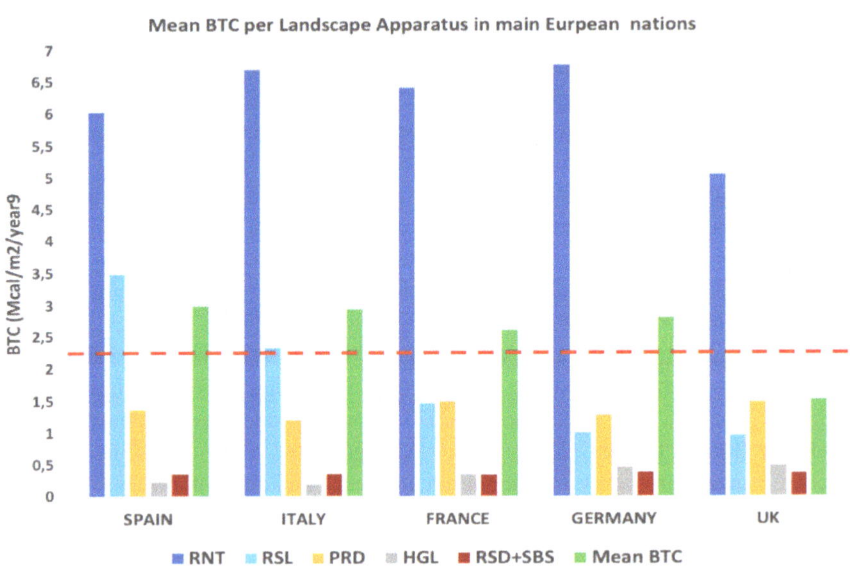

Fig. 8.2 Distribution of average BTC values in the major sets of landscape systems in the most important Western European nations, with a gradient from the South (Spain) to the North (UK). Surveys in the years 2018–20. The red dotted line shows the average BTC of the Emerged Lands in 2015

Table 8.1 BTC Estimation per Landscape Apparatus/Eco-countries Functional Systems in Western Europe (2015)

Nation	RNT	RSL	PRD	HGL	RSD+SBS	BTC
SPAIN	6.03	3.48	1.36	0.231	0.355	2.99
ITALY	6.70	2.33	1.20	0.201	0.355	2.94
FRANCE	6.41	1.46	1.49	0.350	0.339	2.61
GERMANY	6.77	1.00	1.27	0.453	0.374	2.81
UK	5.05	0.96	1.48	0.480	0.362	1.52
BTC mean	6.19	1.85	1.36	0.340	0.360	2.57

RNT Resistent L. app. (forest), *RSL* resilient L. app., *PRD* productive L. app., *HGL* hydrogeological L.app., *RSD* + *SBS* residential and subsidiary L. app., *BTC* Mcal/m^2/year

Spain, Italy, France, Germany, and the United Kingdom with a gradient from South to North.

The figure shows that only the UK has an average BTC (green) lower than that of the Emerged Lands (red dotted line). This result is mainly due to forest cover and the percentage of anthropogenic plantations, which are particularly relevant in the UK.

All the considered nations show an insufficient level of BTC in agricultural landscapes (yellow): before agricultural industrialization, several studies show that the average BTC was equal to 1.6–1.8 Mcal/m^2/year, while today, the average is BTC = 1.36, for Italy even 1.20. Even the BTC of the urbanized areas is very low, about 0.36 Mcal/m^2/year, which should instead reach values of at least 0.60–0.70 for HH >90%.

Given the inverse presence of Mediterranean scrub, the BTC of shrublands clearly decreases in relation to the average latitude. We add Table 8.1 to express the values more precisely.

8.2.2 *Forests in Western Europe*

As expressed by EFI (European Forest Institute, 2025), and contrary to the ongoing deforestation at global level (remember Fig. 8.1), forest area and growing stock in the EU (27) have expanded continuously since the 1950s and impacted on the carbon sequestration and storage dynamics of forests.

However, 85% of forests are available for wood supply and wood harvesting, while only 9% are protected (EU Natura 2000). In the last 30–35 years, the forest area is growing of about 14 million ha: but a wide part of this grown forests is mixed with allochthonous species, especially near the most urbanized districts.

As we can see from Table 8.2, the average BTC of the resistant apparatus RNT in the most important European nations is equal to 6.19 Mcal/m^2/years, decidedly lower than BTC = 8.6–9.8 of natural mature forests.

Table 8.2 Western European Transect of typical forests examples, from Poland to Sicily

Latitude	Place	*Tree composition*	Hc	vPB	BTC	bQ
52°44 N 23°53 E	**Poland**, Bialowieza, Ts3	*Quercus robur 85, Tilia cordata 10, Carpinus betulus 5.*	29.4	768	10.34	90.1
52°39 N 23°50 E	Poland , Polana, Ts1	*Pinus sylverstris 85, Picea abies 9, Quercus robur 6.*	24.0	257	5.89	62.1
50°54 N 04°12 E	**Belgium**, Asse, Ts2	*Acer pseudopl. 30, Prunus avium 22, Fraxinus exc. 17, Alnus glut. 8, Q. robur 4.*	26.9	584	7.60	67.2
50°50 N 04°10 E	Belgium, Asse, Ts4	*Fraxinus exc. 37, Fagus syl. 31,Quercus robur 14, Prunus avium 10.*	29,7	625	7.70	67.4
47°18 N 10°01 E	**Austria**, Schoppernau	*Picea excelsa 77, Abies a. 6, Fagus s. 17*	34.8	953	9.33	86.1
46°58 N 10°11 E	Austria, Galtur, Ts3	*Picea abies, 100*	26.5	667	6.89	62.4
46°42 N 10°06 E	**Switzerland**, Zernez	*Picea excelsa 72, Larix d. 26, Pinus syl.2*	26.2	688	7.53	71.7
46°18 N 11°44 E	Trentino, Paneveggio, Ts4	*Picea abies, 100*	32.4	992	9.11	80.92
46°15 N 10°28 E	Lombardia, Val Paghera, Ts1	*Picea abies 100*	26.8	636	7.78	74.4
45°40 N 09°21 E	Lombardia, Colli Briantei, Ts3	*Quercus robur 60, Q. petr. 25, Platanus hyb. 13, Robinia psac. 2.*	30.1	551	7.64	68.2
41°54 N 14°21 E	Abruzzo, Rosello, Ts2	*Q.cerris 40, Abies a.24, Carp.bet. 13, Acer cam. 10, Acer obt. 4*	19.5	359	7.68	75.0
40°30 N 03°40 W	**Espagna**, Madrid, wood of Botanical Garden Ts.5	*Q.ilex, Q.pirenaica & Bosque Escleròfilo sp.*	14.5	400	6.19	58.3
40°09 N 15°47 E	Campania, Casaletto S.	*Quercus ilex, 80; Fraxinus ornus 7, Carpinus betulus 10*	14.6	416	7.67	69.0
38°23 N 16°11 E	Calabria, Aspromonte, Piano Limina	*Fagus Sylvatica 100*	27.1	596	7.69	67.9
37°30 N 13°22 E	Sicilia, Parco della Ficuzza	*Quercus gussonei 90, Q. pubescens 10*	14.0	210	5.75	53.5

These last BTC values are sometime present in natural Parks, as shown in Table 8.9, where 4/5 of the case study forests are in protected areas with a mean BTC = 7.65, that is 23.61% more elevated. In this transect, only three forests presented BTC >9.00 Mcal/m^2/year, in Poland, Austria and Trentino.

8.3 The State of Vegetation in Lombardy

8.3.1 Forest State in Lombard Parks

As seen, albeit in summary, the LaBiSV method is precise and sensitive; therefore, it allows to arrive at significant assessments of the state of the vegetation. For a preliminary overview, some studies in Lombardy that I consider particularly interesting will be reported. Figure 8.3 summarizes the state of the forests in the Regional Parks of Lombardy. The data was derived from the study of 60–70 forests in Alpine, plain and river areas. Even moving away from the Milan Hinterland that we have just shown, the situation does not change much: in fact, it is even more disappointing, given that these are predominantly Natural Parks.

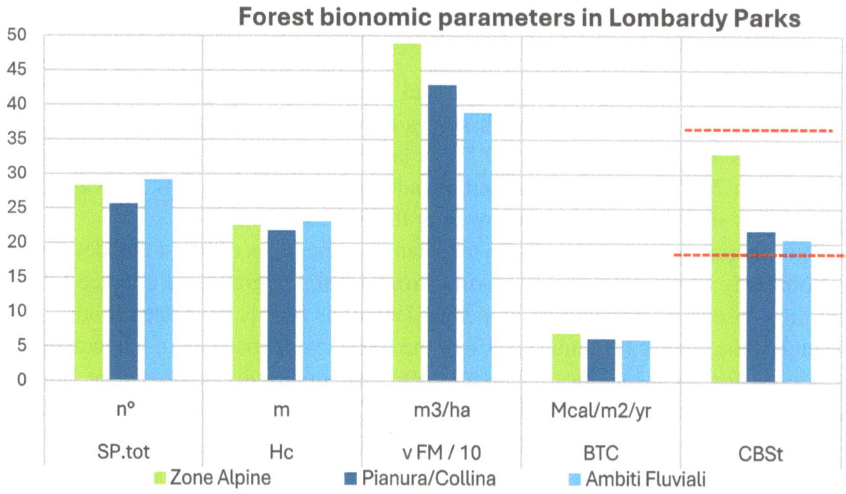

Fig. 8.3 From around sixty surveys carried out in the Regional Parks of the Alpine, Plain/Hill and River areas, a situation emerges that is not very different from the Milan Hinterland. Hc = canopy height, vFM = phytomass volume. Bionomic state: normal if CBSt >36, altered if CBSt <18. Therefore: 86% riverine, 82% plain and 55% alpine are altered forests

Table 8.3 Bionomic state of forests in Lombardy (LaBiSV)

Forest types	Land cover %	For. %	SP. tot n°	Hc m	v FM m^3/ha	BTC $Mcal/m^2$/year	CBSt
Boreal-Alpine	8.0	24.0	27.31	22.09	517.33	6.73	32.13
Temperate Deciduous	2.5	10.0	27.42	25.12	448.92	6.33	23.01
Coppice Woods	16	64.0	26.17	14.34	231.73	4.76	10.62
Boreal+Temp.	0.5	2.0	29.33	18.82	348.79	6.21	20.98
Weighted Average	25	100.0	25.40	17.40	326.40	5.42	17.23

Note that not even in the Alpine Parks, whose forests have a CBSt 1/3 higher than in the other Parks, it does not exceed the average value of 33%. The average BTC is still clearly <7.00 $Mcal/m^2$/year and the volumes of phytomass are also low (from 385 to 485 m^3/ha). To try to understand better, we report Table 8.3 which distinguishes forests by type: Boreal-Alpine, Temperate-Deciduous, Mixed Boreal-Temperate, Coppiced Woods.

The average values are always low (BTC = 6.30 $Mcal/m^2$/year; FM = 435.4 m^3/ha; CBSt = 23.75), but it must be admitted that between the coppice forests and the boreal-alpine forests the difference is marked: however, if we make a comparison within the Boreal-Alpine forests of Trentino (Italy), we will find a BTC that goes from 6.73 to 7.03, a phytomass that goes from 517.3 to 594.8 and, above all, a CBSt that goes from 30.1% to 35.0%.

8.3.2 *Forest Alterations in Lombardy*

To understand what happened, we must add another figure (Fig. 8.4) [(Agnoletti & Martegani, 2003), reworked], which shows how forests in Lombardy have grown over the last 50 years, going from 19.5% of land cover in the post-war period (1970) to 24.5% in 2005, mainly due to land abandonment in the Pre-Alps. However, the degradation shown previously by lowering the BTC is extreme.

This alteration is primarily due to the change in the types of forest that developed in Lombardy during this period. Starting from 1890, the growth of coppice forests and the consequent decrease in temperate-deciduous forests have been robust. If at the end of the nineteenth century, 54% of the forests were made up of tall temperate-deciduous forests, while the remaining 46% was equally divided between Boreal-Alpine and coppiced, today we have 67% of coppiced, 23% of Boreal-Alpine, and only 8.5% of high-stem Temperate!

Therefore, forest management must carefully change, especially in park areas (Bonizzi et al., 2019), where it is not acceptable to have such an unhinged state of vegetation! An in-depth study of the most significant formations and planning capable of considering the principles of landscape bionomics in safeguarding vegetation, especially forestry, will be essential.

We also report Fig. 8.5, which shows the chronological spectrum of the agricultural landscapes of the Milan hinterland. The comparison with the most typical

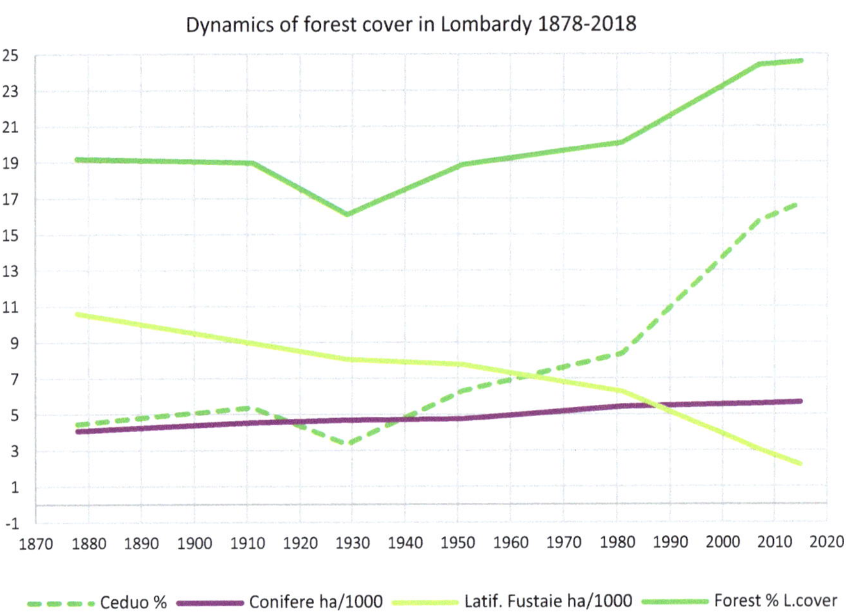

Fig. 8.4 Dynamics of forest cover in Lombardy 1878–2018. At the top (green) the total forests; below: the trends of the Broad-leaved High Forests (light green) and of the Coppices (dashed), while in blue the growth of the Boreal-Alpine ones. (From Agnoletti & Martegani, 2003, reworked)

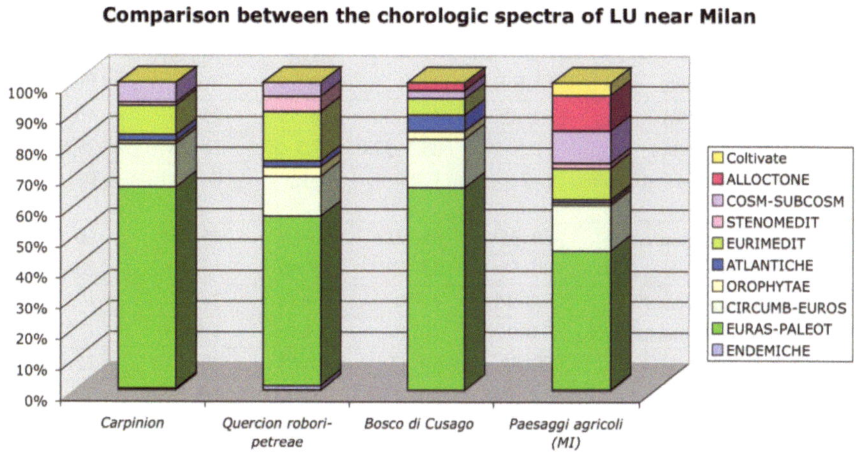

Fig. 8.5 Chorological spectrum of agricultural landscapes in the Milan hinterland. Note the alteration of agricultural landscapes which reaches around 45% in West Brianza (Coltivated, Allochthone and Cosmopolite. (From Ingegnoli, 2015))

alliances of the area allows us to understand better the extent of the alteration, which reaches almost 45% (Allochthone, Cultivated and Cosmopolitan) of farm landscapes in West Brianza. What's more, we are mainly in protected areas because the agricultural landscapes have been surveyed in South Milan Park and Lambro Park!

8.4 Bionomic Analysis of Agricultural Landscapes

8.4.1 Main Types of Agrarian Landscapes

For the restoration of agricultural landscapes, action must be taken at various levels, considering that the ecological efficiency of ecotopes managed with Organic Farming methods is much higher than that of Conventional Agriculture, as it has also been directly experienced, and which requires a diagnosis at different spatiotemporal scales performed with bionomic methodologies before any therapeutic intervention.

The evaluations through the bionomic function of ecological efficiency are such as to be able to distinguish different types of agricultural landscape. Ecological Efficiency is much more complex than productivity and is measurable through the systemic state function called Territorial Biopotentiality (BTC) (remember: measurement of the negentropic energy flux linked to the degree of order and metastability of a system, expressed in Mcal/m^2/year, see Sect. 3.3.2). The BTC allows us to evaluate the order achieved by an ecological system, which explains why it is a significant function.

As it can be seen in Table 8.4, therefore the main types of agricultural landscapes, from the least to the most efficient, are:

Table 8.4 Normal ranges for the main types of agricultural landscapes according to bionomics

Bionomic parameters	SUB.R	AGR	AGR. PRT	Agr-Syl-Tour	AGR. FOR
RNT, forest area (%)	8.0–25	10.0–30	20–55	25–65	40–75
CBSt of forest patches	22.5–30.5	23.5–32.0	27.5–37.5	28–38	29.5–40.5
Allocthonous forest formations (%)	1–2.0	1–2.0	0–1.5	0–1	0–1
pCA, potential core area (%)	20–35	35–50	50–65	45–70	60–75
PRD, agricultural areas (%)	30–55	45–75	35–60	15–40	20–50
RSD, urbanized areas (%)	8.0–15	4–9.0	2–6.0	2–7.0	1–3.0
SBS, industrial & transport (%)	7.0–12	4–7.0	2–4.0	1.5–3.0	1–2.0
HH, human habitat (%)	70–86	60–82	42–68	20–40	22–48
HS/HS* carrying capacity	0.6–3.0	1.5–9.0	3.0–12.0	1.2–12.0	3.5–15
BTC, LU (Mcal/m^2/a)	0.80–1.60	1.00–2.00	1.70–3.50	1.8–4.5	3.00–5.00
LTpE = 10 (HNd/HU) + BTC_{A+F}	5.7–11.0	10.2–24	22–40	34–55	45–88
IFF, river functionality	180–200	180–200	200–250	200–270	250–270

Landscape types: *SUB.R* suburban-rural, *AGR* agricultural (productive), *AGT.PRT* agricultural protective, *AGR-SYL-TOUR* Agri-sylvo-touristic, *AGR-FOR* agricultural-Forest

Table 8.5 Comparison between some types of agricultural landscapes according to bionomy

Agricultural landscapes	Conventional		Organic		Alpine	
Main components	%	BTC	%	BTC	%	BTC
Meadows	12.00	0.75	15.00	0.80	60.00	0.90
Grasslands	17.00	0.45	8.00	0.50	6.00	0.50
Crops	60.00	0.78	58.00	0.82	14.00	0.80
Rows	1.00	2.85	3.00	3.20	1.00	3.00
Forest	2.00	5.60	9.00	6.10	17.00	7.20
Urbanized	8.00	0.15	7.00	0.15	2.00	0.15
Total	100.00	0.79	100.00	1.29	100.00	1.94

(a) Suburban-rural L., with optimal BTC from 0.80 to 1.60 Mcal/m^2/a, which, however usually does not exceed values equal to 0.70–0.90 Mcal/m^2/year,
(b) Traditional agricultural-productive L., with optimal BTC from 1.00 to 2.00 Mcal/m^2/a, but which today stops often at values of 0.90–1.40,
(c) Agricultural-protective L., with optimal BTC of 1.70–3.50 Mcal/m^2/a, which today however stops at values of 1.60–2.80,
(d) Agro-forestry-tourist L., with optimal BTC of 1.80–4.50 Mcal/m^2/a, in the Alpine valleys richest in forests.

These values represent conditions of optimal normality (absence of syndromes) for the landscapes of the temperate zone. Minor adjustments are to be considered necessary in all cases involving particular conditions. We now present Table 8.5,

Paesaggio Agrario Protettivo				Paesaggio Agrario Produttivo				Paesaggio Agrario Industriale			

HH	BTC	BF	gLM	HH	BTC	BF	gLM	HH	BTC	BF	gLM
% LU	Mcal/m^2/a	% Norm	bit x BTC	% LU	Mcal/m^2/a	% Norm	bit x BTC	% LU	Mcal/m^2/a	% Norm	bit x BTC
68,0	1,72	0,90	24,2	77,2	1,28	0,80	14,6	91,7	0,76	0,45	3,65

Fig. 8.6 Comparison between three types of agricultural landscapes: Protective, Productive, and Industrial. Note that the bionomic parameters used are very precise in the ecological evaluation of these landscape units, which are located (from left) in the Oltrepò Pavese, in the South-Milan Agricultural Park, and south of the Po near Voghera

which shows the diversity of bionomic-ecological conditions between conventional, organic and alpine agricultural landscapes. The values marked are deduced from the average case studies in Ingegnoli (Ingegnoli, 2002, 2011, 2015) and Ingegnoli & Giglio (2005). Note that in the Po Valley itself, conventional farms (still the vast majority) have ecological efficiency values that are 35–40% lower than those of organic farming.

These differences must be associated with other more well-known data, such as the lower biodiversity of species in crops and meadows, the lower Omega-3 content in milk, the disappearance of an adequate ecological network of tree-lined rows, excessive use of antibiotics, with consequences of *increased AMR* (*Antimicrobial resistance*), pollution of water networks, etc. In Fig. 8.6, you will see how different protective, productive, and industrial agricultural landscape types can be. Only four parameters are used: human habitat (HH, %), vegetation biopotential (BTC, Mcal/m^2/year), bionomic functionality (BF, %Norm.), and information content (g-LM, bit × BTC).

The latter value goes from 24.2 for the protective landscape (PRT) to only 3.65 for the industrial one, which, thus, has lost approximately 85% of its potential ecological efficiency and information content (see Sect. 7.4.1). Neuro-psychic imbalances can, therefore, also arise in people who must live in these non-landscapes, with further repercussions of stress and increased morbidity, as we will see further ahead. Finally, industrialized agriculture reduces the so-called Ecological Services to forced production because the precious services that can only occur on a landscape scale are eliminated, such as protective aspects.

These observations can also influence the economy because the identification of the values of the assets (agricultural and forestry lots) with the labor component alone is strengthened: as known, it leads neoclassical economics to forget Nature entirely and to not understand the equivalence between economic costs and bionomic-ecological costs, which would bring the values of the lots mentioned to an order of magnitude greater than today's distorted market value (Ingegnoli et al., 2018), as we will see later (see Chap. 12).

8.4.2 Agricultural Landscapes in the Alps

A reference to how the significant eco-bionomic parameters change in relation to the altitude ranges in the Alps is shown in Fig. 8.7. The agricultural land cover ranges from 60–70% in the plain around Milan to zero at around 2000 meters above sea level in the Valley of Aosta. Most of the parameters are mentioned in this volume. The differences among these main types of landscape are due to the different qualitative and quantitative presence, spatial arrangement and structural and/or functional role of heterogeneous natural and anthropic components, measurable with other state functions of the complex system under examination, such as the landscape apparatuses/eco-countries functional systems (L.App), network circulation and connection, human habitat (HH), carrying capacity (SH/SH*), vegetation efficiency (CBSt), etc.

Although Fig. 8.8 provides the best example of the Alpine agricultural landscape regarding BTC level, it is essential to clarify that agricultural landscapes in the Alps have substantially altered in the last century. Contrary to the trends at the Trentino-Alto Adige regional level, where in 77 years the arable land has decreased by around

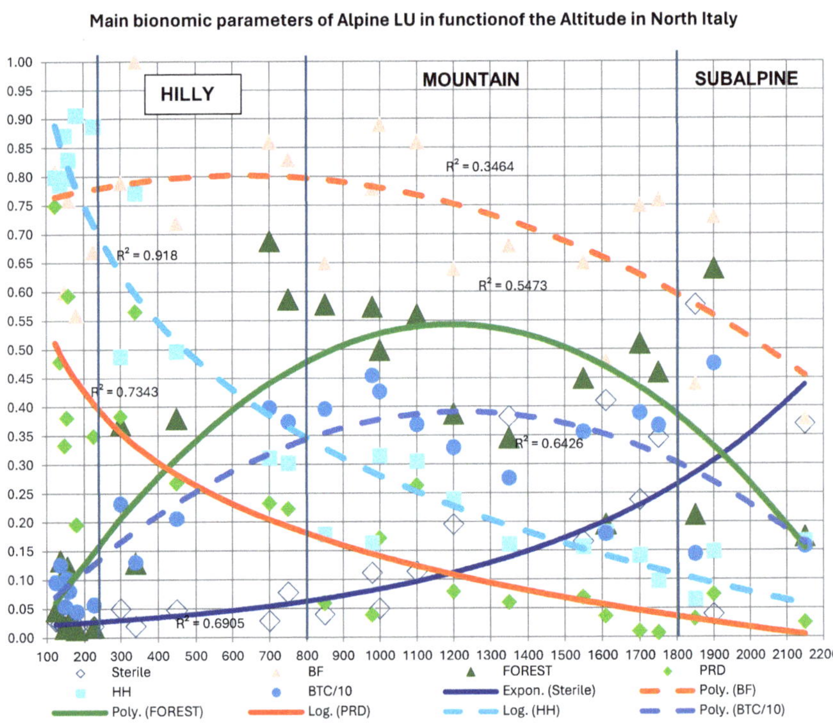

Fig. 8.7 Modification of ecological and bionomic parameters in relation to the Alpine altitudinal ranges. Note that in the Subalpine sector (>1700 m above sea level) as the sterile components (rocks and ice) increase, the BF and BTC values also decrease

Comparison: Trentino Region/Gresta Valley in the period of agricultural decrease		
Years: 1927-2007	PRD (%)	decrease
Trentino Region	8,9-3,9	56,18%
Region arable land	8,16-0,61	92,25
Gresta arable land	10,5-8,16	22,30
Gresta Valley	28,5-23,5	17,54%

Fig. 8.8 Terraces in Val di Gresta, at an altitude of approximately 650 m. The Valley was among the first to adopt predominantly organic farming, so much so that it created a specific District. (Watercolor by V. Ingegnoli, painted in 2012)

90%, as it can be seen for the Val di Fiemme, the Val di Gresta has managed to maintain a good level of agriculture with a decrease in arable land of no more than 22%, reduced to 17.5% for agriculture in general (i.e. including vineyards). All this has not prevented a tourist development, not massive but still significant, of the order of 1135 presences/km^2 already in 1997–98, only 33% less than the much more famous Val di Fiemme (Prov. Trento, 2010).

Its ecological-bionomic state has meanwhile improved, with an increase in the territorial biopotential of the vegetation, even if the ecological functionality has not yet reached full normality, achieving a value of 80%: an implementation of organic agriculture can contribute to increase this value, provided that it grows in parallel with an ecological improvement of the entire valley, which is also a useful driver of attraction for quality tourism. Studies in this sense would be desirable.

8.5 Alteration of Agricultural Landscapes in Lombardy

8.5.1 Framing at Different Scales

To have a framework and a yardstick for comparison, consider that for a couple of centuries the Lombard agricultural landscapes (fields, rows, and woods) had been remaining constant with a BTC equal to 1.60–1.80, while in the last 50 years the average BTC of these landscapes fell significantly below the sustainability limit (equal to 1.20), recording values of BTC = 0.80–0.90 Mcal/m^2/year (Fig. 8.9). Let us now return to the conditions of the alteration of agricultural landscapes to delve deeper into this vital aspect.

A study in the South-Milan Regional Agricultural Park in Albairate shows the local transformations of an agricultural landscape. Figure 8.10 is displayed based on the HH/BTC diagnostic model. From 1900 to 1990, the local transformations occurred within the tolerance band (dotted yellow line), equal to BF = 0.88. After

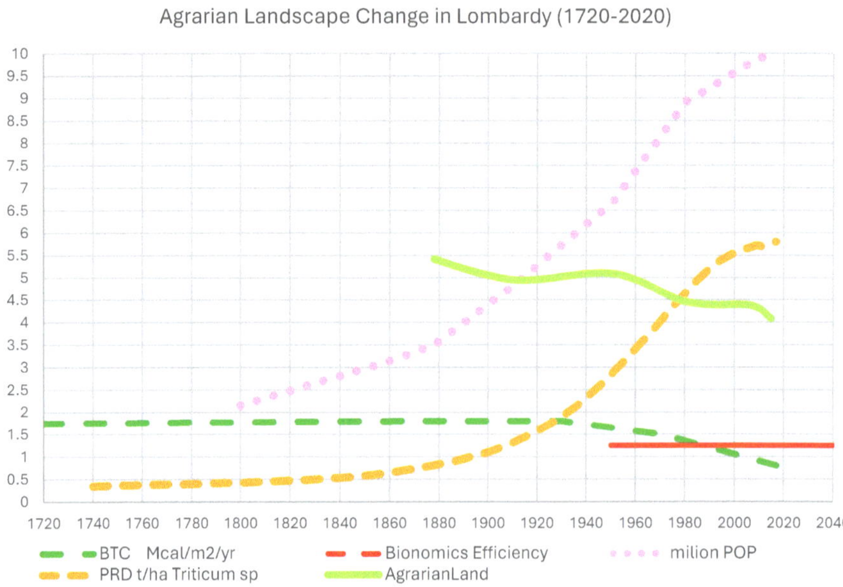

Fig. 8.9 Alteration of agricultural landscapes (cultivations and woods) in Lombardy 1720–2015. With the start of the Green Revolution (1930) degradation also began, today below the minimum level of normality (red)

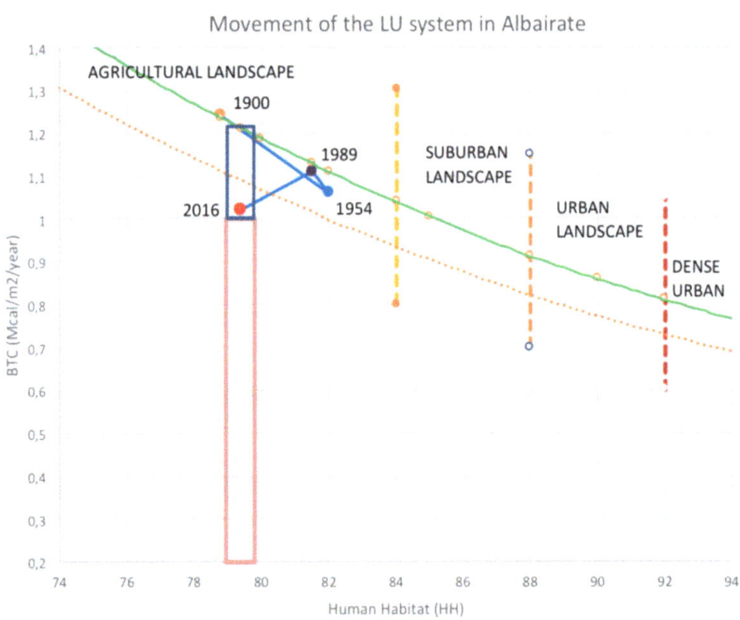

Fig. 8.10 South-Milan Regional Agricultural Park, municipality of Albairate (1900–2016). Movement of the system in the landscape unit (LU) examined. Note that until 1990 the local transformations were all within the tolerance range, while the latest one (2016) comes out almost perpendicular

this period, the transformation of the agricultural landscape drastically changed direction, leaving this band perpendicularly. The result is an altered bionomic functionality equal to BF = 0.80 due to the alterations of industrialized agriculture, which, among other things, has destroyed the network of tree-lined corridors. As we can see, the BTC of the Albairate LU does not exceed 1.03 Mcal/m^2/year, while the Lombard agricultural landscapes have been having BTC values = 1.25–1.30 Mcal/m^2/year for at least two centuries, a value that could even reach BTC = 1.70–1.80 also considering the patches of forest in these landscapes, as shown in Fig. 8.10. One must, therefore, give more weight to Agroecology, trying to restore these agricultural landscapes.

8.5.2 Networks of Tree Corridors

These corridors, which have been characterizing the agricultural landscapes of Lombardy for over two centuries, have indeed been greatly affected by agricultural industrialization (Fig. 8.11). In the example of Albairate, they show a decrease not only in surface area (from 115 ha in 1954 to 50.7 ha today), but also in BTC (from 3.20 to 2.90 Mcal/m^2/year).

The calculation of the ecological network's state, conducted according to planar graph theory (Forman & Godron, 1986), gave disastrous results. The connectivity $\gamma = 0.22$ in 1954 has halved today, while the *circuitry* $\alpha = 0.05$ is even negative today. The width of the fields has consequently changed, going from around 100 m in 1954 to over 350 m today.

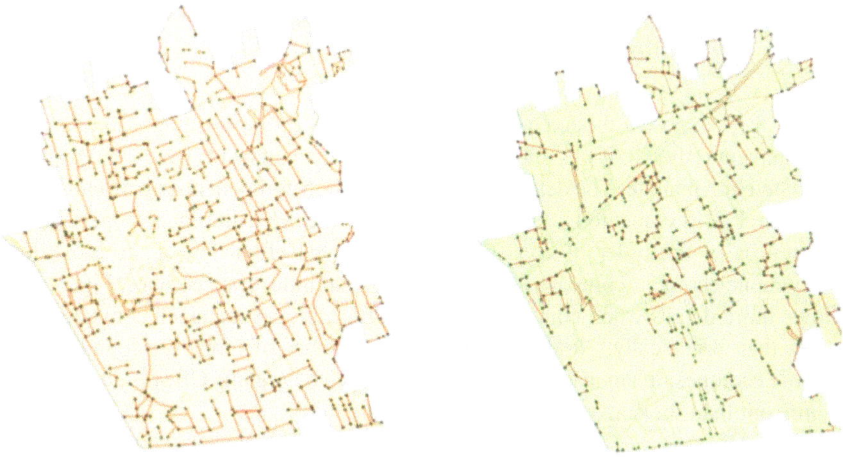

Fig. 8.11 The tree-lined corridors in Albairate (South-Milan Agricultural Park) surveyed in 1954 and 2016. Connectivity has halved, circulation is now negative

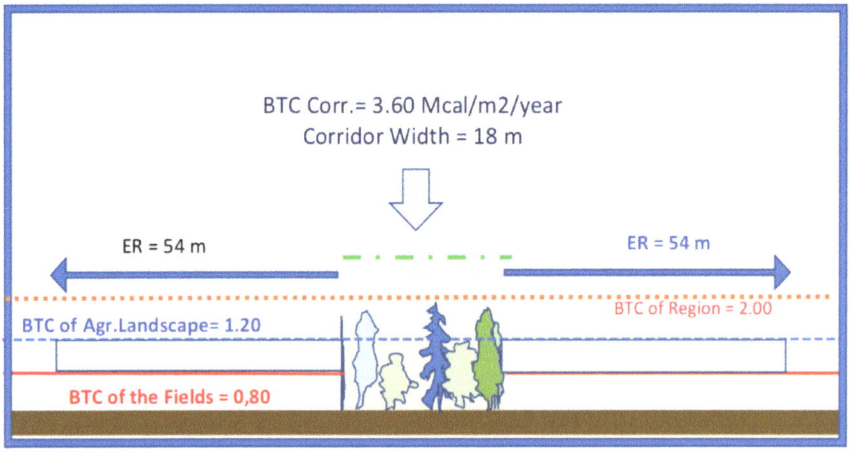

Fig. 8.12 Ecological radius given by a tree-lined corridor 20–25 m high and 18 m wide, with BTC = 3.60 Mcal/m^2/a. It follows that the optimal width of the fields between 2 rows would be approximately 110 m

A basic theoretical check following the concept of ER (ecological radius, see Sect. 7.4.7), with the data relating to Albairate, shows that the optimal width of the fields is around 110 m, as seen in Fig. 8.12. The ecological radius is given by a tree-lined corridor with dominant *Quercus robur*, 20–25 m high and 18 m wide, with BTC = 3.60 Mcal/m^2/year, reaching 54 m. The optimal width of the fields between 2 rows is approximately 110 m.

8.5.3 Comparison Between Conventional and Organic Agricultural Ecotopes

Figure 8.13 exposes about comparing agricultural ecotopes under Conventional management (BF = 0.68) vs. those under Organic management (BF = 0.85), always evaluated in Albairate. The detection of the vegetation components, again carried out with the LaBiSV method, which characterizes the agricultural ecotopes, shows differences of the order of 25% of more ecological efficiency of the biological/biodynamic ecotopes compared to the conventional/industrialized ones. Figure 8.14 shows more details when comparing agricultural ecotopes managed conventionally rather than organically. Note that the residual patches of forest are found only in organic ecotopes; in any case, for each component, the organic CBSt is always more significant than the conventional one.

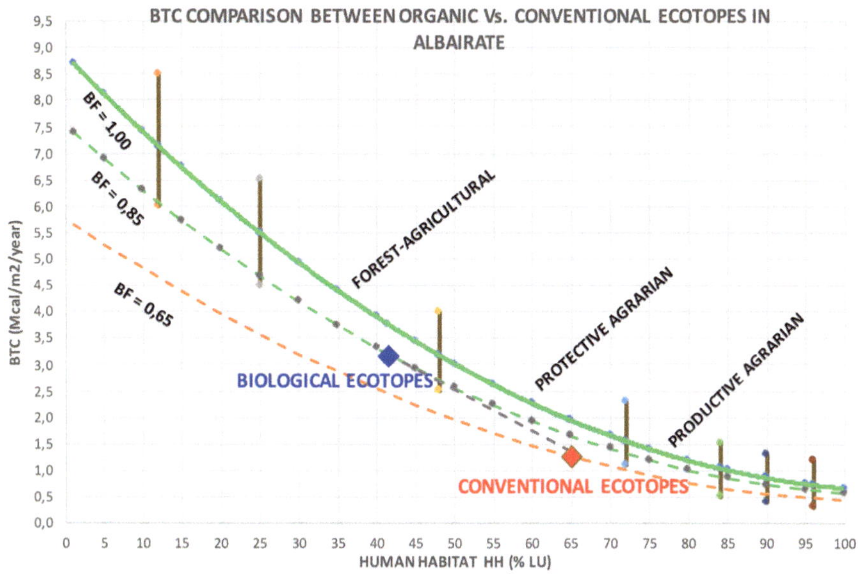

Fig. 8.13 South-Milan Regional Agricultural Park, municipality of Albairate. Comparison between conventional agricultural ecotopes (red) BF = 0.68 and organic (blue) BF = 0.85. See Fig. 7.7 for BF reference values

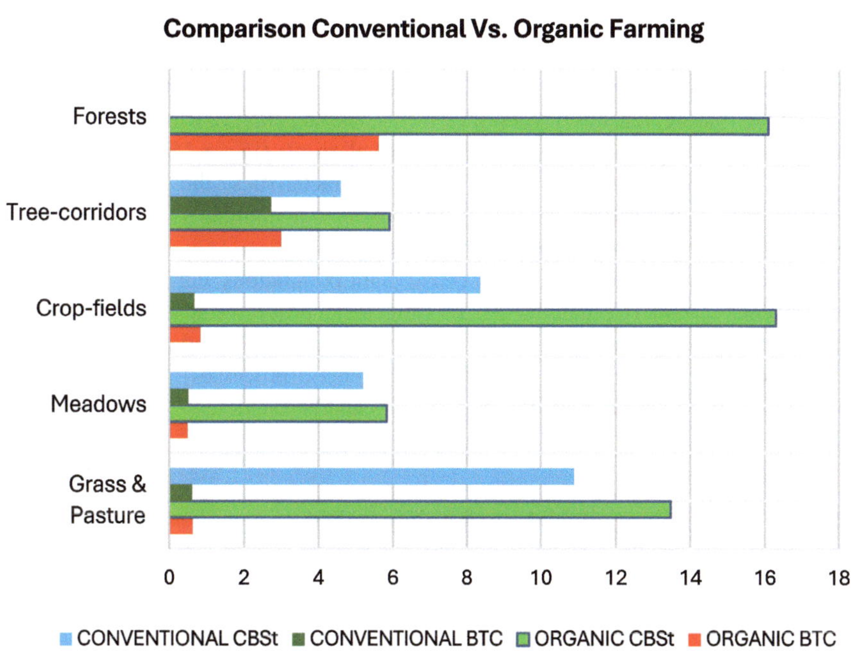

Fig. 8.14 Comparison between the main bionomic parameters (BTC and CBSt) relating to conventional vs. organic agricultural ecotopes, in two contiguous landscape units within the Municipality of Albairate (Lombardy), detected with the LaBiSV method. Patches of forest are present only at the edges of organically farmed ecotopes. Note that the ecological efficiency of all the tesserae with organic farming is higher than that of conventionally cultivated fields.

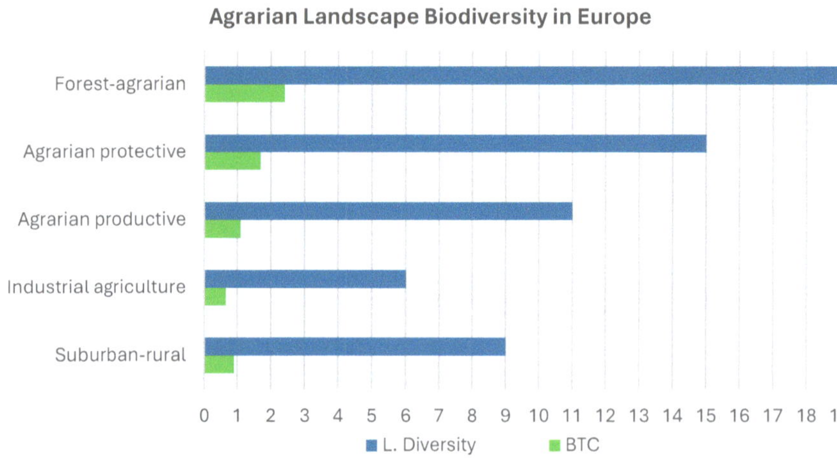

Fig. 8.15 Biodiversity of the elements of the agricultural landscape in the main types of formations. Note that industrialized agriculture (barely) reaches 6 types of elements, less than half of a mixed productive-protective landscape

8.5.4 Other Limits of Industrialized Agriculture

The *biodiversity of the elements of the agricultural landscape* in the main types of formations is of great importance. Remembering what was said in paragraph 2.2.4, this diversity must be neither insufficient nor exaggerated. Note that industrialized agriculture (barely) reaches 6 types of elements, less than half of a mixed productive-protective landscape, which usually has 13–14 elements (Fig. 8.15).

Due to cultivars with the same genetic characteristics, the homogeneity of monocultural industrialized crops is dangerous for our immune system and mental health, disrupting the diversity of our gut microbiota.

Agroecology suggests cultivating cereal plant populations, e.g., mixed varieties of *Triticum durum* (Ceccarelli, 2013). Epigenetic processes lead to the famous ‘Evolutionary Plant Breeding’ (EPB): local environmental characters drive DNA expressions and make them transmitted (epigenetic processes). Note that these populations need not be treated with pesticides and herbicides.

8.6 The Concept of Urban Region (UR)

8.6.1 Structure and Functions of Urban Landscapes

The city is the most typical of the ecological systems formed by man and one of the most recent in the natural evolution of the biosphere. The town was born from the integration of neighborhoods with complementary functions, as a residence with protected gardens and allotments, as a market, as a public management and defense, as a religious center, as a transport crossroads: that is, as an interacting set of human environments. We should speak more properly not of urban ecosystems but of urban landscapes, formed in a natural supporting landscape following the presence of various agricultural landscape units, necessarily complementary to the urban ones. For these reasons, it is necessary to combine territorial sector studies, which follow an ecosystemic logic, with landscape bionomics studies (Forman & Godron, 1986; Ingegnoli 1980, 1993, 2002, 2015; Ingegnoli & Giglio, 2005; Naveh & Lieberman, 1984; Zonneveld, 1995).

Furthermore, it is essential to reiterate that a landscape, even an urban one, is much more than a spatial context of different environments because it is a specific level of biological organization.

The origin of the structures of the urban landscape depends on the high specialization of roles in the population, which is unthinkable in a village (where each family produces a large part of its own needs). For these reasons, the city has a structure analogous to a heterotopic organism, therefore precisely to a landscape. In this complex system, the road network (which also includes squares) represents the most evident aspect of the structure in the plan. It can be stated that urban structures are part of biological structures, as many scholars have understood, starting from Marcus Vitruvius (n.d.) up to the Finnish Saarinen (1962). Forman and Godron (1986) also express a similar opinion and demonstrate that, in summary, the urban structure is made up of two elements of the landscape: the streets and the "city blocks" (i.e. the *insulae* in the Latin sense). Neighborhoods are characteristic sets of the previous elements.

The basic structural models for urban landscapes are, according to Forman and Godron, at least three: (1) concentric zone model, where neighborhoods surround a business center similarly in each direction; (2) segment model, in which diversified neighborhoods surround a center in a radial pattern, (3) multiple core model, in which an asymmetric mosaic of different neighborhoods surrounds one or more central cores. Other structural schemes can be added to these models, as the history of urban planning suggests, e.g., (4) the citadel and market model, one of the oldest (early Iron Age), (5) the square mesh model of Roman origin, in which the different neighborhoods are set up regularly around an intersection of main roads (*cardus* and *decumanus*). New structures are also appearing today: (6) the large urban landscapes of "megacities" which include suburban landscapes and form ecological systems of less stability and difficulty to manage. In ecology, it is known that the more the flow of inputs and outputs, storage and distribution are enlarged, the less stable the system is (Zonneveld, 1995).

8.6.2 The Urban Region According to Forman

The structure of urban landscapes is very complex, even if an urban region can be characterized by many structural parameters, according to Richard Forman (2014,

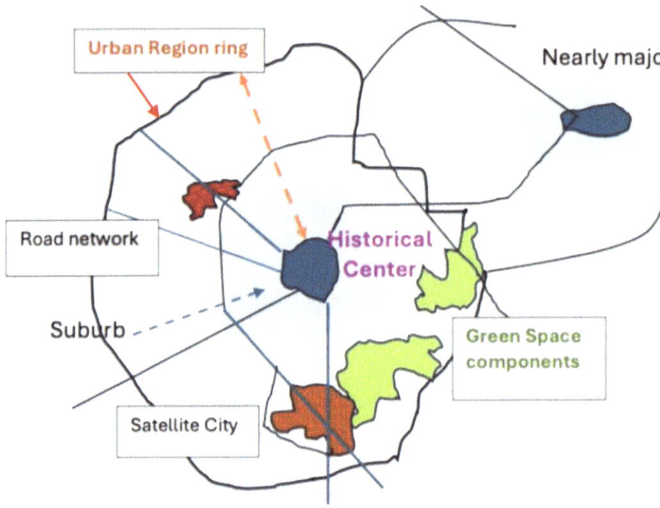

Fig. 8.16 Concepts and terms concerning an Urban Region (UR), according to RTT Forman (2008). The definition of UR acquired a deeper significance following Landscape Bionomics principles and methods

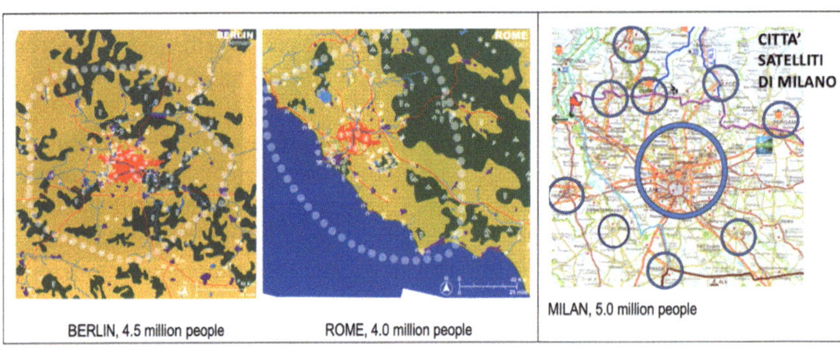

Fig. 8.17 The cities of Berlin, Rome, and their urban region according to Forman (2008) and Milan according to Ingegnoli (2011). Note that Milan is the largest city but has a smaller Urban Region than the other two, which however are the capitals. The Northern satellite city is Lugano, CH

2008). The urban region is the territory in which the dominant town exerts its influence both as economic and ecological operator.

Figure 8.16 distinguishes: (i) a Metropolitan Area with the inner Historical City, (ii) one or more satellite cities, (iii) a vast mobility network, (iv) some areas of natural reserves or parks, (v) a supporting agricultural matrix. Forman compared 38 large and medium-sized cities located on each continent, considering some concepts of landscape ecology (e.g. number of natural landscapes per urban region, urban parks, water resources, ecological networks).

It is necessary to report in Fig. 8.17 the comparison between the Urban Regions (UR) of Berlin and Rome, proposed by Forman (2008) and that of Milan cited above

(Ingegnoli, 2015). Note that the largest city is Milan (5.25 million inhabitants) which appears to have a decidedly lower UR than the other two, which however are capitals.

Milan also has an Urban Region of the same order of magnitude as Rome and Berlin, considering the complete Lombardy Region, which does not coincide with the current one, but in fact also includes the Canton of Ticino. The reference metropolis for Locarno, Bellinzona, Mendrisio and especially Lugano is in fact Milan, of which they are satellite cities, dating back to Roman times. Speaking of which, it must be remembered that the concept of urban region has a historical reference in the Roman world. In fact, the real definition of City for Vitruvius and Columella (Columella, 1955; Vitruvius, n.d.) is the following: *Civitas* = *Urbs* + *Territorium*, showing that they have already understood the definition of city in a systemic sense!

On the contrary, today traditional ecology is unable to identify the phenomena that characterize the territory: it's interesting to compare the Metropolitan Area of Milan (that is the more densely urbanized and crowded one) (Fig. 8.18) with the Metropolitan City of Milan, which includes a large portion of agricultural fields to the south and west of the city proper. The difference is dangerous because it inevitably leads to ever more ruining the territory; the North urbanized areas (more than two million people) are not comprised, while the S-W rural one are considered as urban.

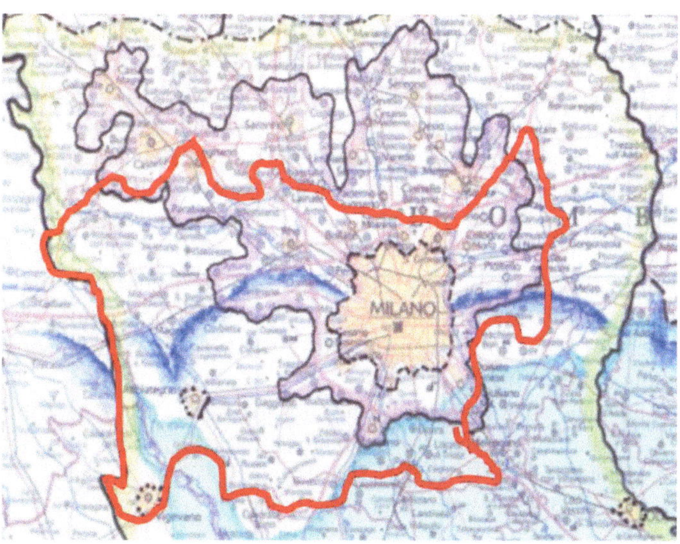

Fig. 8.18 The more densely urbanized metropolitan area of Milan, the fourth largest conurbation in Western Europe after Paris, London, and Madrid, is marked in pink. The comparison with the current political-administrative "Città metropolitana"/Metropolitan City (red line) shows how political-administrative choices can be in contrast with scientific ones

8.7 An Integrated Approach to Detect Alterations of Urbanized Landscapes

8.7.1 From Survey to Diagnosis: Shifting LaBiSHH into LaBiDHH

The concept of the urban region comes to be valid and precise; however, there is a lack of connections to theories and methods capable of reasoning even quantitatively. Landscape Bionomics also comes in handy in this field. Minding the definition of Human Habitat (cfr. 3.1.4), as the set of areas in which the human population lives or the ones they permanently manage (i.e., in which the self-regulation capacity of natural systems is limited and subsidiary energy is added), Landscape Bionomics has already proposed a method of biological survey of Human Habitat, the LaBiSHH (Ingegnoli, 2011): here the next step is proposed, that is the LaBiDHH (Landscape Bionomics Diagnosis of Human Habitat) to diagnose it. It is a new systemic approach for the evaluation of urbanized landscapes by integrating the bionomic dimensions with the structure and functions of urban planning and infrastructure.

As enlightened in Sects. 7.4.2 and 7.4.3, the first diagnostic frame for a landscape or for a LU is the relationship between the HH and the BTC, that is the degree of Bionomic Functionality, which also revealed to what type of landscape its current operation/functioning belongs to. To check if this surveyed functionality is the proper one, the calculation of the landscape typology using the LTpE index (Ingegnoli, 2015) is useful: after, we may access the reference range of parameters to be used for diagnosis (see Table 7.2).

Alteration of an ecological system is also revealed by the absence, reduction, configuration and characteristics of the Landscape apparatuses or, at least, by the % presence of the major categories of urbanization, agriculture, forest.

Figure 8.19 frames the relationships among Human Habitat and the five most important bionomic apparatuses/ functional systems of HH as they appear from the study of 57 LU within Europe. Note that the RNT (Forest land cover) is expressed by a straight line, while the other correlations assume high degrees polynomial lines.

The dispersal accounts of the great diversification of landscape types. Note that between 25% and 75% of HH we find the increase of PRD (agriculture), from 75% to 96% the increase is due to urbanization (RSD + SBS). The BTC function presents two opposing but contained inflections.

Deepening some other integrated indicators are useful such as carrying capacity (SH/SH*). The shifting between the simple LaBiSHH (Landscape Bionomic Survey of Human Habitat) to the LaBiDHH (Landscape Bionomic Diagnosis of Human Habitat) is presented through different case studies.

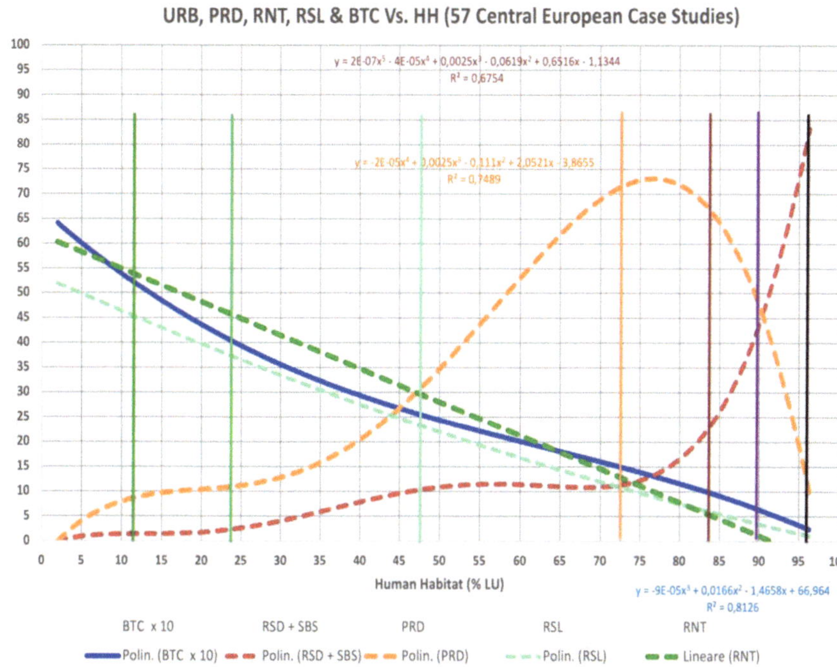

Fig. 8.19 Structure of the landscape types (see Fig. 7.7) in function of their human habitat (HH). We can appreciate the main landscape elements (L.App) whose interrelation characterize the landscape. Note the difference between the Semi-Natural (HH = 18.0%) and Sub-Urban (HH = 87.0%) dispositions of the L.App. (BTC = 4.60 vs. 0.80 Mcal/m^2/year)

8.7.2 *The Diagnostic Meaning of Bionomic Functionality of Urbanized Landscapes at Different Scales*

A study was done for the hinterland (Metropolitan Area) of Milan, and then similar studies were carried out for both the Historic Center and the Metropolitan Area. The comparison graph on the environmental state of Rome and Milan concerning Bionomic Functionality is in Fig. 8.20. On the left, the values of the Metropolitan Area and Historical Center of Rome are marked in yellow circles on the HH/BTC diagnostic model (remember Fig. 7.7), while those of Milan are in red. We immediately notice that: (a) the Historical Center of Rome pertains to the same landscape type of the Metropolitan area of Milan, as it is historically equipped with an adequate extension of qualitative green areas; (b) Rome remains with positive values (not lower than the norm, i.e. the green curve), while Milan has both negative values. On the right you find the incidence of tumors and cardiovascular diseases in the two cities: Milan records 14–15% more cancer. Since the incidence of cancer in Lombardy (2012) is equal to 5.88/1000, this difference is very high, corresponding to the 5.2 million inhabitants of the Milan Metropolitan Area, with 4604 more sick people/per year than in Rome.

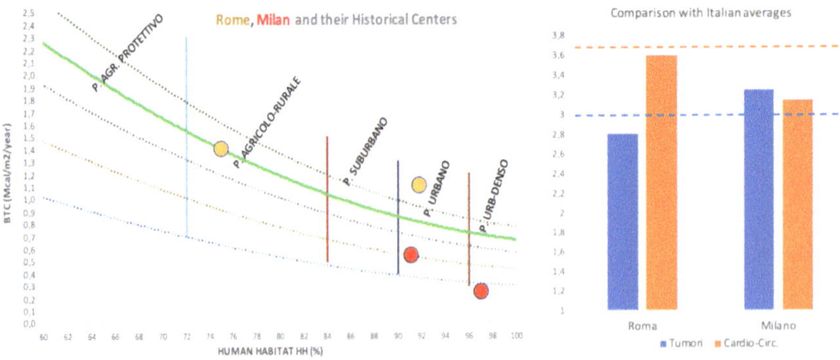

Fig. 8.20 Bionomic-ecological state of the cities of Rome and Milan (referred to 2015). On the left, the values of the Metropolitan Area and Historic Center of Rome are marked in yellow on the HH/BTC diagnostic model, and those of Milan in red. We immediately notice that Rome remains with positive values (not lower than the norm), while Milan has both negative values. On the right, the incidence of tumors (blue) and cardiovascular diseases (orange) in the populations of the two cities in 2012.

8.7.3 *The Diagnostic Meaning of Landscape Apparatuses of Urbanized Landscapes*

We must now show the applicative utility of Bionomics in diagnosing the most significant landscape functional subsystems, an essential operation, e.g., for planning capable of going beyond a Park's zoning. The tool of Territorial Coordination Plans (PTC) is insufficient to establish the accurate diagnosis of the landscape units (LU) that compose it:

We have shown that systemic indicators are essential.

A suitable *method* for studying the multifunctional system of landscape apparatuses (L.App.), thanks to their ability of recording with good sensitivity the state of health of the landscape under examination, includes the following operations:

1. *grouping of landscape elements by apparatuses: residential (RSD), subsidiary (SBS), protective (PRT), productive-agricultural (PRD), resilient (RSL), resistant (RNT), hydrogeological (HGL);*
2. *measurement of the surface area by element and by apparatus;*
3. *measurement, as above, of the percentage with respect to the entire LU;*
4. *estimation of the human habitat (HH') for each individual element (cfr.* (Ingegnoli, 2011; Ingegnoli & Giglio, 2005))*;*
5. *calculation by weighted average of the human habitat (HH);*
6. *field study of the most significant values of the vegetation formations using the LaBiSV method (cfr.* (Ingegnoli, 2011; Ingegnoli & Giglio, 2005; Ingegnoli & Pignatti, 2007))*;*
7. *assessment of the vegetation biopotential (BTC) for each element;*
8. *calculation by weighted average of the BTC;*

9. *calculation of the standard habitat per capita (SH);*
10. *estimate of the minimum theoretical standard habitat for climatic and altitudinal zone (SH*)*
11. *assessment of the Carrying Capacity SH/SH**

8.7.4 Difficulties in Data Collection and Processing: The Example of Robecco

An *example*, carried out in the municipality of Robecco sul Naviglio with the help of a student of prof. Ingegnoli (Grassi & Ingegnoli, 2006), is shown in the Figs. 8.21, 8.22, and 8.23 (data from (Dusaf-ESRSAF, 2015)).

Note the sharp difference between Figs. 8.22 and 8.23, where forests are reported in only three generic and incorrect typologies, vs. a forest bionomic analysis. Figure 8.22 is a near casual example of the land cover collection by the technical offices of Lombardy Region, while Fig. 8.23 expose the data ordered following the previous methodology (1–11)

1. *We must add a synthetic example on the forest evaluation.*

We can see that about 5% of land cover (101.4 ha) is made by high forests, and we can report at least seven types of woods (Ingegnoli & Pignatti, 2007; Pignatti, 1998):

1. *Quercus Carpinetum* (*Q. robur, mesophilic*)
2. *Quercus Carpinetum* (*Q. robur and rubra*, degraded)
3. Alder grove (*Alnus glutinosa)*
4. Poplar forest (*Populus alba and P. canescens*)
5. Willow forest (*Salix alba*)

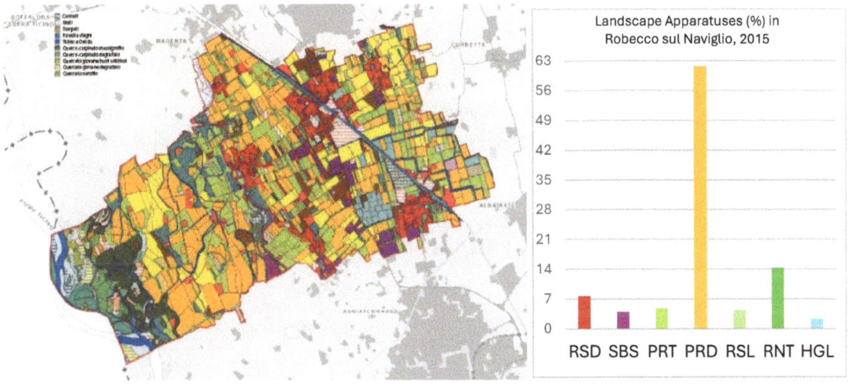

Fig. 8.21 The municipality of Robecco sul Naviglio (Lombardy, Italy; 20.5 km^2 and 6700 inhabitants). (left) Land use tesserae (Grassi & Ingegnoli, 2006). (right) The main landscape systems in percentage (2015). Note the strong prevalence of agricultural production (PRD) and the discrete presence of forests (RNT)

Robecco sul Naviglio, Land-use Data, ERSAF-Dusaf, 2015

134	degraded areas that are not used and not vegetated	0,52
213	rice fields	66,30
222	orchards and minor fruits	2,53
331	beaches, dunes and gravelly riverbeds	24,84
411	vegetation of internal wetlands and peat bogs	0,07
511	riverbeds and artificial watercourses	23,07
1112	continuous residential fabric of medium density	24,68
1121	discontinuous residential fabric	85,84
1122	sparse and nucleiform residential fabric	17,43
1123	sparse residential fabric	15,68
1221	Road networks and accessory spaces	0,42
1411	Parks and gardens	28,99
1412	Uncultivated green areas	11,32
1421	Sports facilities	4,98
2111	Simple arable land	1118,25
2112	arable land with trees	8,50
2115	family vegetable gardens	1,72
2241	poplar groves	101,98
2242	other woody agricultural plants	5,45
2311	permanent meadows in the absence of tree and shrub species	66,12
2312	permanent meadows with the presence of scattered tree and shrub species	1,33
F	rows of trees	38,52
3113	riparian formations	101,36
3222	vegetation of riverbeds	15,34
3223	vegetation of raised banks	2,07
3242	shrubs of abandoned agricultural areas	5,37
5121	Natural water basins	0,98
5122	water basins artificial	0,51
11231	farms	23,86
12111	industrial, artisanal, commercial settlements	46,93
12112	agricultural production settlements	25,95
12122	public and private service facilities	1,08
12123	technological facilities	7,84
12124	cemeteries	3,19
21131	open-field horticultural crops	0,39
21141	open-field floriculture crops	0,44
31111	medium and high density deciduous forests managed by coppice	193,06
31121	low density deciduous forests managed by coppice	3,59
	Total (ha)	**2041,98**

Fig. 8.22 Land-use and Land-cover data (ha) on the territory of Robecco sul Naviglio (Milan hinterland) surveyed by ERSAF-Dusaf Section of Lombardy Region, 2015. Note the traditional ranking criteria, quite chaotic

6. Robinia forest (*Robinia pseudoacacia*)
7. Xerophilous thicket

Moreover, the description of "low density deciduous forests managed by coppice" is insufficient to evaluate the bionomics efficiency of a forest formation, which needs at

	ROBECCO sul Naviglio POP. 67.60 ab (2017)		ha	%	HH'	HH_w	BTC'	BTC_w
1	Continuous residential fabric of medium density		24,68	1,21	99	1,196	0,1	0,001
2	Discontinuous residential fabric (80%)		68,74	3,37	97	3,265	0,15	0,005
3	Sparse and nucleiform residential fabric		17,43	0,85	96	0,820	0,15	0,001
4	Sparse residential fabric		15,68	0,77	95	0,730	0,2	0,002
5	Sports facilities		4,98	0,24	98	0,239	0,3	0,001
6	Cemeteries		3,19	0,16	99	0,155	0,25	0,000
7	Farms		23,86	1,17	92	1,075	0,2	0,002
		RSD	*158,56*	*7,77*		*		*
8	Degraded unused and unvegetated areas		0,52	0,03	98	0,025	0,05	0,000
9	Road networks and ancillary spaces		0,42	0,02	98	0,020	0,1	0,000
10	Industrial, artisanal and commercial settlements		46,93	2,30	99	2,275	0,2	0,005
11	Agricultural production settlements		25,95	1,27	98	1,245	0,25	0,003
12	Public and private service facilities		1,08	0,05	97	0,051	0,05	0,000
13	Technological facilities		7,84	0,38	98	0,376	0,05	0,000
		SBS	82,75	4,05				
14	Parks and gardens		28,99	1,42	85	1,207	2,4	0,034
15	Private gardens (20% RSD discontinuous)		17,10	0,84	89	0,745	1,3	0,011
16	Rows of trees		43,30	2,12	50	1,060	2,75	0,058
17	Uncultivated green areas		11,32	0,55	50	0,277	1,5	0,008
		PRT	100,71	4,93				
18	Rice fields		66,30	3,25	90	2,922	0,75	0,024
19	Orchards and minor fruits		2,53	0,12	87	0,108	1,4	0,002
20	Simple arable land		1.074,90	52,64	88	46,325	0,7	0,368
21	Tree arable land		8,50	0,42	84	0,350	1	0,004
22	Family vegetable gardens		1,72	0,08	90	0,076	0,75	0,001
23	Poplar groves		101,98	4,99	75	3,746	3	0,150
24	Other woody agricultural crops		5,45	0,27	80	0,213	1,9	0,005
25	Open field horticulture crops		0,39	0,02	90	0,017	0,75	0,000
26	Open field flower and nursery crops		0,44	0,02	90	0,019	0,9	0,000
		PRD	1.262,21	61,82		*		*
27	Permanent meadows		66,12	3,24	60	1,943	0,58	0,019
28	Permanent meadows with scattered tree and shrub species		1,33	0,07	45	0,029	0,8	0,001
29	Scrublands of abandoned agricultural areas		5,37	0,26	30	0,079	1,4	0,004
30	Vegetation of internal wetlands and peat bogs		0,07	0,00	10	0,000	2,5	0,000
31	Vegetation of riverbeds		15,34	0,75	10	0,075	2,9	0,022
32	Vegetation of raised embankments		2,07	0,10	50	0,051	0,7	0,001
		RSL	90,30	4,42				
33	Riparian formations (high forests)		101,36	4,96	12	0,596	6,5	0,323
34	Medium/High density deciduous coppice forests		193,06	9,45	15	1,418	5,6	0,529
35	Low density coppice forests		3,59	0,18	20	0,035	4,4	0,008
		RNT	298,01	14,59				
36	Beaches, dunes and gravel beds		24,84	1,22	10	0,122	0,1	0,001
37	Riverbeds and artificial watercourses		23,07	1,13	30	0,339	0,02	0,000
38	Natural water basins		0,98	0,05	10	0,005	0,1	0,000
39	Artificial water basins		0,51	0,03	40	0,010	0,05	0,000
		HGL	49,40	2,42				
		Total	2.041,9			73,24		1,59
	MAIN FOREST TYPOLOGY		character	%	HH		BTC	
	Quercus Carpinetum (*Q. robur, mesophilic*)		High forest	1,50	11		7.3	
	Quercus Carpinetum (*Q. robur and rubra*, degraded)		coppice	4.00	16		5.8	
	Alder grove (*Alnus glutinosa*)		High forest	1.10	10		7.8	
	Poplar forest (*Populus alba and P. canescens*)		coppice	2.45	15		5.1	
	Willow forest (*Salix alba*)		High forest	2,36	12		6.0	
	Robinia forest (*Robinia pseudoacacia*)		coppice	3.00	17		4.9	
	Xerophilous thicket		coppice	0.18	20		4.3	

Fig. 8.23 Elaboration of the Dusaf data 2015, following the Landscape Bionomy. Note the division in Landscape sub-systems (7 Landscape Apparatuses), the separation of small private gardens from discontinuous residential buildings (p.t 2 & 15), the bionomic analysis of forest typologies, the evaluation of HH' and BTC' per L. element and their weighted sum (HH_w, BTC_w)

least a value of its average height of the canopy and possibly a range of phytomass aboveground.

In summary, we can note that the systems in this municipality have a fair component of forest patches (RNT = 14.59%), even if poorly distributed because almost all in the floodplain part; on the contrary, the urbanized systems (RSD + SBS = 11.82%), are almost all in the terraced part. This municipality could be divided into two complementary LUs (see Fig. 3.7) but, considering the systems, Robecco can be viewed as a whole LU. The average BTC of this LU is equal to BTC = 1.59 Mcal/m^2/year.

As regards the values of HH and BTC, it can be observed that these values are normal for an agricultural landscape, but for a *protective* agricultural landscape, like the one to which the territory should belong being a LU of a park, they should change: HH should not be higher than 68% (today it is 73) while BTC should be higher than 1.70 Mcal/m^2/year (today it is 1.59).

We can now take into consideration the individual systems. It should be noted that the examination of the individual systems requires a composition of landscape elements that is partly different from that studied on a general scale. In fact, from the colors shown in Fig. 8.24 the protective (PRT) and resilient (RSL) systems, which are typically functionally shared with other systems, are composed of parts that must be considered by both the residential (RSD) and agricultural (PRD) systems.

Paying attention to the Tables 8.6 and 8.7, the values of HH entail corresponding values of BTC that, in the HH/BTC model (Fig. 7.7), can be read according to a correlation curve that provides the values of BTC-n (normality). As in Medicine, the analyses require a comparison with normality values for each element (or groups of elements) examined and this comparison provides the values of BF, that is, of the bionomic functionality.

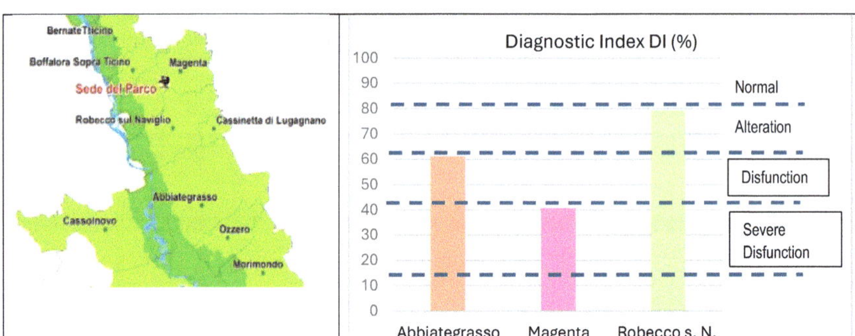

Fig. 8.24 (right) Exemplary comparison of three municipalities of the Ticino Lombardy Park evaluated according to a clinical-diagnostic method specific to Landscape Bionomy in reference to 2015. The three case studies consequently require different interventions, which a Systemic Territorial Plan must foresee and indicate. (Left) The municipality of Magenta, Robecco sul Naviglio, and Abbiategrasso. The territory of the Ticino Park (central part) is green. Dark green the fluvial bed

As seen in the tables, if BF-PRD = 0.88 is slightly altered, for BF-RSD = 0.82 the situation indicates a more serious syndrome, although not yet dramatic. These tests lead to a first general diagnosis, which should be followed by further investigations according to the multiple methodologies of Landscape Bionomics.

For examining the subsidiary apparatus (SBS), which mainly concerns industries and mobility, a table summarizing ecological-bionomic analyses already tested (Ingegnoli, 2011; Ingegnoli & Giglio, 2005) relating to industries can be referred to.

Table 8.6 Residential apparatus in Robecco sul Naviglio (2015)

RESIDENTIAL APPARATUS (RSD)	ha	%	HH'	HH	BTC'	BTC	BTC-n	BF
Continuous residential medium dense	24.68	13.65	99	13.513	0.10	0.014		
Discontinuous residential (RSD)	68.74	38.02	97	36.881	0.15	0.057		
Sparse and nucleiform residential	17.43	9.64	96	9.256	0.15	0.014		
Sparse residential	15.68	8.68	95	8.242	0.20	0.017		
Sports facilities	4.98	2.75	98	2.700	0.30	0.008		
Cemeteries	3.19	1.77	99	1.749	0.25	0.004		
Parks and gardens	28.99	16.03	85	13.628	2.40	0.385		
Private gardens (20% RSD)	17.10	9.46	89	8.418	1.30	0.123		
Totali	180.79	100.00		94.387		0.623	0.758	0.82

BTC-n, value of normality (mod. HH/BTC); BF, bionomic functionality = BTC/BTC-n

Table 8.7 Agricultural productive apparatus in Robecco sul Naviglio (2015)

PRODUCTIVE APPARATUS (PRD)	ha	%	HH'	HH	BTC'	BTC	BTC-n	BF
Rice fields	66,30	4,88	90	4,391	0,75	0,037		
orchards and minor fruits	2,53	0,19	87	0,162	1,40	0,003		
Simple arable land	1.074,90	79,10	88	69,609	0,70	0,554		
Arable land with trees	8,50	0,63	84	0,525	1,00	0,006		
Family vegetable gardens	1,72	0,13	90	0,114	0,75	0,001		
Poplar groves	101,98	7,50	75	5,629	3,00	0,225		
Other woody agricultural crops	5,45	0,40	80	0,321	1,90	0,008		
Open field horticultural crops	0,39	0,03	90	0,026	0,75	0,000		
Open field flower and nursery crops	0,44	0,03	90	0,029	0,90	0,000		
Farmhouses	23,86	1,76	92	1,615	0,20	0,004		
Permanent meadows	66,12	4,87	60	2,920	0,58	0,028		
Permanent meadows with tree and shrub species	1,33	0,10	45	0,044	0,80	0,001		
Shrublands of abandoned agricultural areas	5,37	0,40	30	0,119	1,10	0,004		
Totali	1.358,89	100,00		85,503		0,870	0,988	0,88

BTC-n value of normality (mod. HH/BTC); *BF* bionomic functionality = BTC/BTC-n

8.7.5 The Diagnostic Evaluation of the Ecological-Bionomic State

Remember again that we can arrive at a general systemic diagnostic assessment of a Landscape Unit by comparing the detected values with the normality intervals for each parameter, normality that depends on the type of landscape and that has been already evaluated and proposed by Ingegnoli (2011) for Europe, even if limited here to only 11 parameters: they are (1) Landscape typification (LTpE), (2) Forest biopotentiality (BTC-For), (3) Carrying Capacity (SH/SH*), (4) Biopotential of the landscape unit (BTC-LU), (5) Human habitat (HH), (6) Land Cover (%) of the 6 main landscape systems (RSD, SBS, PRT, PRD, RSL, RNT). Note that the reference landscape for protected areas must be at least an agricultural-protective landscape (AGR-PRT), as already mentioned.

To the study of the municipality of Robecco, the two municipalities of Abbiategrasso and Magenta (Fig. 8.24, left) are added as a further example of diagnostic evaluation (Table 8.8): three municipalities more different from each other than it appears from conventional planning considerations.

The red values in Table 8.8, which indicate alterations, are in fact different both in number and distribution: (1) Diagnostic Index DI for Abbiategrasso = 61.40%; (2) DI-Magenta = 38.64%; (3) DI-Robecco = 79.50%.

As established in the Bionomics texts (Ingegnoli, 2011, 2015) the main diagnostic evaluation bands are marked in Fig. 8.24 (right): we see a first degree of alteration for Robecco, a greater level of dysfunction for Abbiategrasso and a very severe degradation for Magenta (see Sect. 7.4.2). These different degrees of alteration of the health of the environmental systems considered that must be followed by appropriate

Table 8.8 Diagnostic systemic evalutation of the municipalities of Abbiategrasso, Magenta and Robecco (2015)

Parameters	AGR-PRT	Abbiategr	scores %	values	Magenta	scores %	values	Robecco	scores %	values
BTC – LU [Mcal/m²/yr]	1.7–3.5	1.59	−6.5	2	1.28	−24.7	1	1.59	−6.5	2
BTC – Forest [Mcal/m²/yr]	6.5–7.0	5.47	15.9	1	5.42	−16.6	1	5.89	−9.4	2
SH/SH*	3.0–12.0	0.68	−77	0	0.5	−47	0	1.38	−54	0.5
LTpE	22–40	13.95	−36.6	0.5	10.56	−52	0.5	15.86	−27.9	1
HH [% LU]	42–68	69.7	2.5	2	78.7	15.7	1	73.3	7.8	2
RSD [L. cover %]	2.00–6.00	8.9	48.3	0.5	16.6	176.7	0	7.7	29	1
SBS [L. cover %]	2.00–4.00	5.6	40	0.5	12.9	222.5	0	4.0	ok	2
PRT [L. cover %]	3.00–6.00	4.0	ok	2	6.4	6.7	2	4.9	ok	2
PRD L. cover %]	35–60	51	ok	2	47.9	ok	2	61.8	3	2
RSL L. cover %]	3.00–9.00	7.8	ok	2	2.0	−33.3	0.5	4.4	ok	2
RNT L. cover %]	20–55	17.6	−12	1	13.4	−33	0.5	14.6	−27	1
Totale				13.5			8.5			17.5
DI [%]	*> 85.00*			*61.40*			*38.64*			*79.50*

Evaluation Scores: ***0 – 0.10 =2****; – 0.10-0.30 = 1; – 0.30–0.60 = 0.5; > –0. 60 = 0;*

remediation interventions, which in turn can be controlled with the same methodologies.

Concerning Robecco, we will add that, since SH* = 1455 m^2/inhabitant, and SH = 2012.8 m^2/inhabitant, the Carrying Capacity is equal to SH/SH* = 1.38. This landscape is therefore autotrophic for its population, as it is also shown by the indicator LTpE = 15.86 which corresponds to an agricultural landscape type; however, since it is located in a Park, it would have been better if it were an agricultural-protective landscape, with LTpE >22.00 (Ingegnoli, 2015).

8.8 Territorial Anamnesis and History

8.8.1 History of the Territory/Landscape, Fundamental Analysis

In Fig. 2.6a, we have shown that the historical method, which is much neglected in land use planning, becomes mandatory again for reasons of advancement in the scientific paradigm. Having already recalled the importance of non-equilibrium thermodynamics in studying complex systems such as landscapes, we refer to the example of Ilia Prigogine et al. (1972) to better underline the historical aspects. In common language, the term history is used to connote the temporal transformations of a system or, more generally, of an entity. At the same time, scholars refer to the etymology of the Greek-Latin term *historia*, which properly means "research, investigation, cognition", derived from an Indo-European root, from which the Greek ιστωρ is also "he who knows".

It should be remembered that the development of a landscape as a living entity cannot be reduced to a simple event. Consequently, extremist visions of scientism and fanatical anthropocentrism must be overcome to study a complex system such as the landscape seriously. History is the research on the evolution of natural systems, that is, the 'occurrence' of phenomena in the past. It must be clear that only an 'eventful' conception of nature can overcome the dichotomy between an ultra-anthropic and a scientistic sense of history, being able to appropriately develop historical studies on complex systems such as the landscape.

By following this conception, it will be possible to avoid the occurrence of distortions that lead:

(a) to the arbitrary fragmentation of landscapes (Forman & Godron, 1986) as if they were indifferent geometric portions of space assumed to support events and political-administrative choices of man, opportunistic or journalistic; or on the contrary
(b) to the deterministic conception of the landscape, according to a descriptive meaning of the succession of typifications (Langé, 2008), formalized abstractly in conformity with some rules that are presumed valid in the absence of human interference.

As Luigi Zanzi (1995) wrote, speaking of "environmental" history is necessary today. This well-known historian, who not by chance was in contact with Ilya Prigogine, states that the concept of environmental history is to be adequately understood as local history, that is, as the history of a place (for example, of a landscape unit) considered in its autonomous and differential individuality starting from its environmental consistency. 'Natural history' must be replaced by a 'history of nature', attentive above all to the process of formation and continuous metamorphosis of individual 'local landscapes'. Every place has its history, completely unrepeatable, which must be reconstructed according to the traces that territorial transformation events leave locally. In this perspective, there is no longer a nature that can be universally assumed as an object of history. However, there are many natural places whose history of nature is a theoretical compendium.

The history of an evolutionary entity (e.g. a landscape unit) consists mainly in the story of the events through which it was formed. It follows that finding the data needed to reconstruct landscapes in the recent past becomes, often, a difficult task. As is known, the main sources of information on the past of human and natural components constitute five sets. These sources are: (a) bibliographic; (b) cartographic; (c) iconographic; (d) real direct; (e) real indirect (see (Ingegnoli & Giglio, 2005) with updates).

It should be emphasized that the critical, integrated, comparative and reasoned reading of the information derived from the sources necessary for the study in question, a reading aimed at the main objectives of the research, constitutes a basic methodological scheme that is always valid. It should be remembered that the reconstructions of at least two past states are necessary for the study of the dynamics of transformation of a landscape unit and this indication is fundamental for orienting the historiographical analyses. In fact, at least three points are necessary to recognize the dynamic trend of a curve (Fig. 8.25). The distance from the current state of the two reference periods in the past is rather variable and essentially depends on two factors: (1) the main scale of interest; (2) the availability of the sources for finding the data.

It is known that the larger is the spatial scale of a landscape unit (LU), the more it is necessary to move back the historical periods of reference. It should be added that a general methodological framework is not sufficient to address our topic: it may be necessary to have a further outline starting from the main types of sources.

8.8.2 *The Critical Reading of the Maps*

There is no doubt that local maps and geographical and topographical maps remain without doubt the primary source of information for the historical reconstruction of a landscape unit. It should be remembered that the critical reading of maps must always start from the present situation, to understand well the modalities of transformations, and to correctly interpret the typologies of landscape elements that have been detected, referring to a series of lines and areas characterizing the locality.

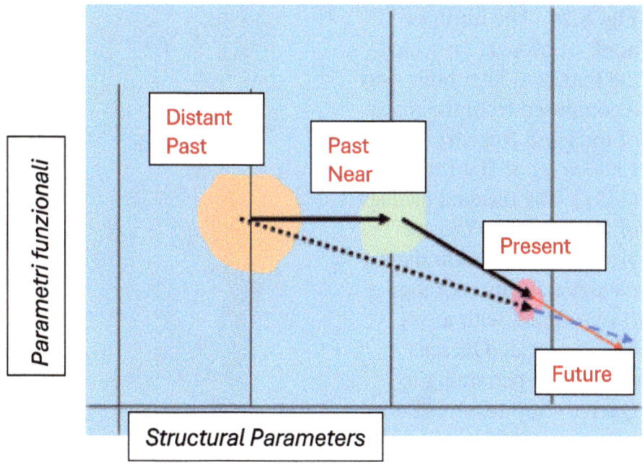

Fig. 8.25 A land use planning project must have at least 4 phases as a temporal reference: (1) distant past, (2) recent past, (3) present, (4) future. Note that the neighborhoods marked in color show analysis errors proportional to the time elapsed. The dotted line demonstrates the insufficiency to refer the dynamic only on two times

It is then necessary to clarify as best as possible the categories of relief mapped in the past, to obtain the correct ecological information and to be able to compare this information with that of today. Not always in the descriptive summaries are specified nomenclatures that are taken for granted, but which today are almost completely modified. As it can be seen in the reconstruction of the territory where the Monza Park arose in the years 1721–22 (Fig. 8.26) (Ingegnoli, 2002) -made using the maps of the Land Registry (1721) (Charles, 1749–60)- coppice is still said today, "soft" and "strong" are instead terms referring to the dominant species of temperate broadleaf trees with soft wood or hard wood, namely: *Populeto-Salicetum* or *Alnetum* (soft wood) and *Querco-Carpinetum* or *Aceri-Tiglietum* (strong wood). On a more synthetic scale (for example, 1: 25,000) it is more difficult to have information on the woods.

It should be remembered that ancient maps only report land use, while the reconstruction of a LU cannot be limited to this first type of information. Consequently, the main themes of land use (for example, forestss, meadows, fields, orchards, urban) must be better specified with distinctions for ecological characteristics, indicating the main physiognomies of woods, meadows, cultivated land, etc. Let us characterize forested and urbanized ecotopes (Table 8.9).

Warning: A further aspect that must always be checked concerning current data is the surface and perimeter measurements of the ancient maps surveyed by hand, whose georeferencing about national or international topographic networks was almost non-existent.

To better understand the type of historical transformation of a territory under examination, the plain between the Seveso and the Lambro was taken into consideration. It should be remembered that this area (Fig. 8.27) includes the municipalities of: Lissone, Desio, Bovisio Masciago, Cesano Maderno, Seveso Meda, Seregno, Macherio, Sovico, Albiate, Carate Brianza, Verano, Giussano and the non-hilly part of Carugo, Mariano and Cabiate (the latter in the province of Como): a considerable surface area of over 100 km^2. In this area only 27% of the land has remained agricultural, approximately 2% is wooded while the remaining 71% is urbanized (built, industrial, road or barren). The reference maps for the

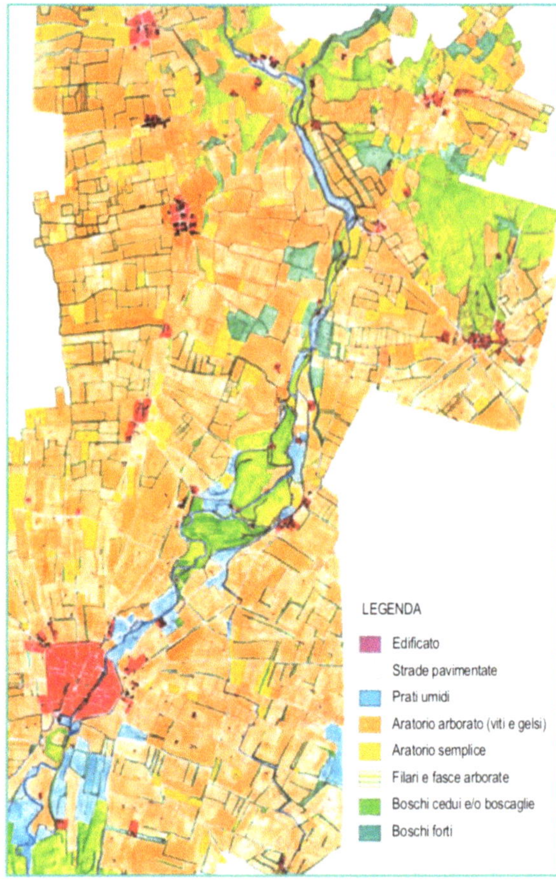

Fig. 8.26 The territory north of Monza, on which the Park was later built, was recomposed from the maps of the Land Registry of Charles VI of Habsburg (1721). The residual patches of "strong forest" (dark green) correspond to the medieval location of the Bosco Bello, with a dominant oak, Quercus robur, thus pertaining to Querco-Carpinetum. (From Ingegnoli, 1993)

historical reconstruction of the transformations are the IGM 1/25,000, while for 2008 the base maps are the regional CTR of Lombardy. Here, the land use was considered.

Table 8.10 summarizes the main ecological parameters for the evaluation of the transformations that occurred in the same Seveso/Lambro plain from 1841 to 2008. These are quantities already defined and used in this paper, except for the *PoBD index* (Potential of Bionomic Development), obtained by considering both the elements of urbanization and the geophysical limits (Ingegnoli, 2015). It is observed that already in the first half of the nineteenth century the general metastability g-LM = 4.53 [bit × Mcal/m^2/year], was low, 42% lower than the optimal theoretical value (g-LM*), therefore beyond any reasonable tolerance (usually around 10–12%).

Today, g-LM has dropped to 2.32 [bit × Mcal/m^2/year], 45% lower than the theoretical value, while the landscape types have changed significantly from agricultural to urban. However, the difference in the value of the bionomic state between 1841 and 2008 is much higher than the ratio 2.32/4.53 = 0.51 since these values of

Table 8.9 Example of land cover distinctions following landscape bionomics suggestions (Central Europe)

Land cover	Ecotope	Condition	Age	HH	BTC
Forestry	Deciduous	Natural	100–150	0–4.0	7.0–11.0
		Coltivated	50–100	5.0–10-0	4.5–7.5
		young or coppice	0–50	5.0–15.0	2.5–4.5
	Conifers	Natural	90–140	0–4.0	7.0–10.0
		Coltivated	30–90	5.0–10-0	4.5–7.5
		young or coppice	0–30	5.0–15.0	2.5–4.5
	D-C Mixed	Natural	90–140	0–4.0	7.0–10.0
		Coltivated	40–90	5.0–10-0	4.5–7.5
		young or coppice	0–40	5.0–15.0	2.5–4.5
Urban	Residential	historical	100–600	80–95	0.20–0.40
		expansion	20–100	50–90	0.30–0.60
	Subsidiary	historical	100–400	85–95	0.05–0.20
		expansion	20–100	80–90	0,10–0.30
	R-S Mixed	historical	100–500	85–95	0.10–0.40
		expansion	20–100	80–90	0.10–0.50
	Monumental	institutional	50–2000	85–95	0.20–1.80
		sacred	60–2000	80–90	0.20–2.00

HH [%]; BTC [Mcal/m²/year]

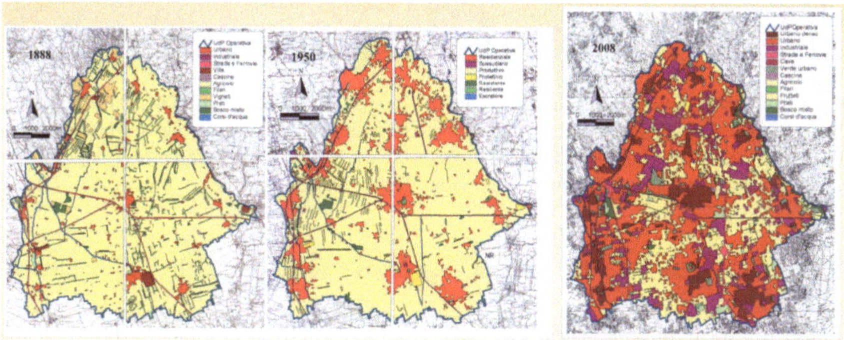

Fig. 8.27 Changes in the territory between Seveso and Lambro from the mid-nineteenth century to 2008. It should be noted that already in 1841, the area was almost devoid of forest patches (green) partially substituted by tree line corridors. Note the expansion of urbanization (in red) at the expense of agricultural areas (in yellow) and the appearance of different industrial components (in purple) after 1950.

"degree of information for energy flux of maintenance of the ordered structure" of the complex system (g-LM) must also consider the component of "potential for bionomic development" (PoBD) of the system itself. This potential is enormous because, in 1841, 94% of the system resources were potentially available to reach the theoretical general metastability values g-LM* (which was equal to 8.79), while in

Table 8.10 Comparison of the main parameters for the evaluation of the transformations of the territory between Seveso and Lambro in the period 1841–2008

LU of Seveso-Lambro	Population	HH (%)	SH/SH*	BTC	LTpE	Landscape	PoBD	gLM	gLM_{max}
BR- 1841	34,500	82.7	1.65	1.67	17,29	Agrarian	0.94	4.53	8.788
BR- 1888	66,750	84.4	0.86	1.55	15,63	Agrarian	0.92	4.16	7.987
BR- 1950	138,600	88.0	0.43	1.20	10,10	Sub-Urb/Rur	0.76	2.89	5.086
BR- 2008	277,325	90.7	0.22	0.65	2,14	Dense urban	0.32	2.32	3.062

HH Human habitat, *SH/SH** Carrying capacity, *BTC ()* Biologic territorial capacity of vegetation, *LTpE* Landscape type evaluation, *PoBD* Potential of bionomic development, *gLM* General landscape metastability, gLM_{max} gLM × (1 + PoBD)

Fig. 8.28 Comparison between the Loppio valley floor (Mori, TN, Italy) in 1900 and in 2007. Note the contrast and complementarity: (**a**) the plain countryside is today almost treeless, while a century ago it was dense with both fruit trees and rows of trees; (**b**) the woods on the first slopes of the mountains are today denser and more continuous. (From Ingegnoli, 2011)

2008, only 32% of the system resources were potentially available, so g-LMmax = 3.06.

Let me show in conclusion Fig. 8.28 (Ingegnoli, 2002) to observe that over the course of a century the changes that have occurred in the municipality of Mori (TN) have been very strong not only in the size of land use, but also in the distribution (and typology) of the woods.

References

Agnoletti, M., & Martegani, V. (2003). *Il paesaggio forestale lombardo, immagini e storia*. Cierre Edizioni.

Bonizzi, L., Campana, F., & Cordini, G. (Eds.). (2019). *Il governo dei Parchi*. Arachne.

Borlaug, N. E. (1974). In defence of DDT and other pesticides. *Pesticides, 8*(5), 14–19.

Buringh, P., & Dudal, R. (1987). Agricultural land use in space and time. In M. G. Wolman & F. G. A. Fourier (Eds.), *Land transformation in agriculture. Scope* (pp. 9–43). Wiley.
Ceccarelli, S. (2013). *Produrre i propri semi. Manuale per accrescere la biodiversità e l'autonomia nella coltivazione delle piante alimentari*. Libreria Editrice Fiorentina.
Columella, L. G. M. (1955). (1st.cent.) *de Agricultura libri X*. New Publ. On Agriculture, Loeb Classical Library, US.
Dusaf-ESRSAF. (2015). *Banca dati dell'uso e copertura del suolo*. Regione Lombardia.
European Forest Institute. (2025). Understanding EU Forest policy governance through a cultural theory lens. *Policy Sciences, 58*(1), 111–144.
FAO (1962) *Global forest resource assessment: Main report*, Alwais FAO, Roma.
FAO (1990) *Global forest resource assessment: Main report*, Alwais FAO, Roma.
FAO (2010) *Global forest resource assessment: Main report*, Alwais FAO , Roma.
FAO (2015) *Global forest resource assessment: Main report*, Alwais FAO, Roma.
Forman, R. T. T. (2008). *Urban regions, ecology and planning beyond the city*. Cambridge University Press.
Forman, R. T. T. (2014). *Urban ecology, science of cities*. Cambridge University Press.
Forman, R. T. T., & Godron, M. (1986). *Landscape ecology* (p. XIX+619). John Wiley & Sons.
Goldsmith, E., & Allen, R. (1972). *La morte ecologica*. Laterza.
Grassi, E., & Ingegnoli, V. (2006). *Caratterizzazione ecologico-paesaggistica del territorio dei comuni di Robecco sul Naviglio e Cassinetta di Lugagnano: Confronto tra l'unità di paesaggio di terrazzo*. Tesi di laurea in Scienze Naturali, Università di Milano.
Hance, J. (2019). *Less rainforest, less rain: A cautionary tale from Borneo*. Mongabay Series.
Ingegnoli, V. (1980). *Ecologia e progettazione*. CUSL.
Ingegnoli, V. (1993). *Fondamenti di ecologia del paesaggio: Studio dei sistemi di ecosistemi*. CittàStudi (poi UTET-Cittàstudi).
Ingegnoli, V. (2002). *Landscape ecology: A widening foundation* (p. XXIII+357). Springer.
Ingegnoli, V. (2011). *Bionomia del paesaggio. L'ecologia del paesaggio biologico-integrata per la formazione di un "medico" dei sistemi ecologici* (p. XX+340). Springer.
Ingegnoli, V. (2015). *Landscape bionomics. Biological-integrated landscape ecology* (p. XXIV + 431). Springer.
Ingegnoli, V., Bocchi, S., & Giglio, E. (2018). Agricultural landscapes rehabilitation suggests 'ecosystem services' updating. *WSEAS Transactions on Environment and Development, 14*, 233.
Ingegnoli, V., & Giglio, E. (2005). *Ecologia del Paesaggio: Manuale per conservare, gestire e pianificare l'ambiente* (p. 685+XVI). Sistemi editoriali SE.
Ingegnoli, V., & Pignatti, S. (2007). *The impact of the widened landscape ecology on vegetation science: Towards the new paradigm* (Vol. XVIII, pp. 89–122). Springer, Rendiconti Lincei Scienze Fisiche e Naturali, s.IX.
Charles VI of Habsburg. (1749–60). *Land register: The situation of the Duchy of Milan close to a Kingdom of Sardinia*. Duchy of Milan.
Langé, S. (2008). *Il paesaggio riconosciuto*. Jaca Book.
Makarieva, A. M., & Gorshkov, V. G. (2012). Revisiting forest impact on atmospheric water vapor transport and precipitation. *Theoretical and Applied Climatology*. https://doi.org/10.1007/s00704-012-0643-9
Marklund, L. G., & Batello, C. (2008). *FAO datasets on land use change, agriculture and forestry and their applicability for national greenhouse gas reporting*. FAO.
Naveh, Z., & Lieberman, A. (1984, 1994). *Landscape ecology: Theory and application* (p. XXVII +360). Springer, New York, Inc.
Pignatti, S. (1998). *I boschi d'Italia: Sinecologia e biodiversità*. Utet.
Prigogine, I., Nicolis, G., & Babloyatz, A. (1972). Thermodynamics of evolution. *Physics Today, 25*, 23–28. 25:38–44.
Prov. Trento. (2010). *Urban reporting of the Province of Trento*. Provincia Autonoma di Trento, Procedura di Rendicontazione Urbanistica

Saarinen, A. (1962). *Eero Saarinen on his work*. Yale University Press.
Vitruvius, M. P. (I cent) *De Architectura libri X*. (Various Publications).
Zanzi, L. (1995). Viatico per una avventura nella storia della Val Grande. In L. Alberti, G. Rainaldi, & E. Rizzi (Eds.), *Val Grande, storia di una foresta*. Fondazione Arch. En- rico Monti.
Zonneveld, I. S. (1995). *Land ecology*. SPB Academic Publishing.

Chapter 9
Altered Landscapes vs. Human Health: Significant Research

Contents

Abstract Since 1999, Ingegnoli has urged us to consider the possible consequences due to landscape degradation with structural and ecological disorders on human health. Theoretically, these involve physiological and ethological behaviors of man through disorders of the hormonal and nervous system. Four research are provided, all conducted by the Author.

The first research concerns Monza-Brianza province, near Milan, and show a remarkable increase in mortality rate MR [× 1000] correlated with an increase in landscape dysfunction: we pass from MR = 7.6 in unaltered landscapes (BF = 1.0) to MR = 9.5 in the landscape with deprivation of 50% (BF = 0.50) of the normal state. The correlation significance (Pearson) is 1.90. The average BF of 72 LU is 0.70.

The second research concerns the COVID-19 contagion related to the bionomic state of the territory. The dynamics of the correlation significance of ecological density (ED), bionomic functionality (BF), and population age (PA) show the dominant importance of BF for ¾ of the first year. At the same time, ED is low,

V. Ingegnoli, *Landscape Bionomics and Planetary Health*, Landscape Series 21,
https://doi.org/10.1007/978-3-032-08908-3_9

and the inversion of these correlations occurs in the second period, when the percentage of the contagion is >5%.

The third research on cancer incidence in Western Europe, 2012–2020, compare 12 parameters (5 traditional and 7 coming from Landscape Bionomics) related to cancer onset. The significance of Pearson's correlation provides clear indications: only five exceed the threshold of complete correlation, while the other five present scarce importance and the remaining two are 'near good'. Note that no one of these parameters can be interpreted to have spurious correlations: GDP (€/capita) is directly linked with the power of landscape transformation and artificialization, Red Meat with agrarian and zootechnical industrialization like Pesticides and Glyphosate.

The fourth research was made comparing European and Indian environmental transformations related to human health from 1880 to 2010. Diagnostic evaluation of the relation HH vs. BTC shows the opposite directions of the land transformations in Europe and India. A focus on the increase of incidence and complex etiology of Diabetes-2 in these two areas is provided.

Keywords Medicine · Mortality rate (MR) · Bionomic functionality (BF) · Covid-19 · Population age (PA) · Cancer incidence · Diabetes-2 · Correlation significance · Organic farming

9.1 First Signs of Health Damage

9.1.1 Not Just Pollution in Vado Ligure

Since 1999, Ingegnoli has urged people to consider the possible consequences on human health due to landscape degradation with structural and ecological disorders. Theoretically, these involve physiological and ethological behaviors of man through disorders of the hormonal and nervous system (Ingegnoli, 2001, 2002). Brain maturation after birth has shown the apparent importance of a link with the environment. Still, weak correlations have been found (Mitchell, 2008) due to the limitations of traditional ecology in assessing the state of complex landscape systems and their relationships with human health.

Even case studies of national resonance, such as Vado Ligure and Quiliano (Savona), where the mortality rate (MR) increased anomalously, and the incidence of tumors in 2013 was double that of Italy (Scarselli, 2013), have been studied exclusively based on pollution, without considering the degradation of the landscape unit (LU). Italian people knew the Coal-Electricity Industry of Vado Ligure to harm the health of the population.

This industry was reduced in 2014, but the MR did not decrease, while urbanization and the port has been continuing to grow (Fig. 9.1), reinforcing the suspicion that the strong alteration of the landscape unit (LU) should be considered a

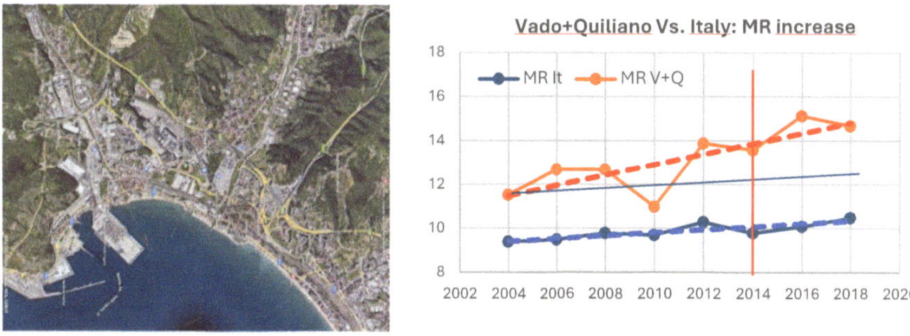

Fig. 9.1 Port and agglomeration (industry and residence) of Vado Ligure and Quiliano (map from Google-Earth, 2019). The graph compares the local mortality rate (MR) with the Italian one

degradation capable of further damaging health. The Google Earth map of this LU is impressive because the built coastline has cancelled out the natural structures for over 12 km.

9.1.2 Similarity Between Lichen Distribution and Lung Cancer in Veneto

A large-scale analysis can help understand the link between a difficult-to-monitor distribution of air pollution (cumulative impacts) and its influence on the aetiology of lung diseases. In Fig. 9.2 is shown a study by Cislaghi & Nimis (1997) on (a) lung cancer mortality in the Veneto region (Northern Italy) versus (b) lichen diversity, the best way to record cumulative impacts in the air pollution, lichens being the best bioindicators of this problem. The comparison of the maps of the distributions (a) and (b) does not need any explanation to express the close correlation between the two phenomena. The green and dark green spots on the northern edges of the cited maps indicate a very low level of cancer mortality and high-quality level of air pollution.

However, we must pay attention to the changes due to the passage of scale because, on a local scale, a municipality presents a green condition (Lichen Index = 36.1), but if we divide its territory into its component landscape units (LU), following bionomic principles, a gradient emerges in which the urbanized LU is orange and red as in the towns of the Po-Adige Valley and/or Avisio one. We can show the case of Tesero (Fig. 9.2) (Ingegnoli, 2011):

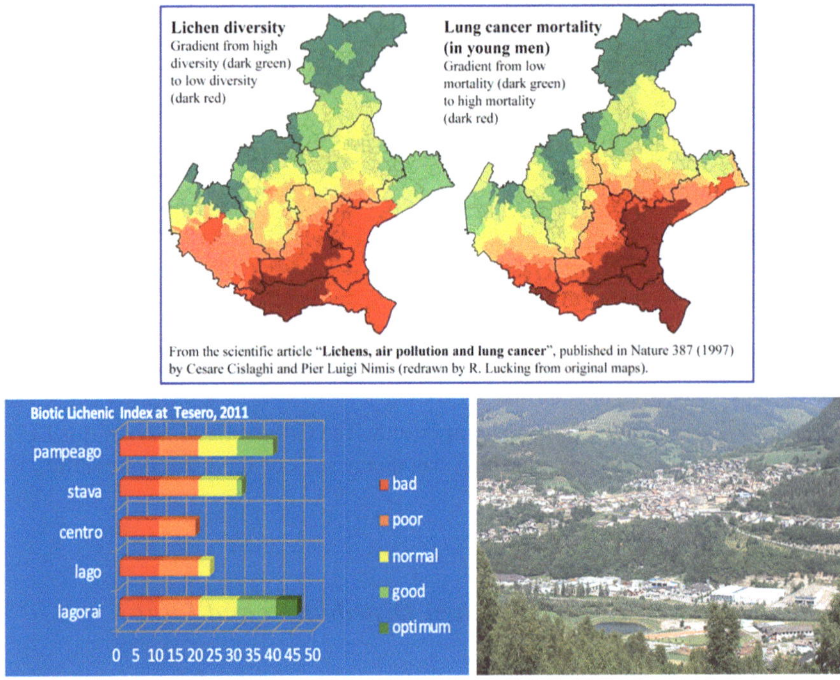

Fig. 9.2 The similarity of lichen/cancer distribution in Veneto demonstrates the need to study macroscale processes in biology. However, we can find a different condition moving from the regional to the local scale, so when the municipality is analyzed in its LU components. The most urbanized LU-1 (in the center) shows a bad lichen index (Ingegnoli, 2011)

9.1.3 A New Field of Study: Health and Environment

The current medical model ignores the complex origin of diseases, medicalizes risk factors, and makes asymptomatic people "sick", thus facilitating treatment with drugs, decreeing the failure of prevention and, at the same time, the supremacy of patentable therapies (Marvasti & Stafford, 2012).

Medicine is moving (i.e., is on a transition phase) from a reductionist criterion of 'strong causality' (e.g., infectious diseases or organic dysfunctions) to a criterion of 'weak causality', i.e., degenerative alterations with cause-effect relationships that are complex and not always understandable. This is why the transition 'from sick to health care' is underlined. Formichi et al. (2022) noted that genetic variability alone could not explain the risk of developing chronic diseases. Genome-wide Association Studies (GWAS) revealed a limited causal effect (estimated less than 20%) of genetic susceptibility on phenotypic variance.

We must underline that environmental exposure and exchanges and their interaction with the body's biological systems and apparatuses play an essential role in disease development. Note that the concept of exposure -e.g., the exposome (sensu

Table 9.1 Environmental alterations capable of influencing human health

(a) ***internal processes*** (e.g., metabolism, hormonal balance, gut microbiota, aging, etc.)
(b) ***specific external factors*** (e.g., infections, pollutants, smoking, drugs, etc.)
(c) ***general external factors*** (e.g., socioeconomic status, technological behaviors, climate change, etc.)
(d) ***landscape alterations*** (e.g., agrarian industrialization, hierarchical relations, the biological territorial capacity of vegetation, the vital space per capita, the ratio human/natural habitats, the inter-scale relationships, etc.)

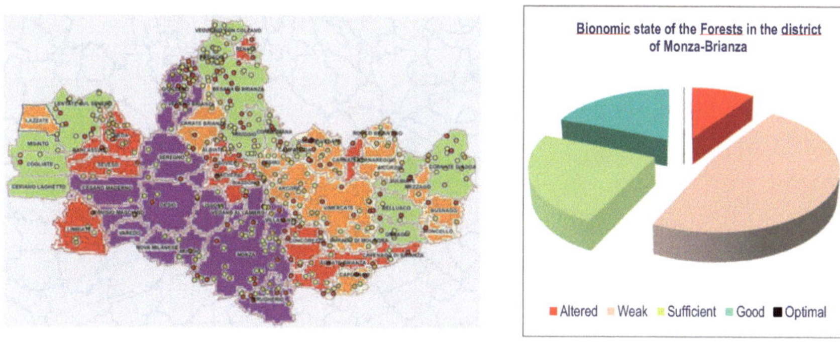

Fig. 9.3 (left) A map of the province of Monza, near Milan, 65,576 ha and about 850,000 inhabitants. The four main types of landscapes are distributed mainly depending on the attraction due to the city of Milan. Colored dots indicate the presence of the main old farms (Ingegnoli, 2015). To the right, the main results of the survey on the bionomic state of the forests, mainly weak

Wild (2012)- may be necessary but *not sufficient* because of the complex structures and interrelations of life. We can distinguish at least four categories of environmental alterations capable of influencing human health through exposure to them. Consequently, the categories of ecological alteration capable of affecting human health must add a new category, as can be seen from Table 9.1.

9.2 Correlation Landscape Health/Mortality In Monza-Brianza

9.2.1 Landscape Alteration Correlated with the Mortality Rate

In 2010–11 all the 55 municipalities of the province of Monza-Brianza (M-B) were studied by Ingegnoli following principles and methods of Landscape Bionomics. The result brought to four main landscape types (Fig. 9.3): (i) agricultural (green), (ii) rural (orange), (iii) suburban (red), (iv) urban (violet). Dividing the three (i),

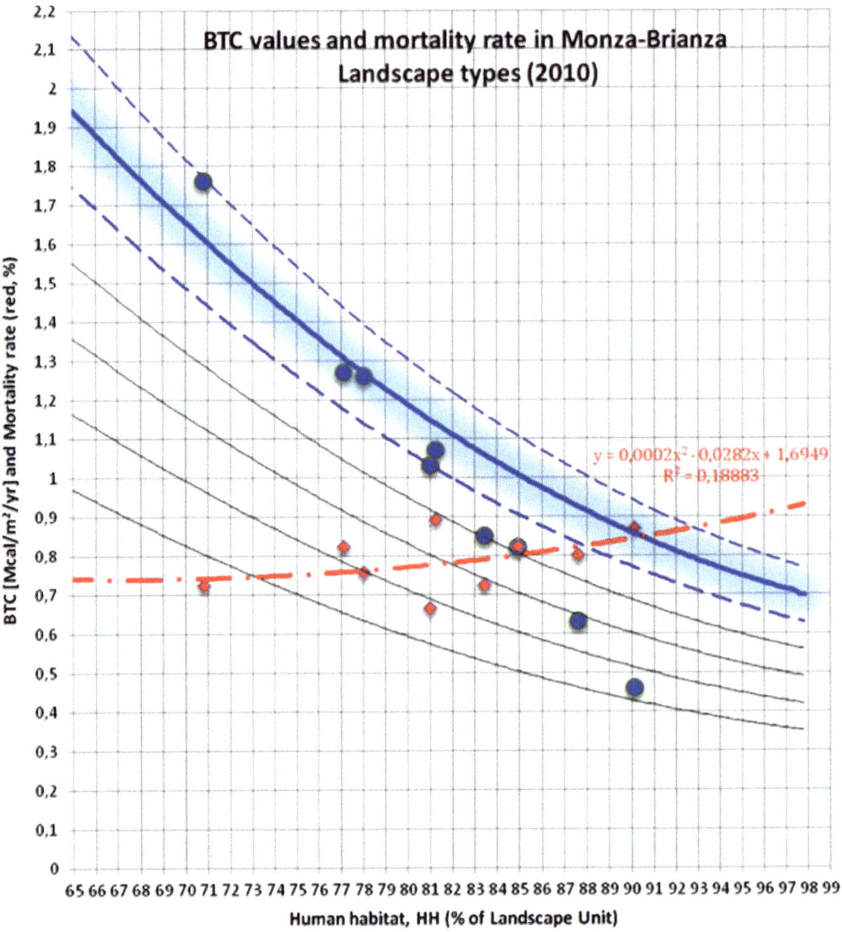

Fig. 9.4 The seven types of landscape show the correlation between the mortality rate (MR) (red dashed curve) and the human habitat (HH): four landscapes are outside the tolerance (dotted blue). The red curve of MR presents low difference in the first part (HH = 65–82), from 0,74 to 0,78 but the difference goes from 0,78 to 0.91 in the second half (HH = 82–99). From Ingegnoli, 2015

(ii) and (iv), into (v) agricultural-protective, (vi) rural-suburban and (vii) dense-urban, *seven types of landscape* were considered.

Ingegnoli (2015) evidenced the relationship between landscape alteration and human health in Fig. 9.4. It shows the correlations of these seven types with the mortality rate MR = f (HH, BTC), plotted in the HH/BTC model (blue line). Only four-five of them can be considered in a normal bionomic state (dotted curves), the other resulting altered, gradually from the blue curve.

Their mortality rate (red) increases from 0.74 to 0.78 before HH = 82% (+0.04) but from 0.78 to 0.91 (+0.13) after this threshold (same HH interval). The Pearson's Correlation Significance is low (0.62) but not near zero. Therefore, a correlation between MR and HH (the red curve) can be seen in this plot because low HH presents low MR. Based on seven landscape types, this evaluation is related to a sample size that is too small, so the statistical significance is too low, but this study pushed towards a deepening.

9.2.2 *Deepening the Research*

The deepening started at the end of the previous first study (2011):

(a) completing the landscape gradient adding, to the cited 7 L-types of M-B (Fig. 9.4), the 8 municipalities of the N-E periphery of Milan (Bollate, Paderno Dugnano, Cusano Milanino, Cinisello Balsamo, Bresso, Cormano, Novate Milanese, Sesto San Giovanni) and the 9 districts of the City of Milan, so reaching the number of 72 municipalities.
(b) considering these municipalities as landscape units (LU) after processing their main bionomic parameters: HH, BTC, HS/HS*, BF (Ingegnoli, 2011, 2015). This elaboration needed to deepen the vegetation survey with the LaBiSV methodology, especially related to forestry (Fig. 9.3, right).
(c) considering the ecological factors capable to increase MR: (1) population age (PA, years), and (2) air pollution, mainly fine dust (FD). Both factors are proportional to the HH but resulted with small local variations (PA = 43.5–46.0) and (FD = about 7,5–10% differ) from Milan to Carate Brianza.
(d) evaluating the distribution of the statistical sample (R^2) and the correlation significance (Pearson) of MR Vs. BF.

As shown in Fig. 9.5, the gradient of the bionomic functionality (BF) decreases from the normality (BF = 1.0) in landscape 1 (AGR-PRT) to 0.5 in landscape

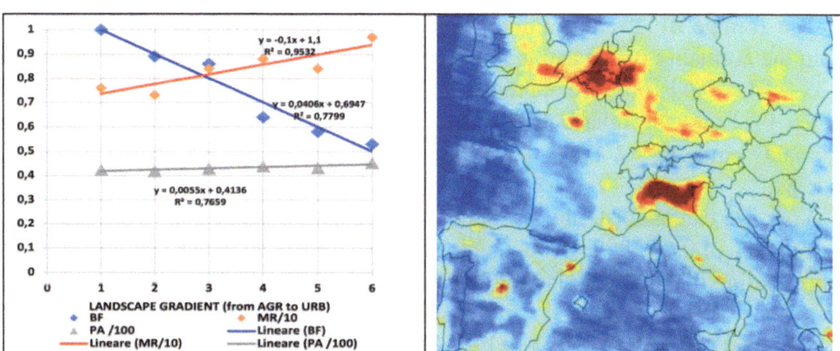

Landscape Types: (1) Agrarian-protective; (2) Agrarian-productive; (3) Rural; (4) Sub-urban; (5) Urban; (6) Dense-urban (see 9.2.1).

Fig. 9.5 The increase of MR (red) and PA was related to the decrease of BF (blue) in the gradient of the six landscapes. Note that PA (grey) shows an increase of about 20% of MR. To the right, are the most polluted areas in Europe, in which Monza-Brianza and Milan are fully inserted

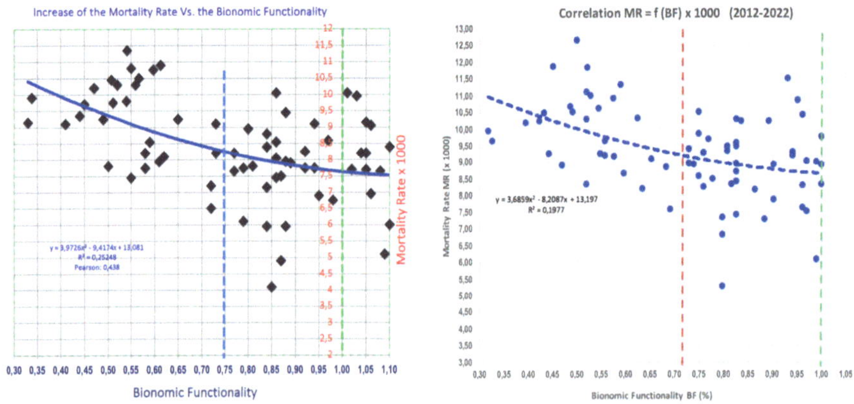

Fig. 9.6 A quantitative demonstration of the correlation between systemic environmental alteration and human population mortality (left). A noticeable increase of mortality rate MR [x 1000] is correlated with the increase of landscape dysfunction: we pass from MR = 7.61 in not altered landscapes (i.e. BF = 1.0) to MR = 9.5 in the landscape with deprivation of 50% (i.e. BF = 0.50) of the normal state. The correlation significance (Pearson) is 1.90. The average BF is 0.75 and the related MR is 8.25 (2010–2012). (right) A recent study (2012–2022) shows a slight worsening of the average BF = 0.72 and of the related MR = 9.2 Vs. the MR = 8.6 at BF = 1.0 and MR = 10.0 at BF = 0.50. The difference, quite contained, is mainly due to the Population Age increase

6 (Dense-URB), while MR passes from 7.4 to 9.4/1000. The Mortality Rate is also proportional to Population Age (which, from the different slopes of the lines, is limited to about 18%) and to fine dust (FD), estimated by WHO to be about 6–7% of mortality. Therefore, about 75–78% of the MR increase is due to BF alterations.

Figure 9.6 shows the plotting of the 72 LU: we can see a distribution of the statistical sample with an $R^2 = 0.25$ and a Pearson coefficient = 0.438 (twice the minimum significance).

The average BF, evaluated in the first research (2010–2012), was 0.75, quite far from the normality of 1.0 and it was related to an MR = 8.25, with a difference of 8.25–7.61 = 0.64/1000. An upgraded weighted mean (2012–2022) gives an average BF = 0.72 and the related MR = 9,2, a bit worse, with a difference from the normality equal to 9.2–8.6 = 0.60/1000, mainly due to the Population Age (PA) increase.

Now, we can estimate the *risk factor* for *premature death* due to the bionomic dysfunction of 0.60/1000. Detracting 0.24% due to population age and fine dust and remembering the total population (2.525 million), we can write:

$$(0.60 \times 0.76) \times 2,525,000/1000 = 0.456 \times 2,525 = 1,151 \text{ death/year}$$

This risk is about 6–7% of the annual death of the territory.

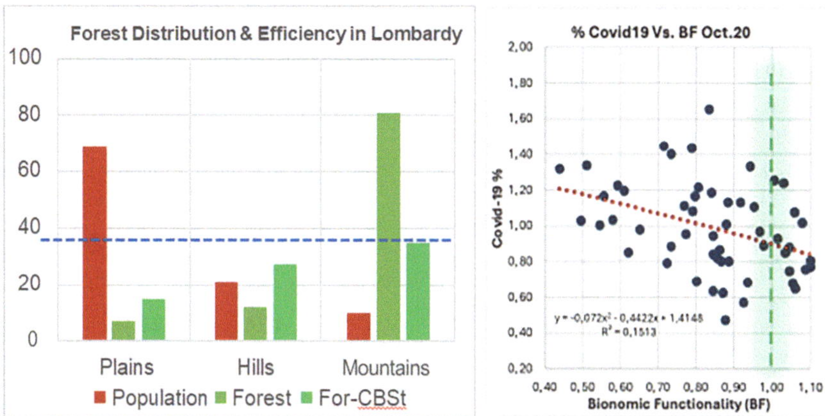

Fig. 9.7 The altered relationships between forests and population in Lombardy, especially in the Plains, where a very insufficient land cover (7%) is linked with very degraded phytocoenosis: CBSt = 17% Vs. a threshold of normality equal to 36%. To the right, the evident correlation Covid-19/BF in the first 8 month of pandemic insurgence in the province of Monza-Brianza (M-B). From Ingegnoli & Giglio, 2020

9.3 COVID-19 Incidence in Monza-Brianza

9.3.1 Incidence Evaluations

In the article on J. Environ. Sci. Public Health 2020, Ingegnoli wrote that both for the first and second insurgence of COVID-19 in Italy, Lombardy was the Region with the most elevated prevalence. The main reasons proposed were: (a) the most populated and urbanized territory, (b) the most congested public transportation networks, (c) the most air-polluted (Coker et al., 2020), (d) a quite old Population Age (PA), not to mention the public health organization, exceedingly depending on nonscientific suggestions. Can other vital reasons due to environmental alteration be added??

Lombardy has dangerous environmental problems, firstly, forest bionomic dysfunction. This problem is not felt. On the contrary, the Region boasts of increasing forest cover in the last 10 years. However, Fig. 9.7 shows a very different situation.

The degraded relationship between forests and population in Lombardy is plotted. The higher alterations emerge, especially in the Plains, which host about 70% of the 10–11 million regional population: we can see a drastically insufficient forest land cover (7%), linked with very degraded phytocoenosis, presenting a concise bionomic state CBSt = 17% Vs a normality threshold equal to 36% (blue dotted line). On the right part of Fig. 9.7 (Ingegnoli & Giglio, 2020) is the evident correlation of COVID-19/BF in the first 8 months of the pandemic insurgence in Monza-Brianza province (M-B, 55 LU). Remember that BF depends on many bionomic parameters,

Table 9.2 Pearson's correlation significance of contagion and the main ecological-bionomic parameters in Monza-Brianza Province, 2020–2022

Covid19 findings	Days	contagion %	ED	BF	PA
APR, 20	30	0,56	0,34	1,40	0,51
JUL, 20	120	0,68	0,12	0,92	0,26
OCT, 20	210	1,03	0,94	1,48	0,53
MAR, 21	330	7,50	1,96	0,71	0,74
MAY, 21	390	8,76	1,70	0,63	0,65
OCT, 21	540	9,51	1,50	0,60	0,62
FEB, 22	640	23,43	1,58	0,86	0,61
Mean			**1,16**	**0,94**	**0,56**
St-Dev			0,71	0,36	0,16
ST.Dev/M			0,61	0,38	0,27

the most important of which are forest efficiency and distribution. So, the answer to the previous question is affirmative.

9.3.2 *Covid-19 In Monza-Brianza: Bionomic Parameter Significance*

To check the significance of the bionomic parameters, let's focus on three of these (Table 9.2): (1) Ecological Density (ED) [people/ha], (2) Bionomic Function (BF) [BTC/BTC_{NORM}], (3) Population Age (PA) [mean years]. The considered period was 640 days (21 months). Remember that the essential bionomic parameters are ED, relating to agrarian fields, urban and human habitat; BF, relating to forest, agriculture, and human habitat; and population age. Note that ED presents an excellent average significance (ED = 1.16 ± 0.71), so the standard deviation is very high. BF presents a bit less average but a better standard deviation: BF = 0.94 ± 0.36. The correlation significance of PA presents a lower average: PA = 0.56 ± 0,16. Moreover, the averages of ED and BF are not significantly different (1.16 Vs 0.94) but they seem to be in opposition: for the first 280 days, ED (mean = 0.70) is low, and BF high (mean BF = 1.23), while for the other period (360 days) is the contrary (ED = 1.63 Vs BF = 0.78). Unlike the different parameters, PA remains lower and more constant.

The Pearson's correlation significances of the parameters ED, BF, PA, are shown in Fig. 9.8. This result is notable because the infection has grown where the environment was altered (BF average significance = 1.23) in the first 278 days (76.2% of the year), leaving the ED contributions as marginal (average significance = 0.70). When the threshold of 5% infected people was exceeded, ED became the dominant correlation, reducing the BF average significance to 0.78 (but not eliminating the correlation). So, a good BF can be considered a defense against infections, slowing down the contagion for ¾ of the first year.

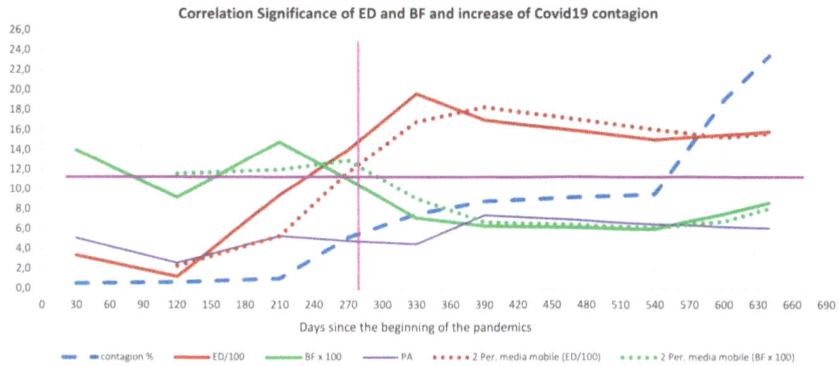

Fig. 9.8 Dynamics of the correlation significance of ecological density (ED), bionomic functionality (BF) and population age (PA) during the Covid-19 contagion in the Monza-Brianza Province. Note the dominant importance of BF for ¾ of the first year, while ED is low, and the inversion of these correlations in the second period, when the percentage of the contagion >5%. The PA remained less prominent

Note that the ecological density ED (inhabitant/ha) is a bionomic parameter related to the concept of the human habitat (HH); so, it has nothing to do with the traditional geographic population density (GD, inhabitant/km^2): being the average human habitat HH = 82%, the ecological density ED is pair to 26.3, while the geographic density GD = 21.5 (inhabitant/ha), with a difference of about +22%.

9.4 Cancer Incidence, a Bionomic European Overview

9.4.1 Widen Correlation Parameters of Carcinogenesis

Today, the study of environmental alteration is generally limited to some pollutants, and often doubtfully, as in the case of glyphosate; diet and behavior have recently increased their importance but remain only marginal, especially in therapy integration. The biological studies on bio-chemical molecules, genetic scales, viruses, and metabolism brought many challenges but also produced insidious errors, such as the DNA "central dogma of molecular biology" (Crick, 1970), because the DNA is not a set of formed characters but a set of potentialities (Bottaccioli, 2014; Bottaccioli & Bottaccioli, 2020). Moreover, we can note the emergence of peculiar paradoxes: tumor cells have and retain the properties to proliferate abnormally and to avoid apoptosis, independently of other genetic factors; this autonomy is well represented by the paradox that emerges in cancer research when accounting for carcinogenesis based on DNA mutations (Bertolaso, 2013, 2018), where quite often the same genes are involved in oncogenetic or tumor reversion processes or, even, they seem to follow, more than anticipated, the neoplastic transformation of the cells.

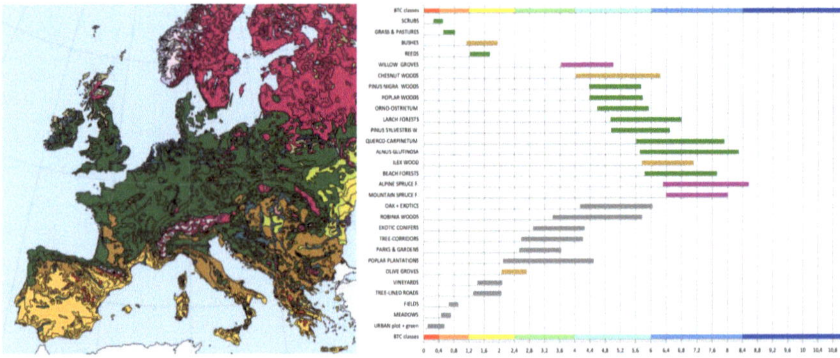

Fig. 9.9 (**a**) Ranges of BTC [Mcal/m^2/year] values from many vegetation surveys in Mediterranean Italy, North Italy, Austria, Belgium, and Poland, in natural and human formations. (**b**) The main types of vegetation in Europe

Another tracking error is the marginalization of biological macro-scales (from landscape to ecoregion); nevertheless, the "rock in the pond" (Urbani-Ulivi, 2019) of the systemic turn in scientific paradigms imposes to change our vision: the alterations of Life at broader scales can damage human health, not unlike small ones! Bottaccioli underlines that health and disease depend primarily on Life's organization. We must affirm 'on the full organization', so the study of biology must be upgraded even to the broader scales, trying to understand their anatomical components (structure), physiological processes and functions, transformation processes and pathologies, and to make a clinical-diagnostic evaluation and to formulate a "sanitation" therapy.

As we can see in Fig. 9.9, the BTC [Mcal/m^2/year] of vegetation formations differ very much: values for grass and crop fields range between 0.50 and 1.40, while vines, shrubs, plantations, and gardens from 1.50 to 3.50; adult/mature forest from 6.50 to 9.50 within Temperate Ecoregions as Europe. Ranges of BTC [Mcal/m^2/year] values derive from about 450 vegetation surveys in Mediterranean Italy, North Italy, Austria, Belgium, and Poland, both on natural and human formations. The bionomic state of 16 EU nations (Austria, Belgium, Denmark, Finland, France, Germany, Greece, Hungary, Ireland, Italy, Nederland, Poland, Portugal, Romania, Spain, Sweden), 90% of EU population and 82.5% of the EU territory, plus the UK and two essential regions, Bayern, and Lombardy has been evaluated using 2012 and 2020 data from EEA and ECIS (EEA, 2011, 2018; ECIS, European Cancer Information System, 2020).

Table 9.3 give an exhaustive frame of all information related to the examined countries: they show the cancer incidence in 2020 (Age Standardized Rate, ASR, evaluated following the EU Journal of Cancer and ECIS- EU (2020) Cancer Information System (Steliarova-Foucher et al., 2014) and the primary environmental parameters to be correlated to HH and BTC for the 19 mentioned "countries". A similar Table was produced related to 2012.

Table 9.3 Data referred to 90% of EU population (16 EU countries and 2 regions) + UK—2020

Country	POP.	Cancer	HH	BTC	Glyphos.	GDP	Org-Farm.	Beef	Forest	Urb-Pop	Pesticide
Austria	8.90	488	39.80	3.45	0.16	38.170	25.30	19.70	46.00	58.30	3.30
Belgium	11.60	640	67.69	1.61	0.30	35.940	6.90	15.00	23.00	98.00	6.70
Denmark	5.80	688	74.76	1.58	0.51	49.720	10.90	18.40	15.00	87.87	1.10
Finland	5.50	558	16.82	4.03	0.31	37.170	13.50	17.00	66.00	85.40	0.60
France	62.30	621	54.08	2.59	0.32	33.270	7.70	23.00	27.00	80.44	3.60
Germany	83.90	570	58.36	2.81	0.27	35.840	9.00	14.50	32.00	77.31	4.00
Greece	10.50	526	43.18	2.92	0.25	17.750	10.30	16.00	30.00	79.06	2.90
Hungary	9.70	623	66.30	2.05	0.31	13.260	5.70	2.40	22.00	71.35	2.20
Ireland	4.90	718	55.60	0.98	0.14	60.170	1.60	35.80	11.00	63.17	6.50
Italy	60.40	549	53.68	2.93	0.30	26.920	15.70	17.10	32.00	70.44	6.10
Nederlands	17.20	655	67.23	1.17	0.39	41.870	3.70	14.00	8.90	91.49	7.90
Poland	37.90	536	58.02	3.05	0.43	13.000	3.50	4.50	30.00	60.06	2.20
Portugal	10.20	510	44.61	3.12	0.35	18.590	8.20	16.00	36.00	65.21	5.40
Romania	19.30	491	53.62	2.65	0.14	9.110	2.90	8.00	29.00	53.97	0.80
Spain	46.80	540	43.36	2.99	0.15	25.200	9.70	8.60	37.00	80.32	3.60
Sweden	10.10	557	18.74	4.51	0.19	43.900	20.40	22.00	64.00	87.43	0.60
U.K.	63.10	590	60.80	1.52	0.11	32.910	2.60	17.90	13.00	83.40	3.20
Bayern	13.10	510	58.00	2.70	0.25	48.300	11.00	14.00	35.00	77.00	3.60
Lombardy	10.10	588	56.10	2.01	0.28	38.600	6.00	18.00	26.00	71.00	5.80

POP = million; Cancer = incidence/100,000; HH = human habitat % territory; BTC = bionomics capacity of vegetation (Mcal/m^2/year); Glyphosate = kg/ha of PRD (agrarian area); GDP = €/capita; Org-Farming = % PRD; Beef = kg/capita; Forest = % territory; Urban Population = % of tot Pop.; Pesticide = kg/ha PRD. Data from: EEA and Eurostat 2018–2020; Index Mundi 2019; Worldmeter, 2018; Antier C. et al. Sustainability Journal, 2020; ISTAT Forest, 2019

Fig. 9.10 The increase of cancer incidence in the last 8 years in Europe was very strong, from an average value of 326 to one of 568/100,000. Note the growing significance of BTC_{20} compared with BTC_{12}, underlined by the increase of the angular coefficient of the correlation line, with the increase of $R^2 = 0{,}530$ Vs. 0,257

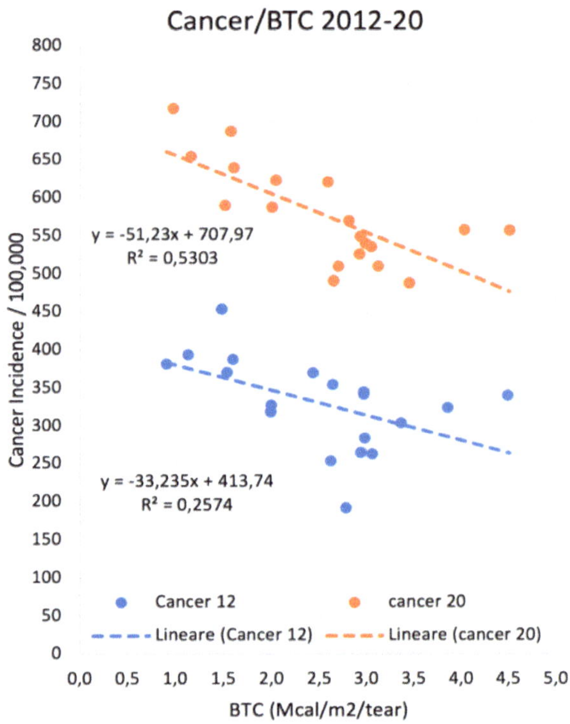

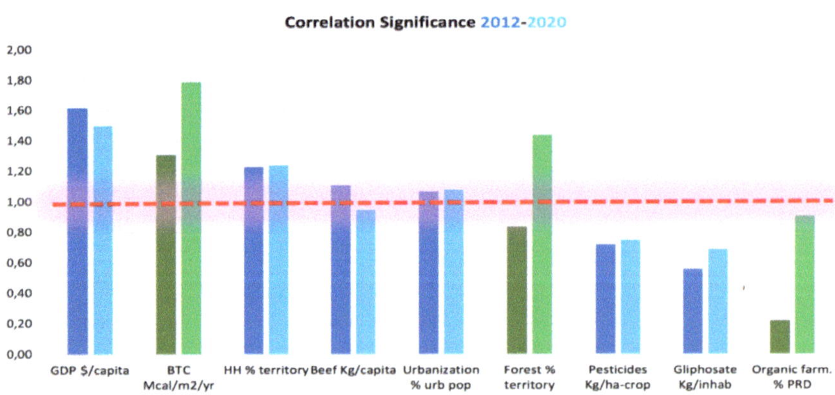

Fig. 9.11 Comparison between the significance of correlations of the studied nine parameters Vs. cancer incidence in European countries 2012–2020. Green bars represent the 'lack of protection' parameters, while the blue p. represent the 'exceeding dysfunctional' ones: the dark colors pertain to 2012, while the light ones to 2020

The first result regards the worrying increase of cancer incidence in Europe, as plotted in Fig. 9.10. In this period 2012–2020, the weak bionomic level of the environment was not able to slow the cancer increase, which grown from an average of 326 to 568/100,000. In this altered environment, the mentioned oncological increase reinforced the correlation significance of BTC, the function better representing the bionomic state of a country.

On Fig. 9.11, the correlation significance Cr-Sg of three parameters in the period 2012–2020 remains near-constant (HH, Urban- Population, Pesticides), two of them decrease (GDP/capita and Beef/capita), but four of them increase: Organic Farming (+ 4.14); Forest (+ 1.44); BTC (+ 1.37); Glyphosate (+ 1.23). So, an evident meaning depended on the alteration of agricultural landscapes emerges.

The correlation significance of the ratio Forest / weighted average BTC of the countries passed from 76,2% in 2012 to 80.5% in 2020. If the forest cover acquires importance (+4.3% in 8 years) concerning BTC (the values of which depend on the two components of Forest and Cropland), this means a further degradation of agrarian landscapes; this alteration is also underlined by the increase of the Cr-Sg both of Organic Farming and Glyphosate. Note that the % increase in 2012/20 of the surface of Organic Farming is equal to 1.5, so the increase, about 2.7 times wider, depends not on the rise in its area but on its effect on cancer onset!

9.4.2 Upgrading Correlation Parameters in Cancer Incidence

The emergence of a correlation between cancer insurgence and environmental alteration is undoubtedly essential. It could be necessary to upgrade it after a more precise description of the health state of the environment. This upgrading is possible following the concept of diagnostic index (DI). As already shown in Fig. 7.10, each of the 12 environmental parameters has been estimated per the 17 European countries, and their resulted range was divided into five classes of importance with the following weight: 5 (green), 4 (yellow), 3 (orange), 2 (brown), 1 (violet). They must be related to the optimum condition established as 80% of the max weight of each parameter, here 0.8 (12 x 10) = 96. So, DI = DI_W / 96 (e.g., Austria = 86 / 96 = 89.58). DI is inversely related to Cancer onset (last two columns). We repeat here Fig. 7.10 = Fig. 9.12 otherwise should be too difficult following these concepts.

The worst environmental state resulted in DI = 45.83% (Nederland). No European nation reached an excellent DI > 89.58%; Austria, Sweden, and Finland express a good DI = 87.5; a fair DI = 79.46 by Poland, Romania, Greece, Portugal, Spain, Italy and Germany; a mediocre DI = 68.75 by Hungary, France, and UK; a scarce DI = 52.60 by Denmark, Belgium, Ireland, and Nederland.

The correlation curve (Fig. 9.13, left) of cancer incidence in 2020 (blue) is plotted on the diagnostic model BTC/DI (orange). We can see the linkage with the bionomic states of the 17 examined countries, classified in 4 levels of health, from the best (green square) to the most altered (red squares). Figure 9.13 shows the average BTC and Cancer Incidence per group of countries.

weight	2	4	6	8	10									
Nation	BTC	RNT (%)	HH	BF	Org. Farm	Urb.Pop	GDP €/cap	RSL(%)	PRD(%)	Red Meat	Pest	Glyph	DI	ancer ons
AUSTRIA	3,45	32,82	39,8	1,08	25,30	58,30	38.170	32,50	23,70	19,70	3,30	0,16	89,58	4,88
ROMANIA	2,65	28,99	53,68	1,08	2,90	54,00	9.110	27,60	35,50	8	0,8	0,14	87,50	4,91
SWEDEN	4,51	66,82	18,74	1,04	20,40	87,4	43.900	10,70	9,00	22	0,6	0,19	85,42	5,57
POLAND	2,98	31,70	57,98	1,34	3,50	60,10	13.000	8,00	52,00	4,5	2,2	0,43	83,33	5,36
FINLAND	4,13	65,80	16,37	0,92	13,50	85,40	37.170	10,00	9,00	17	0,6	0,31	81,25	5,58
SPAIN	2,99	17,47	43,36	0,99	9,70	80,30	25.200	40,90	36,50	8,6	3,6	0,15	81,25	5,40
PORTUGAL	3,12	35,60	44,31	1,05	8,20	65,20	18.590	26,00	31,00	16	5,4	0,35	79,17	5,10
GREECE	2,92	29,84	43,29	0,99	10,30	79,60	17.750	28,50	34,50	16	2,9	0,25	77,08	5,26
ITALY	2,93	30,07	53,34	1,19	15,70	70,40	26.920	14,23	46,60	15,7	6,1	0,30	77,08	5,49
GERMANY	2,81	30,80	56,36	1,22	9,00	77,30	35.840	15,30	41,20	9	4	0,27	77,08	5,70
HUNGARY	2,05	21,60	64,55	1,13	5,70	71,35	13.260	13,50	56,00	2,4	2,2	0,31	75,00	6,23
FRANCE	2,59	26,35	53,93	1,07	7,70	80,40	33.270	24,50	38,50	23	3,6	0,32	66,67	6,21
UK	1,52	12,70	60,45	0,74	2,60	83,40	32.910	41,00	30,50	17,9	3,2	0,11	64,58	5,90
DENMARK	1,52	14,50	72,17	1,05	10,90	87,90	49.720	10,00	63,00	18,4	1,1	0,51	58,33	6,88
BELGIUM	1,61	22,75	67,28	0,97	6,90	98,00	35.940	15,00	50,00	15	6,7	0,30	56,25	6,40
IRELAND	0,98	10,50	53,33	0,42	1,60	63,20	60.170	63,00	8,00	35,8	6,5	0,14	50,00	7,18
NEDERLAND	1,17	8,60	67,26	0,69	3,70	91,50	41.870	34,00	37,00	14	7,9	0,39	45,83	6,55
w-average	2,97	33,43	45,91	1,05	10,01	76,40	30,00	27,16	33,02	15,00	3,00	0,25	72,67	5,65

Fig. 9.12 A set of 17 EU countries are evaluated with 12 parameters, dividing each column in five classes which acquire the weighted values (first row). DI_{SUM} must be related to the optimum condition established as 80% of the max weight of each parameter, here 0.8 (12 x 10) = 96. So, DI = DI_{SUM} / 96 (e.g., Austria = 86 / 96 = 89.58). DI is inversely related to Cancer onset (last two columns). See also Fig. 7.10

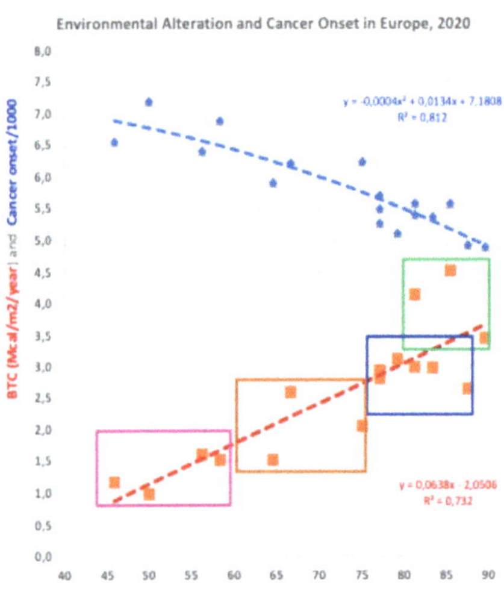

Tab.9.4. Evaluation of DI (%), BTC (Mcal/m²/year), and Cancer Incidence (x 1000) in Western European Countries, (2020)			
W-European Nations	Averages		
	BTC	DI	Cancer
Austria, Romania, Sweden	3.54	87.5	5.12
Poland, Finland, Spain, Portugal, Greece, Italy, Germany	3.13	79.46	5.41
Hungay, France, UK	2.13	68.75	6.11
Denmark, Belgium, Ireland, Nederland	1.32	52.60	6.75

Fig. 9.13 The correlation of BTC Vs. Cancer incidence in European countries is evident, and its significance is more important than traditional carcinogenic parameters (e.g., beef kg/capita, urbanization, pesticides, etc.). The vertical lines indicate Italy (blue, DI = 77.08)

The two curves present a good Pearson Correlation Significance: 1.89 for national environmental health (BTC/DI, $R^2 = 0.73$) and 1.85 for Cancer Incidence (Ci)/ Environmental Alteration (Ci/DI, $R^2 = 0.81$). Correlated with the BTC/DI model, the cancer incidence emerges in undoubtful evidence.

Table 9.4 Pearson's correlation significance (Csg) with cancer onset

Correl.	*Parameters*	*Csg*	Pearson Coeff.
Optimum	BTC	1.55	0.70
Good	BF	1.18	0.53
Good	Urbanized Population	1.11	0.50
Good	HH	1.07	0.48
Good	RNT	1.04	0.47
Near good	GDP	0.88	0.40
Near good	Organic Farming	0.87	0.39
Scarce	RSL	0.44	0.20
Scarce	PRD	0.31	0.14
Scarce	Pesticides	0.29	0.13
Scarce	Red Meat consumption	0.26	0.12
Scarce	Glyphosate	0.25	0.11

Where: BTC = Biological Territorial Capacity of Vegetation (see Sect. 3.4); BF = Bionomic Functionality (see Sects. 7.4.2 and 7.4.3); HH = Human Habitat (see Sect. 3.1.4); GDP = Gross Domestic Product; RNT = Forest Cover (see Sect. 3.2.5); RSL = resilient elements (see Sect. 3. 2.5); PRD = agricultural elements (see Sects. 3.2.5 and 8.3)

In Table 9.4, we expose Pearson's correlation significance (Csg) of the 12 parameters related to cancer onset. Only five exceed the threshold of complete correlation, while the other five present scarce importance and the remaining two are near good. Note that we can interpret these parameters to have spurious correlations: GDP (€/ capita) is directly linked with the power of landscape transformation and artificialization, Red Meat with agrarian and zootechnical industrialization like Pesticides and Glyphosate. This table confirms the effectiveness of DI, which can express the maximum Csg after integrating all the parameters of the state of the landscape, and the crucial importance of BTC in the expression of the health state of a landscape.

9.5 Correlation Landscape Health / Human Health in Europe and India

9.5.1 Opposite Transformations

The diagnostic evaluation of the relationship between HH and BTC shows the opposite directions (dashed lines) of the land transformations in Europe (green) and India (red) from 1880 to 2010. Note that the changes in India were 3.6 times wider than in Europe (Fig. 9.14). Let us examine the situation.

(a) Concerning Europe, the main environmental parameters of the EU have been nearly constant since 1880. Evident increases present the ecological density of the population (red) and Life expectancy (blue). The heavy growth of cancer

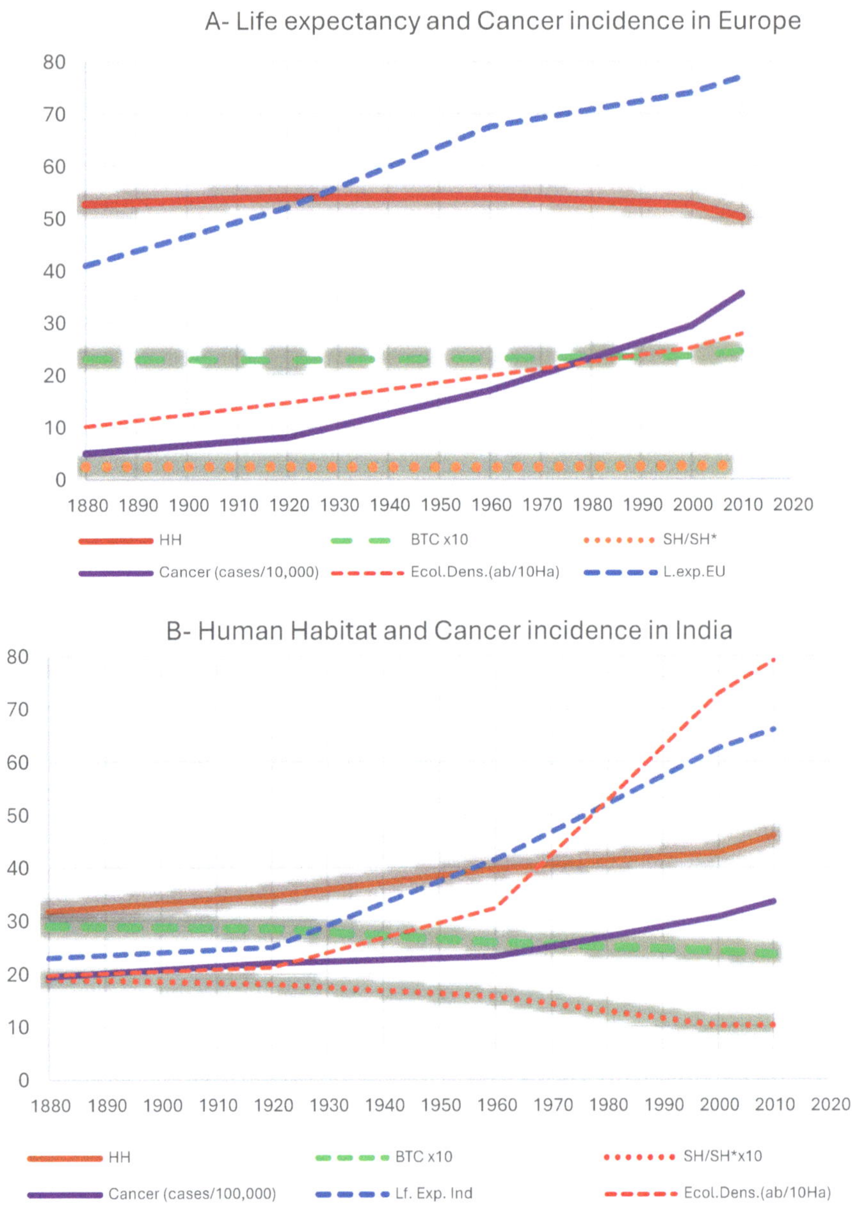

Fig. 9.14 The main environmental parameters of EU and India between 1880 and 2010; are HH (brown), BTC (green), and Carrying Capacity (dotted orange) (Ingegnoli & Giglio, 2021). (From Ingegnoli, Giglio, 2021)

incidence (seven times) is inevitable due to high HH, relatively low BTC, crowded urbanization, and population ageing.

(b) In India, the dynamics of the analyzed land functions present a continuous change, with a detectable change around 1960. The increase in Human Habitat is underlined by the decrease of BTC and carrying capacity SH/SH*, while the ecological density (red) is obviously 2.8 times that of the European one. Life expectancy is lower than that of the Europeans but with a stronger increase. The growth of cancer is limited to 1.7 times, and anyway 1/10 of the European increase.

9.5.2 *Etiology of Diabetes-2 Increase*

Today, another evident increase concerns diabetes-2, so much so many scientists write of epidemics. Contrary to cancer incidence, its growth is higher in India than in Europe. As plotted in Fig. 9.15, the comparison is immediate (blue bars) in urban and rural environments. Furthermore, the difference between low- and high- income is not so vast: 6.1 Vs 8.2%. The problem is that the High-income class (> 12,375.00 $ x capita) in India counts 16 million people, so the prevalence of diabetes-2 may interest 1.3 million (1/1000), while the Low-income (< 1000.00 $) counts as 684 million people, so the prevalence of diabetes interests 41.7 million people. Note that diabetes is known to be linked with the strong growth of overweight people and, consequently, to be a problem for the upper classes.

However, this is not the case because, in the last 10–15 years, even the low-income group has arrived at 75–76% of high-income overweight persons. As High-income (1.3%) presents 10% of people overweight, Low income (56.1%) arrives to show 7–7.5%: so, we will have in India 1.6 million and 47.9 million

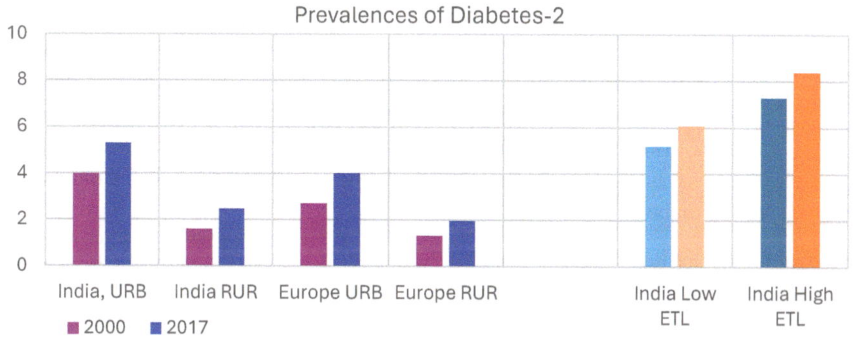

Fig. 9.15 Comparison between the prevalence of diabetes in India and European Union. Note the differences related to urban and rural landscapes, both in India and Europe, and the increase in only 17 years. In the same period, the difference between low and high economic levels (ETL) of population could be meaningful (light and dark blue= 2000; light and dark orange= 2017)

overweight, justifying the prevalence of diabetes, but also the difficulties in its healthcare support.

The reasons for the growth of diabetes are manifold and can also influence low-income people: (1) an increase in caloric food due to industrialized agriculture, (2) Diet modification in modern criteria, (3) an increase in urbanization and consequent food artificialization, (4) Reduction of physical activity. WHO suggested this etiology, but recalling what was exposed in Chap. 5, we must add at least three reasons:

(a) gut microbiota (GM) alterations, due to agrarian landscape banalization.
(b) Inter-scale relationship disruptions, due to agrarian landscape industrialization.
(c) Environmental toxicants acting as endocrine-disrupting chemicals (EDCs).

  (a) The gut microbiome dysbiosis may reshape intestinal barrier functions and host metabolic and signaling paths directly or indirectly related to insulin resistance. Thousands of the metabolites derived from microbes interact with the epithelial, hepatic, and cardiac cell receptors that regulate host physiology, as expressed by (Sapna & Prabhanshu, 2019) and (Sargis & Simmons, 2019). Xenobiotics (e.g., dietary components, antibiotics, and anti-inflammatory non-steroidal drugs) strongly affect the gut microbial composition and promote dysbiosis. Any change in the gut microbiome can shift the host metabolism towards increased energy harvest during diabetes and obesity.
  (b) Bertolaso (2013, 2018) remembers the relationships among a system's components and how such interactions reflect or emerge from the inter-level regulatory processes that are typically the object of inquiry in life sciences.
  (c) Endocrine-disrupting chemicals (EDCs) implicated in diabetes pathogenesis include various inorganic and organic molecules of natural and synthetic origin, including arsenic, bisphenol A, phthalates, polychlorinated biphenyls, and organochlorine pesticides. As underlined by Sargis and Simmons (2019), evidence implicates EDC exposures across the lifespan in metabolic dysfunction; moreover, specific developmental windows exhibit enhanced sensitivity to EDC-induced metabolic disruption, *with potential impacts across generations*. Notably, differential exposures to diabetogenic EDCs also contribute to racial/ethnic and economic disparities. Despite these emerging links, clinical practice guidelines fail to address this underappreciated diabetes risk factor.

As Gandhi underlined, India is a mainly rural nation, so altering agrarian landscapes poses a real danger, even to human health.

References

Bertolaso, M. (2013). *How science works: Choosing levels of explanation in biological sciences.* Aracne.

Bertolaso, M. (2018). Biological robustness – emerging perspectives from within the life sciences. In *History, philosophy and theory of the life sciences* (Vol. 23). Springer.

Bottaccioli, F. (2014). *Epigenetica e Psiconeuroendocrinoimmunologia, le due facce della rivoluzione in corso nelle scienze della vita*. Edra spa.

Bottaccioli, F., & Bottaccioli, A. G. (2020). *Psycho-Neuro- Endocrine-Immunology and science of integrated care*. Edra Ed.

Cislaghi, C., & Nimis, P. L. (1997). Lichens, air pollution and lung cancer. *Nature, 387*, 463–464.

Coker, E. S., Cavalli, L., Fabrizi, E., et al. (2020). The effects of air pollution on COVID-19 related mortality in Northern Italy. *Environmental and Resource Economics, 76*, 611–634.

Crick, F. (1970). Central Dogma of molecular biology. *Nature, 227*, 561–563.

ECIS, European Cancer Information System. (2020). *Incidence and mortality estimates 2020.* European Commission.

EEA. (2011–12). Beating cancer: The role of Europe's environment.

EEA. (2018–19). Beating cancer: The role of Europe's environment.

Formichi, C., Trimarchi, A., Maccora, C., Nigi, L., & Dotta, F. (2022). Environmental factors in the development of diabetes mellitus. In Ingegnoli, Lombardo, & La-Torre (Eds.), *Environmental alteration leads to human disease* (pp. 275–318). Springer-Nature.

Ingegnoli, V. (2001). Landscape ecology. In D. Baltimore, R. Dulbecco, F. Jacob, & R. Levi-Montalcini (Eds.), *Frontiers of life* (Vol. IV, pp. 489–508). Academic.

Ingegnoli, V. (2002). *Landscape ecology: A widening foundation* (p. XXIII+357). Springer.

Ingegnoli, V. (2011). *Bionomia del paesaggio. L'ecologia del paesaggio biologico-integrata per la formazione di un "medico" dei sistemi ecologici* (p. XX+340). Springer.

Ingegnoli, V. (2015). *Landscape bionomics. Biological-integrated landscape ecology* (p. XXIV + 431). Springer.

Ingegnoli, V., & Giglio, E. (2020). Covid-19 incidence and its main bionomics correlations in the landscape units of Monza-Brianza Province, Lombardy. *Journal of Environmental Science and Public Health, 4*(4), 349–366. https://doi.org/10.26502/jesph.96120106

Ingegnoli, V., & Giglio, E. (2021). Environmental bionomic alterations and cancer incidence in European countries: A first overview. *Journal of Nursing & Occupational Health, 2*(3), 266–281.

Marvasti, & Stafford. (2012). From sick care to health care- reengineering prevention into the US system. *New England Journal of Medicine, 367*(10), 889–891.

Mitchell, R. (2008). Effect of exposure to natural environment on health inequalities: An observational population study. *The Lancet, 372*.

Sapna, S., & Prabhanshu, T. (2019). Gut microbiome and type 2 diabetes: Where we are and where to go? *The Journal of National Biochemistry, 63*, 101–108.

Sargis, R. M., & Simmons, R. A. (2019). *Environmental neglect: Endocrine disruptors as under-appreciated but potentially modifiable diabetes risk factors*. Springer.

Scarselli, S. (2013). Relazione sugli inquinamenti di Vado Ligure al Tribunale di Savona.

Steliarova-Foucher, E., O'Callaghan, M., Ferlay, J., Masuyer, E., Rosso, S., et al. (2014). The European cancer observatory: A new data resource. *European Journal of Cancer, 51*, 1131–1143.

Urbani-Ulivi, L. (2019). *The systemic turn in human and natural sciences. A rock in the pond.* Springer- Nature.

Wild, C. P. (2012). The exposome: from concept to utility. *International Journal of Epidemiology, 41*, 24–32. https://doi.org/10.1093/ije/dyr236

Chapter 10
Altered Landscapes Versus Human Health: Processes

Contents

Abstract The studies concerning Biology should be extended to macro-scales, trying to understand their 'anatomical' components, physiological processes and state functions, transformation processes, clinical-diagnostic evaluation, pathologies, and rehabilitation therapies.

V. Ingegnoli, *Landscape Bionomics and Planetary Health*, Landscape Series 21,
https://doi.org/10.1007/978-3-032-08908-3_10

It is possible to rank a list of Landscape Syndrome. It could be helpful to divide well-known traditional processes, e.g., viral infections, water alteration/deficiency, or direct toxicity, and newfound processes. We report an example of how Biological macro-scale alterations and derived physiological processes damage human health. Environmental stress can be registered by the ethological concept of 'value judgment'. The sympathetic nervous system and the hypothalamus-pituitary-adrenal axis mediate the integrated responses of the human organism to stress.

The crucial importance of natural mature forests is vast. In Japan, a 'Forest Bath' was traditionally recommended as a therapy. Note that the psycho-physical relaxation condition enhances the beneficial action of terpenes and monoterpenes, which have antibacterial and analgesic effects and increase the activity of the immune system. A direct relation between psychic stress and inflammation is demonstrated. Moreover, the impact of hormones on the immune system depends mainly on the nervous system, and the haemato-encephalic barrier is considered a mistake today.

Suppose the neuronal 'growth and pruning' process operates in altered landscapes: this operation can result in altered behavior compared to nature, increasing artificial environments, and intensifying landscape pathologies, as a typical positive feedback loop. Therefore, when the environment is significantly altered, artificial environments attract more young people and submit urban green to technology as a sign of a New Babylon.

Note that the Immune System has not only a defense function but many other functions, including Environmental Relationships. The modulation of information linked to the environment (Signaling Molecules, Transcription Factors, DNA, RNA, Protein) is crucial. The new concept of life, at the base of bionomics principles, is not limited to a single organism or a group of species, therefore the linkage due to epigenetic processes is very significant.

In summary, we present a flow chart expressing the complex interrelations among processes linking health and the environment.

Keywords Gut microbiota · Health/environment process · Human health · Landscape syndrome · Neuronal pruning · Transcription factors · Immune system · Information

10.1 Processes in Macro-Scale Biology

10.1.1 Landscape Syndromes and Their Effects on Men

The misunderstanding between Medicine and Ecology is a challenge: we must overcome this impasse! Thus, we cannot discuss the unity of Life, but we must understand better how its scalar interrelations may influence our health. Like small ones, macro-scale life alterations can damage human health (Huttenlocher, 1979; Ingegnoli, 2022). Note that the underestimation of the environment is rooted in

Neo-Darwinian thinking: concepts such as the struggle for existence and natural selection are metaphors (Petanjek et al., 2011; Popper, 1982), not theories, so Darwin's hypothesis becomes "the best-adapted individuals are more likely to have descendants." Thus, other limits of Darwinism appear, first because, in biology, it is possible to demonstrate that the struggle for existence is less significant than cooperation and symbiosis.

At the beginning of *Homo sapiens* development (about 250,000 b.p.), a man/nature *competition*, based on hunting and gathering, has pushed for continuous migrations, allowing the conquest of new continents and interbreeding with other humans (e.g., Neanderthals, Denisovans). However, in the last 10 millennia, man entered a new era, following a *symbiotic* concept, a first *mutualism* with nature, as well recalled by Godron (1984). An alliance with cereals, and vegetables, allowed populations to build villages and first towns. The ability to settle in areas appropriate for the safety and the exploitation of resources has therefore grown in a particular way, together with the capacity of villages planning and first agricultural areas. However, as underlined by E.P. Odum (1971), man is today acting as a *parasite* on his autotrophic environment, taking what he needs, with little regard to the welfare of his host. So, if man does not learn to live in fully mutualism with nature his host may exploit man. Our survival is essentially linked to an era of full symbiosis!

Even a physicist like Federico Faggin (2024) asserts: "Look at scientism: it pushes us to believe that reality is made only of matter, and everything else follows suit. And so, for example, we end up exalting competition, (…) we have accepted the principle of the survival of the fittest, the strongest. In my vision, however, there is only cooperation".

Further considerations concern:

1. Mathematics, which has shown that in a complex system, a random variation always produces damage: see Moser and Kolmogorov's theory (Arnold, 1963).
2. Biosemiotics, which have shown that, in addition to genetic codes, other organic and mental codes are involved in evolution (Barbieri, 1985).
3. Epigenetic, as its control of gene expression due to DNA methylation demonstrates that the phenotype is not directly expressed by the genotype, and part of the methylation pattern of the genome can be inherited in the Lamarckian sense (Bottaccioli & Bottaccioli, 2020).

As affirmed in the previous chapters, health and disease depend on the state of the *entire organization of life*. Consequently, the studies of biology should be extended to macro-scales, trying to understand their '*anatomical*' components, *physiological* processes and *state functions*, transformation processes, *clinical-diagnostic evaluation*, pathologies, and rehabilitation therapies (Par.2.4.1).

Here are some examples of Landscape Syndromes and their effects on men, listed in Table 10.1. As one can see, at least 9 sets of Landscape Syndromes can be ranked (see also Table 7.1). Each section may explicit various pathologic processes, and effects on men, of which some examples are inserted.

Table 10.1 Examples of landscape syndromes and their effects on men

	LANDSCAPE SYNDROMES	EFFECTS ON MEN AND ON HIS HABITAT
1	***Structural alterations***	
	Wrong distribution of energy infrastructures near residential ecotopes	*Acceleration of toxic effects*
	Deficiency of ecological network in LU territory	*Cumulative impacts*
	Increasing fragmentation	*Ethological alarm*
	Improper road networks in natural preserves, etc.	*Obstacle to bionomic functionality*
	Improper Park design and distribution in urbanized LU	*Pollution concentration*
2	***Excess of functional alterations***	
	Disturbance on landscape Apparatuses	*Behavioral alteration*
	Dysfunctional standard habitat per capita and carry-ing capacity	*Behavioral changes & diet alteration*
	Insufficient level of BTC in LU ecotissue	*Alarm sign & environmental stress*
	Dysfunctions in landscape apparatuses	*Exalted life artificiality*
	Heterogeneity destruction	*Food security disruption*
	Hyper-homogeneity of landscapes	*Improper land-use*
	Hyper-trophy of human habitat elements in LU	*Psychological dysfunctions*
3	***Transformation syndromes***	
	Abnormal transformation deficit	*Destructuring of the environment*
	Heterogeneity destruction	*Decrease of health protection*
	Disturbance on landscape Apparatuses	*Increase of artificialization*
4	***Lack of protective agents***	
	Reduced environmental Pool for gut microbiome functionality	*Dysfunction of gut microbiome*
	Lack of protective Ecotopes	*Higher morbidity*
	Difficulty in disturbance incorporation	*Increase of cumulative impacts*
	Deficiency in zoologic proper habitats	*Increase of infections*
	Insufficient natural mature Forest patches	*Insufficient phytoncides*
	Loss of constraints of high level (slow processes)	*Loss of recreational systems*
	Loss of Interface elements (man/nature)	*Zoological reconversion*
5	***Hierarchical deficiencies***	
	Destruction of tree-hedgerows network in agrarian LU	*Decrease of ecological services*
	Loss of scale connections	*information distortions on body life*
	Disruption of agricultural landscapes	*Impoverishment of nutrients*
	Difficulty in disturbance incorporation	*Incapacity of adaptation*
	Difficulty in epigenetic modulation of natural & cereal populations	*Increase of artificial infrastructures*
	Industrialization of agricultural Landscapes	*Increase of water pollution*
6	***Cybernetic alterations***	
		Dysfunction in population structure

(continued)

Table 10.1 (continued)

	LANDSCAPE SYNDROMES	EFFECTS ON MEN AND ON HIS HABITAT
	Wrong influence on animal & human environmental fitting processes	
	Loss of governance of its sub-systems	*Incorrect behavior vs. nature*
	Distortion on Information pools and transmission webs	*Loss of attention to resources*
	Low information Content of a LU	*Incoherent epigenetic code*
	Loss of ecological memory	*Socio-economics alterations*
7	***Infections & invasions***	
	Insect invasions in natural landscapes	*Food contamination*
	Insect invasions in agrarian or forest landscapes	*Vegetational destructions*
	Infections in human populations	*Pandemic onset*
	Infections in natural populations	*Zootic infections*
8	***Pollutions***	
	Fine dusts	*Allergic reactions*
	Complex multiple syndromes	*Epigenetic distortions*
	Chemical pollution	*Respiratory diseases*
	Endocrine disruptors	*Toxic diseases*
9	***Catastrofic perturbations***	
	Wars & terrorism	*Acute pollutions*
	Earthquakes	*Destruction of landscape apparatures*
	Floods & landslides	*Destruction of landscape structure*
	Industrial & Technological Disasters	*Out of scale fires*
	Cyclones	*Water contaminations*

10.1.2 Processes Linking Landscape Alterations and Human Health

These syndromes and their effects can damage human health through many processes linking landscape alteration to human health: it could be useful to divide well-known traditional processes, e.g., viral infections, water alteration/deficiency, or direct toxicity (Table 10.2. column A), and newfound processes (Table 10.2. column B).

Moreover, it is necessary to synthesize some peculiar newfound process, beginning with the ethological one. I try to expose the most important sets capable of understanding the listed processes, being conscious that this is not intended as an organic treatment: (i) ethological alarm signals and their stress influence, (ii) psychic stress, inflammation and meditation, (iii) neuronal fitting brain/environment, (iv) the immune system and the environment, (v) mature natural forests and human health, (vi) information disturbances, and epigenetic code, (vii) scale interactions of landscape pathology/human health, (viii) complex interrelations among health/environment processes.

Table 10.2 Processes linking landscape alteration & human health

A- Traditional processes	B- Newfound processes
A1- Hydrogeologic alterations	B1- Ethologic alarm due to bionomic dysfunctions
A2- Earthquakes and wars destructions	B2- Stress reactions due to ethologic alarm
A3- Viral & Baterial infections	B3- Hierarchical (scale) processes disruption
A4- Fungal & protozoal infections	B4- Environment/psycho-brain correlations
A5- Insect invasions	B5- Environment/meditation / illness correlations
A6- Water alteration or deficiency	B6- Neural pruning / altered landscape information
A7- Micro-climate changes	B7- Gut microbiome / environment relationships
A8- Urban Heath Island	B8- Antibiotic resistance to microbes
A9- Direct toxicity from pollution	B9- Zoonotic disturbances & viral reconversion
A10- Endocrine disrupting chemicals	B10- Environment, food, health relationships
A11- Excessive fine dust	B11- Bionomic carrying capacity SH/SH*
A12- Excessive traffic and noise	B12- Phytoncides and VOC in forests & immune strengthening
A13- Radioactive gas and rocks	B13- Information disturbances and epigenetic dysfunctions
	B14- Lack of Information and cell repairing consequences
	B15- BTC level and health protective services

10.2 Ethological Alarm Signals and Their Stress Influence

10.2.1 Ethology and Stress

Many environmental stressors are due to structural dysfunctions in the landscape, even without pollution. An Ethological Alarm Signal leads to environmental stress, which can be chronic. Let's delve deeper. Remember, first, that a basilar comparative ethological question is linked to information, meaning, and survival:

> The provision and storage of information relevant to the conservation of the species represents for all living entities, a performance as fundamental as the supply and storage of energy. From a phylogenetic point of view these processes are equally ancient: both probably appeared simultaneously with the first life form. As far as I know, Otto Rossler was the first biologist to grasp and clearly express the concept that not only the processes of energy acquisition constitute in themselves a circuit of positive coupled reactions, but they are also in a positive feedback relationship with the information acquisition processes. When, through a mutation or a new combination of hereditary components, the probability of energy acquisition increases in such a way that selection acts to the advantage of the being thus improved, then the number of offspring also increases. (…) This double positive feedback loop of energy and information acquisition processes is characteristic of all living beings, including viruses (...). The selective pressure that pushes to increase the wealth of information relevant to the conservation of the species is so omnipresent that it could be quite sufficient to justify the general direction of evolutionary events... by Lorentz (1973).

Thus, the introduction of the concept of "*values-judgment*" (Lorenz, 1978). This value implies a correspondence between the *innate perception* of values and entropy, thus the passage from 'the most probable and *disordered*' to 'the most *ordered* and improbable' condition.

By applying the scale of values derived from this fundamental process to judge the health of an organism or a landscape, this ethological process can bring, for example, to:

(a) a *Recreational* meaning, if the environment transmits information about good health, thus activating well-being (psychic and physiological) and health prevention.
(b) an *Alarm* meaning, if the environment transmits information about an altered environment, thus activating the *stress responses*. Chronic stress leads to getting sick more easily.

Let us deepen the mechanism: when the ethological process has an alarming meaning, it registers all the *environmental alterations* (Fig. 10.1, photographs) (Ingegnoli, 2022), which remain memorized within our cells. To reach this goal the ethological process must be *innate and unconscious*, because to de-codify environmental signals one must reduce emotions. Remember that the subconscious mind can process 20 x 10^6 environmental stimuli/sec, Vs. only 40 stimuli/sec. of the conscious mind.

Stressors simultaneously activate: (a) neurons in the hypothalamus, which secrete CRH (Corticotropin-releasing *hormone*), and (b) adrenergic neurons. These two responses potentiate each other. The final effect of the activation of neurons that secrete CRH is the increase in cortisol levels, *while* the net effect of adrenergic stimulation is to increase plasma levels of catecholamine (Dopamine, norepinephrine, and epinephrine).

The negative feedback exerted by cortisol can limit an excessive reaction, which is dangerous for the organism. However, when the stress became *chronic*, the circadian rhythm of *melatonin/cortisol is altered* (Fig. 10.1, bottom). Plasma cortisol levels (red dotted curve) bring to a *dominance of the Th2 immune circuit*, with production of typical catecholamine (e.g., IL-4, IL- 5, IL- 13) and the circuit Th17 (Blalock, 2005). Note that the Th2 immune response is not available to counteract *viral infections, neo-plastic cells, auto-immune syndromes,* which need a Th1 response. *So, the premature death risk increases.*

This figure shows the emergence of an ethological alarm signal: people living in cottages and small farms no longer border a protective forest but alienating enormous 10–12 story apartment buildings and very high voltage power lines. The landscape unit BTC changes from 1.42 Mcal/m^2/year to 0.55 and the HH from 75% to 87%. The environmental stress is severe, and once it becomes continuous it alters the level of cortisol.

EXAMPLE OF LANDSCAPE SYNDROMES	MAIN EFFECTS ON MEN	HUMAN HEALTH DAMAGES
FUNCTIONAL ALTERATIONS • INSUFFICIENT level of BTC in LU ecotissue; • INDUSTRIALIZATION of Agricultural Landscapes • ALTERED Ecological Density of Population • DEFICIENCY of Ecological Network	ETHOLOGICAL Alarm CHRONIC Environmental Stress PSYCHOLOGICAL disorders PHYSIOLOGICAL alterations	Chronic stress dysfunctions, Higher morbidity Psychological depression Risk of premature death

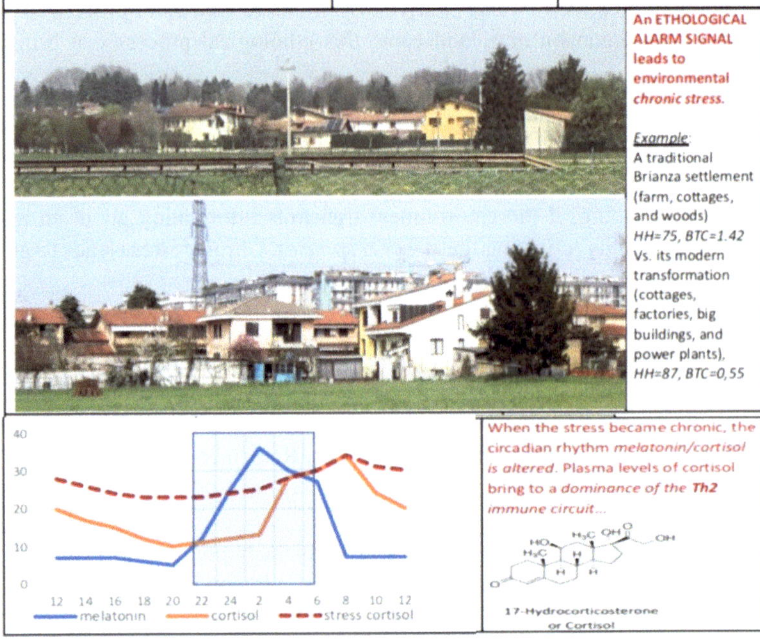

Fig. 10.1 An example of how Biological macro-scale alterations and derived physiological processes are damaging human health. Environmental stress can be registered by the ethological concept of "values-judgment". The sympathetic nervous system and the hypothalamus-pituitary-adrenal axis mediate the integrated responses of the human organism to stress. Note the crucial importance of cortisol (Ingegnoli & Giglio, 2021)

10.2.2 Chronic Stress and Cancer

The above Fig. 10.1 illustrated an example of how Biological macro-scale alterations and derived physiological processes damage human health through the ethological concept of "value judgment" and the integrated responses of the human organism to stress mediate by the sympathetic nervous system and the hypothalamus-pituitary-adrenal axis.

Now Fig. 10.2 [Bottaccioli & Bottaccioli, 2020, modify and replotted] illustrates the complex process of chronic stress related to cancer, showing the biochemical paths which chronic stress increases the cancer risk with. It is an emblematic process extremely serious and not perceived by territorial and policy authorities, because it

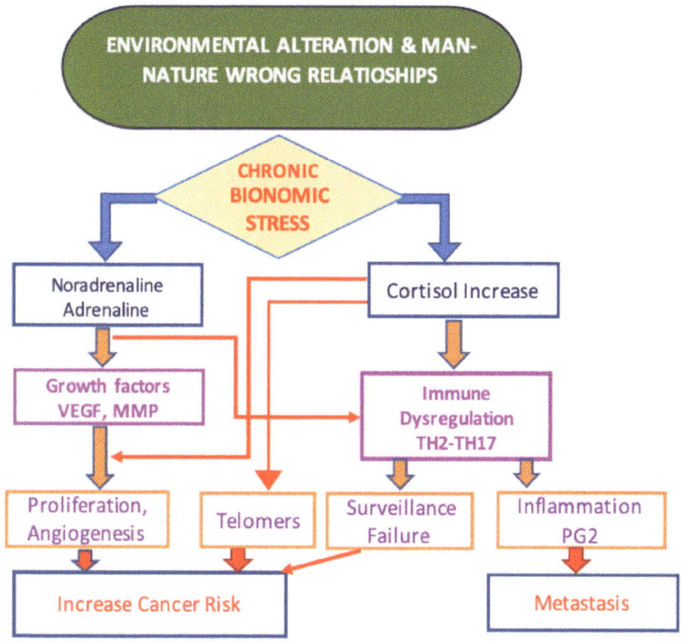

Fig. 10.2 Biochemical paths with which chronic stress increases the cancer risk. Noradrenaline, Adrenaline, and Cortisol induce the formation of VEGF (growth factor of vascular endothelium), and metalloproteinases (MMP), which help the tumor proliferation, facilitated by the immune deregulation with the weakening of TH1. (From Bottaccioli & Bottaccioli (2020); modified and re-plotted)

demonstrates that even independently from pollution the destruction of natural components of a LU may affect human health.

As clearly explained by Bottaccioli and Bottaccioli (2020), Noradrenaline, Adrenaline, and Cortisol induce the formation of VEGF (growth factor of vascular endothelium) and metalloproteinases (MMP), which help tumor proliferation, facilitated by the immune deregulation following the weakening of TH1.

10.3 Psychic Stress, Inflammation and Meditation

10.3.1 Emotional Activation of Brain Signals

The ethological information on the state of our environment can be elaborated following the principles of psycho-neuro-endocrine-immunology. For instance, an emotional or alarm signal registered in the dorsal anterior cingulate cortex (dACC) (i.e., the brain region that subserves cognition and motor control) is correlated with an increase in plasmatic levels of a basilar inflammatory cytokine (type II of TNF,

Tumor necrosis factor) (Fig. 10.3). So, a *direct relationship* between *psychic stress and inflammation* is demonstrated (Simard & Durall, 2004). Moreover, the effects of hormones on the immune system depend mainly on the nervous system, and the haemato-encephalic barrier is considered a mistake today. In add, psychic relationships are not only in negative (stress alarm) but also in positive (relax and benefit) as well known for centuries and acted by St. Francis of Assisi (see next Sect. 10.3.2).

In the face of the materialist theories, scientism, and neoclassical economy, which influence the society on a global scale, humans would need a revolution. Without a revolution seems to be impossible to eliminate the present ecological crisis. What kind of revolution? There are two possibilities: (a) *violence*, e.g., French, or Russian revolution, or (b) *non-violence*, as demonstrated by the Mahatma Gandhi (1962). This direction, underlined by Gandhi, brings to enhance meditation. It is at the basis of the human spirit, so it comes before the same religion. Children can spontaneously meditate in front of emotions detected in the natural environment (Fig. 10.4, left). In the project of his hypothetical house, made by my son when he was 5–6 years old, he had inserted a room "for reflection", placing it on a high floor of the house.

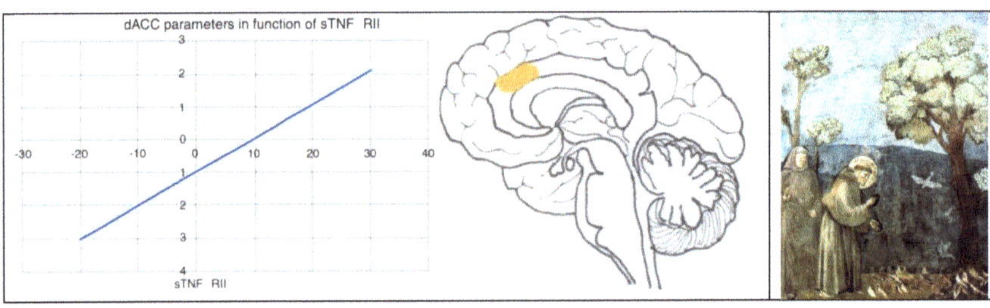

Fig. 10.3 An emotional activation indicated by an increase in signal in dACC, correlates in the graph with an increase in the plasma levels of the soluble TNF-alpha type II receptor, a basic inflammatory cytokine. A similar process is also done in positive, as acted by St. Francis

Fig. 10.4 (Left) Children are spontaneously able to meditate in suggestive natural niches. (Right) Mindfulness-Based Stress Reduction (MBSR) and mental well-being (e.g. contemplation of natural harmony) increase the γ-interferon, produced by the B- and T- lymphocytes, which at the time T4 (1 month) returns to being almost normal, and also manages to counteract any oncological diseases

10.3.2 Meditation, Faith and Human Health

Meditation is a way of connecting with a natural state of mind that is spacious and clear. It is not eliminating thoughts but noticing when our mind is busy or racing. Meditation can help to connect with the breath and bring calmness to the mind. There are many different types of meditation: most involve being still and quiet; some involve movements, such as tai-chi, chi-gong, yoga or walking meditation. It must be underlined how meditation changes the activity of the immune system. Bottaccioli & Bottaccioli (2020) indicate that the activity of the Nk (natural killer cells) increases in the group of meditating patients, so to become like people without cancer: an increase evaluated +50%.

In Fig. 10.4, right, note that the IFN-γ (the primary signal of the immune circuit Th1) increases in 1 month from 7 to 15 ng/mL in the group of meditating patients, while it decreases in non-meditating ones. Andersen et al. (2005) underline the consequences: more than 80% of the patients of the meditating groups did not die of cancer, Vs. 60% of the other; but, even if meditation promoted by Psychic therapies and Oriental philosophies is adequate, following St. Francis suggestions, nothing more than deep religious practice, like spiritual prayers, retreats and fasting, can reach the maximum of efficaciousness in recovery.

10.4 Neuronal Fitting Brain vs. Environment

10.4.1 Central Nervous System Fitting on the Environment

Nature predisposed a biological process to ensure the development of the Central Nervous System fitting with the environment, which is available for higher animals, especially men. Remember that, during the first growth of superior animals and still more of man, the components of the cerebral cortex mature at different times: medicine doctors discovered the possible existence of a process like this on 1909 when Ramon j Cajal (1909) compared the development of the neurons of the brain cortex of babies from 3 to 24 months; but neuroscientists have been believing that neural pruning ended a few years after birth for decades. However, in 1979, Peter R. Huttenlocher (1979), a neurologist at the University of Chicago, demonstrated that this excess production and pruning strategy has been continuing for synapses long after birth, and recently, Petanjek et al. (2011) extended this neurologic process during the entire life (Fig. 10.5).

As shown in this figure, the incredible growth of neural spines and their networks, from 8 to 38/50 μ in only 6–7 years, resulted in 50% of them pruned in the following 15 years, while the other 50% being eliminated in the last 60 years. So, the age of majority begins only after a compulsory adaptation to the environment. It is credible to suppose that this long process is done to reach a proper fit with the environment. The environment is necessary for developing brain capacities in every mammal,

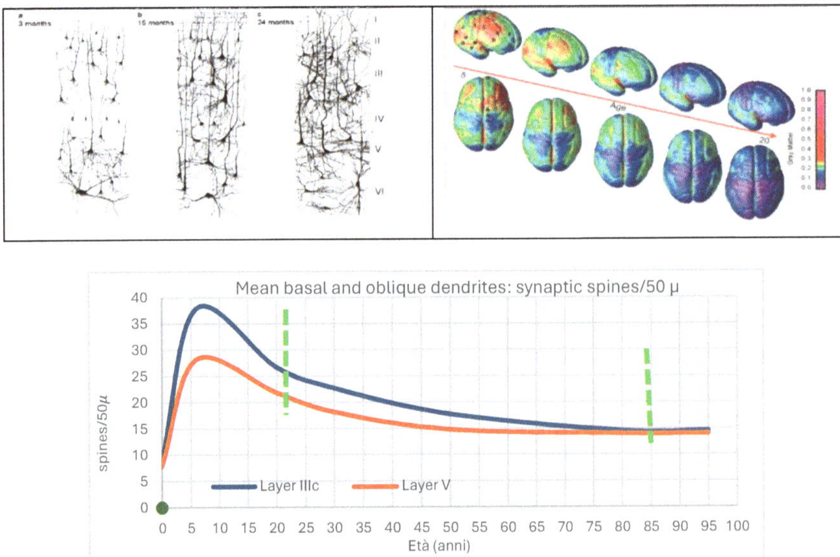

Fig. 10.5 (Above) Studies from Ramon j Cajal (1909), and the group of Gogtay et al. (2009) brought to the Neural Pruning effect. (Below) Synthesis of regression curves (Petanjek et al., 2011; modified) elaborated at the linear scale during the 100-y human lifespan

even in man. The property of having some non-determined cerebral structures at birth, which we must specify in relation to the environment, has been emphasized in ethology (Lorenz, 1973, 1978).

10.4.2 Exalted Feedback from Artificial Environment

In our brain, areas that mediate "executive functioning" mature later than areas responsible for basic functions (Gogtay et al., 2009). Therefore, artificial environments seem to attract more and more young people and to submit urban green to technology as a sign of a New Babylon. Let observe that the two processes exposed before, the 'growth & pruning neural synapsis' and the 'environmental stress alarm', must be obviously integrated in relation to landscape pathologies and human health. In Fig. 10.6 (Ingegnoli, 2015) a logic flow diagram tries to explain their main linkages and loops.

If the neuronal growth and pruning process operates in altered landscapes, all this may result in changed behavior. First, versus Nature, bringing to increase artificial environments, then intensifying landscape pathologies: a typical positive feedback loop; another behavioral feedback loop appears when an increase in cortisol increases arousal and aggression; learning and social disturbances lead to acting against natural lows, thus degrading ecological components and producing a new

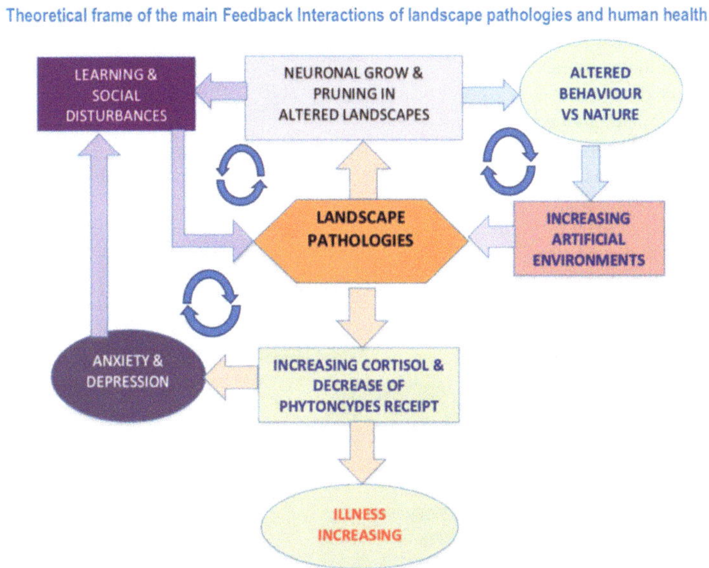

Fig. 10.6 Landscape pathologies increase artificial environment and produce exalted feedback, which in turn generates other exalted feedback. Note the formation of at least three self- exalting feedback loops, which can aggravate the landscape syndromes, therefore the consequences on human illness

intensification of landscape pathologies. Chronically elevated cortisol levels have been linked to problems including abdominal fat gain, cognitive decline, and compromised immune function. Note that learning and social disorders are reinforced by altered neuronal growth & pruning.

As a result of the experience of bionomic applications, the brain/environment adaptation following neural growth and pruning process depends on the bionomic state of the landscape unit (LU). When the state of the environment in a LU is significantly altered, that is the potential of bionomic rehabilitation PoBR <20% (see Sect. 8.8.2 and Table 8.2), bionomic functionality BF <0.85 (see Sects. 7.4.2 and 7. 4.3), environmental cumulative stress ECS >25 (see Fig. 11.3), it can lead to the degradation of the person who reacts with technical sustainability and an increase of artificiality (Fig. 10.7).

The examples of the recent green deal in Milan seem to confirm this hypothesis (Fig. 10.8): the hyper-tech green tower and the new park in the same Business District were designed with no possibility of ecologic and bionomic compensation.

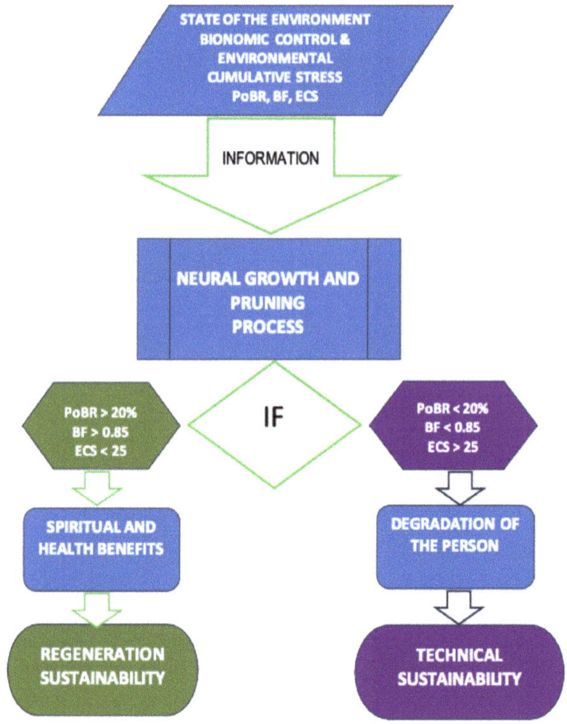

Fig. 10.7 Brain/environment adaptation depending on the bionomic state of the landscape unit (LU), strictly related to the previous figure. In this hypothesis, when the LU is very altered (PoBR <20%, BF <0.85, ECS >25) we can arrive to the degradation of the person (i.e. people denying the qualities that characterize a man as a person) which react going towards a technical sustainability

Fig. 10.8 It is necessary to underline that today the whims of a designer can reach hallucinating formalisms: man creates for himself, in total freedom (understood as the absence of constraints) the criteria for his life, and for its landscape, aided by technology

10.5 The Immune System and the Environment

10.5.1 Immune System: Not Only Defensive Function

Giving the definition of the Immune System as a complex network of cells (Fig. 10.9) and biochemical markers—developing in an organism coordinate systemic actions to control the variations of its internal and external allostasis, providing

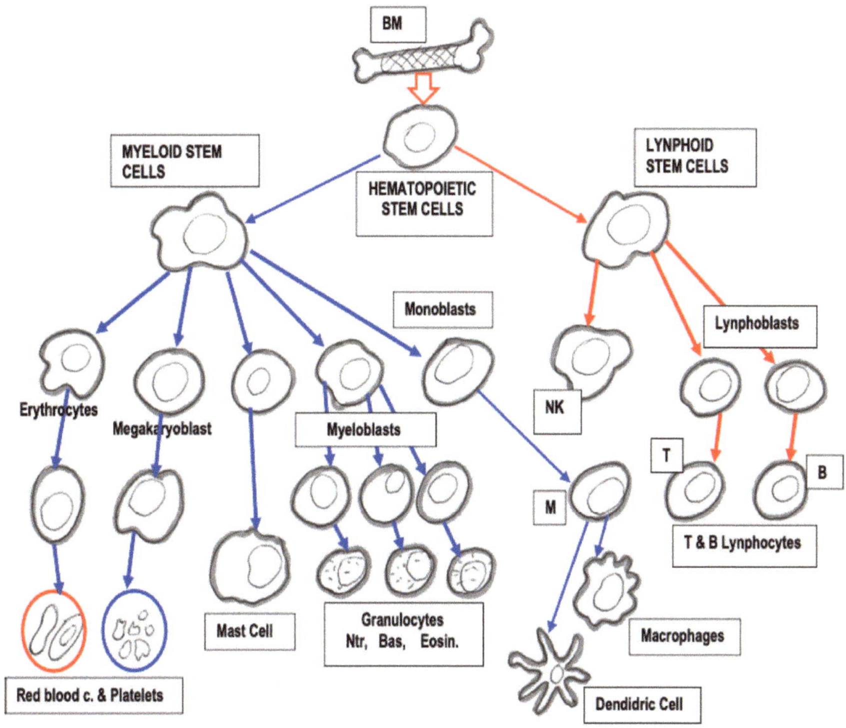

Fig. 10.9 Hematopoietic stem cells develop into two branches: Mieloid stem cells, and Lynfoid stem cells. BM bone marrow, Nk Natural killers, M Monocytes, Ntr Neutrophils, Bas Basophils, Eosi Eosinophils, T T lymphocytes, B B lymphocytes. Note that M, T, B can be named agranulocytes

adequate responses to stimuli that often depend on the contest—the Immune System is generally associated with the Nervous System (central, CNS, and peripheral, PNS), which has a convergence of functions and exchange of biochemical markers and electromagnetic fields.

Blalock (2005) spoke of the "sixth sense" and immune-regulation activity from the CNS. Anyway, information can pass bidirectionally between nervous and immune systems. Moreover, as demonstrated by Besedovsky and Sorkin (1977) and Besedowsky and Del Rey (2007), during an immune reaction humans witness a neuro-endocrine modifications. The inflammatory cytokines (IL-1, IL-6, TNF-α) can induce relevant biological modifications to the main neuroendocrine axis.

So, the Immune System has not only a defence function, but many others too:

1. *Defence function,*
2. *Clearance function,*
3. *Repair function,*
4. *Development function,*
5. *Homeostasis maintenance,*
6. *Environmental relationships.*

10.5.2 Immune System and the Environment

Moreover, the Immune System is even related to the Endocrine System, as underlined by Psycho-Neuro-Endocrine Immunology (Bottaccioli & Bottaccioli, 2020), which indicates that the gut-immune system (hosting 70–80% of immune cells) is linked to the MALT (Mucose-associated lymphoid tissue), divided into BALT (bronchial-associated lymphoid tissue) and other subsystems (reproductive, nasopharyngeal). So, let me list some biological macro-scale processes where the Immune System is concerned:

1. Défense against infections and zoonoses
2. Micro-biota exchange
3. Stress responses from psychological-allostatic tensions
4. Stress responses from an ethological alarm process to chronic changes in the environment
5. Neurological modulation of the biological-structural state of the environment (pruning effect)
6. Reactions to phytoncides and forest bath
7. Reactions due to environmental pollution
8. Response to alteration of edible crops that do not consider EPB (evolutionary plant breeding)
9. Reactions to alterations of the scale-dependent processes of the bionomic organization

As set out briefly, the Immune System can have a potent role of integrator among all these processes that link environmental alterations to human health. The most frequent environmental alterations lead to damage the immune system: the processes listed here seem to confirm this crucial relationship.

10.6 Mature Natural Forests and Human Health

10.6.1 Gut Microbiome and Vegetation

Gut Microbiome can change and be influenced in diversity and composition by the diet, the neurotransmitters and metabolites, by the immune functions, and by the *bionomic-ecological habitat* with its vegetation too (Fig. 10.10).

Here, the crucial importance of direct and indirect contact of men with natural vegetation systems to maintain and enrich their GM, even independently from nutrition must be underlined. The Environment plays a much more significant role than host genetics in determining the composition of the human GM, according to a study recently published on Nature (Rotschild et al. 2018), e.g., living within farms and near forests offers protection against developing allergic diseases in childhood. The crucial importance of maintaining GM exchange areas near our landscape unit

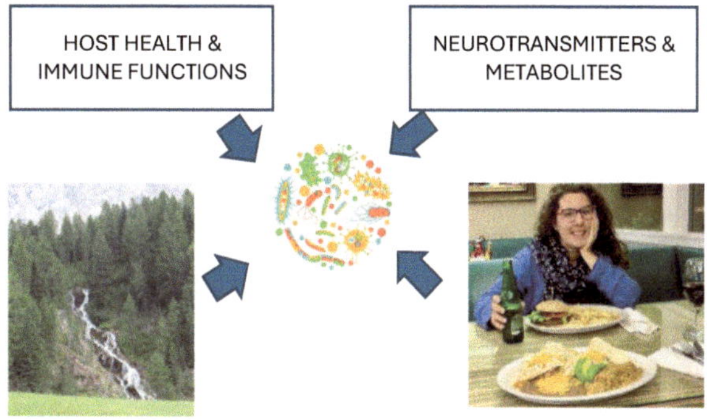

Fig. 10.10 Main groups that influence the Gut Microbiome, to which the importance of the environment, both for food production and for the presence of natural ecotopes must be added

(LU), where we live and work, is demonstrated by the many components of a forest patch that show the presence of high microbial biomass.

10.6.2 Forest Bathing and Human Health

As can be deduced from the volume "Forest Therapy" (Meneguzzo & Zabini, 2022), a study by Meneguzzo and Zabini recalls that among B-VOCs of forest, researchers have identified more than 1000 different compounds, including isoprene (ca. 50% of total emissions) and various terpenes, mostly monoterpenes (MT: α- and ß-pinene, ß-ocimene, d-limonene, sabinene, myrcene and camphene constitute approximately 35–40% of total emissions), with smaller quantities of other substances, including some potentially harmful (10–15%). Therefore, it makes no sense to speak of trees as polluting entities, given the proportions of emissions and the results of various experiments on the health benefits of "forest immersions" and the role of B-VOCs (Bioactive-Volatile Oxygen Compounds) (Fig. 10.11).

The crucial importance of natural mature forests is vast. In Japan, a Shinrin-yoku (a Forest Bath, i.e., a path in a natural forest landscape for a couple of days) was traditionally recommended as a therapy, as follows: 1st day = 2 h of walking; 2nd day = 4 h of walking.

Note that the forest bath's psycho-physical relaxation condition enhances the beneficial action of terpenes and monoterpenes (α-pinene, β-pinene, γ-pinene, linalool, limonene), which have anti-bacterial, analgesic effects and increase the activity of the immune system.

Quing Li et al. (2008) and Li (2009) has shown that phytoncides (forest essential oils) have beneficial effects on natural killers (Nk) because they increase the number and effectiveness of the cytotoxic enzymes granulysin, perforin and A and B granzymes expressed by peripheral blood lymphocytes cells (PBLs), i.e. CD8 lymphocytes (T effectors), strengthening the immune system (Fig. 10.12. top right).

Let me summarize the result of the studies of Li et al. (2008) and Li (2009) on the measurement of the proportions of NK cells, T cells, granulysin, perforin and

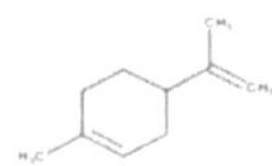	d-Limonene	Anti-inflammatory, analgesic, antioxidant, anxiolytic, antidepressant, antiproliferative.
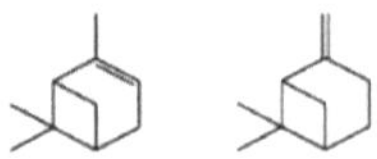 α-pinene β-pinene	a-Pinene e ß-Pinene	Anti-inflammatory, analgesic, antioxidant, anxiolytic, antidepressant, sedative, antiproliferative.
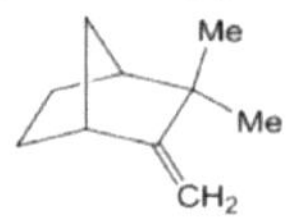	Camfene	Lipid-lowering, with stimulation of metabolism, antioxidant, analgesic, antiproliferative.
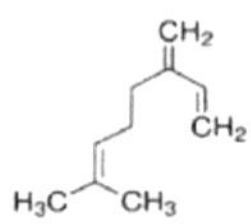	ß-Myrcene	Anti-inflammatory, analgesic, sedative, muscle relaxant, gastroprotective, antiproliferative

Fig. 10.11 Health effects of terpene emissions in forest

granzymes A/B in Peripheral Blood Lynphocytes, on 12 males whose age were 43.1 ± 6.1 years for a Forest bathing lasting 2 days: Control = normal routines. No crossover. NK activity higher after the trip (increase of about 50%) than before and between days 1 and 2. Significantly increased of the score for vigor and decreased of the scores for anxiety, depression and anger (Psychological tests POMS, i.e. Profile of Mood States).

Moreover, Peter Wohlleben (2016), in "The Hidden Life of Trees", describes how trees feel and communicate. His research shows that trees form highly communicative social networks (remember 5.4.2). They care for their sick, nurse their young, and share valuable information and resources. Wilson (2019) and Wohlleben's research is consistent with Simard's findings (Simard, 2018, 2021; Simard & Durall, 2004), showing that trees communicate through a vast fungal network around their roots. Wohlleben found that trees use scents to communicate with each other. While the understanding that trees communicate is now supported by scientific evidence, some Native peoples have always known this. According to Native American wisdom, trees "talk to each other, and they'll talk to you if you listen" (Nesburn & Mengelkock, 1991).

People must be directed to consider natural environments (if not contaminated or structurally and functionally altered) as a source of well-being, integrative therapies and prevention, gut biome enrichment and exchange, and psychic recreation & meditation (Slavich et al., 2009). For example, we underline the mature forest as the best environment to adapt natural activities but leaving them in their natural conditions. A forest transmits information to man both through light and sound and

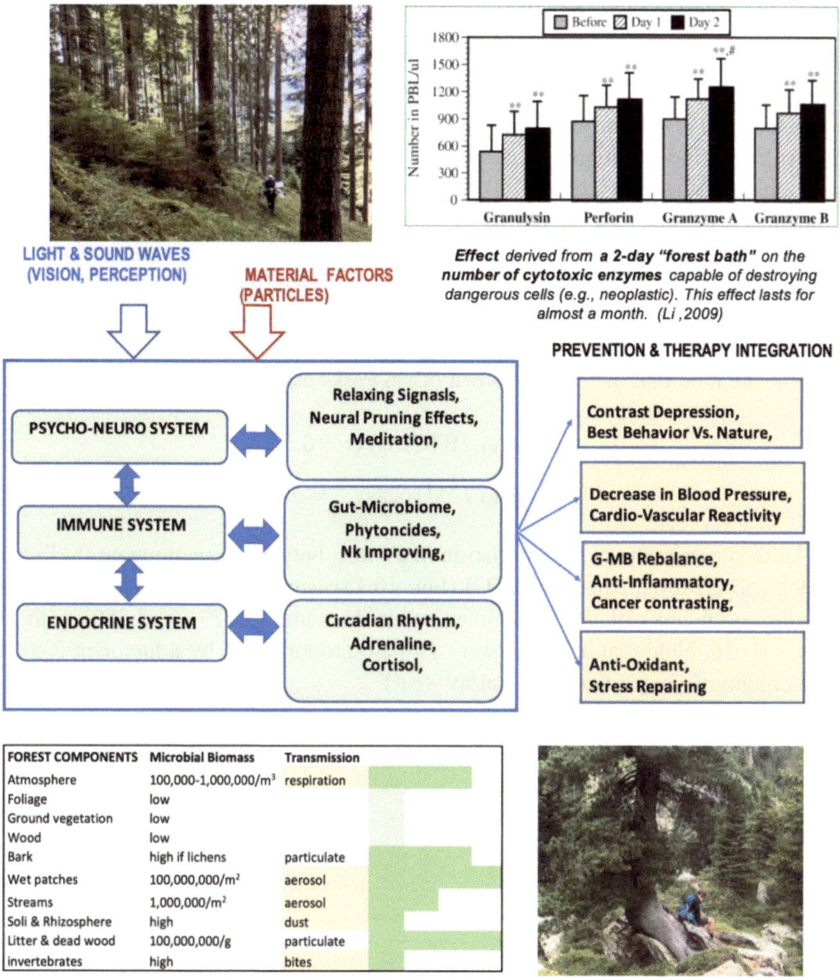

FOREST COMPONENTS	Microbial Biomass	Transmission
Atmosphere	100,000-1,000,000/m³	respiration
Foliage	low	
Ground vegetation	low	
Wood	low	
Bark	high if lichens	particulate
Wet patches	100,000,000/m²	aerosol
Streams	1,000,000/m²	aerosol
Soli & Rhizosphere	high	dust
Litter & dead wood	100,000,000/g	particulate
invertebrates	high	bites

Fig. 10.12 Complexity of the interrelationships between forests and human health. It is reiterated that health prevention and therapeutic integration, Gut microbiota Enrichment and Exchange, and Psychic Recreation & Meditation cannot well materialize without forests: the table left-down focus on the modalities of transmission

electromagnetic waves and through material factors (Particles): we, therefore, receive information through rational and unconscious perceptions and emotions, microclimatic changes, microbial loads, VOCs, and phytoncides (Fig. 10.12. center and down left).

To this information, one must add *epigenetic information*, as it will be shown in Fig. 10.16.

10.6.3 Proportionality Between Biological Territorial Capacity of Vegetation and Monoterpene Values

Therefore, the effects on our body and soul are multiple, as underlined by Psycho-Neuro-Endocrine Immunology, from neural pruning and meditation to endocrine reactions and the activation of Nk. Monoterpenes (MT) emissions that are most influential on health (see Fig. 10.4) vary by forest type, as it can be seen in Fig. 10.13. Averaging across various studies, a significant emission of MT is noted above all in Boreal-Alpine forests (17.5–22.5%), followed by Mediterranean sclerophyll forests (16–21%) and temperate broadleaf forests (15–20%), while urban parks drop to values that are more than halved (7–9%).

One can note that the ratio between urban Parks and Forests:

$$BTC_{PARK}/BTC_{FOREST} = 0,38$$

$$MT_{PARK}/MT_{FOREST} = 0,40$$

So, let us measure the close correlation in a forest between monoterpene (MT) and Biological territorial capacity (BTC) (Fig. 10.13): simply multiplying by two, the angular coefficients of the trend lines result the same (3.475 vs. 3.350) with an error <0.036. Note that MT emissions (%) are proportional, by a *factor of 2.07* to forest bionomic functionality (Mcal/m^2/year).

$$MT = 2.07 \times BTC$$

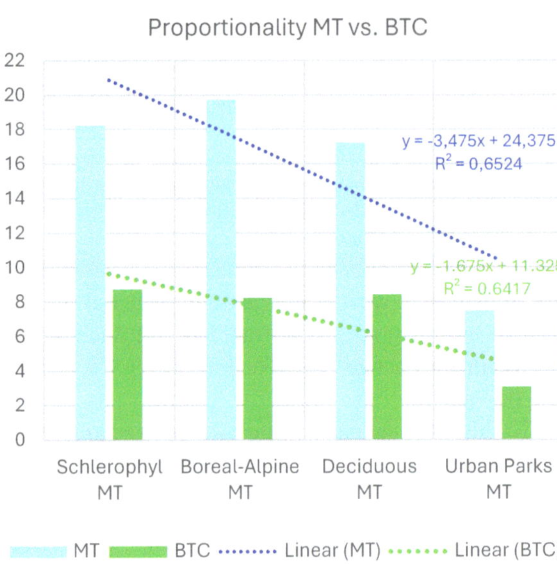

Fig. 10.13 Maximum estimates of monoterpene (MT) emissions (%) in different forest types of temperate and cold-temperate zones in relation to the average BTC values (Mcal/m^2/year) of mature forests. The proportionality between BTC and MT values of urban parks in relation to forests is evident

Abies alba 53.5 %; Picea excelsa 37.8 %; Larix decidua 8.7 % *Canopy Height, Hc = 33.00 m* *Phytomass Vol. = 960 m³/ha* *Biol. Terr. Capacity of Vegetation, BTC = 6.87 Mcal/m²/year* *Concise Bionomic State of vegetation, CBSt = 27.05 %* *Humification (Ell.) R x N = 7.23* *Species Biodiversity = 36* *Oldest Tree: Larix d., Diam.=70.7 cm; High = 39.2 m*	*Picea excelsa 81.0 %, Larix decidua 17.5 %, Abies alba 3.1 %* *Canopy Height, Hc = 32.90 m* *Phytomass Vol. = 732 m³/ha* *Biol. Terr. Capacity of Vegetation, BTC = 8.26 Mcal/m²/year* *Concise Bionomic State of vegetation, CBSt = 50.55 %* *Humification (Ell.) R x N = 16.58* *Species Biodiversity = 41* *Oldest Tree: Picea exc., Diam.= 115.0 cm; High = 51.0 m*

Fig. 10.14 Comparison between two Boreal-Alpine formations in Fiemme Valley at the altitude of 1000–1200 m a.s.l. Note the sharp difference in ecological efficiency CBSt, independently from phytomass and canopy height, but related to the presence of older mother trees (remember Fig. 5.11). The phytoncide effect is near proportional to the biological territorial capacity of vegetation (BTC) and to the concise bionomic state (CBSt) of the forest

This emerged correlation is of fundamental importance in contributing to the calculation of the Health Protective Index HPI of a landscape (see 11.5).

The LaBiSV methodology (Ingegnoli, 2002, 2011, 2015, 2022; Ingegnoli & Giglio, 2005; Ingegnoli & Pignatti, 2007) is needed for a correct evaluation of Forest really performing for a Bath Therapy. See for example Fig. 10.14 concerning the visual vs. the bionomic comparison of two Alpine Forests which seem very similar, while their BTC difference (6.87 vs. 8.26 Mcal/m^2/year) may signify MT = 13.74% vs. MT = 16.52%.

10.7 Scale Interactions of Landscape Pathology vs. Human Health

10.7.1 Hierarchical Systems and Scales

In the previous paragraphs, we have dealt with the environment as a not well-defined entity, even though recalling attention to the environment as a hyper-complex system of complex systems (Par.2.1). Thus, to deepen the relationships between human health and the health of the environment one must go back to the System Hierarchical Theory and its central concept concerning the organisation of a system which results from differences in process rates, which change with the scale.

Boundaries, which are not only the physical ones, separate the set of processes from components in the rest of the system. So, the inferior level components explain the origin of the level of interest (i.e., allow an inner description). In contrast, a superior level system explains the significance of the level of interest (i.e., enables the characters, derived from transferred conditions, to be understood).

This premise is necessary because only studying the relations among different scales one can eliminate the dominance of reductionism.

***Territorial Scale (Landscape Unit)*:** The extension of Life to large scales, as the landscape ones, permits a first understanding of the linkages among hierarchic levels of biological organization, because constraints are imposed from above. Some retrievals:

(a) the characters of the Landscape Unit are indispensable to the evolution of populations;
(b) the zoological reconversions, due to LU structure changes, lead to an exacerbation of viruses and parasites influencing man and animals;
(c) within the correlation between Mortality Rate and Bionomic Functionality, the health damages depend on the dysfunctions of living complex system (i.e. LU), independently from pollution;
(d) the mentioned 'pruning effect', which improves the environmental fitting in the development of the brain, extends the influence of large scale on organisms, especially on man;
(e) the interactions 'landscape apparatuses/vital space per capita' express other meaningful linkages among scales, permitting the assessment of Carrying Capacity and the measure of heterotrophy of a LU;
(f) the *protective* effects of landscape structures, such as tree corridors networks and forest patches with high BTC, have just influenced the detailed scales.

Focusing on the importance of nutrition and food for every living entity, let's examine the scale interactions of landscape pathology with human health through the problem of the health state of agricultural landscapes and obsolete agroecology interventions.

Agroecology only updates farming systems without considering the entire agricultural landscape while, from the interactions among vegetation, fauna, and bacteria at different space-temporal scales, it is possible to avoid a broad input of fertilizers and pesticides. But not only that! The reasons by which agroecology must rehabilitate the entire agricultural landscape are mainly concerned with two disciplines, Epigenetics and Bionomics, and their relationships with the field-crop scale, i.e., ecocoenotope, ecotope (Fig. 10.15).

***Local Scale (Ecocoenotope)*:** Population and the environment. As the complete homogeneity of monocultural industrialized crops is dangerous for our immune system and mental health, therefore, agroecology suggests to plant populations of cereals: they evolve under different types of agronomic management and in the face to specific combinations of *landscape unit characters* (LU) as microclimate, insects, weeds, trees, and shrubs, BTC, HH, etc. In this way, the frequency of genotypes that

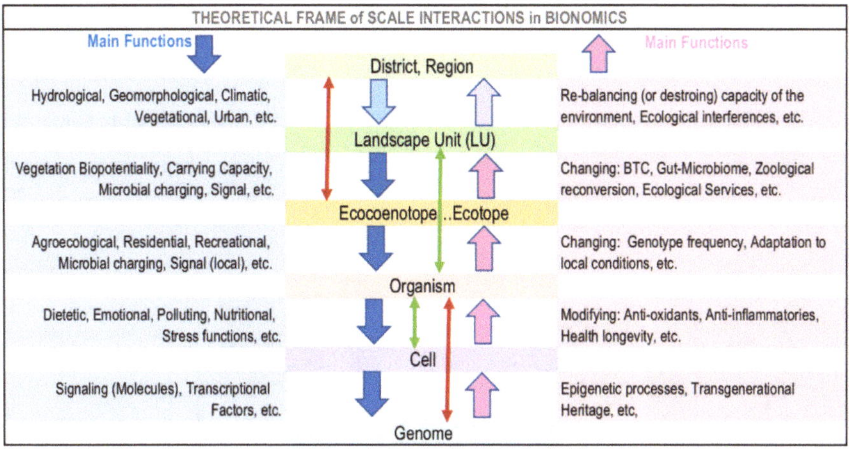

Fig. 10.15 The main links between the scales from the cell to the landscape unit show the focal reasons why the rehabilitation of the entire agricultural landscape is mandatory for agroecology. Red arrows represent bottom-up interrelations; blue arrows represent top-down interrelations and constraints

have adapted to local conditions gradually increases. Epigenetic processes lead to the mentioned "Evolutionary Plant Breeding" (EPB). Note that these populations do not need to be treated with pesticides and herbicides (Ceccarelli & Grando, 2020).

Cellular Scale**:** Epigenetics and the environment (Fig. 10.16). The information written in the chromosome bases is insufficient, a modulation of information given by the environment and our behaviour is also needed. So, the interaction with the environment is crucial. This fact implies a direct linkage between cell scale and environmental scale. Signalling molecules and transcription factors modulate genetic information, producing anti-inflammatory cytokines, antioxidants, anti-mutation, a decrease of cortisol, etc. Very important are nutritional factors and environmental parameters, both exciting to agroecology.

The information derived from our environment is shown in 4 sets: (A) positive bionomic-ecological characters, (B) diet & nutrition, (C) sport & exercise, (D) emotion & spirit. This information is transmitted by signalling molecules and transportation factors, which modulate the genome, promoting health and protection effects on our body and diseases in case of environmental degradation. So, by modifying the environment, men also act towards themselves.

Fig. 10.16 It is not sufficient the information written on the chromosome bases; humans also need the modulation of information given by the environment and by the behavior. Note that the link between micro- and macro-scales presents today robust scientific bases

10.8 Information Disturbances and Epigenetic Code

10.8.1 Environmental Information and Cell Repairing Processes

The necessity to interrelate all the levels of biological organization (see Par.2.4.1) confirms the bionomic principles. However, these interrelations were possible only after acquiring the importance of epigenetics. The modulation of information linked to the environment (Signaling Molecules, Transcription Factors, DNA, RNA, Protein) is crucial (Figs. 10.16 and 10.17). Based on bionomics principles, the new concept of life is not limited to a single organism or a species group. Therefore, the linkage due to epigenetic processes is very significant. Even active bionomics functions are linked with environmental signals and transcription factors. Note that their context addresses the information (Fig. 10.17). Food, health, and environment

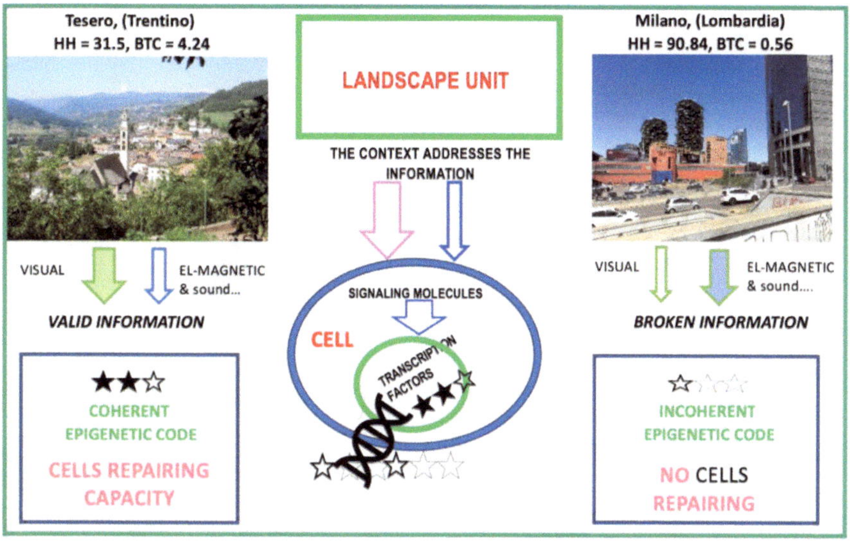

Fig. 10.17 Landscape unit alterations affect information, and a broken information can degrade the epigenetic codes affecting the repair capacity of cells

are strictly connected with gene expression and the bionomic state of landscape units, especially where man lives.

Transcription factors represent the epigenetic code. Regulatory factors are transferred to cells as different information packages contained in vesicles, called *exosomes*, which have specific cells as targets, based on the information contents, i.e., the specific regulatory factors that they transport. Remember what was affirmed by Lorentz (see Sect. 10.2.1) and by Laszlo & Biava (2019):

> Diseases must be considered as an imbalance of information. In other words, we must define diseases as pathologies of information that can be classified according to different types of disruption of information. Diseases are pathologies of information that can affect the symbolic and semiotic codes of meaning. The therapist must consider both the semiotic and symbolic levels. When decoded, the languages of the two codes provide the correct information to the psychosoma.

10.8.2 *Cancer and Cell Reprogramming*

In his book "Cancer and the Search for Lost Meaning", Pier Mario Biava (2009) wrote that cell reprogramming also contains fascinating conclusions about what the lessons of this work can offer, regarding how we view human life and existence. Cancer is a by-product of the sense of loss and lack of meaning in modern society, and healing cancer involves healing that sense of meaninglessness. This process can lead to a deeper understanding of -and connection to- life.

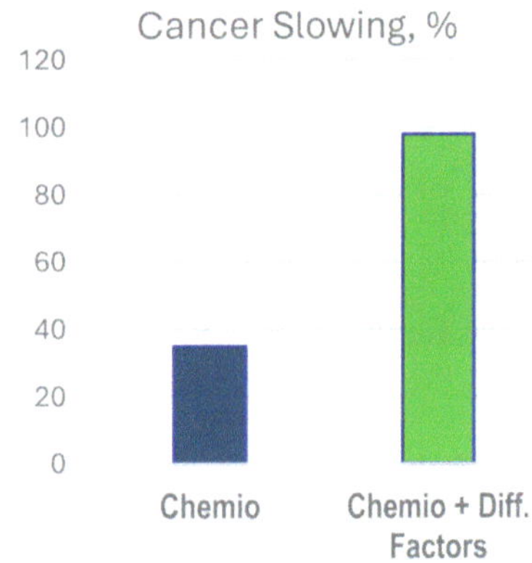

Fig. 10.18 Cancer slowing obtained adding differentiation factors to chemotherapy increased the slowing results of about three times. (From Biava & Laszlo 2019)

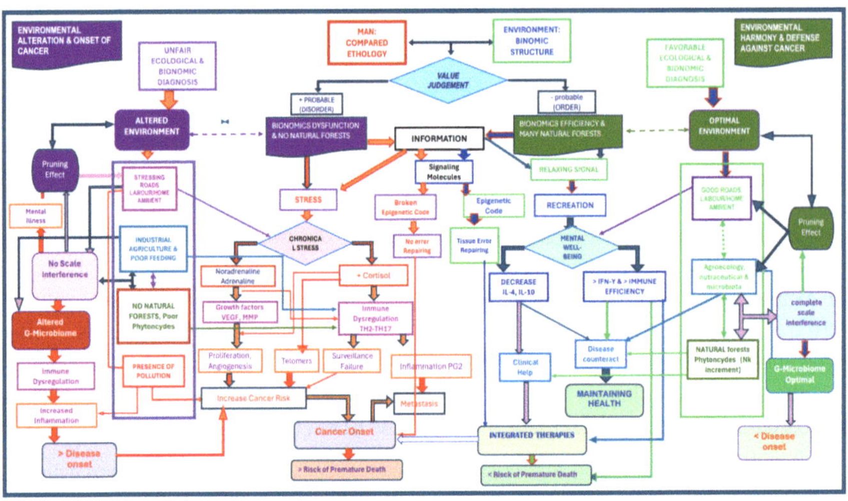

Fig. 10.19 An inauspicious diagnosis of the syndromes of a landscape unit (LU) shows various possible pathways of morbidity: e.g., (1) Chronic stress (yellow), (2) presence of pollution (red), (3) lack of defensive contributions (green), (4) lack of scale interference (blue grey). Complex interrelationships and convergences exacerbate health changes

The synergy between chemotherapeutic agents and differentiation factors: in 2011, research conducted at the La Sapienza University of Rome was published in Current Pharmaceutical Biotechnology. The slowing of the growth of colon tumor cells was observed in vitro: a 35% slowing was achieved with treatment with 5 Fluoro-Uracil alone, and a 98% slowing was instead obtained with the simultaneous administration of 5 Fluoro-Uracil and differentiation factors (Fig. 10.18).

10.9 Complex Interrelations Among Health vs. Environment Processes

10.9.1 Negative and Positive Influence of the Environment on Health

Beginning with the environmental alteration and the emergence of cancer (left part of the flow chart, Fig. 10.19), we find a violet frame containing four agents of modification:

(a) Stressing roads, labor, and home environment,
(b) Industrial agriculture and poor feeding,
(c) No natural forests and poor phytoncides,
(d) Presence of pollution.

These agents present many linkages with other surrounding agents and processes derived from environmental stress (to the right). For instance, left: (e) the Pruning effect, strictly linked with mental fitness; (f) no scale interference, related to the destruction of agrarian landscapes and inadequate food; and (g) altered gut microbiome, leading to immune deregulation and increased inflammation, so to major disease onset and cancer risk.

Complex linkages may characterize some processes of carcinogenesis due to environmental alterations. Let us consider the agrarian landscape syndromes: the dominance of industrialization in agriculture destroys the landscapes and produces mediocre food, and the lack of phytoncides reinforces these alterations. This process influences the chronic stress sequences, but also it blocks or alters the scale interrelations, which affect mental illness and the pruning effect, cutting the sensibility toward nature and enhancing the artificialization of the environment. Simultaneously, the alteration of the gut microbiome leads to the depression of the immune regulation, which can worsen the analogous immune deregulation due to chronic stress (cortisol increase, Th2 dominance, etc.).

Moreover, the presence of pollution based on pesticides, which is not among the most important agents of carcinogenesis, could reinforce the other mentioned factors. An extraordinary road and its disturbances and pollution may have a similar effect on the agrarian landscape. Finally, the risk of premature death is increasing.

On the other hand, the possibility to evaluate the bionomic state of the landscape units and consequently correlate its bionomic functionality (BF) with the mortality rate gives the possibility to control environmental syndromes, reduce the impacts of transformation, and advise the local Authorities on the necessity of ecological rehabilitation. This bionomics consulting acquires a specific function of health protection and prevention. One can also add clinical functions brought by harmonic natural LU related to integrated therapies, as proposed by Psycho-Neuro-Endocrine-Immunology.

10.9.2 Complexity of the Processes Related to the Environment

That is why a complete frame of the primary relationship between an altered environment/reduced health must be presented with the addition of a healthy natural environment (Fig.11.19, right). The right part of this broad logic flow chart shows the interrelations semi-specularly with the alarming signals. Thus, the relaxing signals, leading to psychic well-being, may develop favourable biochemical processes. For instance:

1. decreasing inflammation enhancing IL-10, an immunomodulatory cytokine with a critical role in limiting inflammation in immune-mediated pathologies, or
2. increasing IFN-γ, which activates macrophages and makes them better able to mount an effective immune response,
3. enhancing antigen processing and presentation through upregulation of class II MHC, increased ROS and NOS production,
4. inducing autophagy for clearance of intracellular pathogens, etc.

Going on with the environmental rehabilitation and the prevention of cancer (right part of the flow chart), we find again a frame containing four agents of a healthy state:

(a) Good roads, labor, and home environment,
(b) Agroecology and well-feeding,
(c) Natural mature forests and good phytoncides,
(d) Absence of pollution.

The healthy agents are three because of the absence of pollution. Agroecology is particularly essential: plant populations evolve under different agronomic management types and in the face of specific combinations of LU characters (from diseases to insects, weeds, drought, etc.). In this way, the frequency of genotypes that have adapted to local conditions gradually increases. Epigenetic processes lead this EPB; local environment characters guide DNA expressions and make them transmitted. These populations do not need to be treated with pesticides and herbicides. They are deeply linked with natural forest patches, of which we must remember Fig. 10.12: Phytoncides (forest essential oils) have beneficial effects on natural killers (NK).

In this positive case, the role of the neural Pruning Effect stabilizes, leading to developing man's sensibility toward nature in the correct direction.

References

Andersen, et al. (2005). Meditations and cancer. In F. Bottaccioli & A. G. Bottaccioli (Eds.), *Psycho-neuro- endocrine-immunology and science of integrated care*. Edra Ed.

Arnold, V. I. (1963). Proof of a theorem by Kolmogorov AN on the invariance of quasi-periodic motions under small perturbations of the Hamiltonian. *Russian Mathematical Surveys, 18*, 13–40.

Barbieri, M. (1985). *The semantic theory of evolution* (p. 188). Harwood Academic.
Besedovsky, H. O., & Sorkin, E. (1977). Network of immune-neuroendocrine interactions. *Clinical and Experimental Immunology, 27*, 1–12.
Besedowsky, H. O., & Del Rey, A. (2007). Physiology of psychoneuroimmunology: A personal view. *Brain, Behavior, and Immunity, 21*(1), 34–44.
Biava, P. M. (2009). *Cancer and the search for lost meaning: The discovery of a revolutionary new cancer treatment* (p. 168). North Atlantic Book.
Biava, P. M., & Laszlo, E. (2019) *Information Medicine*. Healing Arts Press, USA
Blalock, J. E. (2005). The immune system as the sixth sense. *Journal of Internal Medicine, 257*(2), 126–138.
Bottaccioli, F., & Bottaccioli, A. G. (2020). *Psycho-neuro- endocrine-immunology and science of integrated care*. Edra Ed.
Ceccarelli, S., & Grando, S. (2020). *Participatory plant breeding: Who did it, who does it and where?* Online Cambridge University Press.
Faggin, F. (2024). *Oltre l'invisibile. Dove scienza e spiritualità si uniscono*. Mondadori.
Gandhi, M. K. (1962). *Village swaraj*. Jitendra T. Desai.
Godron, M. (1984). *Ecologie de la végétation terrestre*. Masson.
Gogtay, N., Giedd, J. N., Lusk, L., et al. (2009). Dynamic mapping of human cortical development during childhood through early adulthood. *Proceedings of the National Academy of Sciences of the United States of America, 101*(21), 8174–8179.
Huttenlocher, P. R. (1979). Synaptic density in human frontal cortex: Developmental changes and effects of aging. *Brain Research, 163*, 195–205.
Ingegnoli, V. (2002). *Landscape ecology: A widening foundation* (p. XXIII+357). Springer.
Ingegnoli, V. (2011). *Bionomia del paesaggio. L'ecologia del paesaggio biologico-integrata per la formazione di un "medico" dei sistemi ecologici* (p. XX+340). Springer.
Ingegnoli, V. (2015). *Landscape bionomics. Biological-integrated landscape ecology* (p. XXIV + 431). Springer.
Ingegnoli, V. (2022). *Landscape bionomics dysfunctions and human health*. In Ingegnoli, Lombardo, & LaTorre (Eds.), *Environmental alteration leads to human disease. Chapter 5* (pp. 95–130). Springer.
Ingegnoli, V., & Giglio, E. (2005). *Ecologia del Paesaggio: Manuale per conservare, gestire e pianificare l'ambiente* (p. 685+XVI). Sistemi Editoriali.
Ingegnoli, V., & Giglio, E. (2021). Environmental bionomic alterations and cancer incidence in European countries: A first overview. *Journal of Nursing & Occupational Health, 2*(3): 266–281.
Ingegnoli, V., & Pignatti, S. (2007). The impact of the widened landscape ecology on vegetation science: Towards the new paradigm. Springer. *Rendiconti Lincei Scienze Fisiche e Naturali, s. IX, XVIII*, 89–122.
Laszlo, E., & Biava, P. M. (2019). *Information medicine: The revolutionary cell-reprogramming discovery that reverses cancer and degenerative diseases*. (English edition). Simon and Schuster.
Li, Q. (2009). Effect of forest bathing trips on human immune function. *Environmental Health and Preventive Medicine, 15*(1), 9–17.
Li, Q., Morimoto, K., Nakadai, A., Inagaki, H., Katsumata, M., Shimizu, T., Hirata, Y., Hirata, K., Suzuki, H., Miyazaki, Y., Kagawa, T., Koyama, Y., Ohira, T., Takayama, N., Krensky, A. M., & Kawada, T. (2008). Forest bathing enhances human natural killer activity and expression of anti-cancer proteins. *Journal of Biological Regulators and Homeostatic Agents, 22*(1), 45–55.
Lorenz, K. (1973). *Die Ruckseite der Spiegels* (p. 419). Piper & co.
Lorenz, K. (1978). *Vergleichende Verhaltensforschung: Grundlagen der Ethologie*. Springer.
Meneguzzo, F., & Zabini, F. (2022). *Terapia Forestale*. CNR-CAI.
Nesburn & Mengelkock. (1991). *Quoted in Stout 2013:21*.
Odum, E. P. (1971). *Fundamentals of ecology*. WB Saunders.

Petanjek, Z., et al. (2011). Extraordinary neoteny of synaptic spines in the human prefrontal cortex. *Proceedings of the National Academy of Sciences, 108*(32), 13281–13286.
Popper, K. R. (1982). *The open universe: An argument for indeterminism.* WW Bartley III.
Popper, K. R. (1983). *Realism and the aim of science.* WW Bartley III.
Ramon Cajal, S. (1909). *Histologie du Système Nerveux de l'Homme et des Vertébrés.* A. Maloine Editeur.
Rotschild, D., et al. (2018). Environment dominates over host genetics in shaping human gut microbiota. *Nature, 555*(7695), 210–215. https://doi.org/10.1038/nature25973. Epub. 2018 Feb 28.
Simard, S. (2018). Mycorrhizal networks facilitate tree communication, learning and memory. In Baluska, Gagliano, & Witzany (Eds.), *Memory and learning in plants.* Springer.
Simard, S. W. (2021). *Finding the mother tree, discovering the wisdom of the forest.* Borzoi Book, Publ. by A Knopf.
Simard, S. W., & Durall, D. M. (2004). Mycorrhizal networks: A review of their extent, function and importance. *Canadian Journal of Botany, 82*, 1140–1165.
Slavich, G. M., Thornton, T., Torres, L. D., Monroe, S. M., & Gotlib, I. H. (2009). Targeted rejection predicts hastened onset of major depression. *Journal of Social and Clinical Psychology., 28*(2), 223–243.
Wilson, R. (2019). *Trees and the human spirit.* Cambridge Scholar Publishing.
Wohlleben, P. (2016). *The hidden life of trees: What they feel.* Greystone Books.

Chapter 11
Widening Anamnesis and Prevention

Contents

Abstract Note that the traditional point of view related to the interrelations Ecology/Medicine does not consider the concept of "One Health" but an "assessment of the response to environmental pressure", which had nothing to do with the "care of the patient" of medicine. The physician's responsibility is to repair the damage. Still, in doing this, a doctor should contact internal specialists and external ecoiatra, at least to avoid reintroducing the patient into the same environment that contributed to the insurgence of the disease.

The assessment of the impact of the man/environment relationship concerning health is based (a) on the cumulative effect of the stress due to an improper man-environment relationship (cWR), physio-environmental parameters (WR-Phys.) increased by the impacts of the stress due to the "Psycho-social" environment (WR-PSoc.); (b) cWR receives a further cumulative effect from the bionomic conditions of the landscape unit where the patient lives, synthetically measured by the bionomic parameter of the Bionomic Functionality (BF), as stated in the introduction, leading to their Environmental Cumulative Stress (ECS). An

V. Ingegnoli, *Landscape Bionomics and Planetary Health*, Landscape Series 21,
https://doi.org/10.1007/978-3-032-08908-3_11

anamnesis able to consider ECS can confirm that the deficiencies of natural vegetation infrastructures influence morbidity, leading to anomalies from environmental stress and altered environment-human body information.

Listing the main sets of prevention expressed today in Western civilization, we can see (A) stress reduction, (B) periodic medical checkups, (C) adequate physical-sporting activity, (D) control of pollution and smog, and (E) correct nutrition and diet. However, following bionomic principles and methods, at least three new sets must be added: (F) landscape syndrome evaluation, (G) environmental conditioning of people, and (H) adequate natural activity.

This could be useful for preventing an Indicative Morbidity per each municipality considered as a landscape unit (LU). The FIM (Function of Indicative Morbidity) is proportional to the DI (Diagnostic Index) and the average disease ratio. An evident proportion of DI/Morbidity must be investigated.

Keywords One health · Environmental anamnesis · ECS (environmental cumulative stress) · FIM (function of Indicative morbidity) · Lung cancer · Adenocarcinoma · Disease ratio

11.1 Widening Anamnesis and Therapy

11.1.1 Ecology and Medicine

Doctors appreciate the problems that come with environmental degradation but generally see them as someone else's problem to solve while they focus on repairing the damage.

Similarly, better environmental protection for human health is recognized as a benefit, but again, medics do not 'do' preventative health in this way—public health policymakers do. It is not entirely clear what the medical profession or students are meant to 'do' with this Planetary Health knowledge or how they can use it to help patients.

In Fig. 11.1, we underlined the differences between the mentioned position of doctors related to ecology (a) when the systemic paradigm was not recognized, so the reductionism maintains Ecology and Medicine strictly separated (Lenzi, 2022), and the new one (b) when they admit the new definition of life as a unitary biological spectrum (see Sect. 2.4.1).

Note that the first point of view did not consider the concept of "One Health" (Gregory et al., 2024; Istituto Superiore di Sanità, 2019) but an "assessment of the response to pressure" of ecology had nothing to do with the "care of the patient" of medicine. New links between the two disciplines emerge following the new systemic paradigm, both in the diagnostic and therapeutic fields and especially in etiology (Bonizzi et al., 2022). One Health can be defined as the study of integrated ecology/medicine expanding the body's internal life systems to external ones.

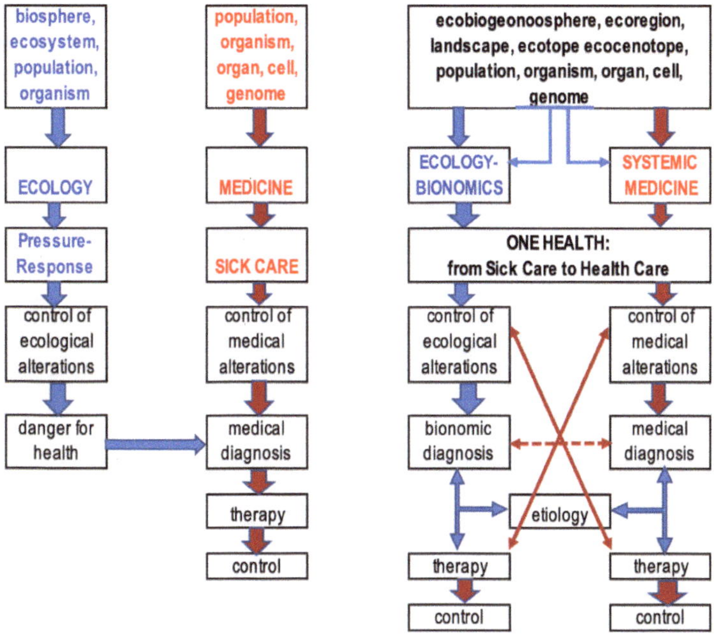

Fig. 11.1 Comparison between the ecology/medicine relations (left) as separated discipline, where the ecological pressure/response leads to health damage treated following reductionist medicine, and (right) if we consider the integration ecology/medicine, where the concept of One Health leads to systemic medicine

11.1.2 Body Internal and External Life Systems

We will see that Landscape Bionomics (Ingegnoli, 2011, 2015), while sustaining the listed physiological processes, opens new perspectives to etiopathology, health prevention and therapy integrations, and anamnesis. So, new linkages between the two disciplines, Landscape Bionomics and Medicine, emerge following the new systemic paradigm in diagnostic and therapeutic fields in etiology and anamnesis. We can indicate an answer to what is not entirely clear to the medical profession/students (see Introduction): what they are meant to 'do' with the ecological problems and how they can use them to help patients.

Figure 11.2 shows that the primary landscape syndromes can be grouped into six categories. These processes, frequently cumulative (at least partly), lead to health damage with an interchange between the body's external and internal life systems. The physician's responsibility is to repair the damage (Bottaccioli & Bottaccioli, 2020). Still, in doing this, a doctor should contact internal specialists and external ecoiatra, at least to avoid reintroducing the patient into the same environment that contributed to the insurgence of the disease. Moreover, we need to add a wider anamnesis and then an integrated therapy.

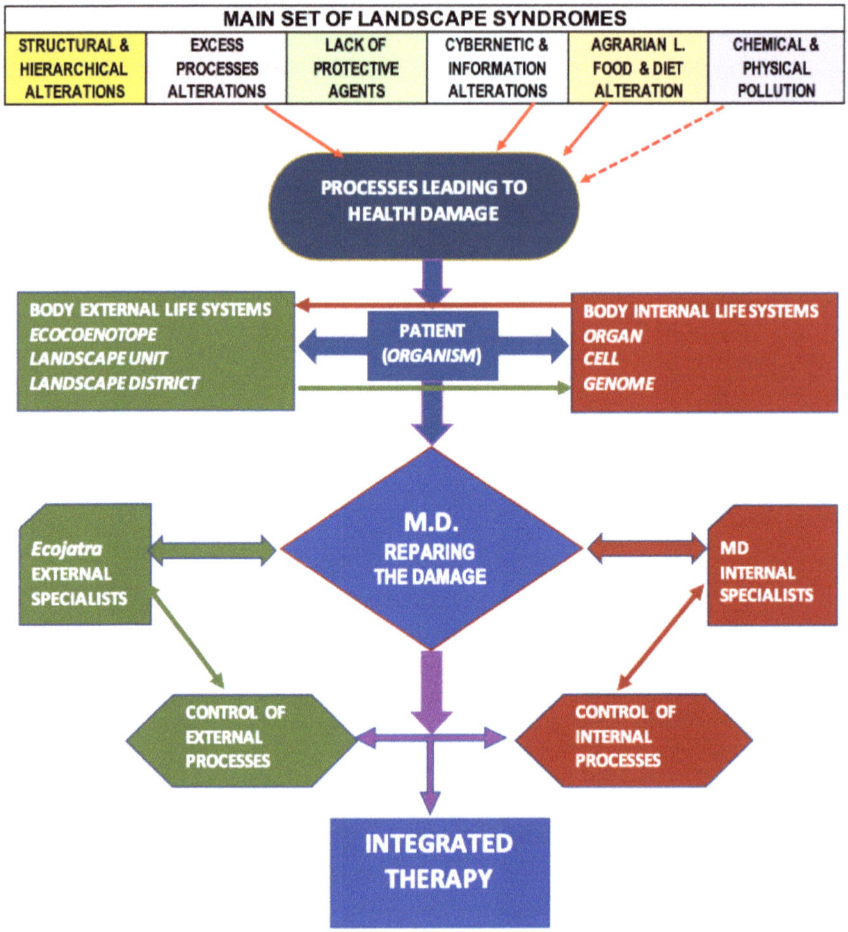

Fig. 11.2 This flowchart tries to explain that the medicine doctors (MD) must repair the damage to human health but, since most of this damage is due to environmental alterations and since the organism is linked to the internal and external life systems of the body, the MD must collaborate with internal and external specialists, arriving at an integrated therapy. (From Ingegnoli & Giglio, 2021)

Remember that a synthesis of the complex interrelations of processes linking the environment to cancer disease is shown in Fig. 10.19 (left): an inauspicious diagnosis of the syndromes of a landscape unit (LU) shows various possible pathways of morbidity, e.g., (1) chronic stress (yellow), (2) presence of pollution (red), (3) lack of defensive contributions (green), (4) lack of scale interference (blue grey). Complex interrelationships and convergences exacerbate health changes. The emergence of ethological alarm signals, capable of activating the adrenaline and hormonal processes, can be reinforced by feedback due to the neuro-cerebral "pruning" effect (Gogtay et al., 2009; Petanjek et al., 2011) (see Fig. 10.5), leading to increased

anthropized technology environments (see Figs. 10.6 and 10.8). "Scale interference" indicates the lack of- Vs the presence of- the correct hierarchical interactions.

11.2 Environmental Anamnesis

11.2.1 Patient/Environment and Landscape/Environment

As summarized in Fig. 11.3, the assessment of the impact of the man/environment relationship concerning health is based -firstly- on the cumulative impact of the stress due to an improper man-environment relationship (cWR), detected through physio-environmental parameters (WR-Phys.) increased by the impacts of the stress due to the "Psycho-social" environment (WR-PSoc.). Secondly, cWR receives a further cumulative impact from the bionomic conditions of the landscape unit where the patient lives, synthetically measured by the bionomic parameter of the Bionomic Functionality (see Sects. 7.4.2 and 7.4.3), as stated in the introduction, leading to their Environmental Cumulative Stress (ECS).

Pilot research coordinated by Ingegnoli was done within the Pulmonology Department of the Vimercate Hospital (province of Monza-Brianza, Lombardy, Northern Italy), directed by Dr. Paolo Scarpazza, through two summary questionnaire tables to record the patient's responses, submitted to 70 cancer patients, who decided to collaborate freely. These tables were compiled by patients, with the help

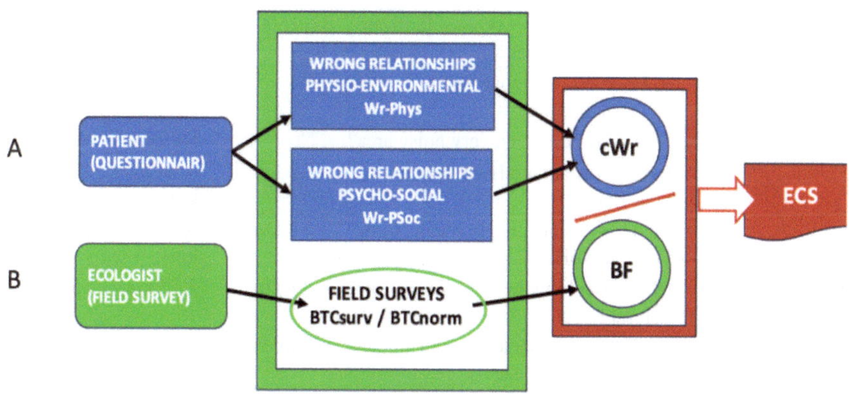

$$\text{Cumulative wrong relationships } (\mathbf{cWR}) = \text{Wr–Phys} \cdot \left(1 + \frac{\text{Wr–PSoc}}{100}\right)$$

$$\text{Cumulative environmental stress } (\mathbf{ECS}) = \mathbf{cWR} \cdot \frac{1}{\boldsymbol{BF}}$$

Fig. 11.3 Diagram of the relationships between man and the environment necessary to propose specific indicators that can be developed into two integration scales: patient scale and ecologist scale

of an assistant, Dr. Silvia Pagani. A final revision was done by prof. Luigi Bonizzi (Milan University) and dr. Elena Giglio, assistant of Ingegnoli.

The first table investigates the history of their improper relationships with the physical and natural environment (WR-Phys). In contrast, the second concerns stress evaluation due to the psychological and social environment (WR-PSoc). The answers to all the questions also require a series of specifications and examples that can guide the detection of patient responses, given that only some of them have the same level of cultural education. The present research was carried forward thanks to the assistants who helped the patients answer the questionnaires correctly. A seminar was held precisely before the start of the surveys.

The two mentioned tables (Tables 11.1 and 11.2) are reported, to better understand the method. Note that some questions will be better express.

Table 11.1 Anamnesis of wrong relationships with physical environment

Patient: Name…Surname… Sex…, Age …, Smoker…, Municipality…		I (0.1)	II (0.5)	III (1.0)
Structural and hierarchical alterations	1. Where did your mother spend the time, she was pregnant with you?	Country	City	SubUrban
	2. Which is the distance between your home and recreation area/park?	<600 m	600–1200	>1200
	3. Do rivers or canals near you present riparian vegetation?	Yes	partial	no
	4. Does your neighborhood/village have a lot of green areas?	Yes	partially	no
	5. Do you live in a tree-lined avenue?	Yes	partially	no
Excess processes alterations	6. Is your home/workplace far from the industrial areas?	>1500 m	500–1500	<500
	7. Do you often move within congested traffic?	No	seldom	yes
	8. Is your home in a very noisy area?	No	partially	yes
	9. How distant are Highways from your home?	>2 km	0.5–2 km	<0.5 km
	10. How distant are Highways from your workplace?	>2 km	0.5–2 km	<0.5 km
Lack of protective agents	11. Is your home/workplace far from a natural forest?	<1 km	1–5 km	>5 km
	12. Does your home have a garden or at least a green balcony?	Garden	balcony	nothing
	13. How many times in a year do you take walks for a few hours within natural forests?	>3	1–3	0
	14. Can you compensate for the environmental alteration by often going to naturally healthy places?	Yes	seldom	no
	15. Do you often take antibiotics even without a prescription?	No	seldom	yes

(continued)

Table 11.1 (continued)

Patient: Name…Surname… Sex…, Age …, Smoker…, Municipality…		I (0.1)	II (0.5)	III (1.0)
Cybernetic and information alterations	16. Could you understand if the environment is degraded?	Yes	partially	no
	17. Does it bother you to stay for hours in a degraded environment?	A lot	little	no
	18. Did you often play in natural environments as a child?	Frequently	seldom	never
	19. Do you play sports or walk in public parks?	Frequently	seldom	never
	20. Do you like living in new artificial/technologic environments?	No	seldom	yes
Agrarian landscape, food and diet alterations	21. How often do you eat red meat or sausages?	1–2	3–4	5–6
	22. Is the agrarian landscape in your municipality of industrial type (monocultural) or heterogeneous (tree corridors and wooded patches)?	Tree corr. & woods	Only few tree corridors	Agro-industrial
	23. Are the cattle and sheeps whose product you consume raised in an environmentally sound way?	>30%	15–30%	<15%
	24. Do you eat a lot of fruit and vegetables?	Yes	seldom	no
	25. How many times a week do you feed on organically grown food?	5–7	3–4	0–2
Chemical and physical pollution	26. Is your residence area crossed by high voltage power lines?	>150m	50–150m	<50m
	27. Do you live or work in an area with high presence of fine dust in the air?	No	marginal	yes
	28. Are there radon gas emissions within your residence area?	No	little	high
	29. Is the water within the municipal plant hard and with a high ionic charge?	No	little	yes
	30. Do you spend more than an hour a day in traffic within a car?	No	1–2 h	>2h
	Score:	X = (A × 0.1)	Y = (B × 0.5)	Z = (C × 1.0)
	WR-Phys = (0.1 × A+ 0.5 × B+ 1.0 × C)/30 [%]	**WR-Phys = (X+Y+Z) /30 [%]**		

Table 11.2 Anamnesis of psychosocial stress levels

Evaluation of psychosocial stress levels (WR-PSoc.)					
Patient: Name…Surname… Sex …, Age …, Smoker: …, Municipality: …					
Stress	Stress evaluation (last 10 years)	Very low	Low	Medium	High
Family stress	1. disagreement with wife or children or divorce				
	2. chronic illness of a relative or of the same patient				
	3. death of a close relative				
	4. housing problems				
Work stress	5. lack of recognition of merit				
	6. insufficient economic recognition				
	7. unjust career impediment				
	8. workplace & office problems				
Social stress	9. lack of medical center requirements				
	10. lack of public transport				
	11. conflicts or persecutions due to religion				
	12. socio-cultural marginalization				
	Score:	A	B	C	D
	WR-PSoc= (A × 1 + B × 3 + C × 5 + D × 7) /84 = %	**WR-PSoc = (A+B+C+D)/84 [%]**			
Very low: incorporable stress, which does not affect behavior. LOW: anxiety-inducing stress but not lasting, some influence on behavior.					
Medium: persistent anxiety-inducing stress, with certain influence on behavior. HIGH: anxiety-inducing and depressive stress, with noticeable behavioral changes.					

11.2.2 The Questionnaire Tables on Physical and Social Environment

The 30 questions in Table 11.1, concerning the improper relationships with the physical and natural environment (WR-Phys), are divided into six groups, and evaluated through three columns with the environmental stress weights in proportion: (i) acceptable (0.1), (ii) altered (0.5) and (iii) severely impaired (1.0).

$$WR\text{-}Phys = (0.1 \times k + 0.5 \times n + 1.0 \times m)/30[\%] \tag{11.1}$$

The main six groups of landscape syndromes are taken from Fig. 11.2 and may influence human health as man processes, for ethological reasons—firstly the value judgement -, even unconsciously, the lack or the excessive presence around him of

1. structural environmental components, remembering the Emergency Properties Principle
2. significant disturbing factors

Table 11.3 Cumulative impact evaluation of first and second steps (WR and ECS).

Cumulative analysis of the wrong Relationships man / environment			Cumulative environmental stress (ecs) and indicative morbidity		
cWR=WR.Phys x (1+WR.PSoc/100)	Values	Result	ECS = cWR x 1/BF	Stress level	Result
Normal	10–20%		Normal	12–25	
Low	20–40%		Low	25–40	
Medium	40–60%		Medium	40–60	
High	60–80%		High	60–85	
Very high	>80		Very heigh	>85	

3. protective agents, helpful to improve biological and mental human resilience
4. cybernetic linkages and information needed for the improvement of psycho-neural development
5. altered agricultural components and food
6. incorrect relationships with pollutants

The 12 questions in Table 11.3 concerning the WR-PSoc are divided into three groups and evaluated in four columns with environmental stress in proportion: very low (1.0), low (3.0), medium (5.0) and high (7.0).

$$WR\text{-}PSoc = (i \times 1 + k \times 3 + n \times 5 + m \times 7)/84 = [\%] \tag{11.2}$$

The groups concern stress:

(a) due to family and health problems,
(b) relating to work and economic problems,
(c) of a social nature or generated by a lack of infrastructure.

11.2.3 The Environmental Cumulative Stress (ECS)

Table 11.3 also shows the cumulative impact assessment (cWR) according to the following equation:

$$cWR = WR\text{-}Phys \times (1 + WR\text{-}PSoc./100) \tag{11.3}$$

We must now observe that the cumulative impact index cWR is involved in a second level of cumulation, which depends on the bionomic state of the context, i.e. the landscape unit (LU) where the patient lives. A cWR of medium value in a degraded environment context can lead to significantly higher environmental stress, or, in the case of a LU with optimal BF (e.g., 0.95), we can control the cWR value. To arrive at the ECS (Environmental Cumulative Stress) measurement, it is therefore necessary to write a new equation for the second cumulative level:

$$ECS = cWR \times 1/BF \tag{11.4}$$

For example, the patient considered as an example lives in Milan in Zone VI: after the necessary LB processing, this zone appears to have a BF = 0.55, a decidedly low value, given that the normal range for BF ranges from 0.9 to 1.1. As it can be seen, a cWR of average value in a degraded territorial environment has a very high ECS, pair to 98.69 (Table 11.3).

11.2.4 Identification of Alteration Levels

A vital issue emerges from Table 11.3, namely the evaluation ranges of the indices used. The study of the pathology of biological systems becomes essential, a discipline that is not easy to apply, as the meaning of pathology is inextricable from that of physiology. As Konrad Lorenz (1978) underlined, to understand a function in physiology, it is necessary to consider its alteration and vice versa in a continuous postponement of answers.

System Theory says that the scale capable of maximizing the importance and the quantity of the relationships between the components of a system is the scale that allows the different relational forms and configurations to be optimally discriminated. Precisely for this reason, Ingegnoli (2002, 2011) and Ingegnoli and Giglio (2005) have exalted the crucial importance of the landscape, as this territorial scale allows the optimization of the relationships between the anthropic and natural elements. Furthermore, as in Medicine, environmental assessment requires comparing the normal parameters of both ecotopes and landscape units and their indicators.

Consequently, the main problem of diagnostics is knowing the state of normality and, at the same time, the levels of alteration of the system under examination: this applies to the human body and a landscape unit.

Let us establish their evaluation levels in this research by remembering how the indicators used in this study were formed.

WR-Phys A patient whose answers -in Table 11.1- were all in the optimality column (I = 0.1) would reach a value of WR-Phys equal to 3/30 = 10% (minimum threshold): remember that we are testing improper relationships with the physical and natural environment. Since a tolerance margin must always be considered, one can reasonably think of three answers in the second column and two in the third, so Eq. 11.1 becomes:

$$WR\text{-}Phys = (0.1 \times 25 + 0.5 \times 3 + 1.0 \times 2)/30 = 20\%$$

Since the worst threshold is equal to WR-Phys = 100 and considering five intervals, here is the subdivision shown in Table 11.4: Normal (10–20), low (20–40), medium

Table 11.4 Parameter's evaluation and thresholds of normality

Values	WR-Phys	WR-PSoc	cWR	BF[a]	**ECS**
Normal	10–20	14–21	12–25	0.9–1.1	12–25
Low	20–35	20–30	25–40	0.7–0.9	25–40
Medium	35–50	30–40	40–55	0.5–0.7	40–60
High	50–65	40–50	55–70	0.3–0.5	60–85
Very heigh	>65	>50	>70	0.1–0.3	>90

[a]The meaning of BF is the opposite: normal, slightly altered, altered, reduced, seriously insufficient. For both cWR and ECS, it is better to start from 25.00 with a small rounding

(40–60), high (60–80), and very high wrong relationships with the environment (>80). Experience shows that this division corresponds to the case studies, even if the highest interval is usually empty.

WR-PSoc Also in this case, if all the answers of a patient were in the first column, we would have WR-PSoc = 12/84 = 14.3% (minimum threshold). With a tolerance equal to at least one presence in the second and third column, we would have

$$WR\text{-}PSoc = (10 \times 1 + 1 \times 3 + 1 \times 5 + 0 \times 7)/84 = 21.4\%$$

Since the worst threshold is equal to WR-Phys = 100 and considering five intervals, the division into levels is: normal (14–22), low (22–40), medium (40–55), high (55–70), very high wrong psych-social relationships (>70). Also, in this case *experientia docet*.

cWR The normality threshold of cumulative wrong relationships can only depend on the equation (Eq. 11.3:)

$$normality : cWR_{MIN} = 10 \times 1.14 = 11,4 \qquad e \qquad cWR_{MAX} = 20 \times 1.2 = 24.00$$

BF As already mentioned, (Sects. 7.4.2 and 7.4.3), the optimum lies in the range BF = 0.90–1.10 (scientific normality) and the others follow: slightly altered (0.90–0.70), altered (0.70–0.50), reduced (0.50–0.30), seriously insufficient (0.30–0).

ECS Normality of Environmental Cumulative Stress depends on the Eq. 11.4:

$$normality : ECS_{MIN} = 11.4 \times 1.11 = 12.65 \qquad e \qquad ECS_{MAX} = 24 \times 0.91 = 21.8$$

Summarizing the normality levels and the value ranges of the indicators considered (Table 11.4), we will notice that they are all increasingly proportional, except BF, which behaves inversely, as indicated in Eq. 11.4 to find the Environmental Cumulative Stress ECS (second-degree of cumulation).

11.3 A First Application of the Environmental Anamnesis

11.3.1 High Environmental Cumulative Stress in Examined Cancer Patients

The values of the mentioned indicators (WR-Phys, WR-PSoc, cWR, BF, ECS) for each of the 70 patients were calculated according to the method described. Table 11.5 shows the maximum and minimum values, averages, and standard deviation.

As often happens in biology, the second degree of accumulation (ECS) reaches the "high" class despite the previous indicators being in the medium or low classes (compare with Table 11.4). Figure 11.4 shows the distribution of patients examined in the five ECS classes. As it can be seen, this distribution does not follow a classical Gaussian curve but is completely shifted towards a high ECS, leaving the first two classes empty (or almost empty).

Given that the first evaluation class correspond to normality (12–25), no patient shows a good relationship with the environment (Fig. 11.5): all cancer patients

Table 11.5 Frame of the results of the elaboration of 70 questionnaires

Parameters	**WR-Phys.**	**WR-PSoc.**	**cWR**	**BF**	**ECS**
Maximum Val.	61.0	66.7	82.8	1.11	153.3
Minimum Val.	22.3	14.3	26.6	0.51	30.6
Average values	**40.3**	**28.6**	**52.0**	**0.79**	**68.6**
Standard deviation	*8.14*	*12.37*	*12.35*	*0.15*	*25.36*

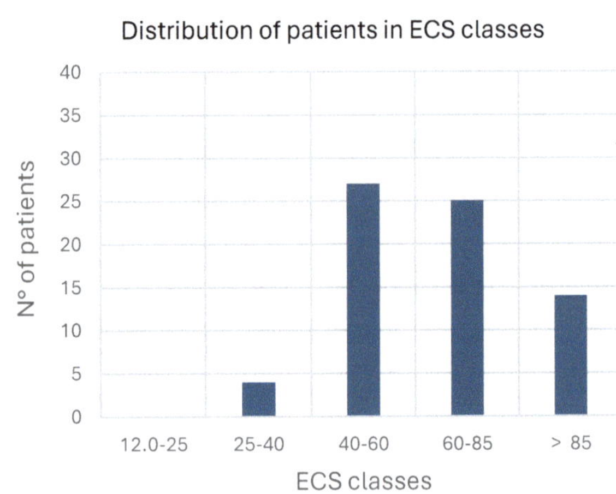

Fig. 11.4 The distribution of patients examined in the five ECS value classes shows a decisive shift towards the high classes. Therefore, high environmental stress is immediately found in cancer patients

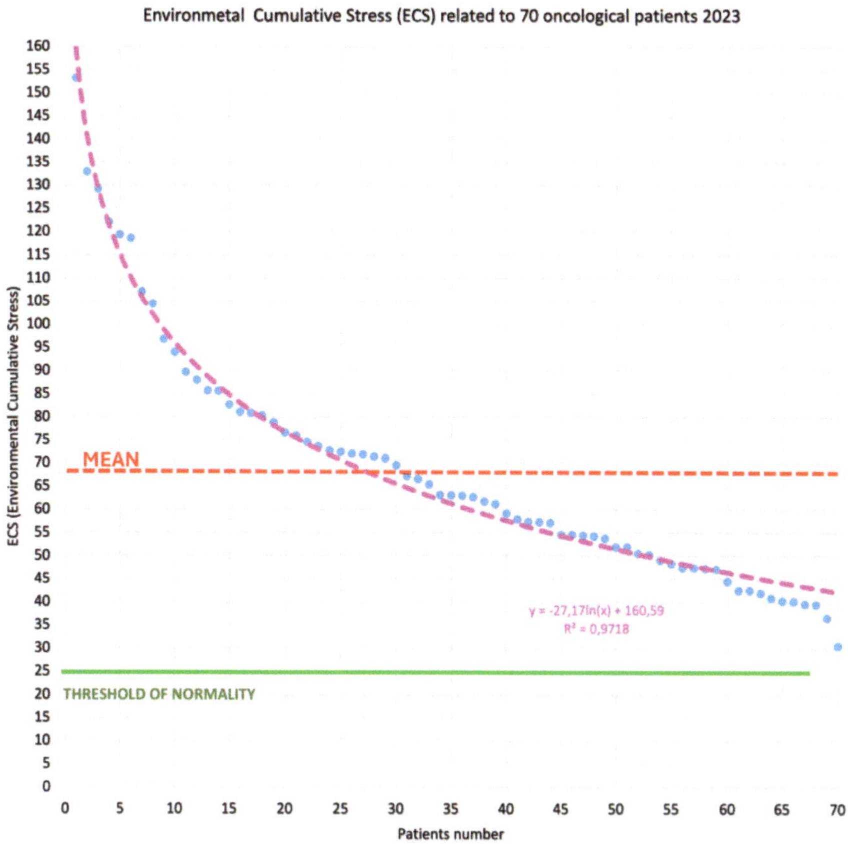

Fig. 11.5 Distribution of the 70-case study ECS. Note the ECS mean = 68.6 and the threshold of normality not over ECS = 25.0; note the logarithmic curve. No one case resulted within the condition of normality

present high environmental stress. As we will see in the discussion, this reality has a non-negligible etiopathological significance.

11.3.2 Environmental Stress in Smokers and Non-smokers

We differentiated the 70 patients among smokers (30%), non-smokers (22.9%), and ex-smokers (47.1%). It is interesting to observe their relationships (a) with tumour histotypes and (b) with environmental stress.

As shown in Table 11.6, ex-smokers maintain similarities with smokers in the frequencies of cancer histotypes: 52.4% and 55.2% of adenocarcinoma vs. 80% of non-smokers (difference of approximately 32.5%). This similarity suggests that, even years later, some lung damage remains in the exes.

Table 11.6 Comparison among smoking patients, their oncological histotypes (Adeno carcynoma; Squamous carcinoma; Non-small Cell Lung Cancer;Small-cell Lung Cancer) and their ECS factors

Smoking	Aden	SP.Carc.	NSCLC	SCLC	WR Phys %	WR PSoc %	cWR	BF %	ECS
Yes	52.40	28.60	14.30	4.80	42.49	36.05	57.79	82.00	78.75
Ex	55.20	31.00	10.30	3.50	39.49	24.63	49.21	81.00	64.16
No	80.00	6.70	13.30	0.00	39.16	26.75	49.64	78.00	61.62
Normality	*60.00*	*25.00*	*10.00*	*5.00*	*20.00*	*20.20*	24.00	*100.00*	*24.00*

Table 11.7 Results for social characterization of all patients

Workers	No Smok. %	WR-Phys	WR-Psoc	BF	ECS
Workers and Labourers	29	40.86	29.97	**73.54**	**78.7**
Employees & Professionals	26	37.95	30.45	**83.26**	**61.85**
Various Pensioners	18	41.36	26.82	**81.24**	**67.32**
Women	34.5	40.92	31.03	80.64	69.39
Men	14.6	39.85	26.73	79.38	68.01

Whether socio-economic and occupational components could have influenced the results was investigated too. Pensioners and Employees present similar BF and ECS values, while

Workers/Laborers vs. Employees/Professionals shown a difference of about 28% of ECS and they were in heavy altered conditions (Table 11.7). This difference implies the value of BF: the survey of the fields where Workers/Laborers appear to live shows more altered conditions. No differences emerge between women and men.

The relationships with environmental stress are various. The former smokers present similarities with non-smokers: 61.6 and 64.2 of ECS vs. 78.75 of smokers (a difference of approximately 20.1%). However, this difference is due to the substantial diversity of WR-PSoc and, therefore, a greater *psychosocial fragility*. Furthermore, smokers get sick even living in a "better" environment, i.e. with a higher BF than non-smokers (+5%). Therefore, non-smokers and ex-smokers are more sensitive to the state of the environment of their Landscape Units (LU), as shown by the WR-Phys/WR-PSoc ratios: 1.17 for smokers vs. 1.46 for non-smokers (+25%) and 1.6 for ex-smokers (+37%). This difference confirms that, for smokers, cancer is due to toxicological issues, while for non-smokers and ex-smokers, it is mainly due to environmental stress factors.

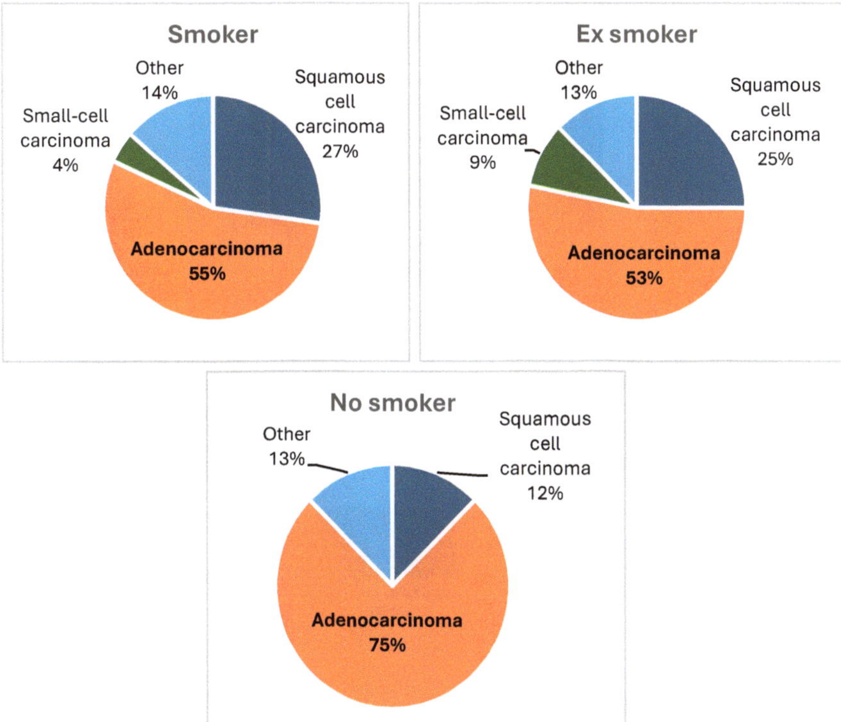

Fig. 11.6 Histotype distribution among non-smokers A, smokers B, ex-smokers C, sum of smokers and ex-smokers

Figure 11.6 summarizes the subdivision of lung cancers according to histotype. Note that our 70 patients present histotypes (Wistuba et al., 2018; Zhang et al., 2023) very near the known general averages: adenocarcinoma = 60%, squamous carcinoma = 25%, Small-cell carcinoma = 5%, and Other = 10%. Adenocarcinoma is confirmed as the most frequent form and even higher in percentage in non-smokers (Fig. 11.6).

Smokers and ex-smoker patients differ from no smokers in oncological histotype independently from BF levels (see also Table 11.6). ECS values of smokers are higher by published data: this difference is due to the substantial diversity of WR-PSoc and, therefore confirming a greater psychosocial fragility.

11.3.3 Discussion

Although environmental factors are essential in the history of lung cancer, we usually limit ourselves to evaluating pollutants in the air (from smoke to fine

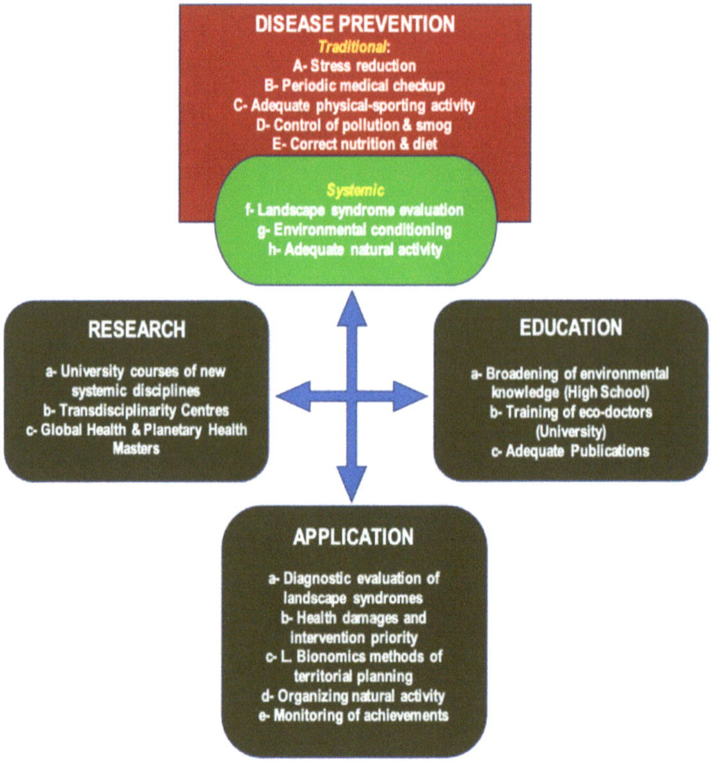

Fig. 11.7 Diagram showing the addition of the three systemic sets to the list of the five main other prevention sets. Note the linkages with Education and Research to reach a correct application

dust). As seen in the previous chapter, we do not understand that the environmental stress that can cause cancer goes well beyond pollution.

Cumulative Environmental Stress (ECS) is quite complex, as shown in Fig. 11.3. and doctors, with the help of advanced ecologists, must understand that a complete environmental history can suggest diagnostic investigations if ECS exceeds typical values (ECS >25). As we have seen (Par. 10.6, 10.7, and 10.8), the deficiencies of natural vegetation infrastructures influence morbidity, leading to anomalies from environmental stress and altered environmental body information.

Medical prevention should be updated, extending the anamnesis to environmental factors, which, as we have said, are complex and linked in a system of cumulative impacts. It is a true *paradigm shift*: the problem concerns the *upgrading in training* for a new specialization: doctors capable to integrate ecological principles to prevent human health alteration and widen integrated therapy (*Ecoiatria*).

Practical steps or frameworks that *clinicians can adopt* immediately? We suggest: (1) to check which one in between WR-Phys and WR-PSoc gives more contribution to cWR; relating to higher WR-Phys (2) to incite patients to improve the frequency of their attendance of really high qualitative environmental places (i.e., Forest baths) and, in general, (3) to focus the attention of each patient to act and take decisions to have the possibility to shift his/her answers to the column of first score (green), (4) try to change the landscape unit (LU) where a patient got sick with a LU with a good BF. However, the most crucial observation is that we need to widen health prevention.

11.4 Widening Health Prevention

11.4.1 Towards a More Systemic Prevention

The British Department of Health & Social Care (2018) asserts that Prevention is about helping people stay healthy, happy, and independent for as long as possible. This premise means reducing the chances of problems arising in the first place and, when they do, supporting people to manage them as effectively as possible. Prevention is as vital at 70 years old as at seven. In conclusion, the cited department shows a synthesis of the main Prevention strategies for young, adult, and elderly people. Now, the British Prevention system is one of the best in Europe; nevertheless, it does not seem to be fully open to the new systemic paradigm. The relationship between man and environment is limited to sports exercises, adequate housing, and healthy eating.

As shown in Fig. 11.7, listing the leading sets of prevention expressed today in Western civilization, we can see (A) stress reduction, (B) periodic medical checkups, (C) adequate physical-sporting activity, (D) control of pollution and smog, and (E) correct nutrition and diet. However, following bionomic principles and methods, at least three new sets must be added: (F) landscape syndromes evaluation, (G) environmental conditioning of people, and (H) adequate natural activity.

Adding the three sets requires a serious commitment to research and education to arrive at a correct application. Without the proper preparation, the new sets cannot develop in good prevention. As in the previous Sect. 5.5, education is the most critical discipline, but it must broaden ecological knowledge with the help of research.

11.4.2 From Traditional to Systemic Disease Prevention

The mentioned addition of systemic sets of new prevention also involves the completion of traditional prevention sets with systemic concepts. Thus, the update of the prevention sets goes beyond the three new sets mentioned: Table 11.8 shows these completions.

Table 11.8 Widening the prevention sets

Traditional:	Systemic expansion:
A- stress reduction	a- to the stress due to LU alterations
B- periodic medical checkup	b- to point (g) and psycho-neuro-environmental relationships
C- adequate physical-sporting activity	c- also to natural trekking
D- control of pollution & smog	d- to point (f) to check cumulative impact
E- correct nutrition & diet	e- to the bionomics of agrarian landscapes
F- systemic	**f- landscape syndrome evaluation**
	g- environmental conditioning
	h- adequate natural activity

(a) Stress reduction cannot remain limited to problems due to economic or work difficulties but must be expanded to the stress due to landscape unit (LU) alterations.
(b) Periodic medical checkup must be linked with environmental conditioning (point G) and psycho-neuro-environmental relationships.
(c) Adequate physical-sporting activities must be related also with natural trekking.
(d) Control of pollution and smog will be linked to point F, to check the cumulative impact with landscape syndromes.
(e) Correct nutrition and diets must be expanded to the bionomics of agrarian landscapes

On the other side, the last three cited sets (F, G, H) need to be better explained. For instance, we report some examples:

(f) ***Landscape syndrome evaluation***

The comparison between mortality rate and Bionomic functionality (MR/BF) (Fig. 11.8) in two agrarian landscapes near Milan, Albairate near Ticino River (Vaglia & Zamprogno, 2017), west of Milan (orange dot), (see Figs. 9.11, 9.12, and 9.14) and Cervesina, near Po River, South of Milan (blue dot), shows the linkage between mortality and landscape syndrome. Despite the blue curve and MR are higher than the orange ones because of the higher population age (PA) of Cervesina, its values need attention: BF = 0.74 and MR = 15.6/1000. The green arrow indicates the necessity to heal the LU to decrease the MR; if we restore the Cervesina land arriving at BF = 0.80, its MR may decrease to 13.7/1000 (VINGEO, 2022).

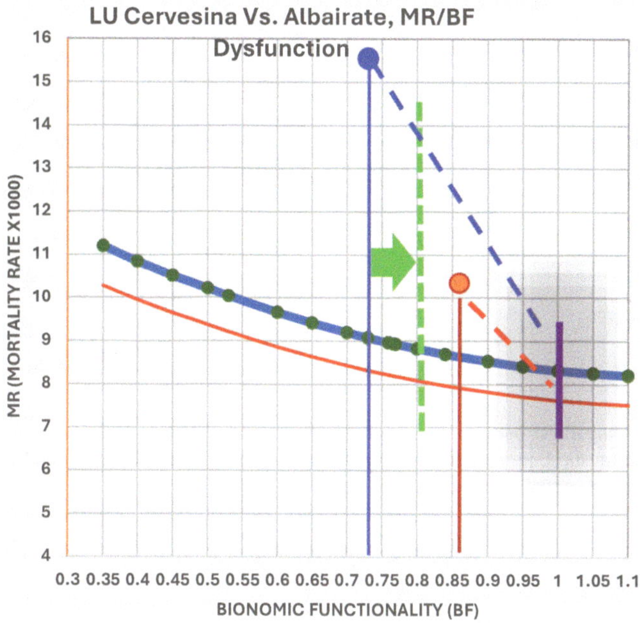

Fig. 11.8 The theoretical model of MR/BF related to two agrarian landscapes near Milan: Albairate (orange) and Cervesina (blue). The blue curve and MR are higher than the orange ones because of the higher PA (population age). The green arrow indicates the necessity to heal the LU. Dashed straight lines indicate the distance between the current real condition and the normal ones (purple line), with a security coefficient (the gray shadow)

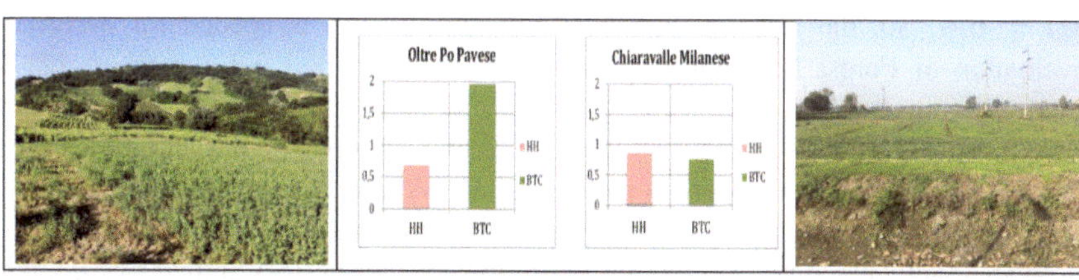

Fig. 11.9 An example of the bionomic comparation of two different agrarian landscapes near Milan: Oltre-Po Pavese and Chiaravalle Milanese

(g) *Environmental conditioning*

People living in the landscape units of Fig. 11.9 do not experience the same conditioning environment as people living in a town or a suburban-industrial LU as we can enlighten even just by applying HH and BTC indicators to these different agrarian landscapes (for the information content evaluation see Figs. 7.4 and 7.5).

Fig. 11.10 Comparison between an historical urban park in Milan (Villa Litta, Affori) and an Alpine pasture in the Cauriòl Valley (Trentino). Both exceptionally beautiful and attractive, but their biomedical potential is at least in proportion of 1.0 to 2.30

We can evaluate their different bionomic states and measure the value of environmental conditioning:

1. Weight of HH, HH/NH >>>
   0.60/0.40 = 1.50 (Oltre-Po) Vs. 0.80/0.20 = 4.00 (Chiaravalle).
2. Local BTC/Regional BTC >>
   1.90/2.00 = 0.95 (Oltre-Po) Vs. 0.70/2.00 = 0.35 (Chiaravalle).

(h) ***Adequate natural activity***

Paragraph 10.6 underlined that MT (monoterpenes) emissions are proportional to forest bionomic functionality: boreal-alpine forest MT = 17.5–22 Vs. urban park MT = 6–9. So, the LaBiSV methodology (see Par.4.2) is needed for a correct evaluation of Forest Bath Therapy because their forest/park ratio may present a sharp difference.

We can prove this statement by comparing two ecotopes of similar surface area (Fig. 11.10): (a) a historical urban park in Milan (Villa Litta, Affori, one of the best of Milan) and (b) an Alpine pasture in the Cauriòl Valley (Trentino, Northern Italy). Both are lovely and attractive grasses, with some patches of trees at their margin, but their biomedical potential is at least in proportion to 1.0 to 2.3. Let us measure the mentioned areas (2 ha each):

Milan: Wood 20% (BTC = 6.4) + Built (20% (BTC = 0) + Grass 60% (BTC = 0.6); average BTC = 1.57; *HH = 54%.*

Cauriòl: Forest 30% (BTC, 8.6) + Creek 10% (BTC = 0.7) + Pasture 60% (BTC = 0.8); average BTC = 3,60; *HH = 29%.*

Their ratio results (BTC ratio): 3.60/1.57 = 2.29, like MT-Forest/MT-Urb.Park ratio (see Sect. *10.6.3) = 2.32.*

11.5 Landscape Unit Evaluation: The Health Protective Index (HPI)

We affirmed that prevention must follow the systemic paradigm. We also affirmed that the Landscape Unit's syndromes lead to human health alterations. Therefore, we can observe that our health is not genuinely controlled if our environment remains dysfunctional. Our Land Authorities must demonstrate to monitor our LU environment, following bionomics principles and methods in territorial planning.

So, here is presented a comparison between two municipalities in the province of Monza-Brianza: Nova-Milanese (NM = 587.2 ha, population = 23,514) and Usmate-Velate (UV = 515,1 ha, population = 10,325). The first presents a landscape diagnostic index DI = 0,65 (quite low), while the second DI = 0,90 (high).

As shown in the next Fig. 11.11, the two LUs (municipal territory) are sharply different and resulting 'altered' (Nova Milanese) Vs. 'normal' (Usmate-Velate) and consequently, their health parameters are different:

(a) MR (mortality rate) /1000 = 9.30 (NM) Vs. 7.48 (UV) ratio UV/NM = 1.243
(b) Covid-19 incidence % = 10.15 (NM) Vs. 6.22 (UV) ratio UV/NM = 1.632
(c) Annual Death due to cancer = 35.5 (NM) Vs. 32.2 (UV) ratio UV/NM = 1.103
(d) Average "disease" = (a + b + c)/3 = 1.326
(e) *L. Diagnostic Index (DI) = 0.65 (NM) Vs. 0.90 (UV) ratio UV/NM = 1.385*

It might be helpful to provide the measure of a provisional indicator, the Indicative Morbidity per each municipality considered as a landscape unit (LU). The indicator is proportional to the DI (L. Diagnostic Index) and can be deduced by noting that the L. Diagnostic Index ratio DI_{UV} /DI_{NM} is pair to 1.385 and the average "disease" ratio equal to 1.326. An evident proportion of DI/Morbidity emerges with a deviation of only 4.23%. So, morbidity of Nova Milanese results to be about 135% bigger than that of Usmate Velate. In case a patient of NM is found to have an elevated ECS (Environmental Cumulative Stress), it would be best for him to move to the UV landscape unit.

However, the situation is more serious, because illness parameters can vary from year to year and are dependent even by sanitary structures. So, we propose to apply the *Health Protective index (HPi)*. This index evaluates the capacity of a landscape unit (LU), for instance a municipality, to present an environmental potential able to protect human health. To reach this goal in a synthetic way, we need to consider two bionomic functions:

(a) The first concerns the necessity to evaluate an average potential of phytoncide in the examined territory, linked to immune system. As underlined in Par.10.6.3, applying MT = BTC × 2.07 [absolute value] to the average BTC it is possible to register this potential.
(b) The second concerns the necessity to evaluate the information content of the LU, linked to the epigenetic processes with our body (e.g., cell reparation capacity). As underlined in Par.7.4.1, the evaluation of g.LM in function of HH has been proposed, with a security coeff. = HH/100. (see Par.7.4.2).

Comparison between two Municipalities with altered Vs. normal environment

NOVA MILANESE	URB	Survey	gap %	score	USMATE VELATE	SUB.R	Survey	gap %	score
LU Surface & Population		587,2	23.514		LU Surface & Population		515.1	10.325	
RNT, forest area (%)	4.0-8.0	0.5	87.5	0	RNT, forest area (%)	8.0-25	12.1	ok	2
CBSt of forest	16.5-22.5	11	33.3	0.5	CBSt of forest	22.5-30.5	27.6	ok	2
PRD, agricultural areas (%)	5.0-15	29.5	Ok	2	PRD, agricultural areas (%)	30-55	50.5	ok	2
RSD + SBS, urbanized areas (%)	25-50	63.9	27.8	1	RSD + SBS, urbanized areas (%)	8.0-15	36.7	245	0
HH, human habitat (%)	85-95	84.9	Ok	2	HH, human habitat (%)	70-86	77.2	ok	2
HS/HS* carrying capacity	0.1-0,5	0.17	Ok	2	HS/HS* carrying capacity	0.6-3.0	0.59	ok	2
BTC, LU (Mcal/m^2/a)	0.60-0.90	0.46	23.1	1	BTC, LU (Mcal/m^2/a)	0.90-1.60	1.23	ok	2
LTpE = 10 (HNd/HU) + BTC_{A+F}	0.35-3.0	1.64	Ok	2	LTpE = 10 (HNd/HU) + BTC_{A+F}	5.7-11.0	9.32	ok	2
Bionomic Functionality, **BF**	0.8-1.0	**0.46**	42.45	0.5	Bionomic Functionality, **BF**	0.85-1.1	**0.94**	ok	2
Ecological Density (ab/ha)	25-50	47.17	ok	2	Ecological Density (ab/ha)	12.5-25	25.96	3.8	2
L. Diagn. Index (DI)	0.85-1.0			0.65	L. Diagn. Index (DI)	0.85-1.0			0,90
Mortality Rate (MR)/1000		9.30		1.0	Mortality Rate (MR)/1000		7.48		0.80
Covid-19 Incidence % of Population (APR-21)		*10.15*		1.0	Covid-19 Incidence % of Population (APR-21)		6.22		0.61
Death due to cancer		*35.5*		1.0	Death due to cancer		32.2		0.90

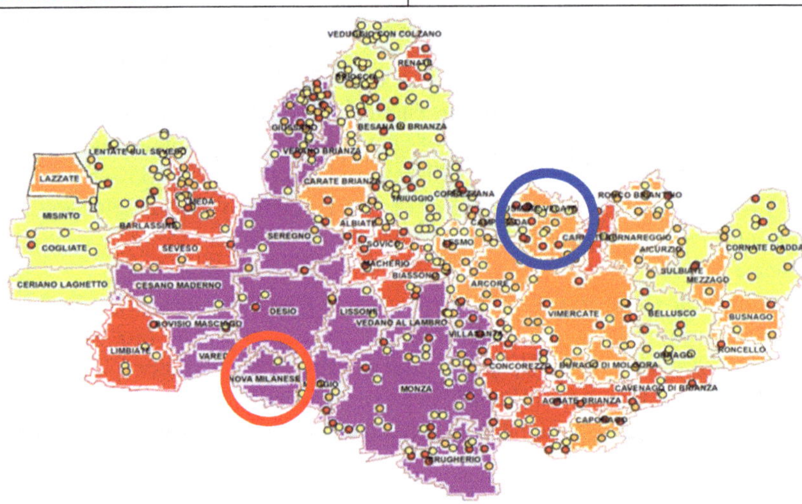

Fig. 11.11 Province of Monza-Brianza (Lombardy). Localization of the municipalities of Nova Milanese (red) in urban landscape (violet), and Usmate-Velate (blue) in a suburban-rural landscape (Orange)

	HH (%)	BTC	lav.MTI	gLM	HH/100	HPi
BRIOSCO	68,13	1,82	3,77	16,81	0,68	43,15
BESANA IN BRIANZA	76,21	1,31	2,71	11,62	0,76	24,02
USMATE VELATE	77,2	1,23	2,55	11,04	0,77	21,70
CARATE BRIANZA	75,85	1,19	2,46	11,84	0,76	22,11
AGRATE BRIANZA	84,24	0,78	1,61	7,37	0,84	10,03
SEVESO	83,1	0,9	1,86	7,91	0,83	12,25
GIUSSANO	86,31	0,73	1,51	6,46	0,86	8,43
NOVA MILANESE	84,93	0,46	0,95	7,06	0,85	5,71
CINISELLO B	93,15	0,52	1,08	4,10	0,93	4,12
MI Z-6 BARONA	94,52	0,41	0,85	3,76	0,95	3,01
AGR-PRT x 0.85	*73,30*	*1,47*	*3,04*	*13,40*	*0,73*	***25,00***

Health Protective Index, HPi = av.MT x gLM x (HH/100)

HH= human habitat (%), BTC= vegetation bio-capacity (Mcal/m²/year), av.MT= average monoterpenes (%), gLM= information content (bit x BTC), HH/100= security coefficient.

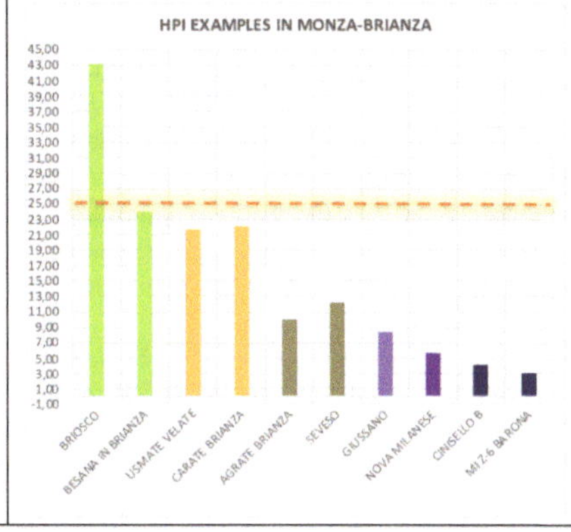

Fig. 11.12 Example of application of the Health Protective index to some municipalities of the Monza-Brianza province, in proportion of landscape types (green = protective agrarian, yellow = rural, grey = suburban, purple = urban, violet = dense-urban). Note the wide insufficiency of HPi in this territory: 84.7% of 72 municipalities present HPi <25.0. This value is the average of the AGR-PRT × 0.85, that is the minimum acceptable value of HPi for this type of landscape, and it remains very distant from suburban and urban landscapes

Therefore, the HPi equation is very simple:

$$HPi = [av.BTC \times 2.07] \times g.LM \times HH/100 \ [bit \times Mcal/m^2] \times year$$

An example of application of this HPi is exposed in Fig. 11.12. Let us observe the interrelations among the parameters: BTC, HH, BF, linked with the registration of human health (Figs. 9.7 and 9.8) and implicated in this index. Note the sharp differences among three LU pertained to the same rural-suburban landscape (see Fig. 11.11) and the yellow line of HPi = 25.0 bit × Mcal/m^2. This value is the average of the AGR-PRT × 0.85 (i.e. even considering a 15% tolerance), and it is a first threshold of normality in that province: note that it remains very distant from suburban and urban landscapes.

References

Bonizzi, L., Campana, F., & Soggiu, A. (2022). Zootechnical systems, ecological dysfuntions and human health. In V. Ingegnoli, F. Lombardo, & G. LaTorre (Eds.), *Environmental alteration leads to human disease: A planetary health approach* (Sustainable development golas series). Springer-Nature. Foreword by Andrea Lenzi.

Bottaccioli, F., & Bottaccioli, A. G. (2020). *Psycho-neuro- endocrine-immunology and science of integrated care*. Edra Ed.

Department of Health & Social Care. (2018). *Annual report and accounts 2018–2019*. House of Commons.

Gogtay, N., Giedd, J. N., Lusk, L., et al. (2009). Dynamic mapping of human cortical development during childhood through early adulthood. *Proceedings of the National Academy of Sciences of the United States of America, 101*(21), 8174–8179.
Gregory G, Roth J, Gensheimer K (2024) *Research from the international symposium on one health research: Improving food security and resilience*. Report ECDC, Brussel.
Ingegnoli, V. (2002). *Landscape ecology: A widening foundation* (p. XXIII+357). Springer.
Ingegnoli, V. (2011). *Bionomia del paesaggio. L'ecologia del paesaggio biologico-integrata per la formazione di un "medico" dei sistemi ecologici* (p. XX+340). Springer.
Ingegnoli, V. (2015). *Landscape bionomics. Biological-integrated landscape ecology* (p. XXIV + 431). Springer.
Ingegnoli, V., & Giglio, E. (2005). *Ecologia del Paesaggio: Manuale per conservare, gestire e pianificare l'ambiente* (p. 685+XVI). Sistemi Editoriali.
Ingegnoli, V., & Giglio, E. (2021). Environmental bionomic alterations and cancer incidence in European countries: A first overview. *Journal of Nursing & Occupational Health, 2*(3), 266–281.
Istituto Superiore di Sanità (2019) One health. Report ECDC, Brussel.
Lenzi, A. (2022). Foreword. In Ingegnoli, Lombardo, & LaTorre (Eds.), *Environmental alteration leads to human disease*. Springer-Nature.
Lorenz, K. (1978). *Vergleichende Verhaltensforschung: Grundlagen der Ethologie*. Springer.
Petanjek, Z., et al. (2011). Extraordinary neoteny of synaptic spines in the human prefrontal cortex. *Proceedings of the National Academy of Sciences of the United States of America, 108*(32), 13281–13286.
VINGEO. (2022). *Indicazioni di Fattibilita' Progettuale Secondo I Criteri Ecologici e Bionomici: Relazione Finale delle Opere di regimazione Idraulica torrente STAFFORA a difesa degli abitati di Oriolo, Cervesina e Pancarana (PV)*. AIPO.
Vaglia, V., & Zamprogno, L. (2017). *Landscape bionomics: Valutazione Diagnostica di un'unità di Paesaggio Agrario con un Metodo Sperimentale Innovativo*. Applicazione ad un Contesto Agricolo-Forestale. Tesi di laurea Corso di Laurea Magistrale in Scienze Agroambienta.
Wistuba, I., Brambilla, E., & Noguchi, M. (2018). Classic anatomic pathology and lung cancer. In H. I. Pass, D. Ball, & G. V. Scagliotti (Eds.), *ISLAC thoracic oncology* (pp. 143–146). Elsevier.
Zhang, Y., Vaccarella, S., Morgan, E., Li, M., Etxeberria, J., Chokunonga, E., et al. (2023). Global variations in lung cancer incidence by histological subtype in 2020: A population-based study. *The Lancet Oncology, 24*(11), 1206–1218.

Part III
Applications and Conclusion

Qui litteris solis confisi fuerunt,
umbram non rem persecuti videntur.

Marcus Vitruvius Pollio
(De Architectura)

Those who trusted in letters alone
they seem to have pursued/chase the shadow, not the thing

Chapter 12
Upgrading Ecology-Economy and Planning Relationships

Contents

Abstract The settlement of a new inhabitant produces not only a deficit (TD) in the energy balance of a district's biological organization, which must be compensated with a bionomic-protective service (PRT) but also a TD in the economic balance because the activities related to the landscape apparatus and minimum living space have costs linked to the operation of the corresponding areas.

Proportioning the basic estimate mentioned to the excursus from BTC = 2.5 up to 9.5 Mcal/m^2/a (current range of BTC in forests), the value of a forest plot can vary from €42 to €160 /m^2. It is also possible to estimate a plot of cultivated land by observing a range of BTC for meadows and crops from 0.5 to 2.4 Mcal/m^2/yr, so the value of an agricultural plot can vary from 8.40 at €40.3/m^2.

The distribution of average BTC values (landscape apparatuses) in the major sets of landscape systems in the most important Western European nations, with a gradient from the South (Spain) to the North (UK) is add as the upper-level framework.

Territorial planning could be a way for therapeutic rehabilitation and care: unfortunately, traditional land use planning scheme lacks of historical investigations and of systemic contributions. Urban planning remains the dominant discipline, even

V. Ingegnoli, *Landscape Bionomics and Planetary Health*, Landscape Series 21,
https://doi.org/10.1007/978-3-032-08908-3_12

beyond inhabited centers. On the opposite, the clinical-diagnostic criterion requires the intervention of an ecologist, who establishes any bionomic-landscape syndromes in progress before considering new possible transformations. Strategic Environmental Assestement must take care of it, widening its methodology. Furthermore, the greater importance of the context scale and systemic historical investigation is noted.

Keywords Transformation deficit (TD) · Global scale · Regional scale · Agricultural landscape · Organic farming · Conventional farming · Economic evaluation · Ecosystem services · Agro-forest lots · Strategic environmental assessment (SEA)

12.1 The Problem of Economic-Ecological Evaluation

12.1.1 Disagreements Between Economy and Ecology

Currently, natural and semi-natural systems provide a wide range of goods and services that are essential to sustain the well-being and quality of life of individuals through ecological processes and functions. This range of goods and services is provided at different temporal and spatial scales and in such a complex and little explored way that they cannot be replaced by technology (Daly, 1999, 2010). In fact, although society has partially emancipated itself from the natural environment through culture and technology, its health and survival are completely dependent on ecosystem services (Millennium Ecosystem Assessment, 2005).

As already anticipated in Par.7.6, we must adapt the so-called Ecosystem Services to the new science paradigms. We must note that the limits to the concept of ecosystem have a particular influence on the critical and more recent concept of "Ecosystem Services" too. As seen in Table 7.9, Health preserving and Protective services are added to the four sets of Supporting, Regulating, Provisioning, and Cultural Services, according to the principles of Bionomy. Furthermore, we should say that the concept of 'services provided free of charge', as SItE (Italian Society of Ecology) writes, is incorrect since, from a systemic perspective, on one side it only refers to the human population, on the other it does not consider that every service has a cost too and, therefore, must be paid for.

It is a general economic principle, and indeed, it is what distinguishes an ecological economy from the dominant neoclassical economy. As we will see, exchange value should always be linked to the role of things in nature; only neoclassical economics demands otherwise. But, still, someone will have to pay for the services. It is known that until a few decades ago, building on a plot of agricultural land took it for granted that urbanization services would arrive sooner or later: but at whose expense? Consequently, urbanization charges had to be imposed because not considering the context in building speculation costs was harmful to society.

Our ability to destroy the planet's natural systems or save them (also saving ourselves) makes us live in a unique generation: we are in a crucial moment in history. Many scholars have underlined that the economy-ecology conflict presents the most significant danger towards the destruction of the environment. Today's dominant economies contain errors that are even serious for an advanced ecologist. We list some of them:

- they claim that exchange value should be the financial measure of everything, regardless of natural functions; instead, we note that the value of everything is also linked to its role in nature, a role that is not easily monetized.
- current culture denies the ethical evaluation of means and ends, so that the end justifies the means: instead means and ends are linked, like the seed to the plant; therefore, they must not be separated, and we must choose between different purposes.
- today absolute scarcity is not seen as a limitation, because low entropy is abundant for our needs; instead, these needs must also contain the adequate functionality of all the ecological systems of the bio-geosphere.

Even the general ecology that is still taught today in academies often makes serious errors, as we saw in Part I, like:

- man believes himself to be detached from ecological processes and in contrast with nature (see EIA), instead human ecosystems are linked and co-evolved with natural ones.
- the concept of ecosystem dominates every level of biological organization, even if it is ambiguous and independent of space-time scales: instead, general ecology must consider the different structural and functional characteristics of each complex environmental system depending on the scales and information content.
- pollution dominates all environmental issues: instead, the alteration of ecological systems (and health) depends first on the structural and dynamic dysfunctions of the systems of ecocoenotopes.

Let us remember that, today, there are two main paradigms in economics: ecological-economic and neoclassical (Fig. 12.1). In the first case, economic systems are part of the ecological paradigm; in the second, ecological systems are part of the economic paradigm. The only limit to growth is technology, which is supposed to have no limits.

12.1.2 Equivalence Between Bionomics and Economics Costs: The Transformation Deficit Criterion

Let us remember that Neoclassical Economics is primarily indebted to nature since the value of goods is measured only in terms of labour (Bjerg, 2016). This is at odds with reality because exchange values derived from human labour cannot be the

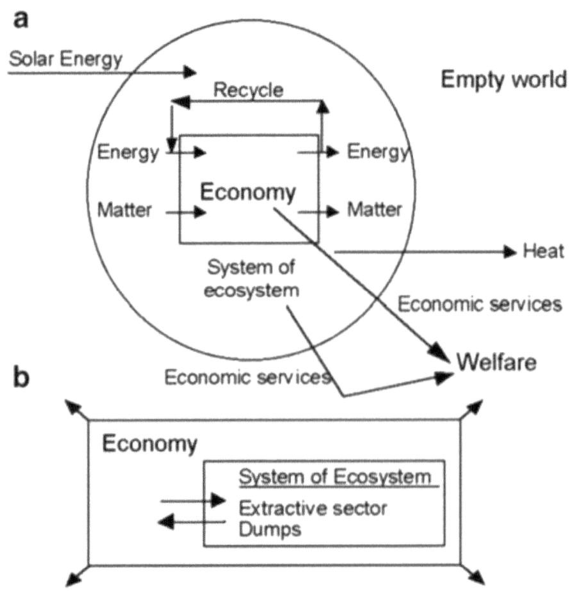

Fig. 12.1 (**a**, **b**) Diagram of the two main types of relationships between economy and ecology. (From Daly (1999), redrawn). (**a**) Ecological Economics, (**b**) Neoclassical Economy. (From: Ingegnoli (2002))

measure of everything. We reiterate: the value of every good is linked to the role assumed in natural systems, so our survival depends on this link; therefore, we must overcome the neoclassical value-work limit. The importance of Bionomic-Ecological Services lies precisely in this: the ability to overcome these neoclassical limitations by recognising that nature also provides services, that is, performs "work".

For example, vegetation formations carry out the following tasks: (a) maintaining the right level of BTC, (b) compensating for transformation deficits (TD), (c) protecting human health, (d) regulating the climate and greenhouse gases, (e) regulating rain, (f) producing food and wood, (g) purifying water, (h) purifying air, (i) slope stabilizing, etc.

In the essential human activities, we can note an equivalence between Bionomic and Economic costs when referring to the same TD (Transformation Deficit). The protective compensation of a TD (e.g., due to the settlement of an inhabitant in a territory), evaluated both as a bionomic cost and as an economic cost, forms a system:

Emergence of an equivalence Between Bionomics & Economics COSTS

$$TD_{BIO} + SH \times PRT_{BIO} = 0 \quad (1)$$

$$TD_{ECON} + SH \times PRT_{ECON} = 0 \quad (2)$$

We can apply these concepts within the Lombardy Region, verifying the transformation deficit (TD) inevitably due to the settlement of an inhabitant in the territory. We know, from research carried out by Ingegnoli, that the average BTC in

Lombardy is today around 2.02 Mcal/m^2/year and that four landscape apparatuses are involved in settlement of an adult person (residential, productive, subsidiary, hydrogeologic, see Sect. 3.3) of which we have the values of the minimum theoretical habitat per capita and of the BTC:

SH^*_{PRD} = 950 m^2/inhabitant and BTC_{PRD} = 0.90 Mcal/m^2/yr;
SH^*_{RSD} = 105 m^2/inhabitant and BTC_{RSD} = 0.30 Mcal/m^2/yr;
SH^*_{SBS} = 65 m^2/inhabitant and BTC_{SBS} = 0.05 Mcal/m^2/yr;
SH^*_{HGL} = 20 m^2/inhabitant and BTC_{HGL} = 0.05

where SH*PRT= Minimum Standard Habitat for the Protective L.Apparatus; SH*PRD= Minimum Standard Habitat for the Productive L.Apparatus; SH*RSD= Minimum Standard Habitat for the Residential L.Apparatus; SH*SBS= Minimum Standard Habitat for the Subsidiary L.Apparatus; SH*HGL= Minimum Standard Habitat for the Hydrogeological L.Apparatus (see 3.2.5 and 3.3.1). The total vital area therefore amounts to 1140 m^2/inhabitant to which the surface of protective elements must be added (high BTC).

These protective components (PRT) are given by forest patches, which in Lombardy have an average BTC = 6.20 Mcal/m^2/yr (Ingegnoli & Giglio, 2005; Ingegnoli, 2002, 2011, 2015, 2019a).

As it can be further deduced from Table 12.1, the calculation of the transformation deficit TD_{BIO} (column 6) is given by the differences in the average territorial bio-potentiality of the devices involved (BTC') compared to the regional average BTC pair to 2.02, as reported in column 5 (BTC'-BTC_R), and by their multiplication by the necessary surfaces (column 2). For the 1140 m^2/inhabitant the result is a TD_{BIO} = −1412.05 Mcal/year. Dividing this value by 4.18 Mcal/m^2/a (i.e. the difference 6.20–2.02) we immediately obtain the surface area necessary to compensate for the TD_{BIO} which is equal to 338 m^2/inhabitant. To settle a new inhabitant in Lombardy we therefore need 1478 m^2/inhabitant.

The settlement of a new inhabitant produces not only a TD_{BIO} in the energy balance of biological organization systems, which we must compensate with a

Table 12.1 Evaluation of the bionomic transformation deficit and the following needed protective apparatus for each new inhabitant in Lombardy

SH*/L.App	m^2/ inhab	% SH*	BTC'*Mcal/m^2/ year*	BTC'-BTC_R	TD_{BIO} *Mcal/year*
SH*PRD	950.00	64.94	0.90	*−1.12*	*−1064.00*
SH*RSD	105	7.18	0.30	*−1.72*	*−180.6*
SH*SBS	65	4.44	0.05	*−1.97*	*−128.05*
SH*HGL	20	1.3	0.05	*−1.97*	*−39.4*
Tot. Transform.	***1140.00***	*77.13*			***−1412.05***
Potential PRT_R			6.2	*4.18*	
TD_{BIO} / potential PRT_R			[1412.05 / 4.18] = 337.8		
Needed SH*PRT/new inhab.	**338**	22.87			
SH* tot/new inhabitant	**1478.00**	**100**	**2.02**	**0**	

Table 12.2 SH* economic cost evaluation for lombardy

SH-PRD	PRD crop-field	PRD farm	PRD canal	PRD roads	PRD edge-row		TD cost €
m^2	900	15	15	10	10	**950**	
%	*94.74*	*1.58*	*1.58*	*1.05*	*1.05*	*100.00*	
€/m^2	3	900	180	200	20	**22,21**	21,100
SH-RSD	RSD flat	RSD garden	RSD roads	RSD services	RSD tech		
m^2	45	25	15	15	5	**105**	
%	*42.86*	*23.81*	*14.29*	*14.29*	*4.76*	*100.00*	
€/m^2	1600	60	250	300	140	**783,33**	82,300
SH-SBS	SBS factory	SBS roads	SBS commercial	SBS technological	SBS energy		
m^2	25	15	15	5	5	**65**	
%	*38.46*	*23.08*	*23.08*	*7.69*	*7.69*	*100.00*	
€/m^2	11,700	250	1000	250	210	**742,32**	50,300
SH-HGL	Water body	well					
m^2	15	5				**20**	
%	75	25				100	
€/m^2	5	200				**53,75**	1075
						Tot.	**154.895**

bionomic-protective service (PRT), but it also makes a TD_{ECON} in the economic balance because the activities related to the functions of landscape apparatus and minimum living space have costs linked to the operation of the corresponding areas. Following a study by the Polytechnic of Milan (Cerea & Re Cecconi, 2010) on the parametric estimates of construction costs, integrated with the agricultural costs for the PRD areas (Table 12.2), it is observed that these costs (weighted average) can be summarized in $TD_{ECON} =$ €154,895/m^2/inhabitant. Stated that the total cost to make the 1140 m^2/inhabitant operational will be €154,895.00/inhabitant, the amount cost of transformation for m^2 will be 154,895.00: 1140.00 = 135.87€/ m^2.

Since the transformation of 1140 m^2 per capita of productive (PRD), residential (RSD), subsidiary (SBS) and hydrogeological (HGL) habitat appears to have a bionomic cost equal to $TD_{BIO} = -1412.05$ Mcal/year, its economic value is (Table 12.3):

$$€154,895.00(TD_{ECON}) = -1,412.05 \text{ Mcal/year } (TD_{BIO}), \text{ that is}$$
$$109,69€ \text{ lost for each missed/wasted Mcal}$$

From Table 12.3 it can be deduced that the proportion between the PRT compensation area and the total settlement area is equal to 23% (338 m^2/ 1478 m^2 = 0.229), therefore the incidence of the compensation area (SH*PRT) counts economically:

Table 12.3 Evaluation of the economic transformation deficit

SH*/L.App	AREA	BTC	UCT	TD_{ECON}
	m^2 / capita	Mcal / m^2 /yr	€/m^2	€
SH*PRD	950	0.9	22.00[a]	−20,900.00
SH*RSD	105	0.3	783	−82,215.00
SH*SBS	65	0.05	780	−50,700.00
SH*HGL	20	0.05	54	−1080.00
Tot. SH*	1140.00			−154,895.00
SH* PRT	**338**	**6.20**	338/ 1478 = 0.229	154,895.00 × 0,23 = 35,625.85
Tot. transform.	**1478.00**		**UCT-prt** = 35,625.85 / 338 = **105.40**	

[a]22 €/m^2 is the value of fields, roads, canals, silos (i.e. 18 €/m^2) including compensatory reforestation (about 4 €/m^2). (From ERSAF, Regione Lombardia, 2020); ***UCT*** unitary cost of transformation, ***TD_{ECON}*** economic deficit of transformation

$$154,895.00 \times 0.23 = 35,625.85\ €/338\ m^2 = 105.40\ €/m^2$$

Adding the current price of an agricultural lot, pair to about 4.5 *€/m^2* we come back to 109.9 as the previous cost €/Mcal.

12.1.3 Relationships Between Bionomic and Economic Costs

It is immediately apparent that €105.40/m^2 for a naturalistic reforestation plot (BTC = 6.20 Mcal/m^2/year) is at least 3–4 times the market price of a woody lot today. But let's remember that this value includes traditional agricultural production (woody in this case), health protection and naturalistic protection.

The biological evaluation of the needed area of the protective component of the minimum Standard Habitat per capita SH*PRT = 338 m^2/inhabitant depends on TD_{BIO}; therefore, the economic assessment may similarly rely on TD_{ECON}, albeit through a suitable coefficient. To measure this coefficient, we will have to consider Fig. 12.2.

This figure shows the opposite relationship between the amortization period of anthropic transformations (road, house, fields, etc., *orange*), estimated at approximately 80 years, and the development period of natural transformations (reforestation of the SH* area PRT, *blue*), estimated at approximately 180 years.

We will note that the bionomic compensation (green arrows) can only occur 50 years after the destruction of the first decade of settlement because the necessary BTC = 6.20 Mcal/m^2/yr rarely arrives before the *60th year*, corresponding to three-quarters of the depreciation. Since the function of the SH*PRT area is completed only then, its economic value will be equal to 25% of the TD_{ECON} multiplied for a tolerance coefficient of 10%, due to possible disturbances over time:

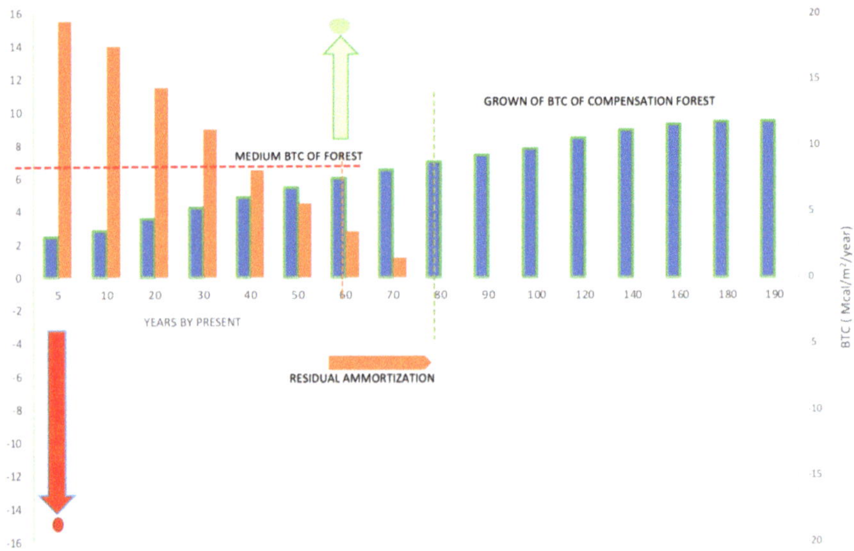

Fig. 12.2 Relationship between the amortization period of anthropic transformations (house, fields, roads, etc., orange) and the development period of natural transformations (reforestation of the PRT area, blue). Note that compensation (green arrows) can only occur 50 years after the destruction of the first decade of settlement and that the necessary BTC = 6.2 Mcal/m^2/y rarely arrives before the 60th year, corresponding to ¾ of depreciation. The function of the PRT area is completed only then, so its economic value will equal 25% of the TD€. Red arrow and dot correspond to the TD_{BIO} value of −1412.05 Mcal/m^2/year

$$SH'_{PRT} = (154,895.00 \times 0.25)/338 = 114.57\,€ \times 0.9 = 103.1\,€/m^2$$

Attention: this value of €103.1 *€/m*^{2} (for a BTC = 6.2 Mcal/m^2/a) is very close to the previous calculation linked to Table 12.3 but it is more significant as it considers the dynamics of the phenomenon detected. We can therefore consider an average value between the two:

$$(105,40 + 103,11)/2 = 104,25\,€/m^2$$

Proportioning the basic estimate mentioned to the excursus from BTC = 2.5 up to 9.5 Mcal/m^2/a (current range of BTC in forests), the value of a forest plot can vary from €42 to €160 /m^2. It is also possible to estimate a plot of arable land by observing a range of BTC for meadows and crops from 0.5 to 2.4 Mcal/m^2/yr, so the new value of an agricultural plot in Lombardy can vary from 8.40 €/m^2 at 40.3 €/m^2.

12.1.4 The Real Value of Agricultural Lots, Considering Eco-Bionomic Services

Remarking the crucial importance of evaluating agro-forest lots in proportion to the ecological systemic function of BTC, it must be reiterated that the economic values of vegetated lots (forest and agricultural fields) are, on average, almost an order of magnitude greater than today's so-called market values (Fig. 12.3). If these values were ignored, society would be forced to shoulder the costs of environmental alterations that inevitably emerge. Furthermore, land consumption would have no limits because today, transforming an agricultural or forestry lot into a built lot is all too easy with such low costs.

Change in values in €/m^2 in the price of land. Today (orange) and considering bio-ecological services (green). The values of BTC × 100 are in blue. Note the greater proportionality in the recognition of services. Constructions: building a = villas with garden, building b = condominiums.

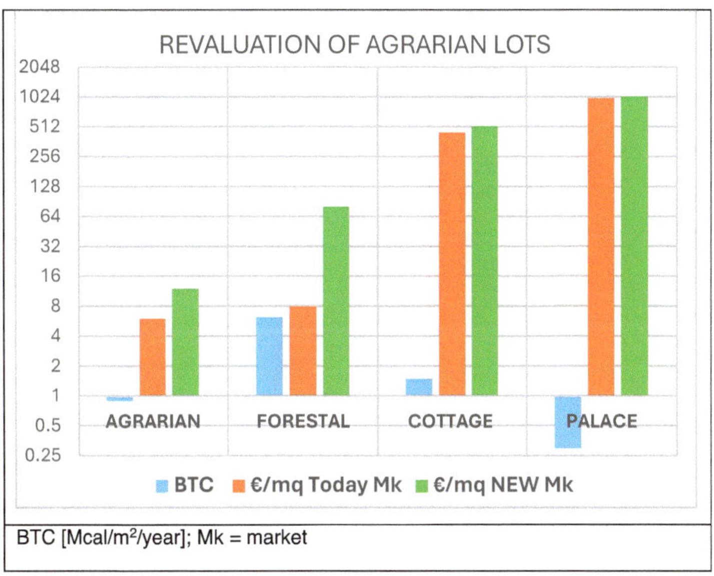

Fig. 12.3 Change in values in €/m^2 in the price of land. Today (orange) and considering bio-ecological services (green). The values of BTC × 100 are in blue. Note the greater proportionality in the recognition of services. Constructions: building a = villas with garden, building b = condominiums

12.2 Ecological Territorial Planning as a Therapeutic Approach

12.2.1 Conventional Planning and Its Limits

In the traditional method, (Fig. 12.4) entirely dominated by the urban planning discipline, even outside the inhabited centers, we still proceed through separate thematic sectors, marked zoning of use, uncritical dependencies on higher-scale plans, technological privileges and road priorities, choices constraints and bureaucracy, standards dominated by the economy (of an anti-ecological economy), captivating social services, banal urban greenery.

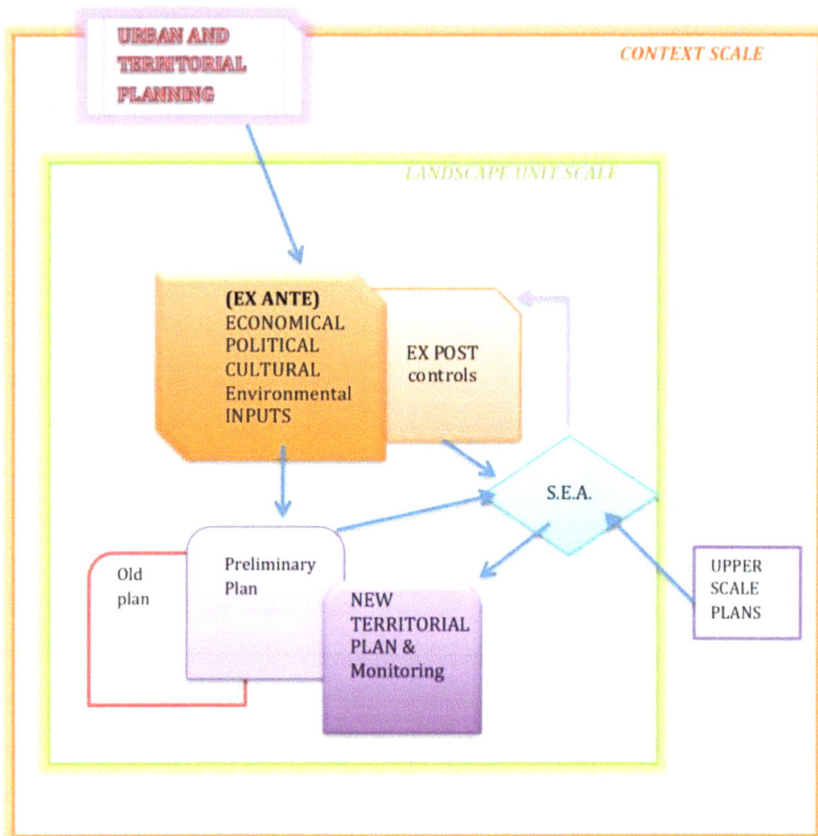

Fig. 12.4 Traditional land use planning scheme. Note the lack of historical investigations and the absence of systemic contributions. Urban planning remains the dominant discipline, even beyond inhabited centers. SEA = VAS. (From Ingegnoli (2015))

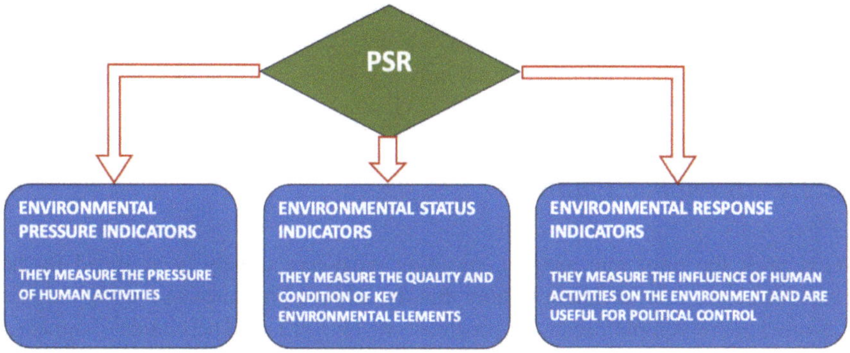

Fig. 12.5 Diagram of the PSR model, the most used today for territorial planning, to arrive at a "multidimensional environmental value"

There is no shortage of references to ecology, but the conventional one, which favors, always in separate sectors, investigations into pollution, rare species, some biotopes to be restricted. Great weight is given to so-called popular participation, but in the sense of use of resources and economic growth and social/recreative objectives or agreement. Historical investigations are usually limited to the state of the previous Plan, plus any in-depth studies only for the monuments to be protected.

Referring to what has been said, for example, are the Territorial Coordination Plans (PTC), which Italian (and European) law imposes for the management of Parks, and the Urban Regulatory Plans, in Lombardy renamed "Territorial Government Plans" (PGT). Today, however, more technologically up-to-date ways to support plan choices are common. The most used model is the "Status/Pressure/Responses" or PSR model (Neri et al., 2016) (Fig. 12.5).

It should be underlined that for Bionomy this model is incorrect. Let's see why:

(i) It does not consider territorial units as dynamic evolutionary systemic entities, as Bionomics claims: in fact, it does not follow the Clinical-Diagnostic methodology, but rather follows a mechanistic approach by applying state indicators "by key elements";
(ii) It considers the landscape only from a historical-cultural and geographical point of view and not from a biological-ecological (i.e. bionomic) point of view.
(iii) It remains predominantly reductionistic, because it considers the components of air, water, soil, geology, and natural factors in a fragmented way.
(iv) It estimates a synthetic index (multidimensional environmental value) obtained with the "Charter of landscape-environmental value", drawn up on sector analysis, in contradiction with the Principle of Emerging Properties, which establishes the impossibility of studying the behavior of a complex system starting from the analysis of its components.
(v) It does not actually consider the dynamics of systemic processes and does not compare the plan choices with the ongoing trends in the state of health of the system as a whole and in relation to the regional context.

The last points (iv) and (v) are particularly critical, as the weight of the socioeconomic issues risks being misrepresented, and they cannot be related to the bionomic-landscape syndromes of the territorial units in a scientifically correct way.

12.2.2 *Bionomic Planning*

As regards the bionomic method, recalling what has already been stated in the previous pages, it is sufficient to note that the scheme presented in the Fig. 12.6 is 2.5 times more complex than the previous one (Fig. 12.4). The clinical-diagnostic

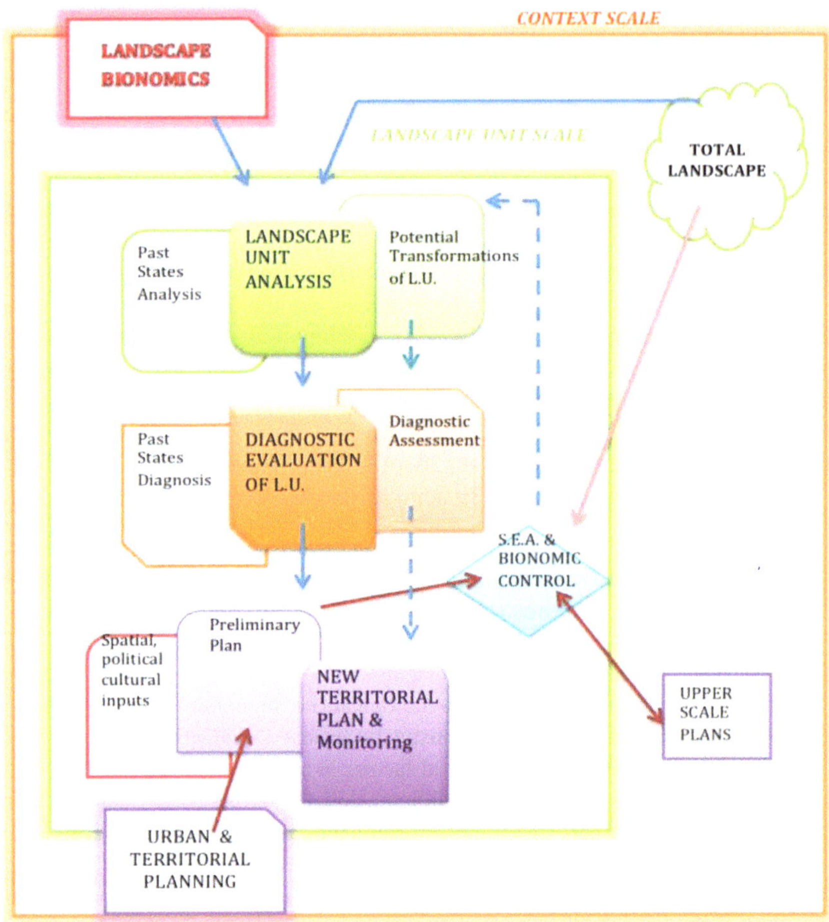

Fig. 12.6 Territorial planning scheme according to a new method, which also involves the discipline of landscape bionomy. Note the greater importance of the context scale, of the historical investigation and above all of the clinical-diagnostic criterion. SEA = VAS. (From Ingegnoli, 2015)

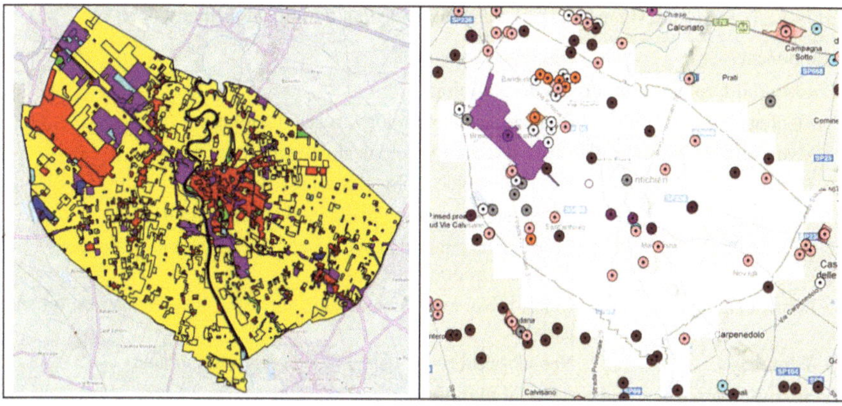

Fig. 12.7 The municipality of Montichiari (Brescia) with its main landscape apparatuses: Red = Residential, Violet = Industrial and subsidiary, Yellow = Agrarian productive. To the right, the localization of pollutant sources, using QCumber program to find the main environmental stressors (Università di Bologna, 2015–16)

criterion requires the intervention of a bionomic expert who establishes any bionomic-landscape syndromes in progress, before thinking about new possible transformations. Furthermore, note the greater importance of the context scale and systemic historical investigation.

Obviously popular participation is necessary here too, but in very different ways, as indicated by the new QCumber type DC-GIS program. These computer programs (Magro et al., 2013, 2014) allow interaction between institutions, businesses, and citizens both in terms of awareness and governance (Fig. 12.7).

QCumber-EnvHealth is a new software tool to quantify urban environmental exposure and health risks under different policy scenarios. It was developed under an Innovate UK collaborative research and development project coordinated by the Italian research society Algebra Srl. that brought together business, universities and the public sector. Decision-making and planning in rapidly growing urban centres require integrated assessment tools to determine impacts due to the exposure to environmental pollution on human health and social inequalities. However, there was a *lack of integrated tools* with which to determine current and future health risks and to evaluate policy options. That is why QCumber is going to be soon completed with the help of Landscape Bionomics principles and methods, becoming an exceptional tool for ecological planning.

An integration of QCumber with bionomic parameters and methodological procedures may help deeply to territorial planning innovations widening health preservation even independently from pollution. This positive kind of AI was presented in the World Congress of IAIA (International Association for Impact Assessment) held in Bologna (Italy) in May 2025, well known as the oldest University of the Western World (Fig. 12.7).

Table 12.4 summarizes the main sectors for a Study of a Territory according to conventional Planning, with the addition of the sectors that Bionomy adds. In the

Table 12.4 Study fields for territorial planning: widening traditional methods

Components	Disciplinary contents
Traditional Ecology and territorial planning	
A- Geology	Geology, geomorphology, rocks, Pedology
B- Hydrology	Water quality, hydrological network, purification
C- Botany	Flora, Geobotany, phytosociology, forests
D- Zoology	Fauna, animal communities, Faunistic sensitivity
E- Agriculture	Crops, companies, cereals, vegetables, irrigation, pesticides, orchards, arboriculture
F- Urban planning	Cities, urban expansions, roads, infrastructures, industries, urban parks,
G- Economy	Neo-classical (work as the value of goods and land)
Landscape bionomics and territorial planning	
1- Anatomical structure of the landscape	Landscape units, Ecotopes, configurations, Ecofabric/Ecotissue, fragmentation, barriers
2- Landscape functions	Human habitat, natural habitat, standard habitat, landscape apparatus, BTC functions, ecological networks,
3- Transformations	Transformation deficit, land use change, landscape interferences,
4- Bionomic state	Strategic evaluation, unit diagnosis, bionomic function (BF), pathology, alterations
5- Human habitat main systems	Green infrastructure, agricultural restoration, urban restructuring, urban ecology,
6- Geo-health/human-health alterations	Premature deaths, landscape-nutrition interferences, ethological alarm, pollution
7- Therapy and planning	Therapeutic criteria, environmental design controls, ecological planning, nature conservation
8- Ecological economy	Ecological economy, ecosystem services, environmental damage, bionomic and economic costs

second part of the table, the studies are extended to disciplinary sectors aimed at understanding the systemic behavior of the landscape in question, also analyzing the relationships between environmental alteration and human health and trying to use economic-ecological principles. The provision of information is much better.

It should also be noted that the bionomic study of the landscape allows an "intelligent design" because it shows the structure, functions, and behaviors of the biological organization that the plan project can follow without too many errors in relation to bio-environmental laws. It is the first rule for any design intervention in a LU: carry out transformations of the environment following its bionomic laws. Today this rarely happens.

We will add Fig. 12.8, which shows the final summary of the study of the municipality of Albairate, in the South Milan Regional Agricultural Park.

This study was finalized to a thesis in agricultural sciences, master's degree, and to theoretical research of application of Landscape Bionomics. Only by applying the principles of Bionomics we were able to give an idea of both the current state and the

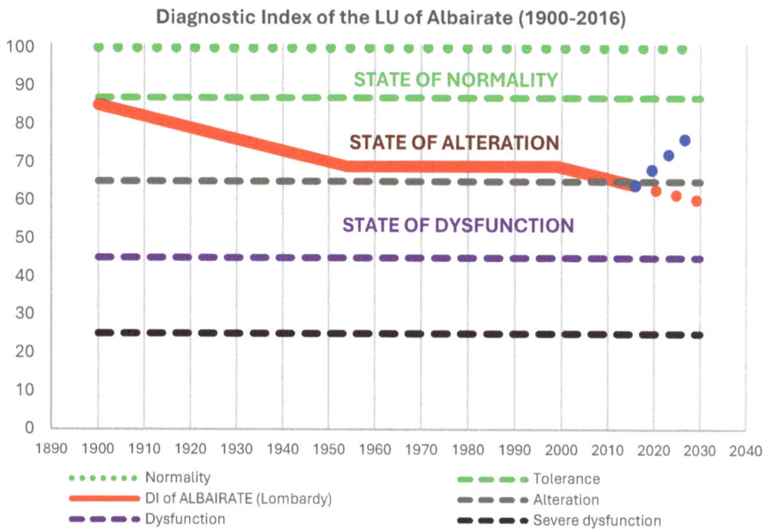

Fig. 12.8 Dynamics of the diagnostic index (DI) of Albairate and current trend (red dotted line). By following Bionomic therapeutic criteria, it is possible to change the ongoing environmental transformation and move towards a state of normality in relatively few years (blue dotted line)

current trend and to contrast (blue) the most correct therapeutic direction to avoid further troubles.

12.3 Planning Criteria

12.3.1 Systemic Planning

Once the systemic analysis and diagnosis have been carried out according to the methodological criteria seen in landscape bionomics, the first phase can be considered complete: "Ecological analysis and diagnosis of the site (site analysis/diagnosis)". Then one can move on to the other phases (Mac Harg, 1969).

- Suitability concerning the intervention objectives (capability),
- Better response (suitability),
- Feasibility choices.

However, be careful: the *intervention* (operation) objectives are *twofold*: (i) eco-bionomic restoration and (ii) socioeconomic use (Fig. 12.9).

According to the principles of Bionomy, the "recovery objectives" consist of the primary and mandatory need to heal the landscapes examined by the syndromes found, through appropriate therapeutic criteria.

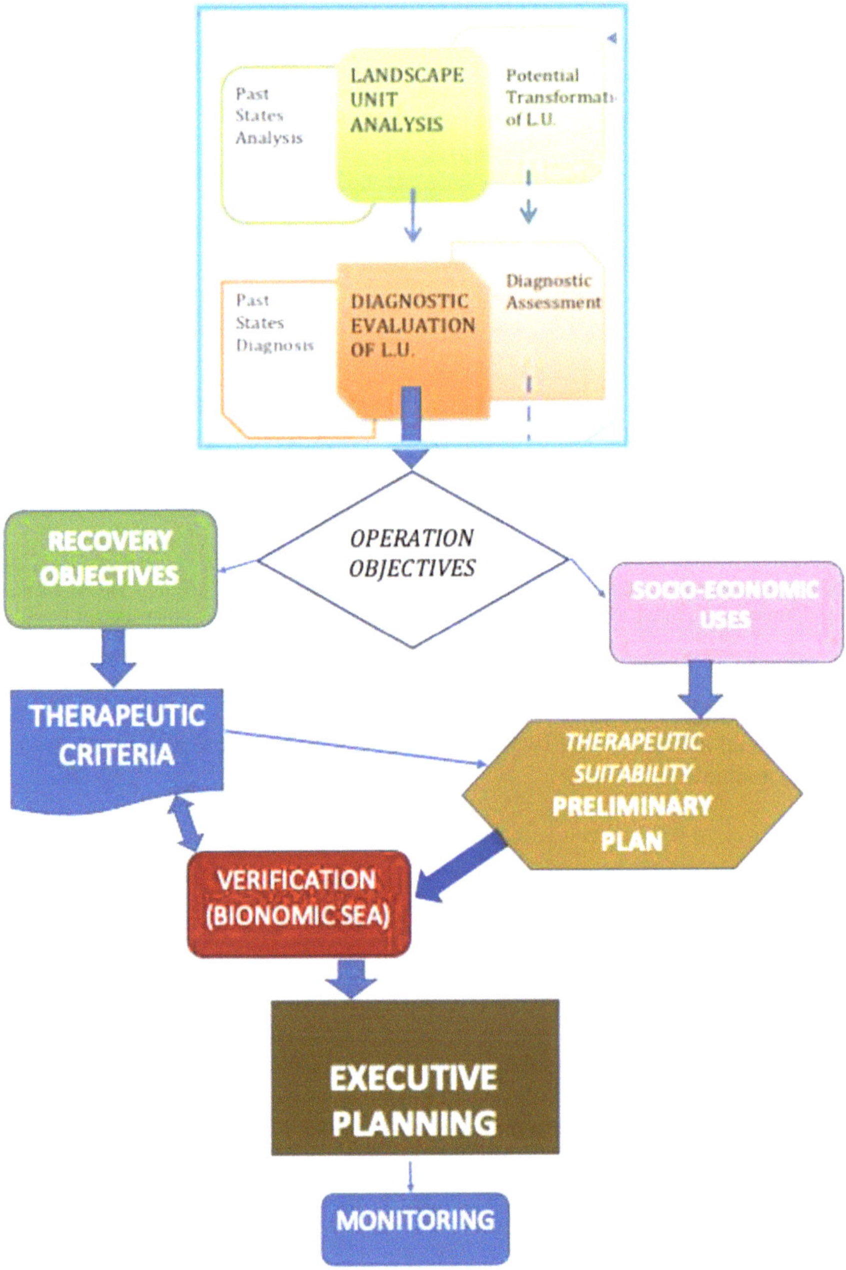

Fig. 12.9 Systemic planning logical scheme. Further information on the planning phases can be found in Figs. 8.16; 8.19; 8.21 (more detailed). After the analysis and diagnosis phases, the intervention objectives are split, maintaining the priorities in the recovery, a sine qua non for the Preliminary Plan, the Verification, and the Executive Plan

Only after the decisions on the redevelopment it will be possible to verify the suitability of the "use objectives" of the territory (health, recreational, road network, tourism, economic, etc.) which constitute the "capability". This is followed by the need to choose the best correspondence (suitability) to the land use objectives, through the preliminary design of the Systemic Plan (e.g., Territorial Plan), to be subjected to checks (always according to the principles of Bionomy). Once the results of the checks have been evaluated (e.g., VAS methods, but always bionomic in nature) we can move on to executive planning (feasibility), which must however also include subsequent actions to monitor the results.

12.3.2 Widening the Strategic Environmental Assessment (SEA)

It should be emphasized that there is no rigid and predefined method of design, which is found in all creative and intentional processes. Like what happens for the construction of models, it is necessary to follow successive phases of approximation, often iterative and with a gestalt method. Here the "Painters' Method" recalled by Lorenz is to be understood almost literally, given that the drawing is the heart of the design and is identified with it.

In concrete terms, the design requires proceeding from the general to particular, organizing the creation and the composition through technical drawings gradually enriched with natural or anthropic thematic parts, to arrive at forming a system capable of responding to the functional needs of the objectives for which one must intervene.

Consequently, one of the most demanding phases is the verification. However, the verifications currently required by the European Union are insufficient, and we must be aware of this.

The *strategic environmental assessment (SEA)* is an environmental governance tool introduced by Directive 2001/42/EC. It constitutes "an important tool for integrating environmental considerations into the preparation and adoption of certain plans and programs which may have significant effects on the environment of the Member States". The methods of carrying it out (art. 11) are the following: (1) verification of subjection; (2) preparation of the environmental report; (3) carrying out consultations; (4) evaluation of the environmental report and results of the consultations; (5) decision; (6) on the decision; (7) possible monitoring.

To better understand how things stand, we report what was established by the Autonomous Province of Trento. In response to Europe's request, the Province of Trento published: "Guidelines for Territorial Planning Strategies and for the Strategic Evaluation of Plans (P.U.P., Annex 2)". The vision that underlies the Trentino document on SEA shows four main strategies: identity, sustainability, integration, competitiveness. The first two also concern the environment, the others are socioeconomic:

- identity = strengthen the recognizability of the territorial offer of Trentino, enhancing its landscape diversity, environmental quality and cultural specificity.
- sustainability= orient the use of the territory towards sustainable development, containing the processes of land consumption and primary resources and promoting urban and territorial redevelopment.
- integration = consolidate the integration of Trentino in the European context, effectively inserting it into the large infrastructural, environmental, economic, and socio-cultural networks.
- competitiveness = strengthen local capacities for self-organization and competitiveness and the opportunities for lasting development of the overall provincial system.

The PUP requires for the SEA the comparison between *ex ante* and *ex post* data and for some evaluations also an ongoing phase; however, it does not require any evaluation understood in a fully historical-evolutionary and clinical-diagnostic sense. On the contrary, it always requires comparison with the PTC, that is with a pre-established planning of a larger scale which however necessarily cannot see any critical ecological-landscape problems, existing or potential, at a local scale that could largely undermine it. The plan is not seen as a means for therapeutic interventions that concern a complex biosystem, but only as an urban planning tool (Table 12.5).

This table compares the summary of the evaluation procedures of the VAS/SEA planning according to the Urban Reporting of the Province of Trento (Prov. Trento, 2010) Vs. the Bionomic Evaluation. It is immediately noticeable that the differences are substantial already from the objectives (Main Strategic Aims of SEA). Subsequently, the methodologies of Assessment (evaluation), Analysis (analysis), Rating (indicators) and Control (control) are even incomparable: what has been said in the comparison between analysis/diagnosis is accentuated in this phase of Evaluation.

12.3.3 Executive Planning and Monitoring

Planning can move to an executive phase when the Preliminary Plan verification is completed, and the necessary adjustments have been made.

Integrating a Landscape Unit (LU) in good ecological status with the basic anthropic requests involves a systemic control of productivity, biodiversity, soil and water about the problems of food, water, health, housing, subsidiary energy and culture (Forman, 1995). As we know, the challenge requires appropriate methods and measures (Ingegnoli, 2011, 2015), such as the eco-bionomic evaluation of vegetation, the BTC (biological capacity of the territory), the SH (standard habitat per capita), the BFF (biotopflächenfaktor, sensu Ermer (Ermer et al., 1996)), the (landscape diversity), the ecological functionality indices of internal waters, the geomorphological characterization, etc. These quantities and methods have already been conceived as integrating multiple factors.

Table 12.5 Comparison between the Strategic Environmental Assessment (SEA) of the Province of Trento (purple) and the Landscape Bionomic one (green)

A- MAIN STRATEGIC AIMS OF SEA	
IDENTITY, enhancing the eco-tourism; SUSTAINABILITY, sustainable development of resources; INTEGRATION, in the European context of infrastructure; COMPETITIVENESS, productive development	HEALTH, of landscape systems and of man; MEMORY, natural history and cultural heritage; REHABILITATION, sustainability of the system and therapeutic indications; CONTROL of RESOURCES, systemic, by landscape systems
B- SEA METODOLOGICAL CRITERIA	
URBAN REPORTING Assessment of consistency with the higher-level planning instruments *Analysis*: indicators, economic and environmental planning and simple decision-making models (e.g. SWOT) *Rating*: action to achieve the strategies are evaluated through impacts by thematic area. The assessment is qualitative, for compliance and indeterminacy. *Control*: Planning, with simple methods of descriptive and effectiveness with urban-economic indicators. The following mitigation measures, compensation and monitoring.	BIONOMICS EVALUATION Assessment of the state of health of the landscape units (LU) *Analysis*: indicators of landscape and ecological models bionomic for complex subsystems *Rating*: structure and functions of each LU are valued through a clinical-diagnostic in relation to dynamic states of normal (3 times). Qualitative and quantitative assessment *Control*: trends with and without plan, by means of models of bionomics and ecological indicators of landscape or ecological-economic factors. The following measures of specific therapy and systemic.

C- SEA EVALUATION CRITERIA	
Following the territorial Planning rules of the Province of Trento	*Following the Bionomic Principles*
URBAN PLANNING REPORTING Verification of Congruence with the superordinate urban planning tools.	BIOMIC ASSESSMENT Assessment of the health status of Landscape Units (LU)
Analysis: Urban and economic-environmental indicators and simplistic decision-making models (e.g., SWOT)	Analysis: ecological-landscape indicators and bionomic models, for complex subsystems.
Evaluation: Actions to achieve the strategies are evaluated through impacts by thematic sector. The evaluation is qualitative, for compliance and indeterminacy.	Evaluation: structure and functions of each LU evaluated in a clinical-diagnostic way in relation to dynamic states of normality (3 times). The evaluation is qualitative and quantitative.
Planning Control: simplified descriptive and effectiveness methods with urban-economic indicators. Mitigation, compensation, and monitoring measures follow.	Planning Control: trends with and without a plan, through bionomic models and ecological-landscape indicators, and partly ecological-economic. Followed by mitigation, compensation, and monitoring measures attentive to the effects on health.
D – TRANSMISSION OF RESULTS	
Summary of consultations and technical and informative synthesis.	Summary of technical-scientific and management consultations and scientific-economic popular synthesis

This allows to quantify and evaluate the transformations of the suburban landscape, of the urbanized environment, to predict the sizing of urban green systems, to guide the expansion of new neighborhoods, to restore agricultural landscapes by applying principles and indicators of landscape ecology and bionomics according to a clinical-diagnostic method. However, this method produces significant changes compared to the concepts of traditional urban planning. Let's try to summarize the major ones:

1. *territory: once defined as a geographical area delimited with a political-administrative intent, it must instead be understood as a local bio-geographic system composed of natural and anthropic elements.*
2. *landscape: defined in planning as a system of territorial assets worthy of protection, with a clear visual and cultural meaning, as repeatedly underlined it must instead be understood as a level of biological organization formed by an interacting system of ecocenotopes* (Forman, 2014)
3. *planning criteria: traditionally technical-urban and socio-economic, mainly for zoning, they must instead be set in an ecological sense, for ecotissue subsystems.*
4. *evaluation criteria: up to now for separate themes, followed by overlapping, scaling up and cluster analysis, they must instead be moved towards an evaluation for integrated systems, with clinical-diagnostic criteria and eco-bionomic controls.*
5. *transformations of an area: it is no longer sufficient that they are locally sustainable, they must be capable of ecologically improving the context (rehabilitation);*
6. *agricultural landscapes today still understood with principles of industrialization, must be rehabilitated according to the principles of bionomics, considering natural infrastructures.*
7. *urban greenery: today ecologically banal, with aesthetic design and many exotic species, it must be designed for vegetation systems and according to principles of landscape bionomics* (Ingegnoli, 2019b)
8. *strategic environmental assessment: the economic-environmental control for thematic sectors in which the concept of impact prevails must be replaced with systemic assessment for landscape systems and diagnosis of eco-landscape syndromes.*
9. *parameters: today almost exclusively based on economic-urban standards, they must be changed with ecological indicators of landscape bionomics.*

Other principles of landscape bionomics are useful in land use planning, such as:

(i) structural congruence: transformations should not erase spatial relationships between geomorphology and vegetation, hydrology and agriculture, old settlements, and landscape grain. Design criteria should instead follow signs due to history, both natural and anthropic.
(ii) patch shape: to support key landscape functions, patch shape should have a large core area, with some curvilinear lobes and orientation according to the main energy flow (Forman, 1995);
(iii) aggregation of functions: bring together land uses while maintaining small patches and natural corridors and at least one large patch of forest (Forman & Godron, 1986);
(iv) local compensation: environmental compensation for urban transformations should be designed following eco-landscape principles and methods but within the same landscape unit, not elsewhere.

(v) basic pattern: protection and renaturation of the river network, ecological networks and forest patches must be pursued even in the most anthropized territories.
(vi) complementarity: the basic law of ecology (not too much, not too little, just enough) underlines the need to avoid any excess or deficiency, such as e.g., a minimum of HN even in areas of prevalent HU, complementarity also indicated by the ecotissue model.

The ecological network may require interventions to relocate in areas of anthropized landscape (agricultural, urban, or suburban). In such circumstances, it is necessary to provide forms of equalization with appropriate incentives to improve the environment in concrete terms. It is immediately evident that the territory in question must be subjected to various controls, therefore, serious monitoring, starting from the enforceability of the Plan. The elaborations that have allowed us to arrive at the diagnoses (phase I) also serve this purpose.

References

Bjerg, O. (2016). *Parallax of growth, the philosophy of ecology and economy*. Polity Press.
Cerea, A. P., & Re Cecconi, F. (2010). *Stima Parametrica del Costo di Costruzione Individuazione di un metodo di stima, in fase di progettazione preliminare, del costo di costruzione tramite l'uso delle regressioni lineari*. Tesi Magistrale al Poiitecnico di Milano.
Daly, H. E. (1999). Uneconomic growth and the built environment: In theory and in fact. In C. J. Kibert (Ed.), *Reshaping the builtenvironment* (pp. 73–86). Island Press.
Daly, H. E. (2010). *Ecological economics: Principles and applications*. Island Press Books.
Ermer, K., Hoff, R., & Mohrmann, R. (1996). *Landschaftsplanung in der Stadt*. Ulmer.
Forman, R. T. T. (2014). *Urban ecology, science of cities*. Cambridge University Press.
Forman, R. T. T. (1995). *Land mosaics: The ecology of landscapes and regions*. Cambridge University Press.
Forman, R. T. T., & Godron, M. (1986). *Landscape ecology* (p. XIX+619). John Wiley & Sons.
Ingegnoli, V., & Giglio, E. (2005). *Ecologia del Paesaggio: Manuale per conservare, gestire e pianificare l'ambiente* (p. 685+XVI). Sistemi Editoriali SE.
Ingegnoli, V. (2002). *Landscape ecology: A widening foundation* (pp. XXIII–357). Springer.
Ingegnoli, V. (2011). *Bionomia del paesaggio. L'ecologia del paesaggio biologico-integrata per la formazione di un "medico" dei sistemi ecologici* (p. XX+340). Springer.
Ingegnoli, V. (2015). *Landscape bionomics. Biological-integrated landscape ecology* (p. XXIV + 431). Springer.
Ingegnoli, V. (2019a). Infrastrutture ecologiche e Diagnosi dell'ambiente. In Bonizzi, Campana, & Cordini (Eds.), *Il Governo dei Parchi* (pp. 173–214). Aracne Ed.
Ingegnoli, V. (2019b). Protezione e valorizzazione della vegetazione. In Bonizzi, Campana, & Cordini (Eds.), *Il Governo dei Parchi* (pp. 273–305). Aracne Ed.
Mac Harg, I. (1969). *Design with nature*. Natural History Press.
Magro, G., et al. (2014). Worldwide social for participation in 2.0 planning. In *Annual conference IAIA 2014, impact assessment for social and economic development*. Vina del mar. 8-11/04/14.
Magro, G., et al. (2013). Q-cumber: The environmental-social network for IA. In *Annual conference IAIA2013, impact assessment: The next generation*. 13-16/05/13.
Millennium Ecosystem Assessment. (2005). *Ecosystems and human well-being: The assessment series*. (4 vol + Summary). Island Press.

Neri, C., Dupin, P., & Sanchez, L. E. (2016). *A pressure–state–response approach to cumulative impact assessment*. Elsevier J. of Cleaner Production.
Provincia Autonoma di Trento (2010) Valutazione ambientale strategica dei Piani urbanistici - VAS. Assessorato all'Urbanistica.
Università di Bologna. (2015–16). *Analisi mediante tecniche di Ecologia del Paesaggio di situazioni di degrado ambientale: il caso Montichiari*. Tesi di Laurea di Daniel Rossi, coordinata dai prof. V. Ingegnoli, G. Magro e M. Sumini in Alma Mater Studiorum.

Chapter 13
Urban Parks and Health Protection

Contents

Abstract An introduction on the items concerning the concept and evaluation of the Urban Green through the systemic approach given by Landscape Bionomics is provide at the beginning of the chapter, which enlightens the quantitative values of some ecological parameters too. We deal with the problem of the correct localization, design and planning in the territory also, in relation to the Principle of Emergent Properties.

The same relations are tackled concerning ten masterplan models, in the research of their ecological congruence with the new quarter in Milan. Each model was composed by six landscape elements, having the same standard land cover but different mosaics. After a landscape bionomics analysis, the best model was chosen for the development of the executive project. As predicted by the deterministic chaos, a small difference in the starting parameter of two projects can led to a sharp difference in the behavior of the Parks: a wrong choice means a delate of about four decennia for ecological benefits.

The possibility of the restoration of the environment of a Region trough the introduction of a *compensative ecotope*, for instance a park area in an urban

V. Ingegnoli, *Landscape Bionomics and Planetary Health*, Landscape Series 21,
https://doi.org/10.1007/978-3-032-08908-3_13

landscape is shown subsequently: an example concerning the Chiaravalle Abbeys Park (South of Milan), with a semi-natural park ecotope is provided. Today, being the average BTC of the Abbeys Park = 1.10 Mcal/m^2/year, the influence range of the park is very short (Ecological Radius ER = 0.75 Km), so that only the peripheral margin of the town can benefit from this Park. If the design of the new seminatural park could reach an average BTC = 2.25 Mcal/m^2/year, ER = 3.60 Km, its influence should arrive at the margin of the central City.

Finally, a study is proposed that intends to verify the opportunity of a plan of reforestation in an urbanized territory to improve our health. This aim needs: (a) to follow the discipline of Landscape Bionomics, (b) to apply a risk analysis concerning zoonosis and premature death, and form function design, and (c) to improve human habitat by adding semi-natural parks. The Study Area is presented: the Province of Monza-Brianza linked to the City of Milan. The main principles of Landscape Bionomics (LB) are shortly presented with some of LB state functions and diagnostic models. If we increase forests from 7.46% to 8.20% (+393.5 ha) we go from 296 cases of zoonosis/year to 325. Even if Mortality Rate should grow near to 3–4% (e.g., antibiotic resistance) the increase of death/year due to reforestation will remain near 570 in the next 40 years. But reforestation at the same time may reduce premature death by about 2150 units, of course if carried out according to the criteria reported.

Keywords Masterplan · Urban Park · Urban green · BTC of the Quarter · Regional Park South of Milan · Chiaravalle Abbey · Environmental compensation · Green Belt · Form-function relationships · Deep vein thrombosis

13.1 Evaluation of the Urban Green

13.1.1 Urban Green and Bionomic Parameters

Let us try to summarize some of the most common considerations that are still made today regarding urban greenery. This concerns the backwardness of the design and control criteria of urban green systems, such as:

1. the green surface per inhabitant (m^2/inhabitant),
2. the formal criteria judged on an aesthetic-visual level,
3. the diversity of floristic species per park (Species/Park),
4. the "green belts" in the suburbs…
5. plant trees (!) in the sense of measuring the number of plants per inhabitant…

These are environmentalist banalities that are often counterproductive and that, in any case, have nothing to do with the scientific considerations proposed by Landscape Bionomics.

Table 13.1 shows some of the main bionomic parameters concerning this item. For each of the ten types of urban greenery, the table provides the average values of the following parameters: (a) Built or paved components (Built, %), (b) Non-native

Table 13.1 Ecological parameters characterizing main types of urban green

Types of urban green		Ecological parameters					
		Built (%)	Allocthonous plants (%)	Soil over built (%)	Tree cover (%)	Permeability BFF	BTC Mcal/ m^2/yr
1	Sport green	65	35	50	10–20	0.30	0.8–1.5
2	Road green	30	30	50	5–20	0.50	0.4–1.5
3	Square green	75	40	60	10–20	0.40	0.4–2.0
4	Vegetable garden	15	20	20	0–10	0.85	0.9–1.3
5	Remnant field	5	15	–	0–10	0.95	0.7–1.0
6	Private garden	20	35	40	20–45	0.80	1.1–3.0
7	Public garden	30	30	50	25–35	0.70	1.3–2.0
8	Urban park	10	20	15	30–40	0.85	2.0–3.0
9	Park with natural core green	2	5	–	50–70	0.95	3.5–4.5
10	Urban forest biotope	0,5	–	–	>60	1	>6.00

BFF index for soil permeability (Ermer, Hoff, Mohman [4])

Table 13.2 Example of Land Cover distinctions following Landscape Bionomics suggestions

Land cover	Ecotope	Condition	Age	HH	BTC
Forestry	Deciduous	Natural	100–150	0–4.0	7.0–10.5
		Coltivated	50–100	5.0–10-0	4.5–7.5
		Young or coppice	0–50	5.0–15.0	2.5–4.5
	Conifers	Natural	90–140	0–4.0	7.0–10.0
		Coltivated	30–90	5.0–10-0	4.5–7.5
		Young or coppice	0–30	5.0–15.0	2.5–4.5
	Mixed Decidous-	Natural	90–140	0–4.0	7.0–10.0
	Conifers	Coltivated	40–90	5.0–10-0	4.5–7.5
		Young or coppice	0–40	5.0–15.0	2.5–4.5
Urban	Residential	Historical	100–600	80–95	0.20–0.40
		Expansion	20–100	50–90	0.30–0.60
	Subsidiary	Historical	100–400	85–95	0.05–0.20
		Expansion	20–100	80–90	0,10–0.30
	Res-subs Mixed	Historical	100–500	85–95	0.10–0.40
		Expansion	20–100	80–90	0.10–0.50
	Monumental	Institutional	50–2000	85–95	0.20–1.80
		Sacred	60–2000	80–90	0.20–2.00

HH [%]; BTC [Mcal/m^2/year]

plants (% total species), (c) soil reported on the buildings (% of urban green surface), (d) tree cover (% as previous), (e) degree of soil permeability (from BFF, biotope flack factor), (f) territorial biopotential of vegetation (BTC, Mcal/m^2/year).

Table 13.2 add a comparison among different forestry vegetation types concerning the Center-European Ecological Region and that of urban green, focusing on their development over time and the types of management.

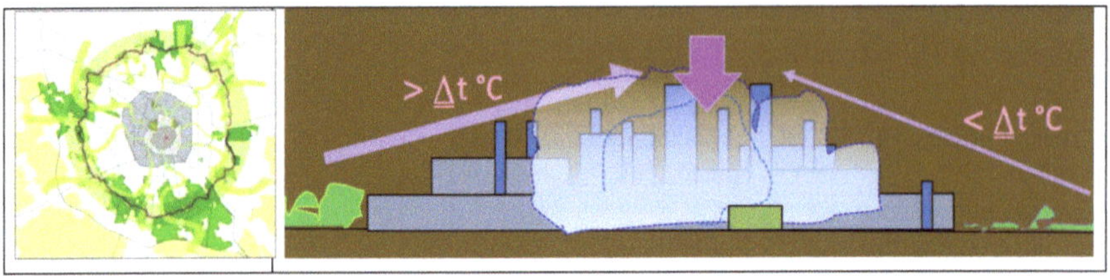

Fig. 13.1 Note that the increase in green parks around the city increases ΔT°C and therefore the transport and deposition of fine dust into the city center. The UHI effect widens this process, which becomes very dangerous for human health

13.1.2 "Green Belt": Incorrect Distribution of Urban Green Space

Creating a high BTC urban park should have notable effects on the health state of a town and allowed the abandonment of plans like the Milan's Green Plan, based on the model of the circular Green Belt, like the London's Green Belt of the 1950s. This model is wrong (Fig. 13.1); as already highlighted by Sukopp and Hejny (1990), the increase in green parks arranged in a circle around the city amplifies the thermal difference (ΔT °C) between the center and the suburbs. It increases the transport and deposition of fine dust into the city center. The UHI (Urban Heat Isle) effect widens the process, which becomes very dangerous for human health.

We recall that in 2008, a study was published entitled Exposure to Particulate Air Pollution and Risk of Deep Vein Thrombosis, mainly based on the Milanese hinterland (Baccarelli et al., 2008), which underlined a clear relationship between the risk of deep vein thrombosis and the particulate matter level <10 μm (PM10) in the year before diagnosis. This process starts at PM10 = 12 $\mu m/m^3$/year, while the EU legal limit is PM10 = 40 $\mu m/m^3$/year. This dangerous process results from the reaction of macrophages that release many pro-inflammatory cytokines, particularly IL-6, which can promote blood clotting.

Despite being much older (mid-nineteenth century), an insertion of Urban Greenery such as that of Central Park in New York (Olmsted, 1870) has greater ecological value, which divides the conurbation bringing an ecological radius effect (see Sect. 7.4.6) much more extensive and effective.

13.1.3 Buildings Compatible with Ecological Principles

Here, we give some hints on the complex problem of developing residential building projects considering the principles of landscape bionomy. Soon, we should make every effort to improve the urban living environment through these interventions too. It is, therefore, important to realize that it is easy to make a mistake, i.e., by

Fig. 13.2 Comparison between tower buildings with advanced technological greening and buildings of another concept, capable of greater integration between ecotopes and much more significant BTC. (Des. by Ingegnoli)

designing tower buildings with green terraces full of shrubs and small trees that may be attractive to look at but of little ecological-bionomic importance. The criticism of these buildings presents a cost/benefit analysis of the following type.

Cost/benefit analysis of a green tower:

1. Abnormal and decidedly high energy consumption.
2. Accumulation point of plants that instead require wide distribution in urban ecotopes.
3. Green spots that are too small and not connected to each other.
4. No direct connection with the surrounding ecotissue.
5. High economic maintenance costs.
6. Aesthetic improvement.
7. Partial protection from heat and fine dust.

It is immediately clear that two attributes are positive, but the other five are not; therefore, it is essential to propose other projects. A useful comparison to understand the difference between green towers and other architectural typologies is shown in Fig. 13.2.

13.2 Form–Function in the Design of an Urban Quarter

13.2.1 Bionomics Methodology for a Masterplan

In wide metropolitan areas, sensu Forman (2008, 2014), the presence of a consistent ecological network is indispensable. The bionomic efficiency of these networks depends on the presence of some "green corridors", linking the central town with rural landscapes, better with a natural park area. In the case of Milan, the *only*

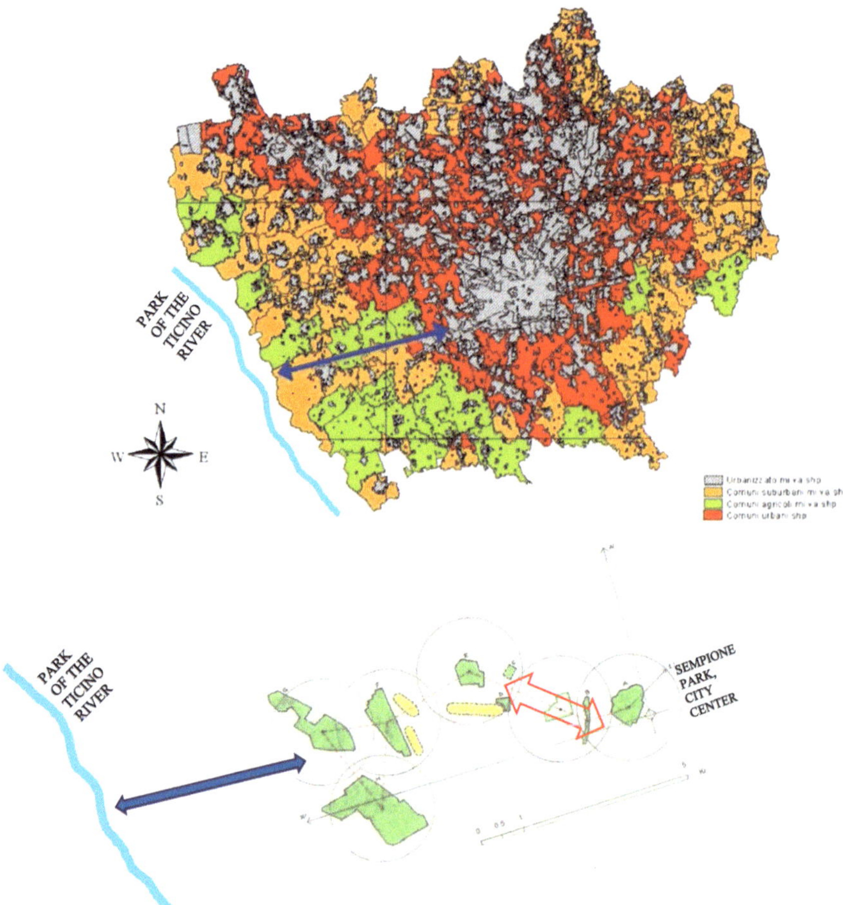

Fig. 13.3 (Above) The metropolitan area of Milan (red and grey) and the green corridor towards the Ticino river (20 km, blue arrow) within the rural landscape nearest to the center of the city (yellow = suburban-rural municipalities; green = agricultural municipalities); (Below) The red arrow indicates the strategic role within the city that the new Cytilife Park should have played in reconnecting the ecological network towards the Ticino Park (blue arrows)

corridor available to give this function is a rural belt from the west side of the city towards the Ticino River (about 20 km far away) and its large Regional Park.

As we can see in Fig. 13.3, creating an ecological network between the central park of Milan (Simplon Park) and the Regional Park of Ticino should be very important. The red arrow in the figure indicates the strategic role within the city that the new Cytilife Park should have played in reconnecting the ecological network towards the Ticino Park, if it had been properly designed and built according to the principles of Landscape Bionomics. Figure 13.4 exposes a transect along the same direction: the urban heath isle (UHI) and the insufficient ecological role of the present urban parks are evident, having a biological territorial capacity BTC

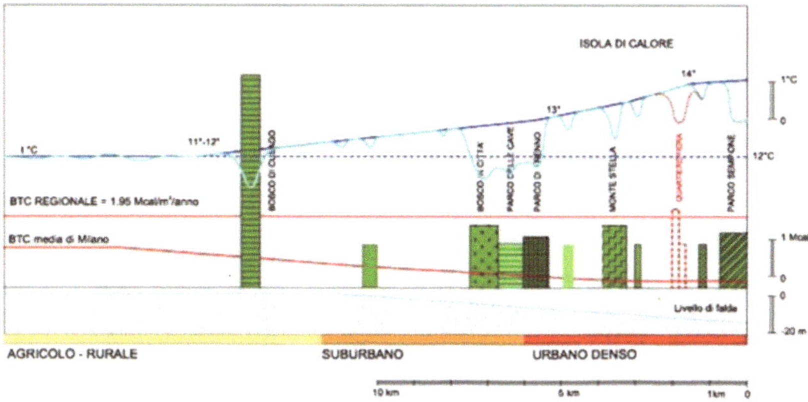

Fig. 13.4 A transect from the regional park of Ticino River to the city of Milan (right). Note the UHI, amplifying the annual average temperature from 12 to 14 °C, and the insufficient BTC of the urban parks remaining below the average BTC of the Lombardy Region (Ingegnoli, 2011)

<1.95–2.00 Mcal/m^2/year, the average regional value, consequently too scarce possibilities to balance the urban desert. The comparison with the remnant forest of Cusago (BTC = 7.67 Mcal/m^2/year) is symptomatic.

To activate the mentioned network, we should have need to restore the vegetation of the urban parks and to insert a new element (e.g., an urban district/quarter) with a good BTC level. The occasion should have started in 2004, with the architectural international competition for the building of a new quarter on the place of the old "*Fiera di Milano*" (35 ha), win by an international group coordinated by Daniel Libeskind (New York) for the company "CityLife". In the mentioned group of Libeskind, Ingegnoli was charged for the ecological aspect of the project (Consigliere, 2004; Ingegnoli, 2011, 2015). To understand the ecological state of Milan, indispensable frame to the correct design of the quarter, see Figs. 8.18 and 8.20. From the figures it can be immediately derived that the most important problem for Milan is the insufficient number of protective elements and the too low level of its vegetation biological territorial capacity (average BTC = 0.52, being the optimum 0.75 Mcal/m^2/year).

The general ecological situation confirms the necessity to insert in the project of the new urban quarter a good park area. That is why in 2006–2007 CityLife charged the Department of Biology of the University of Milan of a research related to the park area of the project. This research was coordinated by the Author (Order of Architects, 2009) and testified that a complex adaptive system (as a landscape unit) is strongly dependent on its configuration even if the components remain of the same extension and with the same function. To evaluate the property of the form is not simple because the visual analysis is generally insufficient.

The target of the research was the choice of the best model (from an advanced ecological point of view) among diverse master-plan scenarios. Each masterplan model reproduces a rectangular area of 25.95 ha, containing: 7.5 ha of residential areas (30% of which green), 5.2 ha of service buildings, 3.25 ha of squares and free

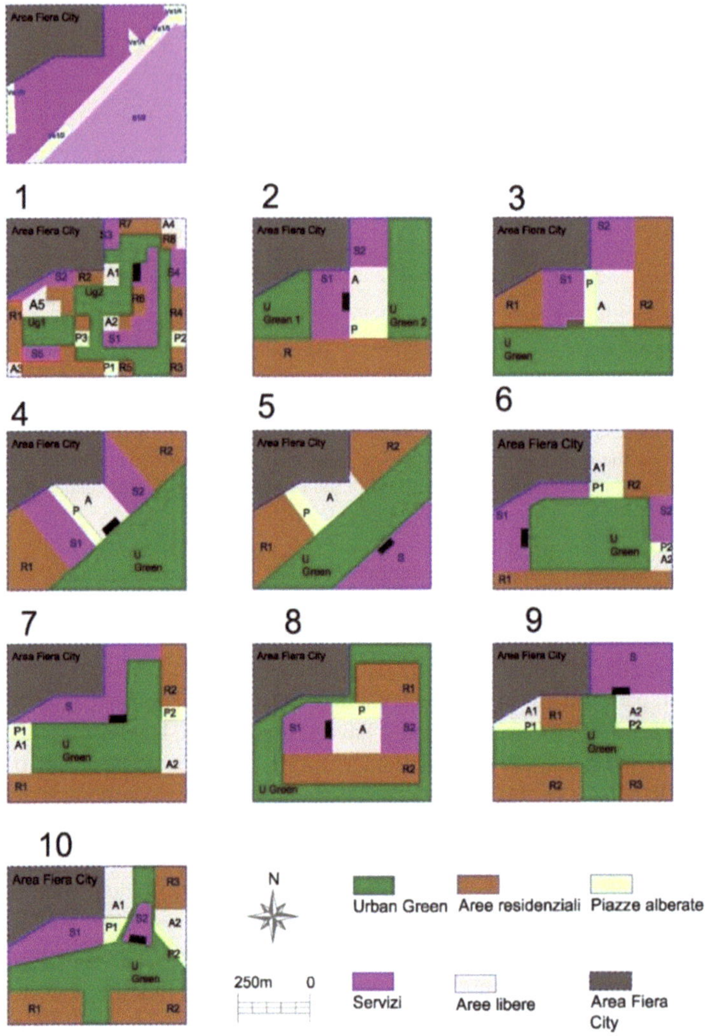

Fig. 13.5 Synthetic diagram of the current state (situation 2004, at the top left) and of the better ten models considered in the research on the ecological congruence of the form for the masterplan of the new quarter. (dark green = Urban Green; light green = tree-lined square; brown = residential with gardens; violet = services and sheds; grey = free areas)

areas, 10 ha of park, the same standard land cover but different mosaics. It was a settlement compatible with the Urban Master Plan of Milan, which prescribes for this area "50–50 % built and open spaces-green". Figure 13.5 shows the synthetic diagram of the status ex ante or Current State (CS), (situation 2004 at the top left) and of the better ten models considered in the research on the ecological congruence of the form for the masterplan of the new quarter. Each model is composed by six

Table 13.3 Vegetation types related to master-plan models

N°	Green types	Composition	%	BTC
1.1	Edge belts [BTC = 1.30]	Meadows or herbaceous formations, both lawn from foot traffic and semi-natural *Arrhenatherion* (about 50 %)	45	0.50
		Patches with trees, only partially exotics, with 50 % of *Quercion robori-petreae* phytocoenosis	25	4.00
		Road green, sport green, etc.	30	0.05
1.2	Core areas [BTC = 2.00]	Meadows or herbaceous formations, both lawn from foot traffic and semi-natural *Arrhenatherion* (about 65%)	45	0.55
		Patches with trees, only partially exotics with 75% of *Quercion robori-petreae* phytocoenosis	20	4.20
		Forest biotopes (e.g. *Populion albae* or *Alnion glutinosae* and *Carpinion* or *Quercion robori-petreae* phytocoenosis)	15	6.00
		Tree lined roads, services, playground and/or life paths	20	0.05
2	Square green [BTC = 1.30]	Patches of trees even with exotic plants (not more than 70%)	35	3.50
		Lawn from foot traffic	40	0.50
		Road green, sport green, etc.	25	0.05
3	Residence green [BTC = 1.40][a]	Traditional garden green, but also about 25 % of *Quercion robori-petreae*	30	4.0
		Lawn from foot traffic	35	0.50
		Road green, sport green, etc.	35	0.05

[a]The residential areas (green = 30%) will have a BTC = 0.40 Mcal/m^2/year

landscape elements with 48.9% of built and 51.1% of open spaces and green (of which 41.6% closely to green); counting the residential green the percentages became: 40.3% built and 59.7% in open spaces and green. The tolerances were of the order of 1–1.5%, for obvious reasons of design of the different configurations.

The models were first designed in number of ten, to which other five were then added. They included the main possible configurations for the area of the new quarter, according to the criteria drawn in Fig. 13.5. Counting the green residential, the percentage of the green area (without open spaces) resulted equal to 50.2, so the photosynthetic area reached about 13.3 ha. In all models, it was meant that the Green should be articulated in the manner summarized in Table 13.3.

The following types of urban green areas were considered: (i) bands margin, consisting of meadows and wooded spots (with the presence of avenues), for which we can estimate an average of BTC = 1.30 Mcal/m^2/year; (ii) core areas, formed by more natural meadows, wooded spots with high coverage and a part in forest biotope, wherein the equipment are more contained, for which we can estimate a BTC = 2.00; (iii) the green parts of squares, which are similar to the values of the group (i), even if they admit some exotic species; (iv) the green residences, with more parts equipped, and BTC = 1.40 Mcal/m^2/year.

The parameters used for the diagnostic evaluation of the models were obviously divided into two groups on (A) the grounds of the neighborhood (Qu), (B) the grounds of the urban operative landscape units (LU), defined based on the principles of landscape bionomics. The area of the former Trade Fair, which already historically corresponded to the military exercise area of the late nineteenth century, was considered with all the blocks surrounding it, to the geometric grid of urbanization next. The normal ranges for each parameter were also set out. The results will be presented in subsequent cards.

13.2.2 New Quarter Models and Evaluation

Let us now review the main case studies, showing maps of the models within their associated Land tables with the findings of the parameters and evaluations. We'll start with the situation (in 2004) before the opening of the construction site (Fig. 13.6). As you can see the urban LU is dominated by the sheds of the former Trade Fair and the evaluation of the status ex ante (SA) model is very low: the diagnostic index DI is equal to 0.162 for the parameters of the quarter and to 0.166 for those of LU. The BTC of LU is approximately 0.35 Mcal/m^2/year, lower than the average of Milan city.

Index of Diagnostic Evaluation and Classification of Results The method is based on the percentage differences from normality of each parameter found in the different models, as described in Sect. 7.4.4. To get a report that you can legitimately call Index of Diagnostics Evaluation (DI) the values of "score" (Y) must be added and the result must be divided by the theoretical value given by the maximum

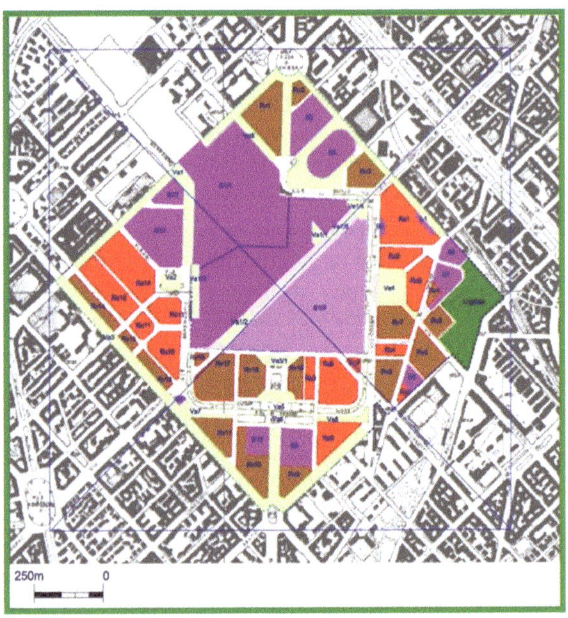

Fig. 13.6 Landscape units (LU) of the former district of "Fiera Milano", the state ex ante (SA). The evaluation of the SA model is obviously very low: equal to 0.162 for the parameters of the quarter and to 0.166 for those of LU. The BTC of LU is approximately 0.35 Mcal/m^2/year (dark green = Urban Green; light green = tree-lined square; brown = residential with gardens; red = residential without gardens; violet = services and sheds; grey = free areas)

Table 13.4 Evaluation of model M4

Ecological parameters	Normal	M4	Δ4	Score
Fractal dimension of green (D)	1.5–1.8	1.22	18.80	0.75
Connection of green (α + γ)	0.75–1.15	0.71	5.30	1.65
Core areas (%V)	45–75	43.39	3.60	1.76
Ratio A/P green	30–60	65.76	9.60	1.36
Contacts of green with the outside (%pV)	20–60	53.69	0.00	2.00
Built area/perimeter V (m^2/m)	45–90	131.78	46.40	0.45
Structural diversity (ψ)	3.6–9.1	6.13	0.00	2.00
Climatic balance (ΔT °C)	0–0.75	1.22	62.70	0.31
Neighbourhood BTC (Mcal/m^2/a)	0.75–0.9	0.79	0.00	2.00
Park BTC (30 year)	1.6–1.9	1.59	0.60	1.96
Scores			*147.00*	*14.23*
Diagnostic index of the neighbourhood	**0.85–1.0**			**0.711**
Fractal dimension of green (D) LU	1.45–1.75	1.17	22	0.647
Connection of green (α + γ) LU	0.5–0.9	0.31	38	0.514
BTC of LU	0.6–0.8	0.494	17.6	0.826
Radius of ecological influence (min)	110–450	103.46	5.95	1.600
Source/sink (% transects)	25–75	13.85	44.6	0.459
Garden grain/general grain	4.0–6.0	4.11	0	2.00
Scores				*6.046*
Diagnostic index of Landscape Unit	**0.75–0.90**			**0.503**

number of ecological parameters multiplied by 2. In our case we have one DI for the quarter/neighborhood (QDI), one for the LU (UDI), plus the more general one (DI gen) measured on all the 16 ecological parameters. DI gen = Y/32, UDI = Y/12; (see Tables 13.4 and 13.5).

We will now show the model M4 (Fig. 13.7, left), characterized by a triangular park.

It should be noted that this model was very well received by the newspapers, however, its ecological assessment (Table 13.4) with diagnostic index is equal to 0.711 for the new quarter, but amounted to only 0.50 for the LU, with average of 0.634, a value certainly not of a good standard.

The model M9 (Table 13.5), instead, propose a park in the shape of a cross. The ecological evaluation results equal to 0.89 for the parameters of quarter and 0.68 for those of LU, with a mean value of 0.784. This assessment is rather high, and it was a surprise, since discarded by many designers (but only based on aesthetic perception).

Note that the latest model that is presented in this chapter, M15, with oblong park (Fig. 13.7, right) and various offshoots, considers one more parameter (the orientation of the park), because we have added the area "pivot" (about 6 ha) to the NW of the old area. The orientation assessment has been carried out with respect to the axis of the ecological network that goes from the park towards the Simplon Park to

Table 13.5 Evaluation of Model M9

Ecological parameters	Normal	M9	Δ9	Score
Fractal dimension of green (D)	1.5–1.8	1.41	5.90	1.61
Connection of green (α + γ)	0.75–1.15	0.875	0.00	2.00
Core areas (%V)	45–75	48.13	0.00	2.00
Ratio A/P green	30–60	52.19	0.00	2.00
Contacts of green with the outside (%pV)	20–60	24.08	0.00	2.00
Built area/perimeter V (m^2/m)	45–90	92.46	2.70	1.82
Structural diversity (ψ)	3.6–9.1	4.96	0.00	2.00
Climatic balance (ΔT °C)	0–0.75	1.10	46.70	0.44
Neighbourhood BTC (Mcal/m^2/a)	0.75–0.9	0.79	0.00	2.00
Park BTC (30 year)	1.6–1.9	1.63	0.00	2.00
Scores			*55.30*	*17.87*
Diagnostic index of the neighbourhood	**0.85–1.0**			**0.893**
Fractal dimension of green (D) LU	1.45–1.75	1.209	19.4	0.706
Connection of green (α + γ) LU	0.5–0.9	0.35	30	0.581
BTC of LU	0.6–0.8	0.494	17.6	0.826
Radius of ecological influence (min)	110–450	295.3	0	2.00
Source/sink (% transects)	25–75	45.6	0	2.00
Garden grain/general grain	4.0–6.0	4.24	0	2.00
Scores				*8.113*
Diagnostic index of Landscape Unit	**0.75–0.90**			**0.676**

Fig. 13.7 Models M4 and M15. The neighbourhood park is shaped like a triangle or in an integrated pervasive oblong way. The evaluation is equal to 0.71 Vs. 0.80 for the parameters of neighbourhood and 0.50 Vs. 0.77 for those of LU, with average values of 0.601 Vs. 0.786

Table 13.6 Evaluation of Model 15

Ecological parameters	Normal	M 15	Δ15	Score
Fractal dimension of green (D)	1.5–1.8	1.69	0	2
Connection of green (α + γ)	0.75–1.15	0.91	0	2
Core areas (%V)	45–75	35.2	21.8	0.65
Ratio A/P green	30–60	26.9	10.3	1.31
Contacts of green with the outside (%pV)	20–60	31	0	2
Built area/perimeter V (m^2/m)	45–90	34.1	24.2	0.63
Structural diversity (ψ)	3.6–9.1	9.5	4.3	1.72
Climatic balance (ΔT °C)	0–0.75	0.67	0	2
Neighbourhood BTC (Mcal/m^2/a)	0.75–0.9	0.75	0	2
Park BTC (30 year)	1.6–1.9	1.53	4.4	1.75
Scores				*16.06*
Diagnostic index of the neighbourhood	**0.85–1.0**			**0.803**
Fractal dimension of green (D) LU	1.45–1.75	1.57	0	2
Connection of green (α + γ) LU	0.5–0.9	0.46	8	1.47
BTC of LU	0.6–0.8	0.49	18.3	0.78
Radius of ecological influence (min)	310–500	194	37.4	0.52
Source/sink (% transects)	25–75	29	0	2
Garden grain/general grain	4.0–6.0	5.6	0	2
Park orientation (% core area)	45–60	55	0	2
Scores				*10.77*
Diagnostic index of Landscape Unit	**0.75–0.90**			**0.769**

Table 13.7 Evaluation of the models

Models	Quarter DI	LU DI	DI mean
SA	0.162	0.166	0.164
M8	0.537	0.457	0.513
M4	0.711	0.503	0.607
M7	0.822	0.671	0.765
M9	0.893	0.676	0,784
M15	0.803	0.769	**0,786**

Ticino. The ecological evaluation (Table 13.6) is equal to 0.80 for the quarter parameters and 0.77 for those of LU, with a mean value of 0.786. This is a decent value, although not optimal.

Table 13.7 shows, finally, the ranking of ecological assessments of the design models (remember Fig. 13.5), with the diagnostic indices of neighborhood, of landscape units and of the average of the whole. The choice made by the technicians of Citylife was for the M15, one of the best, although none of the studied models reached optimality, which would exceed the value of 0.85. In fact, values of 48.9% of built and 51.1% of open spaces and green (of which 41.6% closely to green) are clearly not sufficient to achieve optimality, but only the sufficiency. Sign that there were some obvious shortcomings in the methods of planning by the City of Milan, which included built too!

Some other considerations, deriving from the setting of landscape bionomics, should be added.

We demonstrated that, for the Emergent Properties Principle, even maintaining a standard surface, if the parts of a system are assembled differently, they present functionalities of different value, often in unpredictable ways. This is mainly due to the Form-Function Principle: the interaction between objects is proportional to their contact margin and to the interception of the flow of energy or transport.

13.2.3 *Urban Park Design: Adopted vs. Optimal Project*

The mentioned Form-Function Principles produce very sensible results, as we are showing now. First step: the Park design history. In Fig. 13.8 we present the Masterplan winning the competition (2004), whose Park was due to Ingegnoli, and (right) the design indications of the masterplan Park to be followed by the new competition participants (2010), always due to Ingegnoli.

The result of the second competition on the masterplan Park was won by L. Porter (London) but this project did not follow completely the meta-design of Ingegnoli (compare the Figs. 13.8, right and 13.9, left). No doubt that the project of Carlos Ribas was better, as we can see from this figure: the distribution of urban forest patches does not exaggerate the narrow dimension of the green corridor and chooses vegetation more suited to the Milanese environment.

Fig. 13.8 (left) The Masterplan winning the competition (2004), whose Park had been designed by Ingegnoli; (right) the metadesign of the Park of the masterplan to be developed as an executive project by the new competition (2010), always due to Ingegnoli

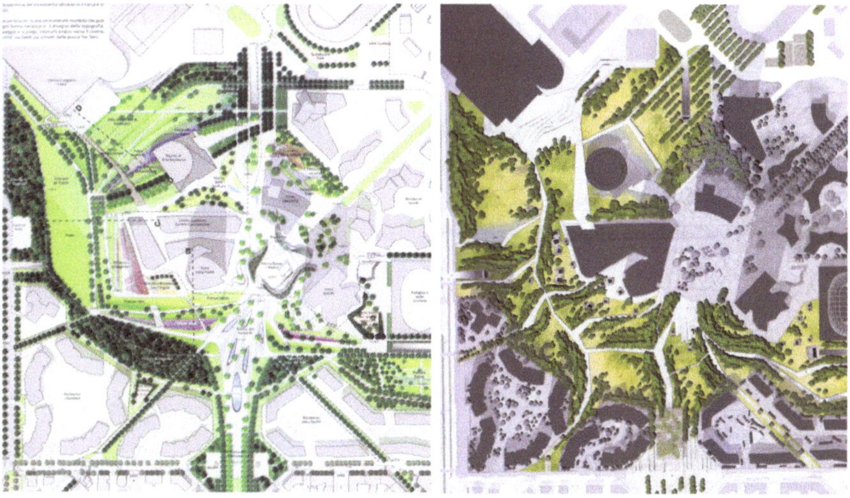

Fig. 13.9 The Park design which won the competition 2010 (L. Porter, London), and the second classified project (Carlos Ribas, Lisboa). The second one features more naturally shaped green spaces, more consistent with the ecological objectives to be pursued and pervasive with respect to the built environment

Table 13.8 Bionomic Control of the Park of CityLife 2050

Components	Adopted project				Optimal project			
	ha	%	BTC'	BTCw	ha	%	BTC'	BTCw
Floorings	4.544	29,02	0,05	0,01451	4.500	28,74	0,05	0,01437
Rows of trees	1.945	12,42	2,5	0,31,052	1.945	12,42	2,5	0,31,052
Lawns,and meadows	3.731	23,83	0,55	0,13,105	3.517	22,46	0,55	0,12,355
Gardens	2.609	16,66	3,2	0,53,325	2.623	16,75	3,4	0,56,962
Woods	2.185	13,96	5	0,69,789	2.429	15,52	5,5	0,85,334
Equipment	0.496	3,17	0,25	0,00792	0.496	3,17	0,25	0,00792
Asylum	0.147	0,94	0,15	0,00141	0.147	0,94	0,15	0,00141
Total	15,659	100,00		**1,70**	15,659	100,00		**1,88**

In Table 13.8 the bionomic evaluation of the adopted project of the Park (Porter design) is now compared with the meta-design (Ingegnoli) of Fig. 13.8 (right). The differences seem to be contained; gardens 2.609 Vs. 2.623 ha, woods 2.185 Vs. 2.429 ha, average BTC after 40 years: 1.70 Vs. 1.88. The adopted project reaches 90.43% of the optimal one, a deviation of about 10% only.

In Fig. 13.10, we show the comparison between the adopted design of the Park (red) and the optimal one (blue). The green line is the average BTC of the Lombardy

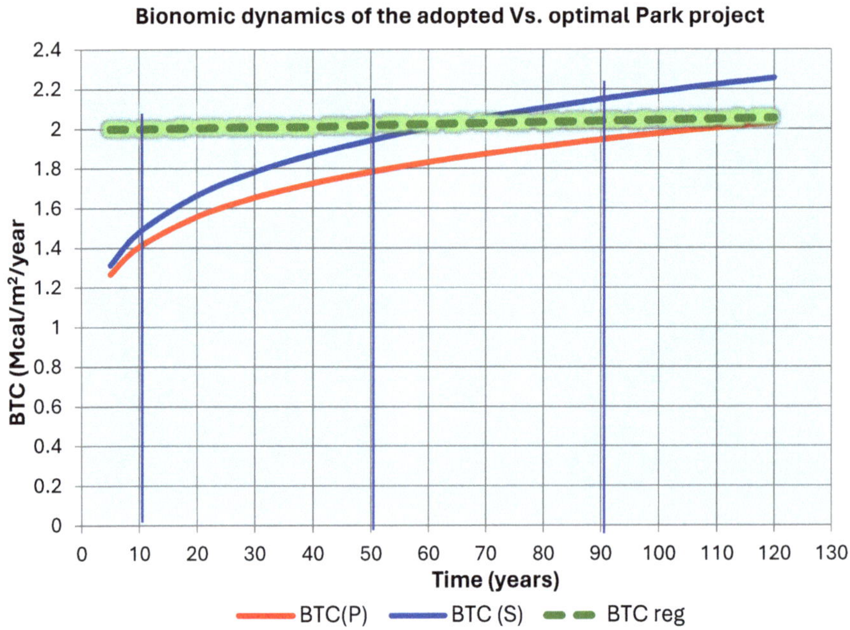

Fig. 13.10 Comparison between the adopted design of the Park (red) and the optimal one (blue). The green line is the average BTC of the Lombardy region. Note that the bionomic efficiency of the park must reach the mean BTC of the Region (10% of security coefficient) in the shortest time possible

region. Note that the bionomic efficiency of the park must reach the mean BTC of the Region (10% of security coefficient) in the shortest time possible. As predicted by the deterministic chaos (Ingegnoli, 2015), a small difference in the starting parameter of two projects can led to a sharp difference in the behavior of the Parks. As we can see in the figure, the minimum goal (arriving near the regional BTC) is reached by the optimal project, while the adopted one delate is about four decennia.

The main reason for the parks to reach the mentioned goal as soon as possible can be seen in Fig. 13.11: a park could -and need to- function reaching a biomedical potential, so a bionomic influence radius able to heal the bionomic functionality of the Zone 8 of Milan, today with a BF = 0.60 (BTC = 0.50–52 Mcal/m^2/year).

The need for our park to reach a biomedical potential in short time reflects the need to decrease the potential morbidity in the complex landscape unit (c-LU) of municipal zone 8 of Milan. Figure 13.12 shows the heal of this c-LU if we increase the Bionomic Functionality from BF = 0.60 to 0.70, so decreasing the Mortality Rate from 8.90/1000 to 8.50/1000: giving that the inhabitants of Quartiere ex Fiera are about 74–76,000, it means 75,000 × 0.0004 × 0.8 = 24 death/year, that is a diminution of 24 unity of premature death risk for year.

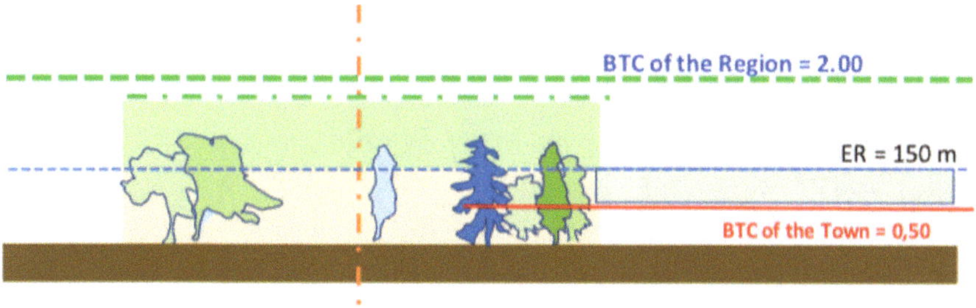

Fig. 13.11 Example of the minimum goal to reach for the examined park, which must go above the regional BTC to function really as a park, a landscape element able to restore the surrounding environment, protecting the human health

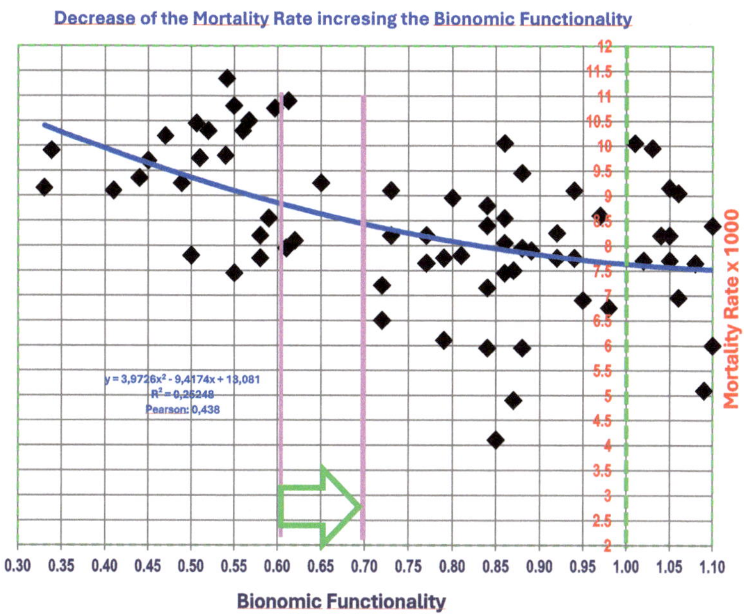

Fig. 13.12 Increasing BF from 0.60 to 0.70 we would have gotten a decrease of MR from 8.90 to 8.50/1000

13.3 Bionomic Strategies for Designing a Suburban Park in Milan

13.3.1 Chiaravalle: The Heart of South Milan Agricultural Park Milano

A Study for the Area Plan of the Abbey Park (2210 ha of surface extension), asked to the writer by the Administration of the Municipality of Milan, was done in 1997

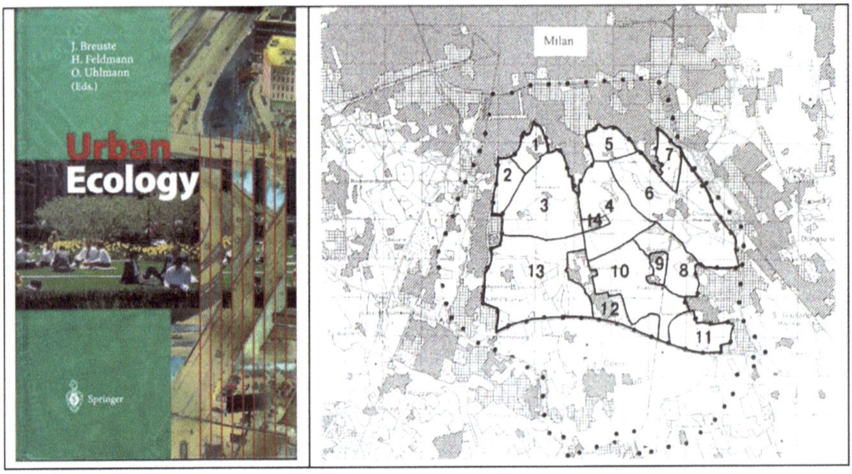

Fig. 13.13 The first study of the central area of South Milan Agricultural Park (The Abbeys District) made by Ingegnoli and published (1998). The map indicates 13 local ecotopes (4.7% of the park) and the local landscape unit (dotted line), 12.6% of the Park total surface (47,000 ha)

(Ingegnoli, 2002, Fig. 13.13). The suggestions given have however never been implemented, and the suburban Park received few restorations, e.g., near the Vettabia canal, on the eastern side of the local landscape, or the Ticinello (on the Western side).

So, a vegetation survey made by Ingegnoli in 2017 near the Vettabia found a wood with BTC = 4.18 Mcal/m^2/yr. and a grass with a BTC pair to only 0.35! The main problem is the misunderstanding of the concept of park: even in Agricultural Park we must consider some indispensable functions/services, as synthetized in Fig. 13.14. In this figure, we underline the importance of bionomics-ecological functions, but the prevalent one in the Agricultural Park South Milan are the social-aesthetic functions.

Let us refer to the parametric transformation of the Lombardy Region (Fig. 13.15). We can observe the modest decrease in human habitat (HH), the limited variations of the average BTC (which now returns to the values of 1870: BTC = 2.00 Mcal/m^2/yr) and the growth of urbanized areas and infrastructures after 1960. The ecological density of population reached 710 inhabitant/km^2. So, the carrying capacity SH/SH* = 0.96; this represents a safety limit, stated that for values around 1.0 (0.95–1.05) the regional landscape system becomes heterotrophic, that is, its agricultural component (935 m^2/inhabitant) would no longer be able to support the needs of the population. In facts, the agrarian components after reaching their max value in 1950 (67%) is today equal to 45%.

Furthermore, the majority of the inhabitants live in the plains (70%) where forests are not 25% of the region, but only 7%, therefore 58 m^2/inhabitant, a value decidedly

	MAINE ECOLOGICAL SERVICES	SPATIAL SCALE		BIONOMICS PARAMETERS		VEGETATION ASPECTS	
	BIONOMICS-ECOLOGICAL FUNCTION	LOCAL	TOWN	$BTC_P \geq BTC_R$	area EFI ≥ PARK area	COMPLEX STRUCTURE	LOCAL SPECIES
1	REDUCTION OF ENVIRONMENTAL STRESS DUE TO BIONOMIC DYSFUNCTIONS						
2	AIR CLEANING FROM FINE DUST						
3	BREAK OF UHI (URBAN HEAT ISLE)						
4	WATER PERCOLATION INTO SOIL (BFF)						
5	OASIS OF NATURAL HABITAT IN THE URBAN DESERT						
6	ELEMENT OF ECOLOGICAL NETWORK						
	SOCIAL-AESTETIC FUNCTION						
7	CULTURAL-EDUCATIONAL ASPECT						
8	CHILDREN AREAS						
9	LIFE-SPORT AREAS						
10	OLD PEOPLE AREAS						
11	DOG AREAS						
12	LOCAL AESTHETIC IMPROVING						

BTC_p = BTC of the park; BTC_R = BTC of the region; EFi = ecological field of influence;

Fig. 13.14 Bionomics, Ecological, and Social Efficiency of an Urban Park

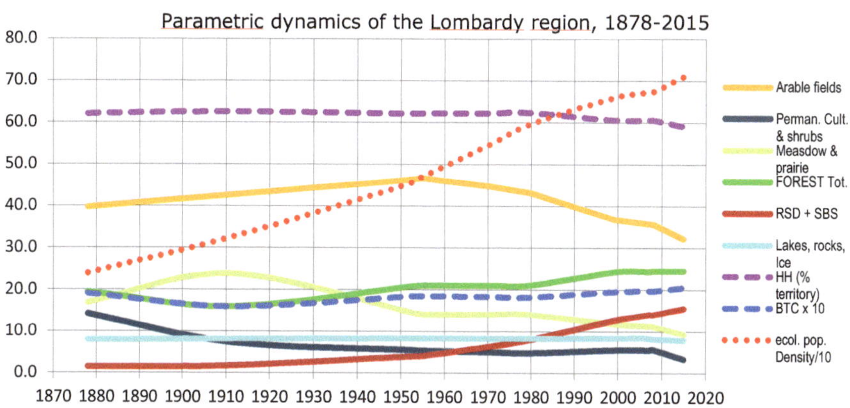

Fig. 13.15 The transformations of the Lombardy Region show the parametric dynamics of its territory, from 1878 to 2015. Note the drastic decrease of agrarian landscapes and the increase of urbanization

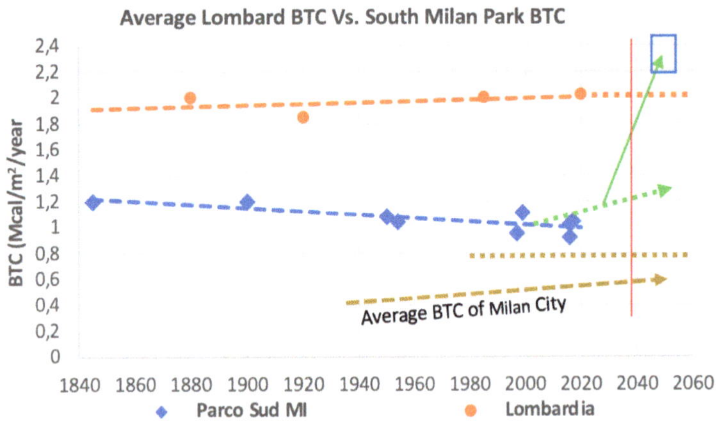

Fig. 13.16 Ecological and bionomic dynamics concerning the biological state of the Abbeys Park (South Milan). Note the decrease of the BTC due to the agriculture industrialization and the building of the Southern edge of Milan, the distance of the present average BTC of the Town from its optimum (BTC=0.80, light brown dotted). Green dotted arrow represents the increase of BTC next 30 years following present tendence, while green arrow indicates the necessary new intervention being the blue rectangle at the right top the correct goal

lower than the minimum necessary for the compensation of the human activities, which is equal to 338 m^2/inhabitant, (see Table 12.1) that is 5.83 times as much: no longer able to provide the health prevention essential for the correct functioning of the PNEI systems (Psycho-Neuro-Endocrine-Immunology, see Par.11.3).

If we want to restore the environment of our Region, we must transform the landscape unit (LU) on which we intervene following the new concept of sustainability, without forgetting a *compensative ecotope*, for instance a park area in an urban landscape. This condition means *going over the average BTC = 2.0 Mcal/m²/ year of Lombardy*, at least in the way able to reach the ecological influence (see Fig. 7.12) necessary to the ecotope sustainability.

Therefore, being the first function of a park its ecological compensation capacity, a Park (even an urban park) involves a design leading to an average BTC > 2.0 Mcal/m^2/year. Therefore, at least one ecotope (5% of the Abbeys Park) in our landscape area must reach BTC = 2.25 Mcal/m^2/yr. in next 25–30 years, as marked in Fig. 13.16.

13.3.2 *Provide a Semi-Natural Park Ecotope Implementing the Range of Ecological Influence*

Suggestions for the design of a semi-natural park ecotope (the mentioned 5% of the Abbeys Park) derive from the volume "Il Governo dei Parchi" (Bonizzi et al., 2019) and the previous chapters 7, 8 of this volume.

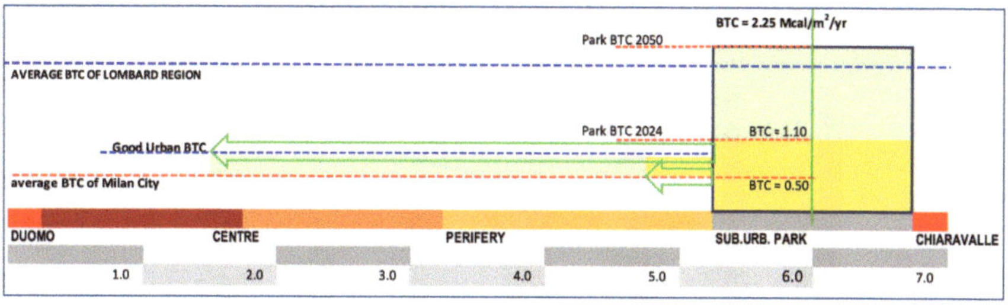

Fig. 13.17 Comparison between the present average BTC of the Abbey Park (1.10 Mcal/m^2/yr., yellow) and the suggested BTC of the new seminatural park (BTC = 2.25). The range of influence changes from 0.75 km (yellow arrow) to 3.60 km

Meta-design of a Semi-natural Park in Chiaravalle

Components	ha	%	HH	HHw	BTC	BTCw
Wood patches	22.40	20	10	2.00	4.80	0.960
Forest area	17.92	16	4	0.64	6.10	0.976
Agroforestry	50.40	45	85	38.25	0.95	0.427
Recreative areas	16.80	15	90	13.50	1.4	0.210
Roads + built	4.48	4	95	4.25	0.1	0.004
Total	112.0	100				2.58
Disturbances		(10)				- 0.26
Results				**58.69**		**2.32**

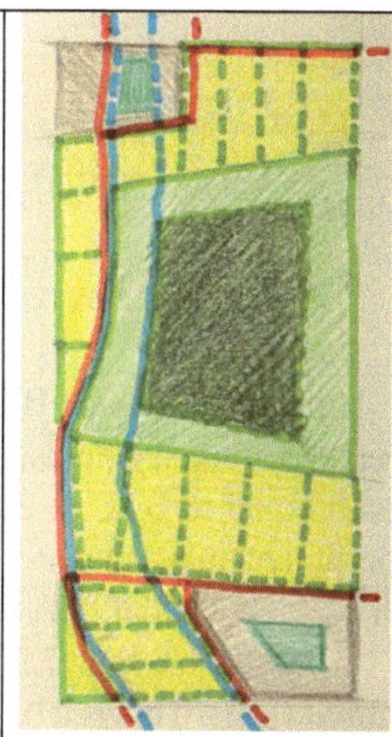

Fig. 13.18 Example of a Meta-Design for the semi-natural new park in the Abbeys Park District of Chiaravalle

As you can see from Fig. 13.17, the South side of Milan is only seven Km wide from the cathedral (Duomo) to the Chiaravalle Abbey, due to the strong productive agrarian landscape. Today, being the average BTC of the Abbeys Park = 1.10 Mcal/m^2/year, the influence range of the park is very short (Ecological Radius ER = 0.75 Km), so that only the peripheral margin of the town can benefit from this park. If the new seminatural park should reach an average BTC = 2.25 Mcal/m^2/year, then ER could achieve 3.60 Km, it influences arriving at the margin of the central City.

So, the proposed semi-natural park (Fig. 13.18) should be a rectangle of 112 ha from Milan periphery and Chiaravalle Abbey, about 1.5 km. Within Fig. 13.18, its values of reference are reported too.

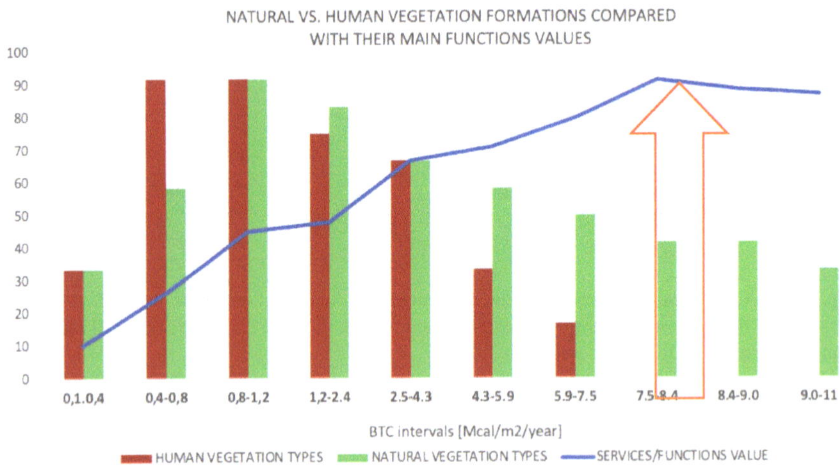

Fig. 13.19 Vegetation formations (human & natural) and their main functions/service values (blue line). See Fig. 7.21. The orange arrow indicates the forest chosen for the formation of the semi-natural park

Human habitat and BTC are linked to the main components: wood patches, forest area, agroforestry fields, recreative areas, roads and built, and their weighted means result equal to BTC = 2.61 Mcal/m^2/year in about 30 years from today. A security coefficient may consider the effect of disturbances (10%), so the result would be BTC = 2.32 Mcal/m^2/year.

In Par.7.7 & Fig. 7.21 we underlined the lack of ecological services in human habitat (HH). This is a big problem related to the disease prevention, because the full forest service's pertain to the three highest class of BTC, reached only by natural formations (Fig. 13.19). That is why we must design at least 16% of the new park area planting a temperate deciduous forest not accessible to citizens (except for integrated therapy reasons) destined to develop at least up to BTC = 8.00 Mcal /m^2 / year.

Note that we should introduce a new semi-natural park ecotope in every district of the Agricultural Park South Milan. In Fig. 13.19 we insert the meta-design of Fig. 13.18 (blue rectangle). Note that we do not design a circular distribution around the town, to avoid the effects shown in Figs. 13.1 and 13.20.

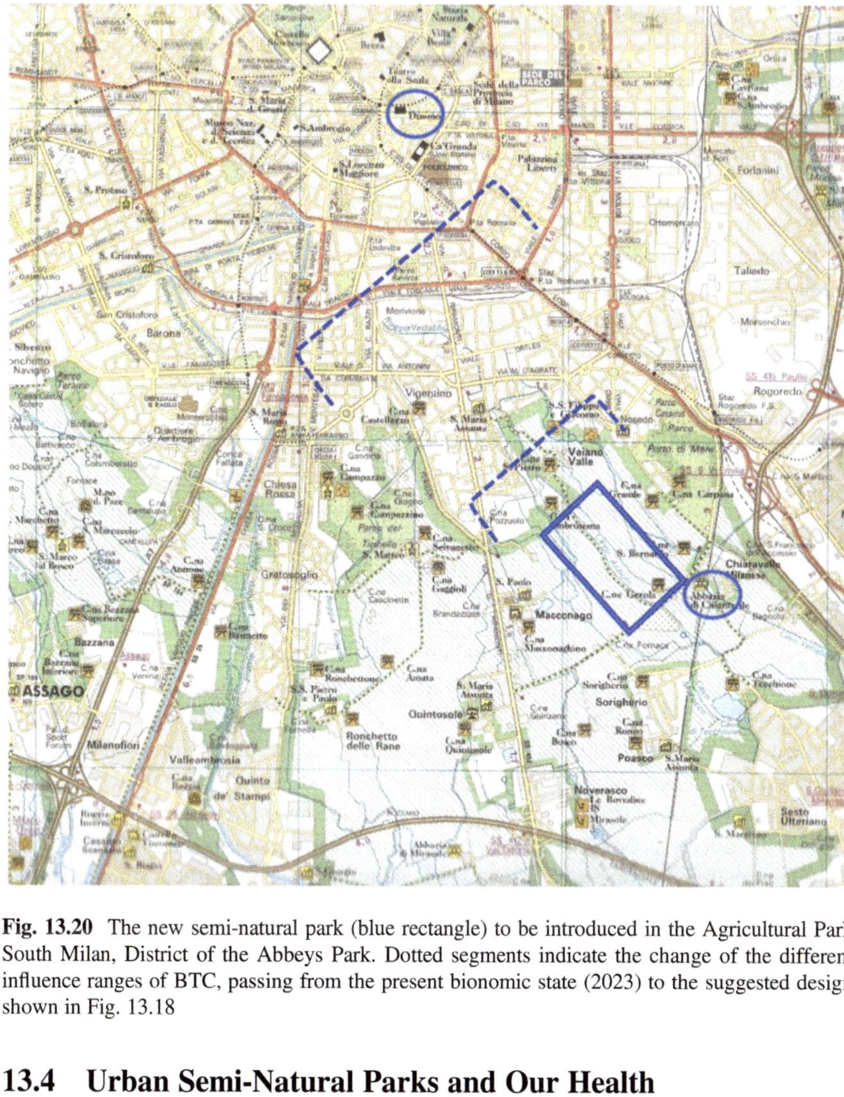

Fig. 13.20 The new semi-natural park (blue rectangle) to be introduced in the Agricultural Park South Milan, District of the Abbeys Park. Dotted segments indicate the change of the different influence ranges of BTC, passing from the present bionomic state (2023) to the suggested design shown in Fig. 13.18

13.4 Urban Semi-Natural Parks and Our Health

13.4.1 Introduction: Forest/Health Correlations and Wildlife

Even before the beginning of the ecological crisis in 1950–60, most ecologists underlined the importance of forests for human health: correct but generic favorable topics, such as (a) oxygen growth, (b) fine dust decrease, (c) mental well-being, (d) micro-climatic improvement, (e) noise mitigation, (f) rebirth of the sacred sense. But, also, dangerous topics, such as sources of allergens and of zoonoses, wildlife habitat (therefore a source of predators, e.g., wolf, bear), and refuge for violent people; let's remember Dante's "dark forest"…

New scientific aspects emerge in the forest/health correlations (Ingegnoli, 2015; Ingegnoli et al., 2022; Ingegnoli & Giglio, 2005, 2017, 2021). The most important seem to be six processes:

1. Detection of *environmental stress* when humans live in degraded environments, with consequences on the nervous and endocrine systems.
2. *Increasing in the mortality rate (MR)* if the *bionomic functionality (BF)* of a landscape unit (LU), linked with forest presence, is far from an optimal state.
3. *Proportionality to Bionomic Functionality of the Pandemic infection,* as demonstrated in Lombardy, Italy, the region most affected by COVID-19 because of lacking of forests where people live.
4. *Inversely correlation of the BTC (vegetation territorial capacity) with cancer insurgence* in Western Europe. BTC mainly depends on forests.
5. *Maintaining and enriching GM* (Gut Microbiome), through men's contact with natural vegetation systems, first forests, even independently of nutrition. It's of crucial importance.
6. *Modulation of information given by the environment* and our behavior, because the information written on the chromosome bases are insufficient. Phytocoenosis alterations affect information, and broken information can degrade the epigenetic codes, affecting the repair capacity of cells.

The positive contribution of forests to human health results is potentially reinforced. These processes widen the relation between advanced ecology and medicine relationships. The doctors must repair thc damage to human health, but since most of this damage is due to environmental alterations and since the organism is linked to the internal and external life systems of the body, doctors must collaborate with internal and external specialists, arriving at an integrated anamnesis and therapy.

If not reduced as banal urban green, for instance, a playground in a grass patch with few scattered exotic trees, urban semi-natural parks and gardens are linked with urban wildlife. The term semi-natural means the inclusion of some forest patches, even in an urban landscape.

As reported in the volume of Urban Ecology (Forman, 2014, 2019) in Western Metropolitan Areas we note the presence of many wild animals, up to 150–200 species. In Table 6.5, it has been the most common 48 species of birds and mammals in EU and USA, underlying that at least 16 species are the same both in European and American towns. Note that 80–85% of this fauna is strictly linked with Urban Green. Moreover, in suburban zones farms with various animal breeding are generally frequent, so we must pay attention to the *zoonosis* when we want to plan new green areas in urbanized territory. Zoonosis has today a small mortality rate (MR $<$ 1.0–1.5%) but it can be dangerous because of unpredictable events (Ist. Sup. di Sanità, 2023; IZSVe, 2023; Collino, 2023; D'Amico et al., 2018). In Table 13.9, we can see an average of 0.73% of mortality (large approximation), the total deaths reaching about 45 to 55 people/year. But, even if the rabies is absent in Italy, we had about 100 bitten in 2023, due to sick animals from Polonia or Romania, a part of which is potentially at risk.

Table 13.9 Main Zoonosis in Italy (2022–23)

Sicks (IT)	%	Disease	Transmission	Mortality	MRw
3.302	50,25	*Salmonellosis*	Food, water, pets	0,08	0,04
1.539	23,42	*Campylobacteriosis*	food, water, chickens, pets	0,05	0,01
723	11,00	*West Nile Virus*	Birds and mosquitos(Culex)	4,5	0,50
345	5,25	*Listeriosis*	Food	0,2	0,01
220	3,35	*Dengue*	Mosquitos	3,6	0,12
125	1,90	*Lyme Borreliosis*	Ticks, from wild in humid/cold environment	0,4	0,01
118	1,80	*Escherichia*	Food	0,05	0,00
89	1,35	*Toscana virus*	Phlebotomus	0,4	0,01
50	0,76	*Neuro-invasiva TBE*	Ticks	1,8	0,01
36	0,55	*Yersinia*	Domestic and wild animals	2,5	0,01
20	0,30	*Brucellosis*	Rural animals	1,2	0,00
15	0,23	*Bovine Tuberculosis*	Aerosols from animals (sneezing, coughing), milk	1,8	0,00
6.582	100,00	ITALY (2022–23)		%	0,73

Moreover, we cannot forget the massive increase in antibiotic resistance, which has reached about 12,000 deaths in 2023 in Italy, one of the worst cases in the Western World.

Supposing we want to estimate an adequate order of magnitude of future zoonotic mortality in Italy, considering unpredictable events and increased antibiotic resistance, we should consider a value equal to 3.0% for the next 35–40 years. Adding new forest patches in an urbanized territory will undoubtedly increase the dangers of zoonosis.

So, we must underlined (a) the potential positive value of forests concerning our health and (b) the increase in zoonosis. The objective of this work is to check the opportunity of a plan of reforestation in an urbanized territory to improve our health. This aim needs:

A. to follow the discipline of Landscape Bionomics.
B. to apply a risk analysis concerning zoonosis and premature death.
C. to suggest a form-function analysis capable of leading an ecological design.
D. to improve the ecological state of human habitat by adding semi-natural parks.

13.4.2 Reforestation: Zoonosis vs Human Health

As shown in Par.9.2 concerning the correlations MR = f (BF), we can see in Table 13.10 the very low forest land cover (7.46%) and the modest average BF = 0.75 in the Monza-Brianza & Milan City. So, we need a *reforestation*, to widen BF and potentiate human health.

Table 13.10 Monza-Brianza + Milan City territory (2015–16)

People	ha	Landscape type	FOR. %	URB. %	AGR. %	HH	BTC	BF
95.377	10.450	Agricultural	16,21	29,56	53,2	73,05	1,46	0,98
123.116	9.150	Rural	7,9	37,39	53,6	80,37	1,02	0,87
173.618	7.553	Suburban	8,55	53,45	35,8	81,5	0,94	0,83
464.100	13.342	Urban	2,33	69,62	25,8	88,29	0,55	0,61
325.373	6.906	Suburbian milan	3,65	74,7	20,2	90,14	0,56	0,66
1.404.431	18.184	City of Milan	6,95	71,22	20,5	94,49	0,43	0,57
2.586.015	**65.576**	**Averages**	**7,46**	**57,85**	33,14	**85,89**	**0,77**	**0,75**

Table 13.11 Proposal of adding 10% of forest cover in the examined territory (2025–30)

People	ha	Landscape type	FOR. %	URB. %	HH	BTC	BTCmd	BF 35
95.377	10.450	Agricultural	2,14	29,56	73,05	1,55	1496	1,03
123.116	9.150	Rural	0,78	37,39	80,37	1,1	1171	0,93
173.618	7.553	Suburban	1,13	53,45	81,5	1,05	1127	0,93
464.100	13.342	Urban	0,23	69,62	88,29	0,65	0,904	0,71
325.373	6.906	Suburbian Milan	0,52	74,7	90,14	0,6	0,854	0,70
1.404.431	18.184	City of Milan	3,44	71,22	94,49	0,55	0,756	0,72
2.586.015	**65.576**	**Averages**	**8,20**	**57,85**	**85,89**	**0,81**	1,05	**0,84**

By reforesting 394.0 ha (adding 10% to present forest cover), we could improve BF, going from 0.75 to 0.84 (Table 13.11; Fig. 13.21). How should the Risk of Premature Deaths change? To answer we must take into consideration the two aspects: (A) the increase in zoonosis; (B) the decrease in the mortality rate.

A. From 2.3 deaths/year (2024–25), we move to a risk proportional to the additional areas (forestation) of greater naturalness equal to future sicks 394 × 0.025 = 9.8 deaths/year However, the increase in mortality is minimal: 7.5 deaths/year.
B. If BF became 0.84 (years 2025–30), MR would fall away from 9.2 to 8.9 (Fig. 13.21), then the distance with optimality (BF = 1.0) would become 8.9–8.4 = 0.50/1000. This value, after subtracting the part due to PA (population age) and multiplied by the inhabitants (2.7 million), would become 0.50 × 0.76 × 2,700,000 = *1026.0 death/year*. In the period 2015–16, with BF = 0.75, the risk cited had higher values: 0.53 × 0.76 × 2,650,000 = *1067.4 death/year.*

The proposed reforestation would *decrease* the risk of premature deaths by 1067.4 to 1026.0 pair to *41.4 fewer death/year.*

So, the comparison of the increase in zoonosis (mortality of +7.5 deaths/year) and the decrease in the mortality rate (mortality of *−41.4 death/year*) are not comparable, with a clear propension for the reforestation. But let us deepen this observation.

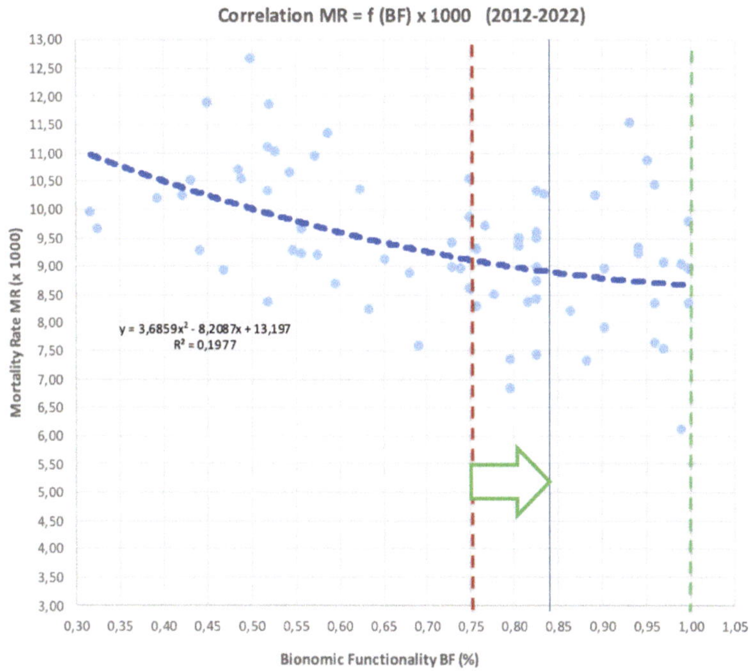

Fig. 13.21 The correlation MR/BF help us in evaluating the change in the risk of premature death within this studied territory

13.4.3 Cost/Benefit Analysis of Urban Reforestation

The Cost/Benefit analysis needs the proportionality of fewer death due to reforestation and more zoonosis death (Fig. 13.22).

The addition of semi-natural parks may cause other deaths if their distribution on the territory is wrong, also recalling the Emergent Properties Principles (see Sect. 13.2.2).

Pay attention also to the capacity of integrative therapy of a forest, which require the forest to be the *most natural* and mature as possible. Remember the positive effects derived from a 2-day "forest bath" on the number of cytotoxic enzymes (perforin & granulysin) capable of alerting Nk cells to destroy dangerous sick cells (e.g., neoplastic) are remarkable (Li et al., 2008; Li, 2009). This effect lasts for almost a month. Below, prevention and therapy integration concern the psycho-neuro, immune, and endocrine systems.

Therefore, the coetaneous and monospecific forests must be avoided, as they are very similar to arboriculture, too frequent forest type even in the Alps: e.g., two layers, Picea abies 95–100%, No ectomycorrhizal web), BTC <7.0–7.5 Mcal/m^2/yr. On the contrary, we must change our forest management, giving more natural

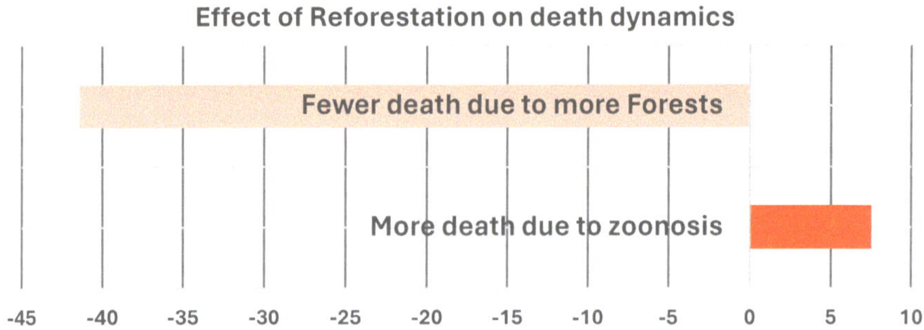

Fig. 13.22 No doubt that RPBD (minor Risk of Premature Death due to a higher BF) despite the heavy safety factors results 3.8 time wide than RPZD (major Risk of Premature death due to Zoonosis)

structures: reforestation should not follow geometric quincunx or equisetaneous plantings; only plantings by age groups and species or following sinusoidal grids are permitted. Remember some important necessities:

1. Presence of mature patches (BTC >8.5 Mcal/m^2/yr),
2. Soil communication networks formed by fungal hyphae,
3. Distinction edge/inner structure, Presence of «Mother Trees»
4. Vertical structure of minimum four layers
5. Presence of large herbivores which, by browsing aboveground vegetation, boost carbon storage into root system while making forests more resilient to global warming-intensified fires.

References

Baccarelli, Zanobetti, Martinelli, et al. (2008). Exposure to particulate air pollution and risk of deep vein thrombosis. *Archives of Internal Medicine*, 920–927.

Bonizzi, L., Campana, F., & Cordini, G. (Eds.). (2019). *Il Governo dei Parchi. Strategie operative per la salvaguardia delle risorse naturali, come fattore di miglioramento della qualità della vita, e di competitività delle imprese*. Aracne editrice.

Collino, A. (2023). *Rabbia: la malattia dimenticata*. Fond. Veronesi Magazine.

Consigliere, I. (2004). *Curare il verde per arricchire il paesaggio in città*. IN: Il Sole-24 Ore, 10 luglio 2004.

D'Amico, W., et al. (2018). *Zoonosi trasmesse da zecche*. Inail coll. Salute e Sicurezza.

Forman, R. T. T. (2008). *Urban Regions, Ecology and Planning beyond the City*. Cambridge University Press.

Forman, R. T. T. (2014). *Urban ecology, science of cities*. Cambridge University Press.

Forman, R. T. T. (2019). *Towns, ecology, and the land*. Cambridge University Press.

Ingegnoli, V. (2002). *Landscape ecology: A widening foundation* (p. XXIII+357). Springer.

Ingegnoli, V. (2011). *Bionomia del paesaggio. L'ecologia del paesaggio biologico-integrata per la formazione di un "medico" dei sistemi ecologici* (p. XX+340). Springer.

Ingegnoli, V. (2015). *Landscape bionomics. Biological-integrated landscape ecology* (p. XXIV +431). Springer.

Ingegnoli, V., & Giglio, E. (2005). *Ecologia del Paesaggio: manuale per conservare, gestire e pianificare l'ambiente* (p. 685+XVI). Sistemi editoriali SE.

Ingegnoli, V., & Giglio, E. (2017). Complex environmental alterations damages human body defence system: a new bio-systemic way of investigation. *WSEAS Transactions on Environment and Development*, 170–180.

Ingegnoli, V., & Giglio, E. (2021). Environmental bionomic alterations and cancer incidence in European countries: A first overview. *Journal of Nursing & Occupational Health, 2*(3), 266–281.

Ingegnoli, V., Lombardo, F., & LaTorre, G. (2022). *Environmental alteration leads to human disease: A planetary health approach.* Springer-Nature. Sustainable Development Golas Series. Foreword by Andrea Lenzi.

Ist. Superiore di Sanità. (2023). Rapporto One-Health sulle zoonosi nel 2022 nella UE.

IZSVe. (2023). Campilobatteriosi e salmonellosi le zoonosi più frequenti in Europa nel 2022.

Li Q (2009) effect of forest bathing trips on human immune function. Environmental Health and Preventive Medicine: 15(1):9–17.

Li, Q., Morimoto, K., Nakadai, A., Inagaki, H., Katsumata, M., Shimizu, T., Hirata, Y., Hirata, K., Suzuki, H., Miyazaki, Y., Kagawa, T., Koyama, Y., Ohira, T., Takayama, N., Krensky, A. M., & Kawada, T. (2008). Forest bathing enhances human natural killer activity and expression of anti-cancer proteins. *Journal of Biological Regulators and Homeostatic Agents, 22*(1), 45–55.

Olmsted FL 1870. Public parks and the enlargements of towns. Cambridge University Press, .

Ordine Architetti Milano. (2009). Grandi Parchi per Grandi Progetti. Il ruolo del verde per la qualità urbana. In *Bollettion dell'Ordine*. Milano 21 maggio 2009, vis Solferino 15.

Sukopp, H., & Hejny, S. (1990). *Urban ecology*. SPB Academic Pub..

Chapter 14
Renaturalize to Preserve Health: Case Studies on the Po River

Contents

Abstract Each territorial intervention, whatever the motivation it starts from, if conducted following the Landscape Bionomic methodologies, thus with a systemic bio-ecological approach, can be an occasion of strategic rehabilitation of the involved territory as well as of improvement of environmental conditions with positive effects on the resident population.

CASE STUDY 1: The Interregional River Agency (AIPo) asked to complete the project of the Staffora diversion, beyond the Margine Maestro del Po (2021)

A bionomic study of the overall LU of floodplain at the current state was applied to obtain the parameters of the human habitat (HH = 58.19%), of the BTC = 2.21 Mcal/m^2/year, of the BF (Bionomic Functionality) = 0.925 and of the information content of the components (gLM = 13.84-bit × Mcal/m^2/a). The type of landscape is agricultural-protective, given the presence of forests that reaches 35% of the LU.

In conclusion, for the floodplain LU the ratio Benefit/Cost B/C = 1.48 (within 20 years from the completion of the work). This is a forecast fully centered in the interval B/C = 1.3–3.0, therefore acceptable.

V. Ingegnoli, *Landscape Bionomics and Planetary Health*, Landscape Series 21,
https://doi.org/10.1007/978-3-032-08908-3_14

CASE STUDY 2 As a re-naturalization of a LU even not contiguous to a town could be important to lower the risk of premature mortality. But, will it be possible to renaturalize the Golena di Gussola (a Floodplain of the Po River)? The question must not be misinterpreted, it is not intended to be captious: it seems, in fact, that at the present time the desired renaturalization should not be done. Both the regions of Lombardy and Emilia-Romagna and the Ministries of Environment and Agriculture seem to give more importance to the local trade associations of farmers and poplar growers. These are political choices that are incomprehensible to the undersigned author of these notes, as it is not so difficult to understand that to help a few hundred people, one can prevent the improvement of the health of hundreds of thousands of inhabitants of the surrounding area

Keywords Floodplain · Fluvial forest · River diversion · Renaturalization · Cost-benefit analysis · Landscape bionomics methodology · Po River

14.1 Final Stretch of the Staffora River Diversion

14.1.1 Examined Landscape Units (LU)

Each territorial intervention, whatever the motivation it starts from, if conducted following the Landscape Bionomic methodologies, thus with a systemic bio-ecological approach, can be an occasion of strategic rehabilitation of the involved territory as well as of improvement of environmental conditions with positive effects on the resident population. In add, as previously highlighted, the new advanced guidelines for the prevention and protection of human health focus on the importance of the rehabilitation of LU going toward an upgrading of biomedical potential through the Health Protective Index (HPI).

We will expose an excerpt of the Investigations to support the Technical and Economic Feasibility Project for the Hydraulic Regulation Works of the Staffora River to protect the towns of Oriolo, Cervesina and Pancarana (PV, Northern Italy) as a practical example of the methods previously exposed (VinGeo, 2021–2022).

To frame historically and geographically the area of intervention, following the ecotissue concept (see Sect. 3.1.5), Fig. 14.1 shows the map of the Pavia District (S. Capsani, XVIII Cent.). Note the presence of three large patches of forest and the sinusoidal shape of the Staffora (red) with its outlet into the Po River. Alongside, the two feasibility projects for a diversion (semi-natural canal): section A (Voghera-Po Master Bank) and B (Po Master Bank-Pancarana), became necessary given the current narrowness of the canal corridor at Cervesina with consequent flooding.

At first, reference was made to the overall Landscape Unit, called Voghera-Po with a surface area of 3246 ha, for general framework reasons, to include all the

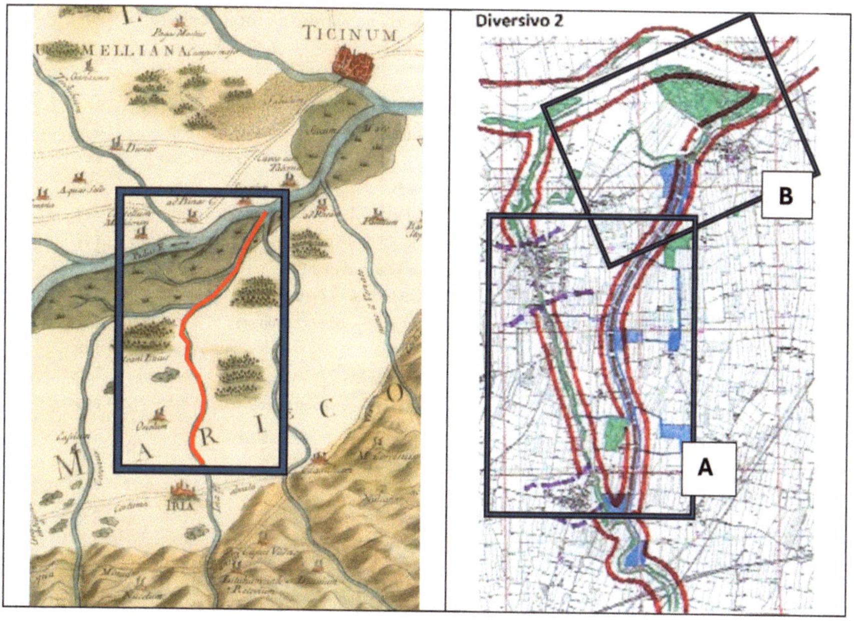

Fig. 14.1 Map of the Pavia District (S. Capsani, 1780). Note the presence of three large patches of forest and the sinusoidal shape of the Staffora (red) with its outlet into the Po River. Alongside, the two feasibility projects for a diversion canal: section A (Voghera-Margine Maestro Po) and B (Margine Maestro-Pancarana-Po), necessary given the current narrowness of the canal corridor at Cervesina with consequent flooding

territories of the inhabited centers involved in the Diversion intervention for the final stretch of the Staffora River: Cervesina, Pancarana, Pizzale and Oriolo (Fig. 14.2).

The assignment on which we will focus our attention referred to an area linked to the development of the new final course of the Staffora (Fig. 14.3), exclusively in floodplain, of approximately 209 ha, equal to 6.5% of the entire general LU but with 65% of its remnant forest cover. However, we had also to consider the fluvial erosion, which has been occurring in recent years and is very significant, as we will see.

The delimitation of the LU of reference, roughly marked in Fig. 14.2, has been specified based on the most recent satellite survey. Its final stretch can be seen in Fig. 14.3, where the North-West bank of the Po has undergone very strong erosion, which has led to a loss of approximately 10–12% of the area in just 5 years (Fig. 14.4). Only the floodplain component is now considered, thus excluding the Po River from the components of the LU.

Along the canal (now dry) that follows the edge of the Po there are tree-lined corridors interspersed with small, wooded patches, which on the south-west side still have stretches of Poplar and Willow trees of considerable size.

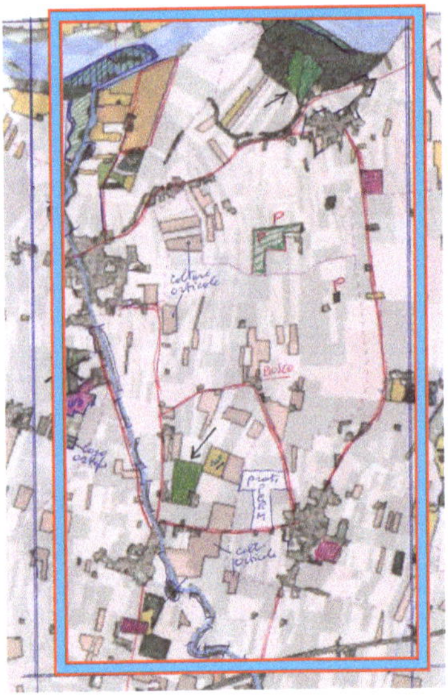

Land use within the total Voghera-Po river LU

LU (A) Voghera-Po, 2020	Sup. ha	%
Urban	135.70	4.18
Farmhouse	14.10	0.43
RSD	**149.80**	**4.61**
Roads	20.80	0.64
Industries	29.20	0.90
SBS	**50.00**	**1.54**
Gardens (15% URB)	24.00	0.74
Staffora Tree-line rows & fields	9.90	0.30
PO Tree-line rows in floodplain	6.60	0.20
PRT	**40.50**	**1.25**
Arable lands	1,908.40	58.79
Forage Meadows and Meadows	828.00	25.51
Vineyards	6.00	0.18
Poplar groves	14.00	0.43
PRD	**2,756.40**	**84.91**
Shrubbery within floodplain	12.40	0.38
Permanent meadows	25.00	0.77
RSL	**37.40**	**1.15**
Willow/Poplar Pancarana	79.10	2.44
Staffora-Po Wood	22.00	0.68
Other woods in the plain of the LU	18.60	0.57
RNT	**119.70**	**3.69**
Po River	82.30	2.54
Staffora creek	7.60	0.23
Other waters	2.50	0.08
HGL	**92.40**	**2.85**
Totale	**3,246.20**	**100**

Fig. 14.2 The total landscape unit (LU) according to the current state surveys. On this survey, after specific inspections, several corrections had to be made, reported on the map

The north-west bank of the Po has seen very strong erosion, which has led to a loss of almost 20% of the area in just 15 years (yellow arrow).

The first photograph (blue arrow at the top left of Fig. 14.4) is a view that allows us to see the main erosion margin of the Po and the considerable size of the forest areas behind it (Fig. 14.5), while the second photo shows a patch of woodland between the main bank of the Po and the trace of the old canal that carried the Staffora into the Po: small patch, but with mature trees of considerable vegetation value, as we will see.

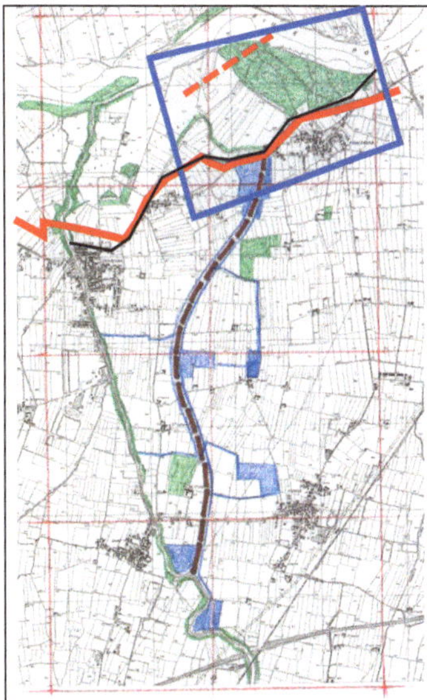

Floodplain LU-B of Pancarana, 2021		
Landscape Elements	ha	%
Arcadia Wood	23.92	11.44
Northern Willow/Poplar Wood	13.09	6.65
Eastern Willow Wood central core	3.00	1.52
Eastern Willow Wood	22.46	11.41
Patch with Tree-line row 1	1.00	0.51
Patch with Tree-line row 2	1.20	0.61
Patch with Tree-line row 3	0.80	0.40
Tree-line rows high BTC	1.92	1.00
Tree-line rows low BTC	1.52	0.77
Agricultural field	115.0	52.35
Gravel and sand	5.60	2.84
Po river	16.00	8.13
Rural roads and clearings	3.30	1.68
Total LU	209.00	100.00

Note: The red curve indicates the Margine Maestro/ Major Bank of the Po river, while the blue trapezoid indicates the floodplain LU of Pancarana.
The dashed red line points erosion.

Fig. 14.3 The part to the north of the Margine Maestro / Major Bank of the Po River concerns the final stretch of the Staffora, dependent on the LU in the blue trapezium (about 210 ha)

Fig. 14.4 Left. Red line: Delimitation of the Golena/Floodplain LU in Pancarana (181 ha). The arrows indicate the photographs shown below to illustrate this LU. Right. A comparison with the map from 15 years ago, to understand the extent of the erosion of the right bank of the Po to the N-W of Pancarana. From the yellow arrow see the Fig. 14.9

Fig. 14.5 (**a**) View of the LU-B, from the blue-green arrow on the previous map; (**b**) View of the LU-B, from the blue-pink arrow on the left of the main margin of the Po

14.1.2 Vegetation Study of the Pancarana Floodplain and Its Therapeutic Significance

The vegetation study was conducted according to the LaBiSV method (Landscape Bionomics Survey of Vegetation (Ingegnoli, 2002; Ingegnoli & Pignatti, 2007; Ingegnoli & Giglio, 2005) already presented in paragraph 4.2. The analysis and bionomic evaluation of the vegetation components (Ingegnoli, 2015) of the forest and tree-lined corridors required several survey inspections and 40 subsequent elaborations, including:

- Number of characterizing species (Herbaceous, Bushy and Trees) (Pignatti, 1982).
- Dominant species and alien species (%) (Pedrotti & Gafta, 1966).
- Ellenberg bio-indicators (L, T, C, U, R, N), sensu Pignatti (Pignatti, 2005).
- Estimation of humus efficiency (R x N).
- Estimation of BTC (Mcal/m^2/y), sensu Ingegnoli (Ingegnoli, 2002).
- Estimation of bionomic quality Qb (%).
- Estimation of apparent and balanced age (Ta and Tb).
- Measure of ecological efficiency CBSt (%) (Ingegnoli, 2013).
- Estimation of basal area AB (m^2/ha).
- Measure of average canopy height (m).
- Calculation of above-ground phytomass FM (m^3/ha).

The dominant trees *Salix alba* and *Populus nigra* often present specimens with good structure, considerable height (even >30 m), therefore mature age: but even they can do very little against the erosion of the Po, especially because it usually occurs below the root network, given the morphological difference in level and the composition of the soil.

Even remembering the importance of Mature vegetation Vs. Human Health (see Sect. 10.6), we now begin to present a summary of the examined woods, from N-W to N-E of Pancarana LU: (1) Arcadia Wood, (2) Spars Poplar wood and (3) Old Willow Wood. Furthermore, we will also have to give some information about the main stretch of the tree-lined corridor (Patch/Tree-line row 1) that develops along the former canal that follows the ancient path of the Staffora (S-W/N-E) (see Fig. 14.3).

A. *Pancarana, Arcadia Wood, ha 23,9*

This is a sparse forest of white willows that can even exceed a century, but with a young average age (41.1 years). For about 15 years it has been being integrated with plantations of poplars of different species. The forest is equipped with areas with benches and maintains various herbaceous patches (Fig. 14.6).

The structure is largely anthropized (planted quincunx), but the alien species are quite limited (15.7%), for a bank forest.

We report Table 14.1 too, to highlight the characterizing trees, of which the dominant ones are green, their phytomass (336.4 m^3/ha) not very high, and all the Ellenberg bioindicators. In add, the bionomic quality values of the detected tiles (bQ = 41.07%), the average territorial biopotential of the vegetation (BTC = 4.42 Mcal/m^2/year) and the concise bionomic state (CBSt = 8.05%), all medium-low values, were calculated.

Fig. 14.6 Plan of Arcadia Woods (red). The inside of the woods has some grassy areas, often with some benches: it is therefore treated as a recreational park, also given the proximity of Pavia

Table 14.1 Summary of the characteristics of Arcadia Wood, Pancarana, 2021

Characterizing trees	Tree N°	Tree h m	A.Bas. m^2	Allom.	FM m^3/ ha	%
POPULUS NIGRA L. ITALICA	33	19.8	12.50	0.65	161.53	45.6
SALIX ALBA L.	28	25.5	8.00	0.65	132.86	37.5
POPULUS ALBA e P. CANESCENS	14	20.3	3.83	0.70	54.371	15.4
ROBINIA PSEUDOACACIA L.	2	17.0	0.33	0.65	**36,465**	1.0
FRAXINUS ORNUS L.	1	15.0	0.16	0.70	1.68	0.5
	78	**21.85**	**24.82**		354.08	100.0
Ellenberg bioindicators					**336.38**	(FM x 0.95)
Humus efficiency	R × N	43.49	A.Bas = basimetric surface			
Temperature	T	6.20	Allom. = Allometric Coeff.; FM = Phytomass above ground			

Fig. 14.7 (**a**) Plan of the Northern Wood under examination (yellow). (**b**) Inside there remain some traces of ancient oxbow lakes

Table 14.2 Summary of the characteristics of the Northern Wood of Pancarana, 2021

Characterizing trees	Tree N°	Tree h m	A.Bas. m^2	Allom.	FM m^3/ha	%
SALIX ALBA L.	25	24.11	14	0.65	219.4	66.5
POPULUS CANADENSIS L. e P. NIGRA	9	33.2	5	0.65	107.9	32.7
ACER NEGUNDO	1	12	0.33	0.7	**2772**	0.8
	35	**26.1**	**19.33**		330.07	100.0
Ellenberg bioindicators					**313.57**	(FM x 0.95)
Humus efficiency	R × N	**42.55**	A.Bas = basimetric surface			
Temperature	T	6.2	Allom. = Allometric Coeff.; FM = Phytomass above ground			

B. *Pancarana, Northern Wood, ha 13,1 (Populeto-Salicetum)*

This is a forest with an average age greater than the previous one (47.1) with a good presence of poplars, including some groups of hybrids (*Populus x deltoides*). The average height rises to 26.1 m, 5 more than the previous one. The herbaceous clearings are a little less frequent, but the alien species are very high here (31.7%). As can be seen from Fig. 14.8, the forest is surrounded by the Po on two sides and still has the remains of ancient oxbow lakes inside it (Fig. 14.7).

Compared to the previous one (see Table 14.1) you will notice (Table 14.2) a smaller number of characteristic trees and that only *Salix alba* is dominant. The phytomass is slightly smaller; therefore, it is immediately clear that the forest is sparser, even if more mature.

C. *Pancarana, Eastern Willow wood, ha 22,46 (Populeto-Salicetum album)*

Sparse forest of Salix alba (Fig. 14.8) with an average age of 61.7 years and trees that can exceed 30–35 m and 125 years. The undergrowth is sparse and grassy due to the floods of the Po. The humus is therefore less rich, about 10–11% less than the

Fig. 14.8 (**a**) Plan of the Willow Eastern wood of Pancarana; (**b**) Inside, the undergrowth is almost absent due to the repeated floods of the Po; some willows exceed 35 m

Table 14.3 Summary of the characteristics of the willow wood east of Pancarana, 2021

Characterizing trees	Tree N °	Tree h m	A.Bas. m^2	Allom.	FM m^3/ ha	%
SALIX ALBA L.	77	26.25	21.5	0.65	366.8	90.6
POPULUS NIGRA L.	6	32.3	1.16	0.65	24.35	6.0
POPULUS ALBA	2	20	0.33	0.7	4.62	1.1
POPULUS CANADENSIS L.	2	32.9	0.33	0.65	**7057**	1.7
ACER NEGUNDO	1	15	0.16	0.7	1.68	0.4
	88	**26.54**	**23.48**		404.6	99.9
Ellenberg bioindicators					**384.3**	(FM × 0.95)
Humus efficiency	**R × N**	**39.73**	A.Bas = basimetric surface			
Temperature	T	6.39	Allom. = Allometric Coeff.			
Light	L	6.19	FM = Phytomass above ground			

previous woods. The alien species are rather limited (12.8%). The 15% of the forest in the central part shows a BTC = 6.0–6.4 Mcal/m^2/year, as already noted the previous year for the study of the LU Voghera-Major Bank of the Po River, to which some general summaries of these woods were added. The low value of the phytomass (384.3 m^3/ha) is striking (Table 14.3), slightly higher than the previous woods in front of groups of trees higher than 34 m.

FORESTE TEMPERATE	1	5	12	22	punteggio
T. CARATTERISTICHE DI TESSERA (Ts) = 51,5					
T1 - Altezza vegetazione (m)	< 9	9.1-18	18.1-29	> 29.1	Chioma dominanti 26,5 m
T2 - Copertura canopy (%)	< 30	> 90	31-60	61-90	Rispetto alla superficie Ts
T3 - Differenziazione struttura	bassa	media	buona	alta	Disetaneità, disp. spaziale
T4- Interno/margine (%)	assente	< 30	31-89	> 90	(% Ts)
T5 - Tipo di gestione	ceduo semplice	ceduo composto	bosco	foresta naturale	O assimilabili (v. nota)
T6 - Permanenza (anni)	< 80	81-160	161-240	> 240	Età degli alberi più vecchi 125 a
F. CARATTERISTICHE DELLA FITOMASSA EPIGEA NELLA TESSERA = 59,1					
F1- Quantità di fitomassa morta	quasi 0	> 10	da 1 a 5	> 5	Calcolata come % sulla Fm viva
F2- Profondità della lettiera	quasi 0	< 1.5	1.6-3.5	>3.5	cm
F3b – Fm in volume (m^3/ha)	< 200	201-500	501-950	> 950	Foresta boreale
F3t – Fm in volume (m^3/ha)	< 150	150-350	350-600	> 600	Foresta temperata, 384 mc/ha
E. CARATTERISTICHE ECOCENOTOPICHE DELLA TESSERA = 52,7					
E1- Specie dominanti (n°)	> 3	3	2	1	Come volume di Fm
E2- Ricchezza di specie	< 15	16-30	31-40	> 40	n° sp./Tessera 34
E3- Specie caratteristiche (%)	< 5	da 6 a 41	41-75	> 75	Associaz. di riferim. 33,7 %
E4- Specie alloctone (%)	> 10	da 10 a 4	< 4	0	ALLOCTONAE 12,8 %
E5- Piante infestanti (% copertura)	quasi completa	> 25	< 25	assenza	Sulla superficie totale della Ts
E6- Individui ammalati	danni evidenti	sospetto o poche	rischio	nessuno	tutte le possibili malattie
E7- Forme biologiche (n°)	< 3	da 4 a 5	da 6 a 7	> 7	Secondo Box 1987, modificato
E8- Stratificazione verticale	2	3	4	> 4	v. nota
E9- Capacità di rinnovo	nessuna	intensa	sporadica	normale	Della/e specie dominante/i
E10- Stato dinamico	degradazione	ricreazione	rigenerazione	fluttuazione	v. nota
U. CARATTERISTICHE DELLA TESSERA IN RAPPORTO ALL'UNITA' DI PAESAGGIO (UdP) = 47,0					
U1- Contiguità vegetazione anal.	0	< 25	26-75	> 76	% del perimetro
U2- Caratteri source o di sink	sink (risucchio)	neutrale	sorg. parziale	sorgente	In Sp e risorse
U3- Ruolo funzionale nell'UdP	ridotto	minore	evidente	importante	Contesto e tipologia
U4- Incorporazione dei disturbi	insufficiente	scarsa	normale	alta	Disturbi locali
U5- Instabilità geofisiche	evidenti	parziali	rischio	nessuna	Legate ai fisiotopi dell'UdP
U6- Interesse da fauna permeante	basso	medio	buono	attrazione	Specie chiave
U7- Motivi di trasformazione d. Ts	forti disturbi	camb. graduali	instabilità temp.	fluttuazione	Ad oggi + tendenza
U8- Patologie dovute al circostante	assai serie	quasi croniche	incorporabili	nessuna	dall'UdP propria o da altre circos.
U9- Permanenza storica (anni)	< 100	100-300	300-1200	> 1200	Presenza storica
RISULTATI DEL RILEVAMENTO					
Punteggio totale Y (= h+j+k+w)	h = 3	J = 9	K = 10	w = ..6	Y = 311
Qualità della Ts		Q t = Y / 616 [700]			bQ = 50,48
Stima della BTC		BTC (t) = 0,01667 (y-28) + 0,13 (FM / 65)			BTC = 5,49 Mcal/m^2/a
Ta = (età apparente)		Ta = $(BTC+0,77)^{1/0,515}$			Ta = 35,2 anni
Tb Età Bilanciata		(Dmd+Dold)/4] / (0,4 to 0,7)*			Tb-Age = 61,7 (years)
Concise Bionomic State		CBSt-2 = (MtL x bQ) (%) where MtL = Ta/125			CBSt = 14,21 (%)

Note: * the increase of wood rings/year (mm) depends on the dominant species of the exained Ts; e.g., 4-5 mm x Quercus robur to 6-7 mm x Populus nigra, In any case, we have to check the behavior of these species to the local environment, to estimate a right wood-ring increase.

Fig. 14.9 Results of the Vegetation Survey of a tessera of *Populeto-Salicetum album* forest detected according to the LaBiSV methodology

In Fig. 14.9 we see that the salient bionomic parameters are also like the previous ones, but a little better: bQ = 50.48 (%), BTC = 5.49 Mcal/m^2/a, CBSt = 14.21 (%), always medium-low values, but on average about 20% better than the previous ones. This wood is undoubtedly the most natural, at least in the central areas.

Concerning the Monoterpene emissions (MT, %) produced by these woods and their utility for recreational scopes or for Forest bath, the detected values indicate MT = 11.36%, a modest value (about double of an urban park). On the other side, the contribution of information content conveyed by them g.LM = 18.70bit × Mcal/m^2/year. This wood presents a quite good structure shaped by flooding.

D. *Tree-lined corridor along the canal*

From the surveys of this corridor (Fig. 14.10) we have seen that the salient bionomic parameters are like the previous ones: bQ = 50.16 (%), BTC = 5.61 Mcal/m^2/a,

Fig. 14.10 (**a**) Plan of the corridor (yellow) and the scrub (orange) within the blue box; (**b**) Corridor: Populus alba with 3.95 m circumference. (**c**) Scrub: Populus nigra with 4.45 m circumference

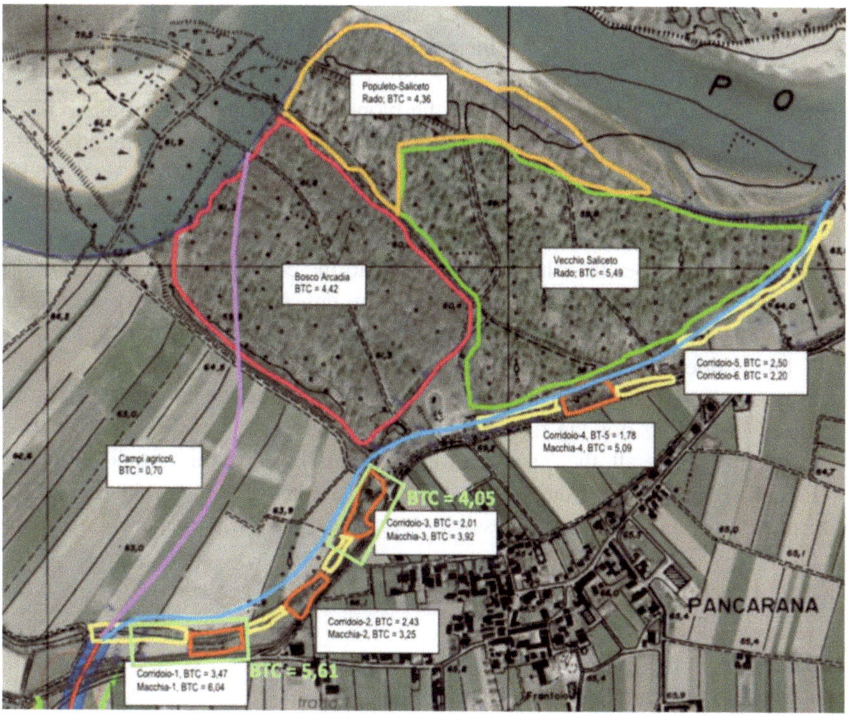

Fig. 14.11 Summary of the surveys of the vegetation components in the Pancarana floodplain LU-B, in which the estimates of the territorial biopotential of the vegetation (BTC) are exposed and the two possible routes for the final stretch of the Staffora diversion are also marked (pink line and light blue line), which will have to be carefully evaluated

CBSt = 16.80 (%), always medium-low values, but on average better, as in the previous case.

E. *Summary of Vegetation Surveys and Possible Routes for Diversion*

Figure 14.11 summarizes the studies carried out on the components of the local vegetation of the landscape unit under examination: the three woods whose bionomic and ecological characteristics we have already seen, the tree-lined corridor with scrub, and other similar but less important presences. Furthermore, the two possible routes for the final stretch of the Staffora are traded on the map: (a) purple = D1 and (b) blue = D2; it also shows the average BTC values to which we refer (BTC= 4,05 and BTC= 5,61), which are significantly higher than those detected.

14.1.3 Bionomic Status of the Landscape Unit

Table 14.4 presents the process of arriving at a correct estimate of the human habitat (HH) and the average BTC of the LU under examination. The table highlight the BTC classes and the percentage of their presence within the LU. As will be seen, the value of the human habitat (HH = 58.19%) is relatively low, given the good presence of woods. At the same time, the biopotential of the vegetation (BTC = 2.21 Mcal/m^2/year) is not entirely optimal, given the alterations found in the vegetation components examined. However, it must be emphasized that the BF measure (Bionomic Functionality of a LU) is acceptable, being only 7.5% far from the optimal state.

We must remember that the BTC is linked to the concept of metastability, and Bionomics allows us to estimate the BTC of an entire LU as we have seen from Table 14.6. This value is significant, especially if measured in a historical perspective, but it does not directly express the level of metastability reached by the landscape unit, because it does not consider the distribution of the diversity of elements in the LU nor the distribution of the energy flow (BTC) between said elements nor, above all, the quantity of information linked to this composition.

Recalling Sect. 7.4.1 from which it can be deduced that an evaluation of the information content of a LU is possible, measuring the g-LM (general Landscape

Table 14.4 Pancarana Golena/Floodplain LU-B, Present State (2021)

Landscape elements	ha	%	*HH*	HH w	*BTC*	BTC w
Arcadia wood	23.92	*13.23*	*20*	2.65	*4.42*	0.58
Northern willow/poplar wood	13.09	*7.24*	*12*	0.87	*4.36*	0.32
Eastern willow wood central core	3.00	*1.66*	*12*	0.20	*6.40*	0.11
Eastern willow wood	22.46	*12.42*	*15*	1.86	*5.49*	0.68
Patch with tree-line row 1	1.00	*0.55*	*9*	0.05	*5.61*	0.03
Patch with tree-line row 2	1.20	*0.66*	*9*	0.06	*4.05*	0.03
Patch with tree-line row 3	0.80	*0.44*	*10*	0.04	*5.00*	0.02
Tree-line rows high BTC	1.92	*1.06*	*8*	0.08	*2.60*	0.03
Tree-line rows low BTC	1.52	*0.84*	*12*	0.10	*1.50*	0.01
Agricultural field	103.05	*56.98*	*88*	52.87	*0.70*	0.42
Gravel and sand	5.60	*3.10*	*10*	0.31	*0.10*	0.01
Rural roads and clearings	3.30	*1.82*	*100*	1.82	*0.05*	0.00
Total LU	180.86	100.00		**58.19**		**2.21**
Optimal theoretical value: (BTC mod)					*2.39*	
BTCmod/BTC = **BF**						**0.925**

Table 14.5 Pancarana Golena/Floodplain LU-B, 2021

BTC class	P (elem)	P*$\log_2$ P
I	0.0492	−0.214
II	0.5698	−0.462
III	0.0084	−0.058
IV	0.0106	−0.070
V	0.3320	−0.528
VI	0.0166	−0.098
VII	0.0001	−0.001
H	Heterogeneity	−1.302
D	Dominance	1.505
T (tau)	Functional diversity	**6.263**
BTC		2.21
g-LM	**g-LM [bit × Mcal/m²/yr]**	**13.84**

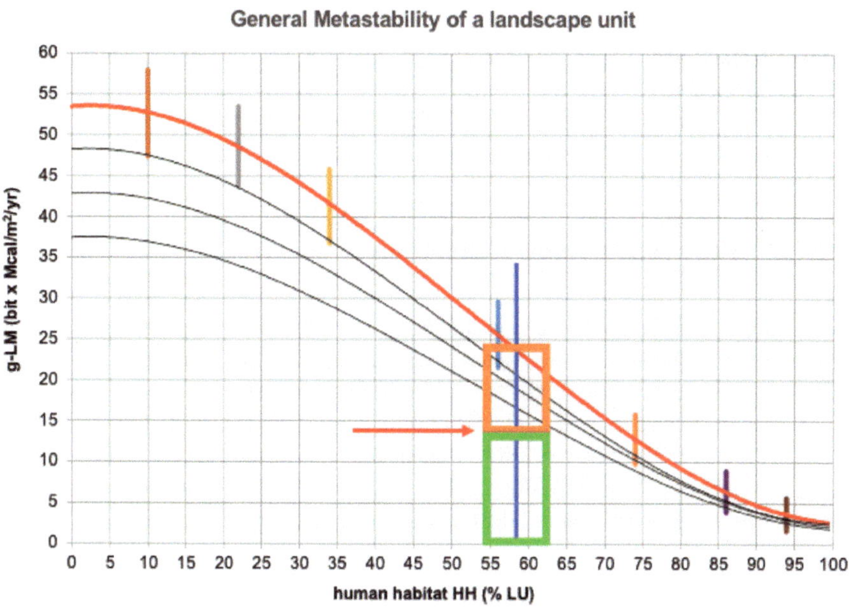

Fig. 14.12 Control of the information content for the LU-B under examination. In green the estimated gLM value, in orange how much is missing to reach the optimum. (Graph from Ingegnoli, 2015)

Metastability), we will report here only the summarizing table of this processing (Table 14.5), observing that the result is equal to g-LM = 13.84-bit x Mcal/m^2/year. This value of g-LM is significantly higher than the g-LM = 3.67 of the LU Voghera-Major Bank/Margine Maestro but lower than it could be, as we will see below.

A precise check of the information content can be obtained from Fig. 14.12, which plots the optimality curve of g-LM (red) as a function of HH.

Being $y = 0{,}0001x^3–0{,}0158x^2 + 0{,}0738x + 53{,}5$ the equation of the optimal g-LM curve, for a HH pair to 58,20 we will have g-LM = 23,99.

Since the current g-LM_{2021} is pair to 13.84, it is equal to 57.7% of the optimal value of the model (see rectangle and arrow in Fig. 14.12), therefore it must be improved with the inclusion of the diversion and its compensation reforestation. The increase should be at least 10–15%. This is an objective to be kept in mind in the project proposal.

We can evaluate the Health Protective Index (HPI) in this LU (see 11.5):

$$\begin{aligned}\text{HPI} &= [\text{av.BTC} \times 2.07] \times \text{g.LM} \times 0.582 = 2.21 \times 2.07 \times 13.84 \times 0.582\\ &= 36.85 [\text{bit} \times \text{Mcal}/\text{m}^2/\text{year}]\end{aligned}$$

14.1.3.1 Control of the LU transformation Dynamics

We will consider the state of the LU under examination in 2016. It is recalled (remember Fig. 14.4) that in 2016 the landscape unit under examination was approximately 14% larger than the current one (209 vs. 180 ha) given the lower erosion of the Po.

However, parameters are not very different from today's ones: a HH = 57.63% (−1.0%), a BTC = 2.15 Mcal/m^2/year (−2.7%), and BF = 0.888 (−4.0%).

14.1.4 PCA Interference Estimation

Continuing with the systemic considerations mentioned above, we highlight the system of PCA (Potential Core Areas, see Sect. 7.6) (Fig. 14.13) and the comparison with the possible valuable routes for the Staffora diversion.

We call the two alternative tracks D1 (North, purple) and D2 (East, red). The loss of PCA in the case of D1 is pair to 9.8%, while in D2 = 3.6%; the increase in the marginal areas (sensu Forman) (brown) consequently reaches 131% for D1 and 31% for D2; the fragmentation in the perimeter of the LU under examination (Fig. 14.13 above) in fact appears to pass from the current 9 spots to 11 with D1, while 9 remain with D2 option.

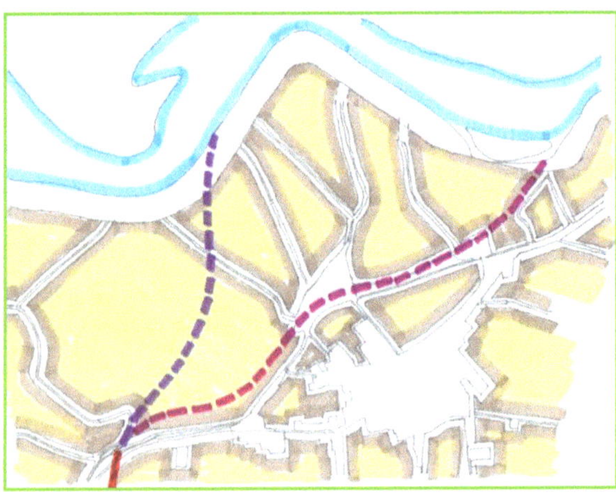

Fig. 14.13 The dotted lines highlight the interferences of the two possible paths D1 (purple) and D2 (red) with the PCAs (yellow). The greater fragmentation produced by the possible diversion D1 is immediately noticeable) (see also Fig. 7.17)

14.1.4.1 The New Tree-Lined Corridor

The design of a tree-lined corridor alongside the new diversion bed and the expansion of the current woodland along the route, with related forestry improvements also at the internal edges of the existing woods, should lead to an increase in the average BTC from the current 2.21 to at least 2.37–2.40 Mcal/m^2/y. Given the plan with the chosen route (D2), Sections 2, 4, 6 proposed by the Geologists are highlighted in Fig. 14.14, with an example of trees to be added.

In the feasibility project, it is possible to hypothesize the planting of a double row of trees on both sides of the Staffora diversion with the reforestation of forest patches.

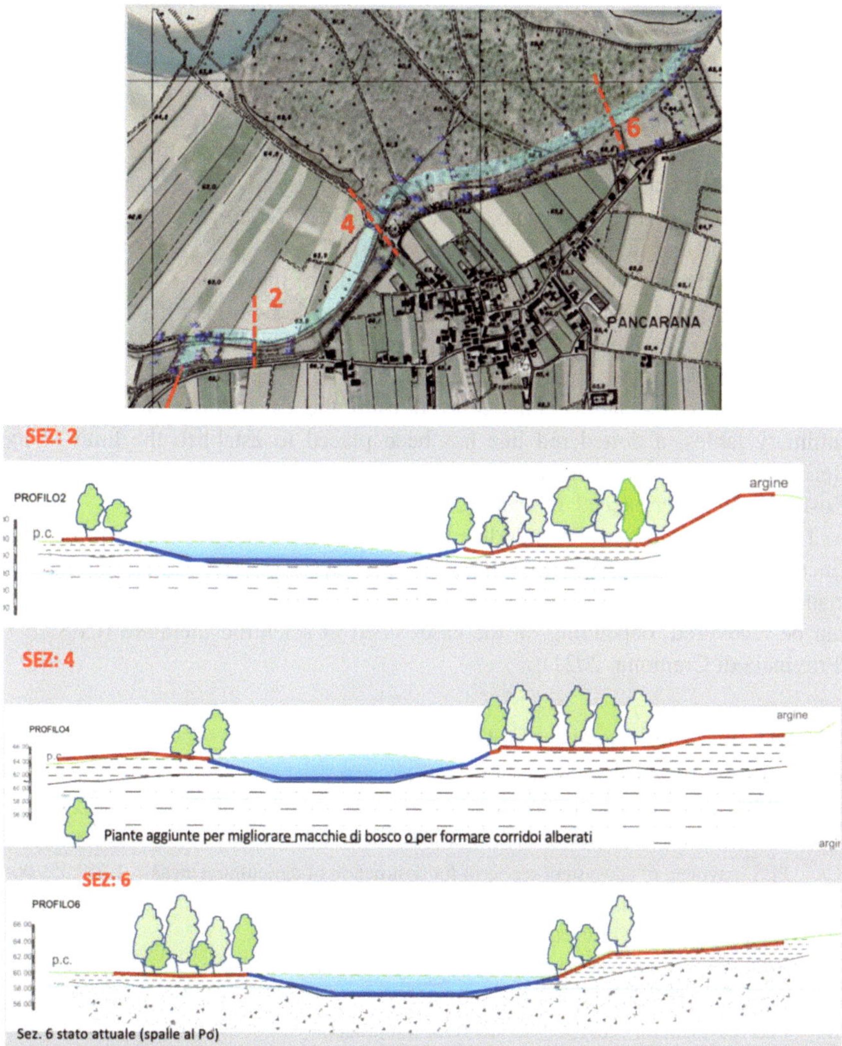

Fig. 14.14 Example of revegetation on the sides of the new course of the D2 diversion. The trees to be added are clearly marked: the shrubs have not been marked for greater clarity of the drawing

14.1.5 Cost/Benefit Forecast

Let us recall once again an important economic principle: what distinguishes an ecological economy from the dominant neoclassical economy is the consideration of exchange value (Bjerg, 2016; Daly, 2007). It should always be linked to the role of things in nature: it is only neoclassical economics that claims otherwise. But in any case, someone will have to pay for the services. In our case study, we must consider the costs and benefits within 20 years, given that the construction is foreseeable in stages over 6–8 years from preparing the final project and expropriations. The costs will be estimated (Table 14.6) through four arguments relating to costs and as many relating to benefits

It should be noted that the PES (Payment of Ecosystem Services) are currently theoretical costs, as our laws have not yet made the "Ruralization Charges" mandatory. However, they are necessary for a correct Cost/Benefit balance.

As reported in the upper part of Table 14.6, the summary table of all Costs for the LU-B, floodplain of the Po, are not too high, being less than €4 million. The highest cost, however, concerns the PES, which, as is known, are higher than the market costs, despite being the most ecologically correct, because they consider: (a) market costs, (b) compensation costs, (c) lower health risks.

Similarly, benefits are shown in the bottom part of the Table 14.6: B1, B2, B3, B4, for LU-B give a result of over €5.5 million, decidedly higher than the Costs.

The Benefits/Costs ratio matters; in this case, this value is 1.475, so the benefits are almost 50% greater than the Costs. As can be seen from Fig. 14.15, which compares the costs and benefits derived from the calculations reported in the summary tables, a dotted red line has been placed to establish the limit of the Benefits: they must be at least 30% higher than the Costs, or, if we prefer, the Costs should never exceed 75% of the Benefits.

This relationship must be underlined, given that trees and green areas provide an "income" that is largely higher than the cost necessary for their planting and maintenance: annually, for every euro invested in greenery, between 1.30 and 3.00 can be recovered, depending on the cases cited in scientific literature (CESAF—(Provincia di Cremona, 2021)).

Table 14.6 List of topics regarding Cost/Benefits Pancarana Golena/Floodplain LU

C1	Diversion Construction + General Expenses	1,429,069.00
C2	Reforestation and grassing plants	493,900.00
C3	Maintenance and unforeseen events (M & I)	210,000.00
C4	PES (payment of ecosystem services) for destruction of agricultural areas	1,669,355.00
	Total expected costs	3,802,324.00
B1	Economic estimate of flood damage elimination	1,588,000.00
B2	Reduction in premature deaths/year (from MR/BF and flood)	1,173,890.00
B3	Absorption of air pollutants by reforestation	122,150.00
B4	PES (payment for ecosystem services) for reforestation	2,846,291.00
	Total expected benefits	5,730,331.00.

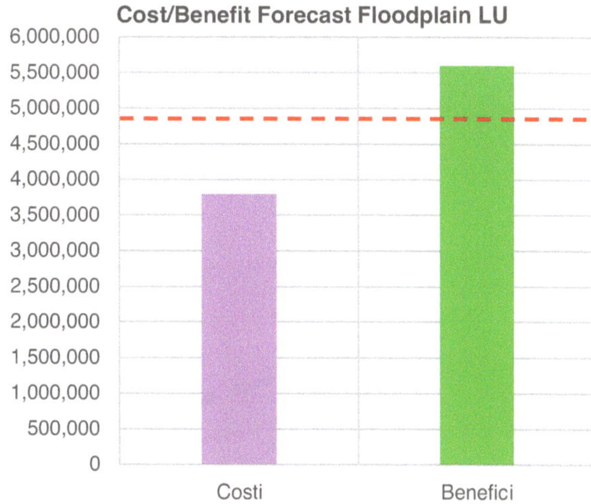

Fig. 14.15 The Cost/Benefit ratio for LU-B is 67.8%. The values are compared with the dotted red line, which shows the minimum value of the Benefits, that is when the B/C index = 1.3

In our case, the ratio B/C = 1.48 (within 20 years of completion of the work) is a forecast entirely in line with the above considerations, as the European Union is keen to underline (Salomoni & Segneghi, 2018; Provincia di Cremona, 2021).

14.1.6 *Reduce the Risk of Premature Death*

Having found an explicit correlation between MR and BF for 72 municipalities in the Milan-Monza area (orange curve in Fig. 14.16) is not enough to make this correspondence valid also in the province of Pavia: it is necessary to adjust the correlation curve with the average age of the local population, which is very high here (MR = 15/1000, blue curve). With a tolerance of 5%, we consider as normal the MR corresponding to BF = 0.95, that is pair to 8.41/1000.

Through the proposed environmental remediation, we could go from BF = 0.73 to BF = 0.80 (green line). The mortality rate should be reduced from MR = 15.55 to MR = 13.72, so we can calculate an estimate of premature mortality (Prm-M) for both the current and future state:

$$Prm\text{-}M = (MR_{BF}–MR_{BF=0{,}95}) \times 76\% = (15,55–8,41) \times 0.76 = 5,43/1000 \times 2.656 = 14,4\ deaths/year(year\ 2020)$$

$$Prm\text{-}M = (MR_{BF}–MR_{BF=0{,}95}) \times 76\% = (13,7–8,41) \times 0.76 = 4,02/1000 \times 2.656 = 10,6\ deaths/year(year\ 2050)$$

As we have seen, since there is (in LU) a population of 2656 inhabitants, a premature mortality due to environmental stress is currently estimated at 14.4 deaths/year. By improving the bionomic state of LU with an appropriate intervention (green line in Fig. 14.16), 3.8 fewer deaths per year are estimated (73,6% of the present). It should

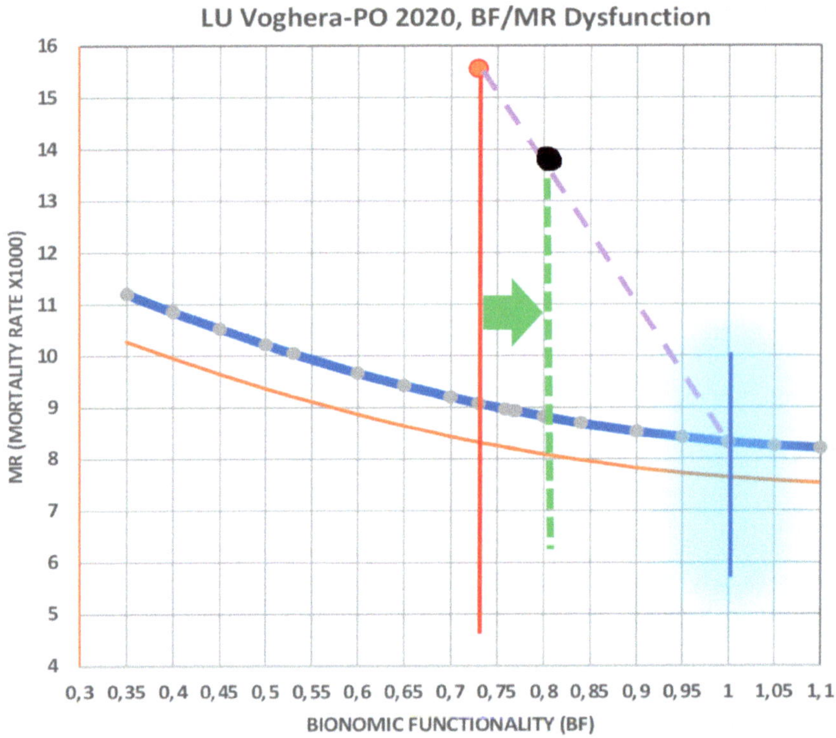

Fig. 14.16 The MR/BF correlation curve (blue) calculated for the LU under examination is related to the local mortality rate (MR = 15.5/1000) and to BF (0.73), which are very altered. By reducing the mean bionomic dysfunction (BF) from 0.73 to 0.80 (green line), the new mortality can be estimated (MR = 13.7 black)

be remembered that each premature mortality is valued (Lattarulo & Plechero, 2005) at least € 600,000–700,000 (medical costs + social costs), therefore even by decreasing by 3.8 units, approximately €2,200,000.00 would be saved and this figure could be considered a valid reason to justify the greater expenses due to the design of the Diversion considering the New Sustainability.

14.2 Environmental Restoration of the Po Floodplain (Parma)

14.2.1 Territorial Ecological Frame

The floodplain section of the Po to the north of Parma is 2 to 4 km wide and about 18–20 km long, from the inflow of the Taro to that of the Enza. The AIPo, Agenzia Interregionale del Po (Ingegnoli, 2023) has identified 6–7 locations suitable for renaturation interventions, mainly on river islands, to be implemented according to the directives of the PNRR. The area closest to Parma, about 12 km directly to the north, is the floodplain section to the south of Gussola, in a predominantly Lombard territory, characterized by the large state-owned Maria Luigia Island, of about 205 ha (in blue).

In the following figures, we will report some historiographical data on the area under examination, starting from the map of 1670 up to 2005 (Fig. 14.17).

Fig. 14.17 Significant Historical maps of the examined territory. From above: 1670, 1880. Note the presence of mani forest patches, covering about ½ of the floodplain. Today's satellite photo, from Google-Earth, 2022. Maria Luigia's Island and some nearby lots appear to be covered by woods, even if they are poplar groves, often hybrids, which have very little to do with the principles of naturalness

In the seventeenth century, the course of the Po made a convex arch towards the North, a few hundred meters from the town of Gussola (Cremona), but under this arch, rather complex braids branched out.

At the time of Lombardy-Veneto, the Habsburg map 1: 25,000 of 1840 shows intense changes, which are very close to the current state of the Po, with a backwater in place of the central arch mentioned above. It should be noted that wooded areas cover more than half of the vast floodplain, while the agricultural space does not reach 1/3 of the area.

Another excerpt from the Brioschi Table from the late nineteenth century (1881) shows that the forest cover has remained almost unchanged since 1840. We can affirm that for centuries, until the first decades of the twentieth century, this stretch of floodplain had remained natural.

Today's satellite photo, from Google-Earth, 2022 (Fig. 14.17, below) shows Maria Luigia Island and some nearby lots that appear to be covered by woods, even if they are poplar groves, often hybrids, which have very little to do with the principles of naturalness.

14.2.2 Territorial Constraints: Zps-It20a0503

Having been a natural area for centuries, as we have seen, and having maintained an ad hoc potential, the Lombardy Region decided in 2005 to constrain a large part of this territory according to the dictates of the Natura 2000 Network, which identifies the areas to be protected in the United Europe (Provincia di Cremona, 2021).

Figure 14.18 shows the Special Protection Area IT20A0503 "Isola Maria Lugia", located within the territories of the Municipalities of Gussola, Martignana di Po and Torricella del Pizzo in the province of Cremona. The SPA protects an area of 558 hectares, which extends to an average height of 24 meters above sea level.

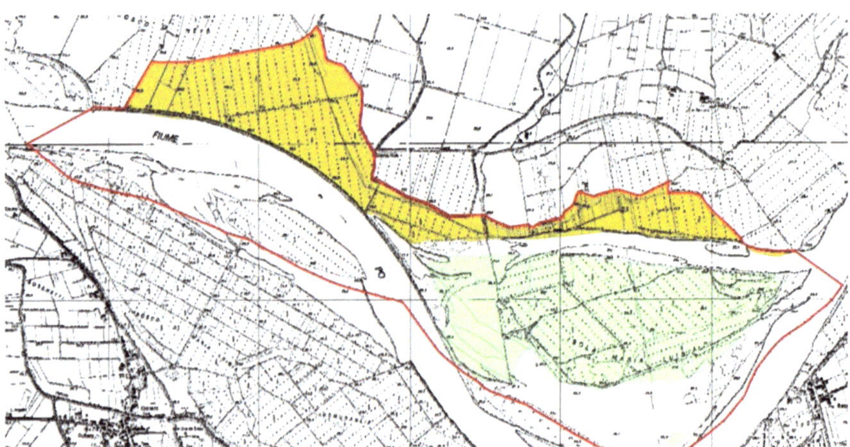

Fig. 14.18 Delimitation of the Special Protection Area (red line) of about 560 ha, in prevalence constituted by state-owned areas (private areas are in yellow)

According to the "Map of Biogeographic Regions" (document Hab. 95/10) the site belongs to the continental region.

The only other site identified by the Natura 2000 form to maintain relationships with the SPA under consideration is a SIC SPA in Emilia, namely IT4020017 "Area delle risorgive di Viarolo, basins of Torrile, fascia golenale del Po", a site that borders the SPA Isola Maria Luigia to the north, defining a single environmental context.

This protected area in Emilia is grafted onto the SPA in Lombardy, forming a link of the interregional ecological network of the utmost importance. It should be immediately underlined that the Lanca di Gussola is also under the protection of Natura 2000, but it should relate to a special green corridor to the two SPAs mentioned.

Also considering these SPA constraints, the renaturation interventions proposed by AIPo take on an undoubted importance.

14.2.3 *Maria Luigia Island: Present Bionomic State*

As can be seen from Fig. 14.19, at least ¾ of Maria Luigia Island is covered by poplar groves, although often integrated with additional plantations of other species.

Fig. 14.19 The vegetation cover of Maria Luigia Island in its current state. The quincunx geometric arrangements of the poplar groves can be seen, although sometimes integrated with additional species. The two areas marked in yellow have been the subject of (quick) surveys using the LaBiSV

The remaining parts show degraded and chaotic vegetation, with many alien and invasive species.

We will report only the most significant data of the vegetation tiles of the island. In the marginal strips and thickets there is a disordered vegetation, with some *Populus nigra* and *Salix alba,* but mixed with *Populus hybrida* and *Morus alba*, with the shrub layer dominated by *Amorpha fruticosa* (North America) with some *Buddleia davidii* (Asiatic) accompanied by cosmopolitan herbaceous plants such as *Fragmites australis* and *Parietaria officinalis*, but with the addition of many often invasive alien species: *Rudbeckia hirta, Sicyos angulatus, Humulus japonicus, Amaranthus deflexus, Phytolacca americana,* etc.

In Fig. 14.20 we see the oxbow lake to the north of the island, surrounded by low scrub vegetation of the *Chenopodium rubri* type, with several alien species.

The surveys carried out by the writer were made (A) on a 5–6 ha patch of *Populus alba* poplar grove (Fig. 14.21) with some presence of *Populus nigra* and *Fraxinus oxicarpa*, therefore with relatively few alien species, shrubs or herbaceous, and

Fig. 14.20 The vegetation cover of Maria Luigia Island is very disorderly in many parts, especially towards the edges, and contains many non-native plants

Fig. 14.21 Survey (A): artificial poplar grove of Populus alba with 10,6% of allochtonous species

(B) on a 3–4 ha patch of *Populus alba* poplar grove, but integrated with *Ulmus minor, Fraxinus oxycarpa* and *Robinia pseudoacacia.*

The previous Fig. 14.22 summarize the findings of the two forest tiles A.

TEMPERATE FOREST	1	5	12	22	Mod. Inflextion = 130 a
T. Tessera characters (Ts)					
T1 – Vegetation height (m)	< 9	9.1-18	18.1-29	> 29.1	Canopy Hc = **25**...m
T2 – Cover of the canopy (%)	< 30	> 90	31-60	61-90	Ts surface
T3 – Structural differentiation	low	medium	good	high	Age, space groups, etc.
T4- Interior/edge (%)	none	< 30	31-89	> 90	(% Ts)
T5 - Management	simple coppice	coppice	wood	Natural forest	Natural must be uneven!
T6 – Permanence (years)	< 80	81-160	161-240	> 240	Old trees (Mother Tr.)
F. VEGETATIONAL BIOMASS (ABOVE GROUND)					
F1- Dead plant biomass	near 0	> 10	0,1-0,5	05,10,0	% of living biomass
F2- Litter depth	near 0	< 1.5	1.6-3.5	> 3.5	cm
F3 t - Biomass volume (m^3/ha)	< 150	150-350	350-600	> 600	pB = **405**....... m^3/ha
E. ECOCOENOTOPE PARAMETERS					
E1- Dominant species (n°)	> 3	3	2	1	as pB volume
E2- Species richness	< 15	16-30	31-40	> 40	n° sp./Tessera
E3- Allocthonous species (%)	> 10	10-4,0	< 4	0	From other ecoregions
E4- Ectomycorhizal Fungi	< 2, no web	3-6,0 part. web	7-12 (web)	> 12 (web)	Species n° & web
E5- Infesting plants %	near all	> 25	< 25	0	Covering area on Ts
E6- Diseased plants	evident	suspect	risk	0	Even acid rain damage
E7- Biological forms (n°)	< 3	4-5,0	6-7,0	> 7	Cfr. Box [15], mod.
E8- Vertical stratification	2	3	4	> 4	traditional
E9- Renewal capacity	none	intense	sporadic	normal	Dominant species
E10- Dynamic state	degradation	recreation	regeneration	fluctuation	Cfr. Ingegnoli [10]
U. LANDSCAPE UNIT (LU) PARAMETERS					
U1- Similar veg. contiguity	0	< 25	26-75	> 76	% of perimeter
U2- Source or sink	sink	neutral	partial	actual source	Species & resources
U3- Functional role in LU	reduced	minor	evident	important	Context & typology
U4- Disturbance incorporation	insufficient	scarce	normal	high	Local disturbances
U5- Geophysical instabilities	evident	partial	risk	none	On the physiotope
U6- Permeant fauna interest	low	medium	good	attraction	Key species
U7- Transformation modall. of the Ts	strong disturbances	gradual changes	temporal instabilities	fluctuation	Today + tendency
U8- Landscape pathology interference	serious	near chronicle	easy to incorpor.	none	from surroundings ecotopes
U9- Permanence of analogous vegetation (years)	< 100	100-300	300-1200	> 1200	Historical presence
RESULTS OF THE SURVEY					
Total score Y (= h+j+k+w)	h = 10	J = ...10...	K = ...7	w = ...1...	**Y = ...166**
Bionomic quality of the Ts (bQ)	**bQ_b = Y / 700 (b) bQ_t= Y / 616 (t)**				**bQ = 26,95 [%]**
Estimation of the BTC Temperate (deciduous) Forest	**BTC (t) = 0,01667 (y-28) + 0,13 (FM / 65)**				**BTC = 3,10** [Mcal/m^2/year]
Ta = apparent age	**Ta = $(BTC+0,77)^{1/0,515}$**				Ta = **13,81** years
Tb Balanced age	**(Dmd+Dold)/4] / (0,4 to 0,7)***				Tb-Age = ...(years)
Concise bionomic state	**CBSt = (Mtl x bQ) %, where Mtl = Ta/125**				CBSt = **2,98** (%)

Fig. 14.22 Results of the Vegetation Survey of a tessera of Artificial Poplar grove detected according to the LaBiSV methodology (october 2023) within the Maria Luigia Isle, Gussola: Surface about 5,0 ha

Phytocoenosis A with:

- height of the canopies Hc = 25.0 m;
- phytomass pB = 405 m^3/ha;
- number of species Sp = 21, of which 10.6% are alien,
- BTC = 3.10 Mcal/m^2/year;
- CBSt = 2.98%;
- soil humification RxN = 39.4.

Phytocoenosis B with:

- height of the canopies Hc = 15.7 m;
- phytomass pB = 200 m^3/ha;
- number of species Sp = 24, of which 21.7% are alien,
- BTC = 2.80 Mcal/m^2/year;
- CBSt = 2.64%;
- soil humification RxN = 31.0.

Although the two surveys should be considered as expeditious, they are sufficient to have an order of magnitude on the bionomic state of the involved woods, which is undoubtedly low.

The floodplain-high forests along the Ticino can reach even higher values, such as in Robecco, Bosco Fasolo (*Querco-Carpinetum*):

- crown height Hc = 24.1 m;
- phytomass pB = 477 m^3/ha;
- number of species Sp = 37, of which 12.2% are alien,
- BTC = 7.94 Mcal/m^2/year;
- CBSt = 30.61%;
- soil humification RxN = 30.6.

Comparing all these results, also with that of the surveys carried out on the promontory in Pancarana along the Po between Voghera and Pavia (Sect. 14.1.2), we can find a range of BTC values ranking from 2.20 up to 3.90.

Having most artificial woods with a value equal to BTC = 2.6–3.8 Mcal/m^2/year, the average BTC of the island can be estimated at around 3.00. The remaining part of the previously identified floodplain section is largely agricultural, with some artificial poplar groves and a few tree rows, therefore with an average BTC estimated at around 1.30–1.40. Including the island Maria-Luigia, it will be possible to go up to BTC = 1.80 Mcal/m^2/year.

14.2.4 Golena di Gussola Renaturalization and Mortality Rate

As expressed in Fig. 14.23, if we want to reverse the trend of the MR (Mortality Rate) in Parma we must increase the radius of influence of the BTC of the Po floodplain from 3 to 12 km. This requires increasing the current value of the average

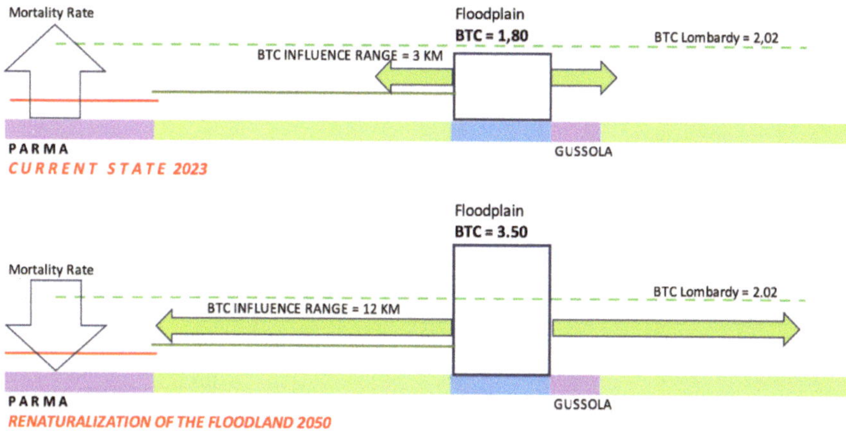

Fig. 14.23 If we want to reverse the trend of the MR (Mortality Rate) in Parma, we must increase the radius of influence of the BTC of the Po floodplain from 3 to 12 km. This requires increasing the current value of the average BTC of the floodplain from 1.80 to 3.50 Mcal/m^2/year

BTC of the floodplain from 1.80 to 3.50 Mcal/m^2/year, re-naturalizing the floodplain.

The distribution of European forests sees in the Po Valley a vast territory almost entirely without forests. Furthermore (Fig. 14.24), Veneto, Emilia and Lombardy have average BTCs lower than the average of the Emerged Lands. Both in Lombardy and Emilia, the population is clearly prevalent in the plains, where over 70% live. But precisely in the plains the presence of forests is not only very low (7% in Lombardy and 21% in Emilia), but very altered, with many alien species and counting the artificial forests of hybrid poplar.

Recalling what was stated in Par.9.4.2 on the onset of cancer in Western Europe and the correlation with the presence of forests (importance of the BF indicator, Bionomic Functionality), we must note that the renaturalization of the Golena di Gussola also takes an aspects of protection and improvement of human health, e.g., following the research of Li and Ingegnoli, Lombardo & La Torre (Ingegnoli et al., 2022; Li, 2009).

Let it be clear that if we integrate the reasons for the naturalistic protection of the territory, reiterated in the SPA area already mentioned, with those for the protection of human health, it turns out that Renaturalization entails a priority moral obligation.

However, bringing back to Golena the forest cover seen in the maps of past centuries may be unsustainable for local farmers. For this reason, it is necessary to renaturalize the forests in state-owned areas and limit us to imposing at least (a) the presence of tree-lined corridors along the fields in private areas and (b) the cultivation of poplar with non-native species.

Even about the forests to be renaturalized (or replanted in a naturalistically correct way), the main rules must be kept in mind, such as:

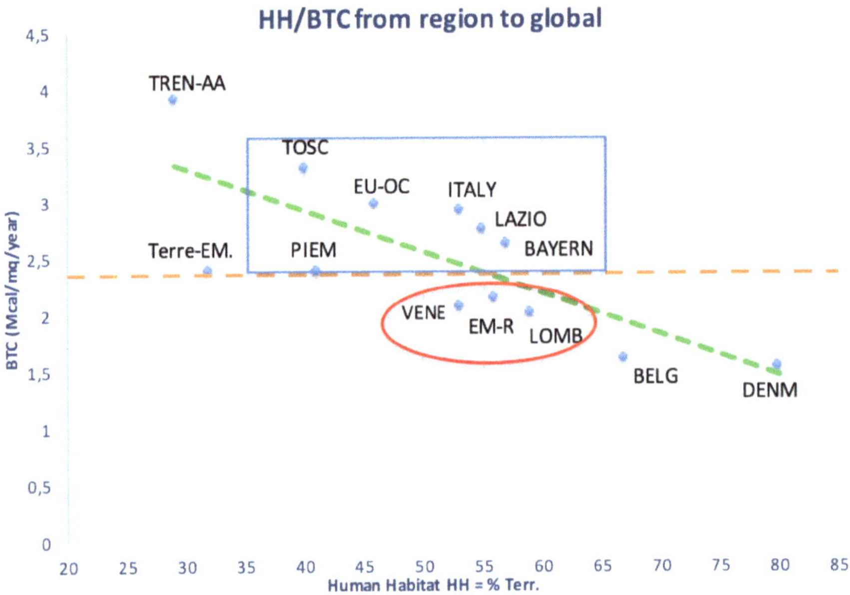

Fig. 14.24 HH-BTC model in which some European nations and regions are highlighted with their average BTC value. These values are compared with the average BTC of the Emerged Lands (orange), equal to BTC = 2.30–2.40 Mcal/m^2/year. Note that Veneto, Emilia and Lombardy have BTC lower than the average of the Emerged Lands. Legend: TREN-AA (Trentino Alto-Adige), TOSC (Toscana), EU-OC (Europa Occidentale), Terre-EM (Terre Emerse), PIEM (Piemonte), VENE (Veneto), EM-R (Emilia-Romagna), LOMB (Lombardia), Belg (Belgium), DENM (Denmark)

1. The Po floodplain must become a source area with respect to the adjacent plain strips.
2. The historical dynamics of the forest areas must be reconstructed, to the extent possible.
3. The interventions can continue to be in spaced-out patches, but with two conditions:

   - be included in an ecological floodplain network with adjacent connections on the margin.
   - have efficient connectivity and circuitry within the network connection.

4. The average floodplain BTC must be higher than the average BTC of the neighboring regions by at least 50%.
5. Some forestry interventions should present a gradient of vegetation types up to the margin.
6. The areas left for agriculture must be restructured with agro-forestry systems according to bionomic principles.

7. The control of the bionomic influence rays must always be carried out for each stretch of floodplain and linked to studies on the PCA (Potential Core Area).
8. Reforestation must not follow geometric and equisetaneous plantations.

14.2.5 Cost/Benefits and Conclusions

It should be remembered that the reduction in the risk of premature mortality is proportional to the level of bionomic functionality of a landscape unit, therefore, primarily, to the average level of the BTC. It should be added that in addition to human tragedies, premature mortality also brings significant economic and social imbalances.

The economic costs of premature deaths can be estimated, inspired by similar calculations regarding the risks of pollution (Lattarulo & Plechero, 2005). This estimate considers both direct costs (hospitalization, drugs, ambulances, etc.) and indirect costs (e.g., loss of work and life expectancy) and the sum of these costs is equal to €75,000–€100,000 + €450,000 = €550,000/death, which reported today reaches €550,000 × 1.321 = €726,000/death.

The current mortality rate of Parma is MR = 11/1000, so the deaths per year are 220,000 × 0.011 = 2420. If, as an order of magnitude, we consider the risk of premature mortality like that of the Milan area (same MR/BF), we will have

$$2,4200 \times 0.06 = 145 \text{ deaths/year}.$$

An improvement of even just 5% for BF regrowth (increase in the bionomic radius of the Golena-Po in question) decreases this mortality by about 8 units within a few years, with an economic saving of about 726,000 × 8 = 5,808,000 €. But be careful: a valid renaturalization can even reach four times as much.

Given that the re-naturalization of about 200 ha (Isola M.Luigia) can cost 12,000 €/ha, this would result in 2,400,000 €. The benefit/cost ratio

$$B/C = 5.80/2.40 \text{ million } € = 2.41 > 1.30$$

value decidedly higher, almost double than the minimum acceptable by the EU ecology sector, equal to 1.30.

The fundamental question remains: will it be possible to renaturalize the Golena di Gussola? The question must not be misinterpreted, it is not intended to be captious: it seems, in fact, that at the present time the desired renaturalization should not be done. Both the regions of Lombardy and Emilia-Romagna and the Ministries of Environment and Agriculture seem to give more importance to the local trade associations of farmers and poplar grovers.

These are political choices that are incomprehensible to the undersigned author of these notes, as it is not so difficult to understand that to help a few hundred people, one can prevent the improvement of the health of hundreds of thousands of inhabitants of the surrounding area.

We will conclude by recalling that we cannot forget environmental ethics with impunity, as stated in the Encyclical of Pope Francis (Pope Francis Ist, 2015).

References

Bjerg O (2016) Parallax of growth, the philosophy of ecology and economy. Polity Press, .

Daly, H. E. (2007). *Ecological economics and sustainable development, selected essays of H. Daly*. Edward Elgar Publ.

Ingegnoli, V. (2002). *Landscape ecology: A widening foundation* (p. XXIII+357). Springer.

Ingegnoli, V. (2013). Concise evaluation of the bionomic state of natural and human vegetation elements in a landscape. *Rendiconti Lincei. Scienze Fisiche e Naturali*. https://doi.org/10.1007/s12210-013-0252-2

Ingegnoli, V. (2015). *Landscape bionomics. Biological-integrated landscape ecology* (p. XXIV + 431). Springer.

Ingegnoli, V. (2023). *Corso di formazione per Tecnici AIPo in materia di "Bionomia del paesaggio"*. Parma.

Ingegnoli, V., & Giglio, E. (2005). *Ecologia del Paesaggio: manuale per conservare, gestire e pianificare l'ambiente* (p. 685+XVI). Sistemi editoriali SE.

Ingegnoli, V., & Pignatti, S. (2007). *The impact of the widened landscape ecology on vegetation science: towards the new paradigm* (Vol. XVIII, pp. 89–122). Springer Link: Rendiconti Lincei Scienze Fisiche e Naturali, s.IX.

Ingegnoli, V., Lombardo, F., & LaTorre, G. (Eds.). (2022). *Environmental alteration leads to human disease: A planetary health approach*. Springer-Nature. Sustainable Development Golas Series. Foreword by Andrea Lenzi.

Lattarulo, P., & Plechero, M. (2005). *Traffico e Inquinamento: i danni per la salute dell'uomo e i Costi Sociali*. Irpet.

Li, Q. (2009). Effect of forest bathing trips on human immune function. *Environmental Health and Preventive Medicine, 15*(1), 9–17.

Pedrotti, F., & Gafta, D. (1966). Ecologia delle foreste ripariali e paludose dell'Italia. *L'Uomo e l'Ambiente, 23*, 1–165.

Pignatti, S. (1982). *La Flora d'Italia*. Edagricole.

Pignatti, S. (2005). Valori di bioindicazione delle piante vascolari della flora italiana. *Braun-Blanquetia, 39*, 1–97.

Pope Francis Ist. (2015). Laudato Sì. Encyclical. Vatican, Rome.

Provincia di Cremona. (2021). Piano di Gestione della ZPS IT20A0503 Isola Maria Luigia.

Salomoni, M. T., & Segneghi, F. (2018). *Gli Alberi e la Città*, in appendice: Valutazione Benefits. In Rebus, Regione Emilia-Romagna.

VinGeo. (2021–2022). Indagini e Studi Integrativi a supporto della redazione Progetto di Fattibilità Tecnica ed Economica delle Opere di regimazione Idraulica torrente STAFFORA a difesa degli abitati di Oriolo, *Cervesina e Pancarana (PV)*. AIPo, Pavia.

Chapter 15
Environmental Bioethics and Education

Contents

Abstract The placing of blind faith in the possibilities of scientific and technological knowledge of man, which results in the relativization of the Truth in the name of progress, denies the dependency that is underlying the concept of Creation. It is not the adoration of progress that will be able to preserve creation, but rather the search for Truth, which is the primary task and essential role of all those endowed with thought. Knowledge implies an act of Truth and at the same time of emotional participation, that is, of "love". Above all, it is noted that Truth and Non-violence demonstrate the maximum concreteness and rationality of the principle, because *ahimsa* (sensu Gandhi) is the only way to approach the Truth, avoiding the possibility that any form of violence could impose the false: Gandhi extended this concept to all life forms and the environment, arguing that the destruction of natural resources was a form of violence.

"The universe is not the product of darkness and unreason. It comes from Intelligence, Freedom, and from the Beauty that is identical with Love" says Albert Einstein. "Being is Supreme Wisdom. Everything that exists in the present arises from the interaction between Purusha and Prakriti, Spirit and Nature, which are the Parents of everything and every being" is affirmed in Chap. 4 of Bhagavad Gita. So,

V. Ingegnoli, *Landscape Bionomics and Planetary Health*, Landscape Series 21,
https://doi.org/10.1007/978-3-032-08908-3_15

Man must obey the laws of nature, which govern biological systems capable of managing many aspects of their components (including man). Let us remember that already Saint Francis, in the *Laudes Creaturarum*, had recognized that *Mother Earth* had the ability to *support* but also *govern* living beings, including man. Bearing witness to the Truth and its consequences, and acting according to it, requires a sacrifice so closely linked to reality, that it constitutes the law placed at the foundations of civilization. Civilization, as Gandhi maintained, does not consist in the multiplication, but in the voluntary and deliberate *restriction* of needs. Following the words of last four Popes, the urgent need for humanity to take seriously into consideration the necessity of an ecological conversion becomes evident.

Antropocene Vs Ecocene. Environmental ethics must be based on at least five assertions: (a) Life is a process of knowledge, able to perceive and process information, to learn, to reproduce itself, to have an history and to participate in the evolution, open, self-transcendent, self-organising, autopoietic, adaptive, dissipative, dynamic; (b) Life cannot exist without its environment; (c) Life on Earth is organized into hierarchical levels, each of which is a complex system; (d) Each level is unique because of its emergent properties; (e) Each level has its own characteristics but, through appropriate constraints, the processes of the higher scale explain the meaning of the level examined, while the processes of the lower scale explain its origin; (f) Man is part of this biological hierarchy and evolves with Nature.

The need for the new figure of the *Doctor of environmental systems and of their relations with human health* (*ecoiatrist*) is reiterated once again: so, in analogy with Hippocrates, an "*Ecoiatrist Oath*", to forcefully indicate the ethical commitment of the ecologist.

Keywords Truth · Non-violence · Justice · Life · Mother earth · Ecoiatrist · Environmental ethics · Environmental doctor · Civilization · Conversion · Oath · Sustainability

15.1 Truth and Justice

15.1.1 The Lie of Self-Referential Freedom

The placing of blind faith in the possibilities of scientific and technological knowledge of man, which results in the relativization of the truth in the name of progress, denies the dependency that is underlying the concept of nature as Creation.

Today, most men do not believe in the created world, but in the world to be created, which no longer needs believe, but only capacity. Transformation becomes therefore the fundamental task of man. Progress is dependent on the change of the truth, a relative truth, which is essential for 'the good of the cause' (Ratzinger, 2006). So, today's version of the myth of Icarus leads to see the man as a force opposed to nature, in the name of a self-referential freedom.

Nature is no longer a prerequisite able to afford and, at the same time, to limit and to direct human actions but it is an object in the hands of man: a man not created but produced by "chance and necessity" (Monod, 1970). The only creator is man, the true creator of himself, hence the unfailing faith in science and technology. Adam and Eve ate the fruit of the tree of knowledge and wanted to replace the Creator.

This absolute confidence in human capacities devoids nature of any practical importance and considers it as a sub-system of the economy (Daly & Farley, 2004). Changing point of view, "we are also witnessing the spreading of the blindness of a purely technical and economic rationality which ignores the profound realities of humanity" (Morin, 2025). This is evidenced by the fact that, when the scarcity of a resource, or an excessive pollution, threaten to cause radical changes in the environment, with negative influences on health, science and technology intervene only with palliatives able to avert the danger, maintaining the status quo.

Relativism, scientific progressivism, and ontological reductionism are among the greatest idolatries of these times, and it is worth remembering how much influence they have on ecology, and medicine, as exposed in previous chapters: in most cases, it means stopping at pseudo-theories without trying to go further, replacing the search for truth with convenient pseudo-truths. Moreover: Progress as ideology is an attack on nature. Unequivocal words were written on the concept of progress to measure the underlying error. Below is a statement of Gandhi, reported by Lanza del Vasto (1980), his greatest disciple in Europe:

> The progress or the acceleration makes more speed necessary, the redundancy more redundancy, the vanity more vanity, the violence more violence: everything is accelerated, magnifying like an avalanche. The avalanche must meet a rock to shatter itself upon, a village to take away and, at its height, it stops choking all life.

John Paul II (1979), taking up the well-known argument of Saint Paul in the Letter to the Romans, underlines the importance of the revelation of the children of God with respect to nature:

> The immense progress, never known, which has taken place, particularly in our century, in the field of man's dominion over the world, does it not itself reveal -and moreover to a degree never achieved – that multifaceted submission "to transience"? It is enough here to recall certain phenomena, such as the threat of pollution of the natural environment [...] or armed conflicts [...] the prospects of self-destruction using atomic weapons [...] the lack of respect for the life of the unborn. The world of the new epoch, the world of cosmic flights, the world of scientific and technical conquests, never achieved, is it not at the same time the world that "groans and suffers" and "waits with eager longing for the revealing of the sons of God"?

Let us observe, however, that the greatest scientists reject scientism and are wary of all preconceptions praising progress, because, as written by Einstein (1953): "in the laws of nature is revealed a reason so superior that all rationality of thought and of human laws is – in comparison – a very insignificant reflection". More recently, Faggin (2024): "Look at scientism: it pushes us to believe that reality is made only of matter, and everything else follows suit. In this way, for example, we end up exalting competition. In the vision I propose, however, there is only cooperation, because we are integral parts of the One [a self-aware Universe], and the One is not competitive, it is cooperative. Instead, we have accepted the principle of the survival of the fittest, the strongest".

15.1.2 Truth and Justice

Indeed, it is not the adoration of progress that will preserve creation, but the search for Truth: the primary task and essential role of all those endowed with thought. Truth who is the essence of all reality, perceivable or not, even if expressed through different cultural shades:

> *Being is Supreme Wisdom. Everything that exists in the present arises from the interaction between Purusha and Prakriti, Spirit and Nature, which are the parents of everything and every being* (Bhagavad Gita, Chap.4, Radahakrishnan & Vecchiotti, 1978).

> *The universe is not the product of darkness and unreason. It comes from intelligence, freedom, and from the beauty that is identical with love* (Einstein, 1953).

> *God made the world to establish a space to communicate his undying covenantal love and draw all creatures to Himself* (Ratzinger, 1996).

> *All that exists is created by One, One of pure light and spirit, dynamic, holistic, conscious. Reality is made of parts-wholes: we exist as parts of a whole, directed by a consciousness that is elsewhere* (Faggin, 2024).

Only in the light of Truth even apparently opposed categories, such as science and metaphysics, can present unpredictable and profitable correspondences that it would be simply wrong to ignore, as Konrad Lorenz (1973) notes.

For the understanding of reality, reason and faith are inseparable, because knowing without believing is like not knowing. This is why Joseph Ratzinger (1996) wrote:

> The reason of the world allows us to recognize the reason of God. The Bible is and remains the true Enlightenment that entrusted the world to reason and not to the exploitation of man, because it opened reason to the truth and love of God.

Knowledge implies an act of truth and, at the same time, emotional participation, that is, "love". This synthesis is evident in the Mahatma's writings and speeches, starting from his well-known autobiography (Gandhi, 1926, 1962). The ancient classical world also intuited this fact: in Latin, for example, the verb "to study" implies a passion for the object one aspires to, as *studere puellam, studere naturalem historiam*.

The heart of the matter lies in the fact that Truth is *ahimsa*, in the sense of *non-violence, charity, and love*. Gandhi said that Truth and non-violence are two sides of the same coin, without effigy and without depth: the first is the end, and the second is the means. Any violence used in presenting a truth ends up prevailing upon it, distorting it and making it so that Truth isn't accepted because of its own merits and its adherence to reality, but because of the violence used to promote it. Therefore, it is noted that Truth and nonviolence demonstrate the maximum concreteness and rationality.

What has been exposed highlights how, in the search for understanding, the greatest danger for biological conservation, landscape planning, sustainable development, strategic environmental assessment, prevention, and integrated health care is represented by the betrayal of the role of man, expressed as violence against life in all its aspects, even in the form of pseudo-scientific arrogance. A betrayal demands a principle of Justice: we speak of Environmental Ethics as a necessary reference in the relationship between man and nature. Marcus Tullius Cicero was already aware of this (De Legibus, I, XXIII) (Cicero, 1973). He recalled the Socratic precepts and a vision of nature capable of bringing man back to his responsibilities:

> Idemque cum caelum, terras, maria, rerumque omnium naturam perspexerit, eaque unde generata, quo recursura, quando, quo modo obitura, quid in iis mortale et caducum quid divinum aeternumque sit, viderit, Ipsumque ea moderantem et regentem paene prehenderit, seseque non unis circumdadum moenibus, popularem alicuius definiti loci, sed civem totius mundi, quasi unius urbis agnoverit: in hac ille magnificentia rerum atque in hoc conspectu et cognitione naturae, Dii immortales! quam se ipse noscet! Quod Apollo Praecipit Pythius; quam contemnet, quam despiciet, quam pro nihilo putabit ea, quae vulgo dicuntur amplissima![1]

The most relevant ethical question arises from the fact that translating 'truth into deeds' (i.e., justice, according to Gandhi) but also acting according to what science has shown to be necessary, usually requires a vision of the causal chains among the laws of nature, pushed to the extreme consequences. Lorenz (1973) writes that what the Truth makes obligatory cannot cancel our free will: however, it can nail us to our responsibilities. For example, in Italy, numerous cases of construction (with regular permits!) of houses in floodplain areas or on landslide slopes, despite the geologists' warnings, often have tragic consequences. Bearing witness to the Truth requires sacrifices, even extreme ones. "Disinformation has a strong advantage over truth, because re-establishing the truth is terribly onerous" says E. Morin (2025). History teaches us that Socrates had to drink hemlock, Gandhi was killed with a gunshot, and Bishop Luigi Padovese was stabbed to death in Türkiye. The martyrs are countless. Even Christ was not spared: history records that He told Pontius Pilate, "I have come to testify to the Truth." The man of power responded: "What is the Truth?"

Even today, we often read that anyone who wants to follow a genuinely scientific conception must be secular, or rather atheist, "irreducibly atheist to the bone" in order not to have to choose "between faith in God and faith in Darwin," or "postulating an agent superior to nature means ceasing to do science". Many secular and atheist scientists, often polemical to the point of becoming spiteful and sometimes violent towards men of faith, do not realize that an objective and impartial research methodology applies even for believing scientists, since faith has nothing to do with magic or fanatism: the very search for truth demands it. The believer has understood what Saint Augustine (386) (Augustine of Hippo, 2005) argues in his

---

[1] Thus, once [man] will have studied heaven, earth, seas and the nature of all the things and will have seen their origin and their possible end, when and how they will perish, what in them is mortal and transient, what divine and eternal, and almost will have touched the Being who governs them, and once [man] will recognize himself as not closed by walls in a limited site, but citizen of the entire world such as a single city: in this magnificence of things, in this spectacle and knowledge of Nature, Immortal Gods, how much he will have known himself such as the precept of Apollo Pytius! How much he will condemn; how much he will despise and consider as nothing what mass reputes as important.

work, entitled (not by chance) *Contra Academicos*: "*Crede ut intelligas, intellige ut credas!*", that means *believe to understand, understand to believe.*

Presumption and unworthy contempt of men of science are therefore at the origin of the anathemas against believing scientists: "obscurantists", "creationists", and "reactionaries", "whose works must be carefully avoided mentioning", even if it is often discovered later that very similar arguments are supported by the most intransigent laymen.

At the opposite are the two mentioned scientists: Einstein (1953) who asserted "Seeing this [The universe ... comes from intelligence, freedom, and ... beauty] gives us the courage to keep on living, and it empowers us, comforted thereby, to take upon ourselves the adventure of life" and Faggin who, recently has become devoting himself to the study of consciousness and free will, arguing in his most recent publications (Faggin, 2022; Faggin, 2024) that consciousness cannot be explained solely in terms of computational or material processes, but is actually a fundamental reality of the universe. The model he proposes is one in which consciousness is an intrinsic property of reality, distinct from matter, and that free will is a direct consequence of it.

15.1.3 Some Misleading Idols

> *Humanity is at a crossroads. Either it returns to believing that it has a different nature than machines, or it will be reduced to a machine among machines.* (Faggin, 2022)

> *If you turn your city into a machine, you will be the ones used to fuel it.* (Lanza del Vasto)

One misleading idol, which we have already had to recall, is the faith in the inexhaustible capacity of human technology. As it emerged, for example, in the debate on nuclear power in Italy, treating this problem not in a scientific way, but almost as in a fairy tale, in which technology would be able to solve everything, as Virginio Bettini et al. (2009) notes with dismay.

Equally dangerous is unbridled capitalism, economic "growth" at any cost: the belief that it always brings benefits and allows all men to be made richer and happier. From international data known for some time (UNDP, 1999), it is by following this ideology that the richest fifth of humanity has come to use 82.7% of the world's resources, leaving the poorest fifth only 1.4%! Note that the distribution of resources between high- and low-consumption countries is in a ratio of 75 to 1. The price of this ideology is, in fact, paid by all of nature's living systems.

On the other hand, falsehood also lies in the so-called green fundamentalism, according to which everything made by man is incompatible with nature, and -at the same time- man is considered on the same level as animals. Thus, environmental protection too often proceeds through political slogans, which are dangerous for living systems themselves or clearly inconsistent and utterly detached from true scientific considerations.

The biblical reference to the god Mammon, an Aramaic word that has different meanings but involves wealth and lies, sums up all the idols well; therefore, they are an explicit reference to original sin. In the painting by Beppe Ingegnoli (1901–-1999) (Ardizzon & Romeo, 2010) the golden serpent symbolizes Evil, which generates all violence and destroys man and the Earth (Fig. 15.1).

15.2 Anthropocene or Ecozoic?

15.2.1 Two Opposing Cultures

Globalization is changing culture; it is evident to everyone, but today two opposing cultural positions are facing each other: the Anthropocene and the Ecozoic.

For us, the Anthropocene can be defined as a descriptor of a historical situation in which humanity's responsibility has been superseded by its invasiveness with respect to natural laws, a situation that must be changed in the direction of ecological-bionomic integration. A vision, the Anthropocene, based on utilitarianism, individualism, competition, extreme anthropocentrism, artificialization, little or no respect for Mother Earth: concepts some of which -it must be said- are also embraced by the Greens. Opposite to it stands the Ecozoic, a term coined by Swimme and Berry (1992) of Columbia University, which instead attempts to align human activities with other forces and with the laws that govern the Earth, to achieve a creative and protective balance. Berry talks about a critical point in Earth's history. As he points out, humanity faces a choice: either embrace an "unsustainable Technozoic-Industrial Era" or create the conditions for the transition to an "Ecozoic-Organic Era" (Berry, 2015).

The first vision, still decidedly dominant today, might seem rational, but it is not. The race for enrichment ignores spirituality and considers ecology dependent on the economy, worshipping Progress. Gandhi's disciple, Lanza del Vasto (1943), had questioned him precisely on the scope of this idolatry of Progress: progress, by its very structure, expresses itself as self-exalting feedback.

From a summary of Agostino Cera's study on the Anthropocene/Technocene (Cera, 2019), it can be seen that the history of the term Anthropocene does not originate with Crutzen's repeated proposals (Crutzen & Stoermer, 2000; Crutzen & Schwägerl, 2011) in the Geosphere–Biosphere Programme, but rather with various antecedents, the most significant of which are due to the American philologist G.P. Marsh (1864), the Italian naturalist and geologist Stoppani (1871–1873, 1876), and, in a psychological sense, also to the Russian geologist Verdnadskij (1926,1945, 1977, 1998; also in Kendall, 1990), at the beginning of the twentieth century.

The concept of the Anthropocene, however, has many facets, and Cera identifies the most appropriate interpretation in the one by J. Baskin (2015), who characterizes it not so much as a scientific concept, but as "the conceptual basis of a particular worldview," even an "ideology" or "a paradigm disguised as an era". However, for

Fig. 15.1 The god Mammon, tempera and pastel painting (2 x 1 m) by Enrico Giuseppe Ingegnoli, known as Beppe (1901–1999), executed in 1991–92 to express the effects of Evil which, through greed that generates violence, attempts to destroy man and the Earth

A. Cera, the spectacle we see today is that of a “techno-nature” (Technature), since nature is conceived, perceived, and enjoyed in entirely technical terms within the Anthropocene framework. The fact that the centrality of technology, its imposition as the current epochal subjectivity, can only occur at the cost of the ‘dehumanization of humanity’ is the reason why he proposes to define this aspiring new era (as a discourse already elaborated and a paradigm already implemented) not as the Anthropocene, but rather as the Technocene. The similarity between Berry’s view of the Technozoic Era and the present discussion of the Anthropocene theory is remarkable. “The task of the Technozoic Era is to extend human dominion over Earth, then the rest of the solar system, and beyond. Although this possibility seems like science fiction, it is increasingly evident that many within the scientific and industrial communities are actively working to realize this possibility” (Henning, 2016).

15.2.2 About the Technological World

As shown before, the impact of the human alteration of Earth’s ecological systems on human health is undeniable. The exemplary case of the historic park of Monza has been debated many times. It preserves forest relics dating back to the Middle Ages. It was designed by an illustrious landscape architect such as Luigi Canonica on behalf of the Habsburg Imperial family in 1805. In 1922, a large racetrack was built, ruining Milan’s most significant green lung.

Since the beginning of the third millennium, strong population growth, from 3.3 to 7.0 billion in 1970–2012, and climate change from +0.01 to +0.60 °C in the same period, have accentuated environmental changes and ecological alarm. Land use changes have been produced in the Emerged Lands in the last 125 years (1890–2015) both by forest cuts from 36.7% to 26.8% and by an increase in agricultural areas from 14.1% to 27.7%. This change has decreased the bionomic efficiency of vegetated landscapes from 88% to 65%; therefore, Gaia (i.e. Planet Earth) has lost 7974×10^{18} kcal: a massive flow of energy no longer helpful for its self-regulation (see Chaps. 8 and 12).

The WHO Environmental Burden of Disease estimated deaths from environmental causes at 17.06% in 2002. However, these values became 21.80% in 2012 (+127.8), while Genome-Wide Association Studies (GWAS) reduced morbidity from genetic causes to below 20% (Dotta, 2022), thus *giving even more importance to environmental causes*! That human activities are causing fundamental biophysical changes at the fastest rates in our species’ history, leading to dangerous health effects, has been demonstrated previously using advanced ecological principles that can communicate with systems medicine (see Chaps. 9 and 10).

Today too many people live in an altered environment. Under the culture of the Anthropocene and the neural ‘pruning effect’ (see Chap. 10), men tend to shape their behavior favoring aspects coherent with the dominion of technology. This process is so current today that even many bioethicists and epistemologists affirm that the

Technological World is the authentic environment of man (Agazzi, 2014). Most people prefer to live in large metropolitan areas and megacities rather than in rural countryside or semi-natural landscapes: on a global scale, urbanization has increased from 30% in 1950 to 50% in 2017. Most people believe that agricultural industrialization and biotechnology are the best way to produce our food. Even many people who admit to the ecological crisis continue to follow technology! Some architects, for example (as already illustrated in Chap. 12), propose "vertical forests" in urban landscapes, using sophisticated technologies and much energy, covering the facades of tower buildings with plants or even 'green walls', convinced that they are acting as ecologists. Nothing to do with the use of climbing plants on the facades of buildings of normal height and of green roofs, left to a semi-natural management.

Warning It must be clear that the Technological World *is not* the real environment of man. "We are also witnessing the spread of the blindness of a purely technical and economic rationality that ignores the profound realities of humanity" (Morin, 2025).

The main reasons are:

(i) Technology cannot solve all problems, as demonstrated by the recent COVID-19 pandemic, especially ones concerning complex leaving systems. But the environment, as repeatedly mentioned previously, is a complex hierarchy of complex leaving systems.
(ii) It is not the processing of the largest amount of collected data or the creation of wider databases that will be able to understand the complexity of reality, not without an integrated criterion capable of linking knowledge and tilling it. "We live under the illusion of quantitative progress in knowledge through the growth of information: but the quality of knowledge is declining because we struggle to connect that information". And more: "thought has been reduced to an aid to calculation" (Bocchi & Ceruti, 2007).
(iii) Knowledge must pass through experienced and judged life. But A.I. and technology cannot do so. "If we don't understand that we are more than artificial intelligence, we will be swallowed up by it. It has no conscience and never will. True creativity comes from us". (Faggin, 2024).
(iv) Not all aspects of life on Earth can be modelled and the thresholds for activating or deactivating processes can vary because of the Emergent Properties of each system.
(v) To manage reality, we need a guiding idea or criterium coming from a deep biological-integrated understanding of the Laws of Nature. Landscape Bionomics has demonstrated the complementarity of human and natural habitats, which must be integrated, and therefore the actual relationship between environmental syndromes and human pathologies (see Chaps. 9 and 10).

All these reasons, and more, can be illustrated following an Ecozoic vision and the principles of bionomics (see also previous Chapters).

15.2.3 About the Ecozoic Vision: From Sora Nostra Matre Terra to the Ecological Conversion

Concerning the Ecozoic vision, instead, we may find Saint Francis as its founder. Let us remember that he, in the *Laudes Creaturarum* (1224), already recognized that *sora* (sister) Mother Earth had the ability to *support* but also *govern* living beings, including man:

> Laudato si', mi' Signore, per sora nostra matre terra,
> la quale ne sustenta et governa,
> et produce diversi fructi con coloriti flori et herba[2]

Man must obey the laws of nature, which govern biological systems capable of managing many aspects of their components (including man). An example of this can be seen in the cultural change in the design of gardens during the Industrial Revolution: from French terraced and structured gardens, when the surrounding landscape was cultivated in a network of natural corridors and forest patches; to English semi-natural gardens, when the surrounding was cultivated in wide monocultures. This unconscious change was guided by Mother Earth, in a time when most Westerners maintained a great sensitivity towards nature. Even today, the spontaneous refuge of a child (our son, although raised in the city of Milan but very often taken to the woods) within the roots of the old tree testifies to the recognition of the Earth as a mother "who sustains and governs us" (Figs. 15.2 and 15.3).

Saint Francis said: "Every day we make use of creatures and without them we cannot live, and in them the human race greatly offends the Creator" (Delorme, 1922). Man's role in relation with nature is an exalting one, but one not always easy to understand. In the book of Genesis (Gen 1, 26) is said that:

> *Then god said, "let us make humankind in our image, according to our likeness; and let them* have dominion *over the fish of the sea, and over the birds of the air, and over the cattle, and over all the wild animals of the earth, and over every creeping thing that creeps upon the earth."*

This quote, which underlines man's dominant role in nature, may seem in line with the view suggested by the Anthropocene and in direct contrast with what we have said about the necessity for man to protect and serve the organization of life that allows him to live. This contrast, however, is nothing but apparent, as John Paul II wrote (1979), since the biblical concept of dominion is to be understood in a loving and responsible sense. Indeed, Genesis 2, 15 says:

> *The Lord God took the man and put him in the garden of Eden* to work it and keep it.

An analogous concept is present in the Indu religion, as expressed in Bhagavad Gita, Chap. 3, Verse 14 (Radahakrishnan & Vecchiotti, 1978):

---

[2]Praised be you, my Lord, through our sister, Mother Earth, who sustains us and directs us bringing forth all kinds of fruits and colored flowers and herbs (from The Canticle of the Creatures, 1224).

Fig. 15.2 Drawing "At the Roots of Life", pencil on paper (Vittorio Ingegnoli, 2008), which symbolises the Earth as mother "who sustains and governs us". The design is a real fact: the son of the Author (3.5 years old) took refuge among the roots of the big spruce tree, in the Natural Park of Paneveggio (Trentino-Sud Tirol)

> The man who lives in harmony with nature, who feeds on what the Earth offers him without exploiting it, is the truly wise.

Dante wrote: "All things have order among themselves, and this is the form that the universe makes similar to God" (The Divine Comedy, The Paradyse, Canto I, vv. 103–105). Of this order, which we have tried to show a fragment of in the previous chapters, man is the guardian, but also the only part of creation who can reject and act against it (free will). Franco Nembrini (2020) recalls that recognizing the laws in which God has placed all reality (realism), taking them into account in the path of knowledge of things (reasonableness), and humbly following them in action (morality) is the way in which freedom is fully exercised. It is Dante again (The Divine Comedy, The Purgatory, Canto XVI, vv. 79–80): "To greater force and to a better nature, Though free, ye subject are, […]".

Following this reflection, the need for humanity to take into serious consideration the necessity of an ecological conversion, as underlined by Pope Francis in the Encyclical Laudato Sì (2015), starting from a note of Josef Ratzinger, becomes self-evident:

Fig. 15.3 The young daughter of the Author spontaneously immersed herself in a flowery meadow

> If "the external deserts are multiplying in the world, because the internal deserts have become so vast (Ratzinger)", the ecological crisis is a call to a profound interior conversion. However, we must also recognize that some Christians who are committed and dedicated to prayer, under the pretext of realism and pragmatism, often mock concerns for the environment. Others are passive, they do not decide to change their habits and become inconsistent. What they lack is an ecological conversion, which involves allowing all the consequences of the encounter with Jesus to emerge in their relationships with the world around them. Living the vocation of being custodians of God's work is an essential part of a virtuous existence; it is not something optional or even an aspect.

The ecological conversion forcefully indicated by Pope Francis must become prayer, in the sense that living the vocation of being custodians of the Work of God means acting according to the laws of Creation in such a way that the same actions become a prayer of thanksgiving for the gift of participation in the evolution of Creation itself.

We must return to the *Laudes Creaturarum* (1224) of St. Francis (Fig. 15.4 and remember what has been already emphasized: Mother Earth sustains and governs us. By the will of God:

> Altissimu, onnipotente, bon Signore,
> tue so' le laude, la gloria e 'honore et onne benedictione.
> Ad te solo, Altissimo, se konfâno
> et nullu homo ène dignu te mentovare[3]

[3]Most High, all-powerful, good Lord, yours is the praise, the glory and the honor and every blessing.To you alone, Most High, do they belong, and no one is worthy to speak your name. (from The Canticle of the Creatures, 1224).

Fig. 15.4 The Ecstasy of Saint Francis, oil on panel, painted by the Italian Renaissance master Giovanni Bellini between 1475 and 1480. It is now housed in the Frick Collection in New York City

Acting according to this role requires a form of sacrifice, the sacrifice in bearing witness to the Truth and its consequences, which itself constitutes the law placed at the foundation of civilization. In the real sense of the term, in fact, as Gandhi maintained, civilization does not consist in the multiplication, but in the voluntary and deliberate restriction or restraint of needs. Indeed, only with self-discipline one can think of accepting and following the prescriptions derived from the study of the laws of nature, rejecting idolatry and concupiscence. Too often, however, us humans pretend not to understand.

15.3 The Ecozoic Vision: Environmental Ethics and Bioethics

15.3.1 A Widening Environmental Ethic

Remember: when enlightened by disciplines such as Bionomics, the *Technological World is a syndrome that reduces Gaia's capacity for self-regulation*. Dealing with a living entity, which man is a part of, completely differs from simply controlling man's activities on a geophysical territory. However, the alternative to the Technological World proposed by the Greens and conventional ecologists, is the return to the state of nature. This proposal is impossible, but not because man's actual environment is the technological one. The reason is that contrary to what the reductionism of traditional ecology affirms, the processes are not reversible, as is the case with the process of growth and life of any living entity.

Indeed, bionomic criteria are concerned with irreversible processes and environmental pathology and with a systemic vision of the problem. So? How do we indicate the main guiding principles capable of reorienting man-nature relations? Environmental ethics must be based on at least six assertions:

- Life is a process of knowledge, able to perceive and process information, to learn, to reproduce itself, to have an history and to participate in the evolution, open (exchange of material and energy), self-transcendent (capable of representing, realizing and transforming itself), self-organizing, autopoietic, adaptive, dissipative, dynamic.
- Life cannot exist without its environment.
- Life on Earth is organized into hierarchical levels, each of which is a complex system.
- Each level is unique because of its emergent properties.
- Each level has its characteristics, but through appropriate constraints, the more extensive scale processes explain the meaning of the level examined. In comparison, the processes of the lower scale explain its origin.
- Man is part of this biological hierarchy and evolves with Nature. Moreover, man is the level of self-consciousness of Nature.

Accordingly, man must take care of the Rhythm and Logic of Gaia and of the systemic approach: this leads us to the necessity to let ourselves be governed by Nature, which co-evolves together with the arrow of time, and to take on the responsibility of rehabilitating the role and functionality of our Mother Earth for future generations. Structural and functional environmental alterations lead to alterations in human health: it cannot be forgotten. These alterations are confirmed once more by the recent birth of studies on Planetary Health, which correlate environmental syndromes with human health pathologies.

Environmental Ethics is closely linked to Bioethics, and a renewed ecology requires an Oath of the Ecoiatrist.

15.3.2 Oath of the Ecoiatrist

At the beginning of this work, it was emphasized that the landscape is a living entity, a specific level of life organization on Earth, and, therefore, subject to the laws that govern every other living entity, even if appropriately declined. Thus, the need arises for the traditional figure of the doctor, who takes care of the human body and the population, as well as the veterinarian and the phytopathologist- to be joined by the new figure of the Doctor of Environmental Systems (*Ecoiatrist*). So much so that in the book "Ecologia del Paesaggio: manuale per conservare, gestire e pianificare l'ambiente" E. Giglio and Ingegnoli inserted an "Ecoiatrist Oath", in analogy with Hippocrates' one, to strongly indicate the ethical commitment of the ecologist (Ingegnoli & Giglio, 2005).

Aware

- That life on Earth is organized at hierarchical levels, and they cannot be defined only within each biological entity but also include each specific environment.
- That the ecologist must care for the health of the ecocoenotope levels and the landscape, just as the doctor is responsible for the state of health of the organism and the population levels.
- Of the importance and solemnity of the act I do and the commitment I take on,

I Swear

- To exercise and practice ecology in freedom and independence of judgment and behavior.
- To pursue the exclusive purposes of defending life, protecting the health and harmony of Nature in all its components (including man), and caring for Nature itself, according to the clinical-diagnostic method and the Principle of Strategic Rehabilitation.
- To inspire every professional act with responsibility and scientific, cultural and social commitment.
- To never commit acts that deliberately cause the destruction of a living system.
- To respect ethical principles in my activity without ever using my knowledge against them, in respect of Life.
- To carry out one's work with diligence, competence and prudence in good faith and observing the laws that are not in conflict with the purposes of my profession, following the most advanced scientific methods.
- Exercise my right to conscientious objection when I am asked to act in violation of scientific evidence.
- Actively oppose all those requests in which the insufficiency of time or available resources may prevent the implementation of studies and interventions commensurate with the definition and care of the state of health of the ecological system considered.
- Entrust my reputation only to my professional skills and study ability and to my moral qualities.
- Avoid, even outside of professional practice, any act and behavior that is detrimental to the prestige and dignity of the profession.
- Avail myself of the help of specialist colleagues in every situation in which my specific skills are not adequate.
- Apply myself to all situations, regardless of the space-time scale and the level of urbanization or degradation, with the same commitment.
- Make myself available to the Competent Authority in the event of a public disaster.
- Keep secret everything I see or have seen, understood or perceived, in the exercise of my profession.

15.4 Sustainable Restraint: Key Actions to Overcome the Environmental Crisis

Increasingly, people are wondering what to do, knowing that the resolution to the problems of the relationship between man and the environment cannot depend only on government authorities and international agreements, but also on their responsible actions. Let us go back to Pope Francis, who was very pragmatic and had a clear idea of what is happening to our planet, reporting some notes written by Callegari and Durastante (2023), two journalists who have studied the contents of the Encyclical *Laudato Sì*.

In the Encyclical he proposed three ways:

(i) care for the *natural* environment, therefore the common home.
(ii) care for *people*, therefore, care for oneself and care for others.
(iii) care for the *relationship* between *God, human beings, and the earth.*

The Pontiff identified well what the environmental, social, and economic issues are and, considering them, he proposed growth, but a growth that is sober and compatible with the limits of our planet. Moreover, he proposed universal solidarity through dialogue, through collaboration and creation: that is an *ecological conversion*. This means being able to change lifestyle and consumption style in the perspective of *integral ecology* (i.e. Landscape bionomics) or, perhaps more broadly, of *sustainability*, intended as *sustainable restraint.*

Sustainability rethinking brought inevitably to change democracy together with economy (Mancini, 2015). Leaving aside the already advanced fight against pollution and climate change, or the problem of waste management and recycling, on which work is already underway; we shall try to list some valid suggestions, dividing the responsibilities of Nations from those of the individual citizen -even if such a separation is not always possible in a clear way-.

(A) *NATIONS, and regional governance*

1. Education and refresher courses at Middle and High School, at Academic levels and at technical-political level on Advanced Ecology and Bionomics
2. Review and updating of legislation regarding environmental assessment, ecological planning and the responsibilities and quantification of damage to human health and the environment as a system, controls with related penalties
3. Conservation and revaluation of cultural knowledge and resources
4. Ecological-bionomic and Systemic Medicine advanced education
5. Definition and monitoring of the Potential of Territorial Health Protection of its proper territory
6. Implementation of bionomic-ecological services with Health protection
7. Diagnosis of the state of health of its proper territories at municipal, provincial, regional and national level following the bionomic methodologies and criteria

8. Regeneration of natural habitats, included all kind of protected areas
9. Rehabilitation of agricultural landscapes and incentive of agroecology
10. Defense against hydrogeological damage and priority protection of water resources
11. Stop deforestation and promotion of the rehabilitation of all forest efficiency
12. Urban-ecological planning focusing on the need for bionomic efficiency of green
13. Adoption of the upgraded bionomic criteria in territorial planning
14. Limits on soil consumption and protection & rehabilitation of soil
15. Inclusion of the bionomic accounting of natural heritage within economic choiches
16. Inclusion of Bionomic-ecological Services and their quantification in the State economy
17. Stimulate the transition to a sustainable ecological economy where production and distribution of goods are aimed at caring for people and the environment and not at economic profit
18. Limits on energy consumption by job category and incentive of renewable energy being aware of the limit of Gaia
19. Ecological rationalization of logistics with respect to goods
20. Renewal of the environmental indicators used with advanced systemic ones

(B) *PEOPLE, both as employers & workers and as private citizens*

1. Relaying importance in true advanced environmental education and sensibility
2. Accept and promote the transition to a mindset open to complexity
3. Abandoning both outdated environmentalism/ecologism and prejudices against Nature
4. Encouraging respect for the genius loci/sense and meaning of a territory/landscape
5. Intelligent recovery and revitalization of ancient local cultural (ecological) traditions
6. Favor and teach life within the real world over the virtual one
7. Stop wasting resources (especially water)
8. Assiduous and respectful frequenting of natural/semi-natural places, such as regular walks on trails through mature natural forests
9. Reduce drastically of red meat in current meals
10. Reduce drastically of exotic and extra-continental food consumption
11. Prefer and look for seasonal vegetables and fruit
12. Increase of consumption of organic foods in current meals
13. Re-evaluate and financially support the work and role of farmers
14. Houseplant care and possibly vegetable growing
15. Stop gardening with exotic plants and flowers
16. Refuse to promote and follow the economy and the behaviors which destroy nature

17. Prefer quality over quantity in purchases of all kinds
18. Significantly reduce air and car travel and encourage the use of trains and buses
19. Stop discriminating woman's work and responsibility
20. Stop considering the advanced green choices as opposed to economy

15.5 Agenda 2030 Global Goals 4: Quality Education

"Ensure inclusive and equitable quality education and promote lifelong learning opportunities for all": so, it is said in n° 4 of the Global Goals. And moreover: "Education liberates the intellect, unlocks the imagination and is fundamental for self-respect. It is the key to prosperity and opens a world of opportunities, making it possible for each of us to contribute to a progressive, healthy society [...]".

But "there is a need for a reform of thought and a reform of education that allows us to face complexities, integrate different knowledge and resist regressions.[…] As long as the sense of the complexity of the world, of life, of human beings and of events remains the prerogative of very few, the world will go from bad to worse" says Morin (2025). And even Pope Leo XIV explained to the youth that "in the face of pollution and climate change, the loss of biodiversity, the deterioration of life and social degradation, global inequalities, the lack of drinking water and access to energy for many populations, an ecological education is essential for everyone to reverse the order of things" (2025).

Despite the growing interest in environmental protection (see Laudato Sì) and Sustainability (see UN-SDG), still today in high schools (in Italy) Ecology is not concretely taught, not even its basic principles and their primary applications, while the study of STEM technological subjects is promoted and encouraged, especially from EU. It is true that the discipline is at least motioned in Civic Education and Science, but, even when the mention goes further than surface level ecologism, only problems of Environmental Engineering are usually considered (energy resources, disposal, recycling and reuse of waste, flood defense, pollution reduction). This, of course, completely excludes Ecology, much less Advanced Ecology or Bionomics (Ingegnoli, 2011, 2015), as man remains convinced that he can solve environmental problems through an increase in the use of technology, despite the recent scientific paradigm shift from reductionism to systems theory (Urbani-Ulivi, 2019).

The topics of Bionomics can be grouped into two parts: (a) upgrading the concepts of Biology and (b) applications to the most social-economics important fields.

Let us see in summary what it is about:

A. Traditional biology remains reductionist and determinist, it marginalizes large space-time scales, scale dependency of structure, functions and transformations, and complexity. Therefore, it must overcome the Neo-Darwinist interpretation of evolution; the concept of the dogma of microbiology, according to which every aspect of life depends on DNA; and the concept of interdisciplinarity in favor of transdisciplinarity. It must also be aware of the ambiguity of the concept of ecosystem and the misinterpretation of biodiversity, of the possibility to apply qualitative-quantitative clinical-diagnostic methodology to territories, and to be involved in the study even of anthropized ones from an integrated point of view.

B. Given that it can be demonstrated that near 80% of diseases are of environmental origin, medicine must know the bionomic processes underlying the relationship between environmental alteration and human health, as well as the relationship with the healthiness of foods and, therefore, the bionomic qualities of agricultural landscapes. Both ecological planning of the territory and the architectural design of outdoor spaces must put the harmony of nature and the compensation of environmental deficits before expressive originality. Finally, the economy must review the concept of sustainability considering the new systemic paradigm.

In summary, we need to propose at least refresher seminars on advanced ecology, useful to help teachers at secondary/High schools and university students to understand the importance of and to follow the new scientific paradigm.

References

Agazzi, E. (2014). *Science, metaphysics, religion*. F. Angeli.

Ardizzon, F. F., & Romeo, R. (2010). *Gli Ingegnoli, una Famiglia di Artisti a Milano dal 1850*. Catalogo della Mostra alla Camera dei Deputati.

Augustine of Hippo. (2005). (386) *Contra Academicos*. Reprint Bompiani Testafronte.

Baskin, J. (2015). Paradigm dressed as epoch: The ideology of the Anthropocene. *Environmental Values, 24*, 9–29.

Berry, T. (2015). *Evening thoughts: Reflections on the earth as sacred community*. Counterpoint.

Bettini, V., Nebbia, G., & Baracca, A. (2009). *Il nucleare impossibile. Perché non conviene tornare al nucleare*. UTET.

Bocchi, G. L., & Ceruti, M. (2007). *La sfida della complessità*. Mondadori Ed.

Callegari, R., & Durastante, E. (2023). *Papa Francesco e la Cura delle Casa Comune*. Secondo Tempo.

Cera, A. (2019). *Dall'Antropocene al Tecnocene. Prospettive etico-antropologiche dalla "Terra incognita"*. Università della Basilicata.

Cicero, M. T. (1973). (52 a.C). *De Legibus libri tres*. Reprint by Mondadori.

Crutzen, P. J., & Schwägerl, C. (2011). Living in the Anthropocene: Toward a new global ethos. *Yale Environment, 360*.

Crutzen, P. J., & Stoermer, E. F. (2000). The Anthropocene. *Global Change Newsletter, 41*, 17–18.

Daly, H. E., & Farley, J. C. (2004). *Ecological economics: principles and applications*. Island Press.

Delorme, F. (1922). Legenda antiqua S. Franscisci du ms. 1046 de la bibliotehèque de Pérouse. *Archivium Franciscanum Historicum, 15*, 23–70. 278–332.

Dotta et al. (2022) Environmental factors in the development of diabetes mellitus. In Ingegnoli V, Lombardo F, LaTorre, G: Environmental alteration leads to human disease, a planetary health approach. Springer-Nature, .

Einstein, A. (1953). Mein Weltbild, ed. C. Seelig Stuttgart-Zurich-Vienna,

Faggin, F. (2022). *Irriducibile. La coscienza, la vita. i computer e la nostra natura*. Mondadori.

Faggin, F. (2024). *Oltre l'invisibile. Dove scienza e spiritualità si uniscono*. Mondadori.
Gandhi MK (1926) *An autobiography: The story of my experiments with truth.* Reprint 2013, Createspace Independent Publ. Platform.
Gandhi, M. K. (1962). *Village Swaraj*. Publ. Jitendra T. Desai.
Henning, B. G. (2016). From the Anthropocene to the Ecozoic: Philosophy and global climate change. *Midwest Studies in Philosophy, 40*(1), 284–295.
Ingegnoli, V. (2011). *Bionomia del paesaggio. L'ecologia del paesaggio biologico-integrata per la formazione di un "medico" dei sistemi ecologici* (p. XX+340). Springer.
Ingegnoli, V. (2015). *Landscape bionomics. Biological-integrated landscape ecology* (p. XXIV + 431). Springer.
Ingegnoli, V., & Giglio, E. (2005). *Ecologia del Paesaggio: manuale per conservare, gestire e pianificare l'ambiente* (p. 685+XVI). Sistemi editoriali SE.
Kendall, B. (1990). *Science and Russian culture in an age of revolutions: V. I. Vernadsky and his scientific school, 1863–1945*. Indiana University Press.
Lanza Del Vasto, G. G. (1943). *Pellegrinaggio alle sorgenti. Viaggio con Gandhi e con l'India.* Jaca Book.
Lorenz, K. (1973). *Die Ruckseite der Spiegels*. Piper & Co..
Mancini, R. (2015). *Ripensare la sostenibiiità: le conseguenze economiche della democrazia.* Franco Angeli Edizioni.
Marsh, G. P. (1864). *Man and nature: Physical geography as modified by human action*. S. Low, Son and Marston.
Monod, J. (1970). *Le hazard et la nécessité*. Edition du Seuil.
Morin, E. (2025). *Semi di saggezza*. Cortina Ed..
Nembrini, F. (2020). *Purgatorio, di Dante Alighieri*. Mondadori.
Pope Francis, I. (2015). *Laudato Sì. Enciclica sulla cura della casa comune*. Libr. Edit. Vaticana, Roma.
Pope John Paul II. (1979). *Redemptor Hominis. Memoria e profezia*. Lateran University Press.
Pope Leo XIV. (2025). Message to french guides and scouts partaking in the "Clameurs" Rally in Jambville.
Radahakrishnan, S., & Vecchiotti, I. (1978). *Bhagavad Gita*. Ubaldini Ed.
Ratzinger, J. (1996). *Relativism, the central problem for faith today*. Libr. Edit. Vaticana, Roma.
Ratzinger, J., & Benedetto XVI (2006). *In principio Dio creò il cielo e la terra*. Edizioni Lindau.
Stoppani, A. (1871–1873). *Corso di geologia. Tre volumi: I. Dinamica terrestre; II. Geologia stratigrafica; III. Geologia endografica*. G. Bernardoni e G. Brigola Ed..
Stoppani, A. (1876). *Il Bel Paese. Conversazioni sulle bellezze naturali la geologia e la geografia fisica d'Italia*. Agnelli.
Swimme, B., & Berry, T. (1992). *The universe story: From the primordial flaring forth to the Ecozoic era—A celebration of the unfolding of the cosmos*
UNDP (1999) Human development report. .
Urbani-Ulivi, L. (Ed.). (2019). *The systemic turn in human and natural sciences. A Rock in the Pond*. Springer-Nature.
Verdnadskij, V. (1926). *Biosfera [the biosphere]*. Nauka.
Verdnadskij, V. (1945). The biosphere and the noosphere. *American Scientist, 33*, 1–12.
Verdnadskij, V. (1977). *Scientific thought as a planetary phenomenon* (Trans. by B. A. Starostin). Nongovernmental Ecological V. I. Vernadsky Foundation.
Verdnadskij, V. (1998). *The biosphere. A Peter N. Nevraumont Book*. Copernicus Springer-Verlag New York, Inc..

Index

V. Ingegnoli, *Landscape Bionomics and Planetary Health*, Landscape Series 21,
https://doi.org/10.1007/978-3-032-08908-3

J

L

M

N

O

P

R

MIX
Papier aus verantwortungsvollen Quellen
Paper from responsible sources
FSC® C105338

If you have any concerns about our products,
you can contact us on
ProductSafety@springernature.com

In case Publisher is established outside the EU,
the EU authorized representative is:
Springer Nature Customer Service Center GmbH
Europaplatz 3, 69115 Heidelberg, Germany

Printed by Libri Plureos GmbH
in Hamburg, Germany